Space Enterprise:
Living and Working Offworld in the 21st Ce[ntury]

Philip Robert Harris

Space Enterprise:
Living and Working Offworld in the 21st Century

 Springer

Published in association with
Praxis Publishing
Chichester, UK

Dr. Philip Robert Harris
Management/Space Psychologist
Harris International, La Jolla, California
Adjunct Professor, California School of International Management
San Diego, California, USA
Associate Fellow, American Institute of Aeronautics and Astronautics
www.drphilipharris.com

SPRINGER–PRAXIS BOOKS IN SPACE EXPLORATION
SUBJECT *ADVISORY EDITOR*: John Mason B.Sc., M.Sc., Ph.D.

ISBN 978-0-387-77639-2 Springer Berlin Heidelberg New York

Springer is part of Springer-Science + Business Media (springer.com)

Library of Congress Control Number: 2008920035

Apart from any fair dealing for the purposes of research or private study, or criticism or review, as permitted under the Copyright, Designs and Patents Act 1988, this publication may only be reproduced, stored or transmitted, in any form or by any means, with the prior permission in writing of the publishers, or in the case of reprographic reproduction in accordance with the terms of licences issued by the Copyright Licensing Agency. Enquiries concerning reproduction outside those terms should be sent to the publishers.

© Praxis Publishing Ltd, Chichester, UK, 2009
Printed in Germany

The use of general descriptive names, registered names, trademarks, etc. in this publication does not imply, even in the absence of a specific statement, that such names are exempt from the relevant protective laws and regulations and therefore free for general use.

Cover design: Jim Wilkie
Project management: Originator Publishing Services, Gt Yarmouth, Norfolk, UK

Printed on acid-free paper

Contents

Dedication xii

Acknowledgments xiii

Foreword by David G. Schrunk, M.D. xv

List of exhibits xix

List of abbreviations and acronyms xxv

About the contributors xxxi

About the author xli

Prologue xlv

1 Toward a global space vision, ethos, and enterprise 1
   1.1   Space visions of humankind 3
        1.1.1   Evolution and space development 5
        1.1.2   Multiple visions of space 6
        1.1.3   Articulating a space vision 7
   1.2   North American space case study: U.S.A., Canada, Mexico 11
        1.2.1   United States space case study 13
        1.2.2   Canadian space case study 25
        1.2.3   Mexico's space vision and ethos 26
        1.2.4   Conclusions about North American space vision and ethos   27
   1.3   Russian space case study 28
        1.3.1   Conclusions about Russian space vision and ethos 35
   1.4   European space case study 37
        1.4.1   France 37
        1.4.2   Germany 38

		1.4.3	Great Britain	40
		1.4.4	Italy	41
		1.4.5	ESA emergence	41
		1.4.6	Conclusion about European space vision and ethos	43
	1.5	Asian space case study		44
		1.5.1	China	45
		1.5.2	Conclusion on Chinese space vision and ethos	47
		1.5.3	Japan	48
		1.5.4	Conclusion on Japan's space vision and ethos	50
		1.5.5	India	51
		1.5.6	Conclusion on India's space vision and ethos	53
		1.5.7	Australia	53
		1.5.8	Conclusion on Asian space vision and ethos	54
	1.6	Other nations in space		55
	1.7	Conclusions on global space vision, ethos, and enterprise		58
	1.8	References		60

2	**Human space exploration and settlement**		65
	2.1	Leaving Earth's cradle: joint venturing	66
	2.2	Emerging space settlement issues	76
	2.3	Interdisciplinary contributions to space habitation	79
	2.4	Redirecting knowledge and workforces upward	86
	2.5	Space is a place for synergy	90
	2.6	Conclusions: synergizing space exploration and settlement	97
	2.7	References	99

3	**Space habitability and the environment**			103
	3.1	Space habitability and life sciences		106
		3.1.1	Emerging life/behavioral science research	108
		3.1.2	Space settlement	115
	3.2	Behavioral analysis of life aloft		120
	3.3	Anthropology and space habitation		121
	3.4	Psychology and space habitation		125
		3.4.1	Emerging contributions of space psychologists	127
		3.4.2	Addressing spacefarers' psychological needs	129
		3.4.3	Space psychologists in the future	130
	3.5	Sociology and space habitation		131
	3.6	Living systems and space habitation		136
		3.6.1	Studying orbital living systems	137
	3.7	Habitability and the space environment		138
	3.8	Conclusions on future research directions		143
	3.9	References		147

4	**Cultural implications of space enterprise**		**153**
	4.1	Culture: a coping strategy	153
	4.2	Emergence of a new space culture	159
	4.3	Cultural influences on aerospace organizations	170
	4.4	Cosmic cultures	176
	4.5	Conclusions on emerging space culture	177
	4.6	References	180
5	**High-performing spacefarers**		**183**
	5.1	Extending human presence aloft	186
		5.1.1 Human orbital expansion	190
		5.1.2 Offworld multiculturalism	191
		5.1.3 High-performance offworld norm	192
		5.1.4 Exceptional ground support services	193
	5.2	Learning from human offworld performance	196
		5.2.1 Improving human factor research	198
		5.2.2 Learning from spacefarers	199
		5.2.3 Advantages of diversity and teamwork	201
	5.3	Crew system productivity	206
		5.3.1 Improving orbital performance	209
		5.3.2 Countering the downside aloft	210
		5.3.3 Human sexuality and family offworld	213
	5.4	Orbital team performance	214
		5.4.1 Orbital team building	215
		5.4.2 Crew-training strategies	217
		5.4.3 Strengthening team culture	220
	5.5	Human/Machine interface aloft	222
		5.5.1 Orbital team mates	222
	5.6	Human resource development of spacefarers	227
		5.6.1 Educating for space	229
		5.6.2 Troubles in paradise	231
	5.7	Space ergonomics and ecology	232
		5.7.1 Space ecology	235
		5.7.2 Habitation ergonomics and ecology	236
	5.8	Space performance research issues	237
	5.9	Conclusions on spacefarers' performance	240
	5.10	References	240
6	**Orbital deployment systems and tourism**		**245**
	6.1	Terrestrial analogs	248
		6.1.1 Polar regions as laboratories for space	252
		6.1.2 European isolation studies	254
		6.1.3 Biosphere 2 (private enterprise)	258
		6.1.4 Design for extreme environment assembly	261
		6.1.5 Mars virtual explorations	263

viii Contents

	6.2	Space personnel deployment strategies	264
	6.3	SPDS stages	265
		6.3.1 Stage One: Assessment	265
		6.3.2 Stage Two: Orientation	275
		6.3.3 Stage Three: In-space services	283
		6.3.4 Stage Four: Re-entry policies and programs	296
	6.4	Transforming space development through tourism	304
	6.5	Conclusions on space personnel deployment and tourism	310
	6.6	References	312
7	**Macrothinking in strategic space planning**		**317**
	7.1	Understanding macrothinking and planning	318
		7.1.1 Large-scale space enterprises	319
		7.1.2 Space strategic planning and management	322
	7.2	Macrothinking examples from the past	324
	7.3	Mini case studies of macrothinking	329
		7.3.1 Spaceport case study by Derek Webber	330
		7.3.2 The European Moon program	337
	7.4	Macrothinking perils	339
		7.4.1 Orion: the original proposal for a nuclear-powered rocket	339
		7.4.2 United Societies in Space, Inc.	339
		7.4.3 First Millennial Foundation	342
	7.5	Current macrothinking illustrations	342
	7.6	Future macrothinking	348
	7.7	Macrothinking conclusions	351
	7.8	References	353
8	**Macromanagement of space enterprise**		**357**
	8.1	Management challenges for the space era	357
		8.1.1 Macromanagement prototypes	359
	8.2	The Apollo heritage of innovative management	361
		8.2.1 Matrix management	363
		8.2.2 Knowledge management	365
	8.3	The impact of organizational culture	367
		8.3.1 Learning from organizational setbacks	368
		8.3.2 Space macromangement culture	371
	8.4	Emerging space roles of Earth-based managers	373
	8.5	Management in orbit	378
	8.6	Macromanagement for/in outer space	379
		8.6.1 Emerging macromanagement	381
		8.6.2 Educating tomorrow's macromanagers	382
	8.7	Space management in the future	384
		8.7.1 Re-education of humanity	386

	8.8	Space Station case study	388
		8.8.1 Mini case study: evolution of the International Space Station	388
	8.9	ISS learnings in macromanagement	399
	8.10	Conclusions on space macromanagement	400
	8.11	References	402
9	**Challenges in offworld private enterprise**	407
	9.1	Challenges and realities in space enterprises	408
		9.1.1 Space entrepreneurialism	408
		9.1.2 Rocket renaissance	410
		9.1.3 Space entrepreneur analysis	412
		9.1.4 *NewSpace* business	416
	9.2	Macrochallenges on the commercial space frontier	421
		9.2.1 Model for astrobusiness analysis	424
		9.2.2 Private/Public-sector cooperation in space commerce . . .	428
		9.2.3 Lowering space transportation access costs	430
		9.2.4 Entrepreneurial view on space access	436
	9.3	Technical space A&R challenges	437
		9.3.1 Space robotic possibilities	438
		9.3.2 Spacebots' prospects	439
	9.4	Legal space frontier challenges	440
		9.4.1 Space laws and regulations	441
		9.4.2 Space settlements and governance	446
		9.4.3 Future space administration and governance	449
	9.5	Political challenges on the space frontier	451
		9.5.1 Influencing space policy	454
		9.5.2 Influencing lunar policy	455
		9.5.3 Influencing commercial space development	460
	9.6	Conclusions on offworld private enterprise	463
		9.6.1 Space education	464
	9.7	References	466
10	**Lunar enterprises and development**	471
	10.1	Reclaiming the Moon: rationale	473
		10.1.1 International lunar agreements and initiatives	478
		10.1.2 Asian lunar initiatives	479
		10.1.3 European lunar initiatives	482
		10.1.4 U.S. lunar initiatives	486
		10.1.5 Private lunar initiatives	489
	10.2	Lunar administration and governance	494
		10.2.1 Background observations	495
		10.2.2 Lunar Economic Development Authority (LEDA) proposal	497

		10.2.3	The strategy of space authorities	499
		10.2.4	Creating lunar social systems	503
	10.3	\multicolumn{2}{l}{Lunar exploration and science}	504

         10.2.3 The strategy of space authorities 499
         10.2.4 Creating lunar social systems 503
  10.3 Lunar exploration and science 504
         10.3.1 The Antarctica Model vs. terraforming 505
         10.3.2 Science role in exploration 506
  10.4 Lunar settlement and industrialization 507
         10.4.1 Lunar start-up enterprises 511
         10.4.2 Earth support enterprises 512
         10.4.3 Lunar base location and expansion 513
  10.5 Conclusions on lunar developments 515
  10.6 References 519

Epilogue 525
  Transforming space visions into realities 525
  Exercising transformational leadership 530
  References 533

APPENDICES

A **Governance issues in space societies** by *George S. Robinson* 535
  A.1 Beyond imperialism: beyond space colonies 535
  A.2 Governance in the absence of human rights and duties 537
  A.3 Transitioning principles for governance of space societies 538
  A.4 Space governance: preliminary conclusions 539
  A.5 References 542

B **Space-based energy: Lunar solar power** by *David R. Criswell* and *Philip R. Harris* 545
  B.1 Introduction 545
  B.2 Rationale for a lunar solar power system (LSPS) 547
  B.3 Lunar solar power system proposal 555
  B.4 Lunar power base technologies 561
  B.5 Conclusions 562
  B.6 References 565

C **Learning from space entrepreneurs** by *Thomas L. Matula* 569
  C.1 Dawn of the Spage Age 569
  C.2 Space Services of America Incorporated 570
  C.3 Orbital Sciences Corporation 572
  C.4 Lessons learned from these space entrepreneurs 572
  C.5 Conclusions 574
  C.6 References 576

D   Health services aloft: Space nurses by *Linda M. Plush* and
Eleanor A. O'Rangers.   579
   Part 1   Healthcare offworld: historical perspectives and current state of affairs relative to medical care knowledge and capabilities in the manned spaceflight program and beyond.   579
   D.1   Dawn of space medicine.   580
   D.2   Space medicine issues in Projects Mercury, Gemini, and Apollo .   580
   D.3   The Skylab program   582
   D.4   Space Shuttle program   583
   D.5   The Shuttle–Mir program and the International Space Station . .   584
   D.6   To the Moon (again) and Mars   585
   D.7   Health challenges of private citizens in orbit.   587
   D.8   Part 1: Conclusions   587
   Part 2   Future role of space nursing in spaceflight arena   590
   D.9   A visionary nursing leader: Martha E. Rogers.   590
   D.10   Role of nurses in a future lunar settlement.   591
   D.11   Emerging personal/commercial spaceflight nursing   592
   D.12   Advancing space nursing.   592
   D.13   Part 2: Conclusions   594
   D.14   References   595
           Part 1   Health services aloft   595
           Part 2   Space nursing bibliography.   596

E   International Lunar Observatory strategy by *Steve Durst*   599
   E.1   Introduction.   599
   E.2   History   600
   E.3   International support   600
         E.3.1   Strategies within states and nations   600
   E.4   Financing the ILO macroproject.   601
   E.5   ILO lunar location   602
   E.6   ILO'S master plan.   602
   E.7   ILO features and benefits   604
   E.8   Conclusions   605
   E.9   International Lunar Observatory Association News   605
         E.9.1   ILOA moves forward with precursor mission, advancing Galaxy forums, and private funding initiatives; SpaceDev completes ILO Payload Study; BOD meeting set for July in Canada.   605
   E.10   References   606

F   Resources   607

Index   613

Dedication
The envoys of humankind: the world's astronauts and cosmonauts

Some of whom gave their lives to advance human migration into space, like the crews below of the Shuttles *Challenger* and *Columbia*!

Challenger crew, 1986 [Back row, l. to r.: Astronauts Ellison Onizuka, S. Christa McAuliffe, Greg Jarvis, Judy Resnik. Front row, l. to r.: Mike Smith, Dick Scobee, and Ron McNair ... Note the gender and racial diversity evident in this group and the one below.]

Columbia crew, 2003 [Back row, l. to. r.: Astronauts David Brown, William McCool, and Michael Anderson. Front row, l. to r.: Kalpana Chawla, Rick Husband, Laurel Clark, and Ilan Ramon. Note: image of the STS-107 crew in orbit was recovered in 2007 from an undeveloped film canister found in the Shuttle's wreckage. Source: NASA.]

May these 14 and those other spacefarers who have lost their lives exploring the high frontier rest in peace! Ad Astra!

Acknowledgments

In writing *Space Enterprise*, the author is grateful for the encouragement of Clive Horwood, my publisher at Praxis; Dr. John Mason, advisory editor; and Neil Shuttlewood of Originator Books for technical production. I am particularly indebted to Dr. Eileen Sheridan Wibbeke without whose computer skills this book may not have been published. This text has also benefited from the peer review of Dr. Albert A. Harrison which led to many revisions.

In writing *Space Enterprise*, the author especially appreciates the illustrations of Dennis M. Davidson, the talented astronomical artist, whose work in three editions of this volume has helped us all to visualize our human future in space. My gratitude also extends to the other space artists, especially Pat Rawlings whose creative renderings appear in these pages, as well as to the National Aeronautics and Space Administration, and the European Space Agency, for the many photographs which they supplied. Furthermore, I am beholden to my friend Dr. David Schrunk who shared his insights in the Foreword. This new release in the Springer-Praxis series of Books in Space Exploration profits from the collaboration of many scholars, but especially the seven contributors. Six shared their thinking in our appendices: Dr. George S. Robinson, attorney; Dr. David R. Criswell, astrophysicist; Dr. Thomas L. Matula, management professor; entrepreneurs Linda M. Plush, M.S.N., nurse practitioner, and Dr. Eleanor O'Rangers, pharmacist; and Steve Durst, M.S., publisher. The seventh colleague, Derek Webber, spaceport consultant, provides a case study in Chapter 7. For more biographical information, consult the About the contributors pages (pp. xxxv–xliii).

I am also obliged to many other esteemed colleagues whom I first met and admired in 1984 at the California Space Institute on the campus of Scripps Institution of Oceanography in La Jolla. There, during a NASA Summer Study, their information and insights contributed to the original subject matter of this text. The author wishes to particularly acknowledge in that regard: Dr. James Arnold, first director of the California Space Institute; Dr. Stewart Nozette, with the U.S. Air Force Phillips

Laboratory; NASA representatives, Drs. Mary Fae and David McKay, as well as Drs. Michael Duke and Wendell Mendell, all then at the Johnson Space Center. At that time, I was a Faculty Fellow in that intensive workshop, planning a lunar base for 2010, about ten years off the mark. This is when I developed the research database for a novel on lunar industrialization, *Launch Out*, plus the first two editions of this professional book, then entitled *Living and Working in Space*.

However, these pages that you read reflect the scholarship of numerous other colleagues, some of whom cooperated in this venture when its publication was initially conceived as an anthology: Roger M. Bonnet, director of ESA's Scientific Programs; B. J. Bluth of NASA headquarters; Joseph V. Brady of The Johns Hopkins University School of Medicine; James D. Burke, California Institute of Technology, Jet Propulsion Laboratory; Julian Christian of Universal Energy Systems; Jacques Collet, head of ESA's Long Term Program Office; Angel and Patricia Colon of the Georgetown University Medical School; Mary M. Connors of NASA-Ames Research Center; Frank Davidson, director of MIT's Macro-engineering Research Group; Ben R. Finney of the University of Hawaii; Nathan C. Goldman, Attorney at Law and adjunct professor at Rice University and the University of Houston; Albert A. Harrison of the University of California-Davis; H. H. Koelle, professor emeritus, Technische Universität Berlin; William E. MacDaniel, professor emeritus of Niagara University; James Grier Miller of the University of CaliforniaLos Angeles/San Diego and Mrs. Jessie L. Miller; Brian T. O'Leary, former Apollo astronaut, author/lecturer; Namika Raby of the California State University-Long Beach; John M. Talbot of the Federation of American Societies in Experimental Biology. Some of their ideas and findings have been synthesized and reported in this present volume.

Without the unfailing support of my late wife, Dr. Dorothy L. Harris, during its 12 years of preparation, and then in this revision, my present wife Janet, this work might not have reached you. In assembling the original manuscript, I owe special appreciation to my former co-workers at Netrologic Inc.: the late Dan Greenwood and Sarah Bode Becker. With this current version, *Space Enterprise*, I also acknowledge the assistance of Dennis Laurie, CEO of Transorbital Inc.

The world of outer space is where imaginations and knowledge combine to become an unlimited human culture!

Philip Robert Harris

Foreword

The "knowledge culture", which consists of knowledge industries, knowledge management, and knowledge workers, now dominates world economic activity. As the fields of science generate new knowledge and develop more sophisticated tools, our problem-solving ability within this knowledge culture increases with time. One result of this optimistic scenario is that humankind is now advancing, on a broad front and at an increasing pace, onto the space frontier. Continuing improvements in computers, communications, pharmaceuticals, robotics, and life support systems, etc., have enabled a growing number of nations, commercial enterprises, and even tourists to participate in space activities. Although relatively few people have ventured into space thus far, it can be confidently predicted that permanent offworld settlements will become a reality by the middle of the 21st century.

Of significance, the establishment of human communities on other worlds, beginning with the Moon, will create a link between the knowledge culture and the unlimited resources of space. When that linkage is secured, humankind will become a true "spacefaring civilization!" Our present "Closed-Earth" mindset, based on limited Earth resources, will be replaced by a much more optimistic "Open-Space" view of unlimited resources and endless frontiers. Large-scale exploration and development projects throughout the Solar System will then be routine and the first robotic emissaries will be launched on missions to nearby stars. In addition, dramatic improvements in living standards and quality of life for everyone on Earth can be expected with the delivery of an abundance of energy and material wealth from space to the home planet.

In anticipation of the challenges and opportunities of becoming a spacefaring civilization, it is essential that we understand the steps the have brought us to this point and formulate a bold vision for the upcoming migration of humans into space. That is precisely what Dr. Phil Harris achieves in this thoroughly researched and comprehensive work, *Space Enterprise: Living and Working Offworld in the 21st Century*. From a behavorial science perspective, the author emphasizes that the future of space will engage not just space scientists and government agencies, but entire cultures. The need for and desire of global cooperation in space among nations, business consortia, universities, artists, and tourists will offer opportunities for synergy. Thus will emerge a *space ethos: the epochal transition to space-based living and the creation of an entirely new space culture!*

In his description of the path to a truly space society, the author provides a wealth of information on the diverse issues involved with living and working offworld, such as habitat design, ergonomics, high performance, physiologic and psychological concerns, protection from space hazards, cultural issues, and medical concerns. Of significance, the coming migration of people into space will involve *macroprojects* requiring *macro* thinking, planning, and management on a scale much larger than the more narrowly focused space missions of the past. These macroprojects will entail competent governance, large-scale financing, and the expertise of private enterprise. In fact, the private sector may play the largest economic role in the development and operation of future settlements in space.

Chapter 10 is devoted to the issues of lunar enterprises and development. The Moon is the closest celestial body to the Earth and it is the logical place to establish the first offworld human settlement. It offers protection from space hazards, it has energy (sunlight) and material resources that can be used to support human activities, and it is an excellent platform for scientific studies, especially astronomy. The transformation of the Moon into an inhabited sister planet of the Earth is a case that this management and space psychologist covers in the first chapters of the book. The "Planet Moon Project" will be a macro undertaking that involves the collaborative efforts from most, if not all, of the nations of the Earth. A new lunar government will regulate macroprojects, and private enterprise will develop industrial parks and infrastructure networks (railroad, communication, pipeline, and solar power systems), as well as provide food processing and banking services, etc. By these means, the space ethos of humankind will emerge and we will truly become a *spacefaring civilization!*

The total effects of the migration of humans to other worlds are difficult to predict, but they will unquestionably be accompanied by advances in every purposeful field of human endeavor. For those who are interested in understanding the origins and future directions of humankind as an emerging spacefaring species, *Space Enterprise* is a must-read, especially for space planners, engineers, entrepreneurs, and students.

David G. Schrunk, M.D., Aerospace Engineer
Senior author, *The Moon: Resources, Future Development, and Settlement*
Chairman, Quality of Laws Institute (*www.glpress.com*),
Poway, California
September 2007

50TH ANNIVERSARY OF THE SPACE AGE: RETROSPECTIVE ON UNMANNED SATELLITES

The launch 50 years ago of Sputnik 1, the first man-made satellite in orbit, marked the beginning of the space race and the Space Age. The little satellite (it was only slightly larger than a basketball) altered the course of history and geopolitics, of U.S. education and culture. It helped spawn new technologies from the Internet to satellite radio. Perhaps most profoundly, it changed the way Americans, Russians, and others saw themselves and their place in the world!

Sputnik opened the floodgate. More than 25,000 satellites have been launched since 1957. Most have since come down or burned up on re-entering the atmosphere. At the last count, there were at least 836 active satellites, according to the Union of Concerned Scientists, each circling somewhere between 39 and 22,356 miles above the Earth's surface. The twin Grace satellites launched in 2002, for instance, monitor the Earth's shifting water masses, ocean currents, and sea floor pressures by precisely measuring the planet's variable gravity field. A net of Global Positioning Satellites allows ground-based receivers to identify their locations on Earth to within 50 feet—or better. This system consists of a planet-cloaking grid of 31 satellites in medium orbit (between 1,243 and 23,236 miles up). These satellites transmit microwaves that allow receivers on the ground in planes, cars, boats, and hand-held devices to triangulate location, speed, direction, and time ... Future satellites will be more complex and capable of smaller sizes at less cost ... At least 10 countries have independently launched satellites into orbit ... A Merrill Lynch study projected that the satellite industry's revenue would be $171 billion in 2008!

The downside is the orbital debris caused by 25,000 plus payloads sent aloft since Sputnik. *The Space Encyclopedia* puts the satellite total now in orbit at 3,211, including non-functioning and abandoned satellites ... The total of man-made objects in orbit has been estimated in the hundreds of millions. These objects, some racing at a speed of 17,000 miles per hour, are a real threat to working satellites, spacecraft, and astronauts. (Source: LaFee, S. "50 Years after Sputnik: Satellite Fever," *San Diego Union-Tribune*, October 4, 2007, pp. A1 and A12.)

AND THE 100TH ANNIVERSARY OF SPUTNIK?

It is safe to say that the 100th Anniversary of Sputnik will also be celebrated on Mars. What else will have happened in space by that date? We will have mastered the thorny problem of nuclear power for interplanetary travel. Decades before, Wernher von Braun, the man who gave us the Moon, recognized that nuclear power is necessary for manned interplanetary travel ... Physics dictates that if you want power, you have to tap it from the place where nature put it, the atom ... Fifty years from now, we will also know we are not alone in our galaxy! (Source: Giovanni F. Bignani, President of the Italian Space Agency, "The Future in Space Will Unlock Wonders," *San Diego Union-Tribune*, October 7, 2007, pp. G1 and G4.)

Exhibits

The author acknowledges with gratitude all who have contributed to these exhibits and who have been noted in the captions. Asterisks (*) in this list of exhibits designate an aerospace illustration by Dennis M. Davidson, former Astronomical Artist for the Hayden Planetarium and founding director of its innovative Digital Galaxy Project.

Front cover   Lunar mining facility: an artist rendering for extracting volcanic soil from the eastern Mare Serenitatus. This rich resource on the Moon includes iron, magnesium, and titanium content. The tailings could be processed as raw materials for a lunar metal production plant (artist, Pat Rawlings of NASA)   cover

Dedication   Space Shuttle *Challenger* crew, Flight 51-L, 1986; Space Shuttle *Columbia* crew, STS-107, 2003   xii

Pi	Libration points/zones in outer space	xlvi
Pii	Our realm of outer space	xlviii
Piii	Offworld human activity	lii
1	Development of Human–Earth–Space, one creative system	3
2	First lunar landing of humans	4
3	Human evolution and space	10
4	Why go offworld?	11
5	Outposts on the Moon	14
6	Launch out: space exploration	19
7	Chinese overseas exploration	22
8	Canadian space vision and ethos	27
9	Energia space launchpad: Kazakhstan	29
10	Europe's spaceport in South America	39
11	A Chinese mission control room	48
12	Creating a solar civilization	55
13	Global space spending	57

Exhibits

14	The Vision for Space Exploration	60
15	Leaving the home planet	67
16	Human challenges in space living	74
17	Transnational, multicultural crew teams	75
18	Offworld settlement challenges	80
19	Human factors in space development beyond the technical*	82
20	Vastness and beauty of space	87
21	Orbital repairs	89
22	Space is a place for synergy	95
23	Russian/American space cooperation on Station Mir	96
24	Russia wants to join NASA Moon program	97
25	Super ecology: future interactive loops	104
26A	Unique space environmment: useful attributes of space living	106
26B	Stressors in the space environment	107
27	NASA perceptions of space life sciences programs	108
28	Shuttle docked at International Space Station	113
29	Radiation challenges offworld	114
30	Space Settlement Act: U.S. Congressional legislation	116
31	Space habitation dimensions	119
32	Correspondences between biological and cultural evolution	122
33	Lunar base activities in the 21st century	126
34	Space factors and human behavior	128
35	Research on living and working in microgravity	132
36	Lunar habitability	135
37	Living systems symbols*	139
38	Space Station: five living systems flows*	140
39	Lunar base: five living systems flows*	141
40	Near-Earth objects impacting the environment	142
41	Lunar dweller with robotic helper transforming Moon/Mars materials	144
42	Biological and behavioral space research challenges	145
43	Work performance in orbit*	152
44	EVA: we are no longer Earthbound!	154
45	Multicultural relations in space	156
46	Transformation of ancient cultures	157
47	Seeking space organizational synergies	160
48	Characteristics of space culture*	163
49	Dress and appearance of humans and robots aloft	164
50	Education on the Moon	168
51	Orbital work habits and practices	169
52	Aerospace organizational culture*	171
53	Learning from Russian spaceflight experience	173
54	Renewing NASA culture	175
55	Work culture on the Moon	177
56	Deploying Eureca from the Space Shuttle	185
57	Peak performance: humans on the Moon!	186
58	Successful space launches 1957–1993	188
59	ESA multinational astronauts	191
60	A team in the Destiny laboratory of the ISS	194
61	High performance offworld	196

62	On-orbit performance research	200
63	Interdependent systems on the Moon	202
64	Crew diversity and ingenuity	204
65	High-performing female spacefarers	205
66	Human factor research areas for space family living	213
67	ISS teamwork	216
68	Orbital team satellite repairs	219
69	Human/Robotic lunar prospects	223
70	Servicing Hubble in orbit	226
71	Can robots be trusted?	227
72	Lunar ergonomics	233
73	Factors in space ecological systems	237
74	Strategies for life science research	239
75	Technauts' lunar deployment	247
76	ESA ground-based isolation studies	256
77	Biosphere 2	258
78	IDEEA conference themes	262
79	Space Personnel Deployment System*	265
80	ESA astronaut deployment strategy	267
81	Future crew prototypes	269
82	Preparing spacefarers	274
83	ESA Astronaut Training Cycle	275
84	Lunar/Mars Excursion Vehicle	289
85	Shuttle Orbiter carrying Spacelab	291
86	Mars mission challenges	301
87	Innovations in space development	305
88A	Macroplanning characteristics	321
88B	Requirements for macrothinking and macroplanning	321
88C	Attributes of macrothinking and macroplanning	321
89	Dynamic management strategic process	323
90	Macrothinking: orbiting spaceport	327
91	Macrothinking: space-based energy from the Moon*	330
92	Launch details	332
93	Tourist spaceport of the future	336
94	ESA utilization of lunar resources	338
95	Space peace symbol	340
96	Living and working in microgravity	343
97	Principles of international coordination for space developments	350
98	Prototype of future macromanagers: John Pierce	359
99	ESA ground-based global establishment	361
100	The management process	366
101	Managing risk offworld	374
102	Managing unmanned space missions: Galileo	375
103	Managing a new world: Mars	377
104	Macromanagement for space enterprises*	380
105	Lunar power plant	385
106	Managing Moon macroprojects	386
107	International Space Station cooperation	389
Case Study Exhibit	ISS at completion in year 2010	398

108	Space business ventures	409
109	Private launch entrepreneurs	413
110	Profile of a *NewSpace* entrepreneur	415
111	Portrait of a *NewSpace* company	417
112	Lunar market prospects	422
113	European space commerce	424
114	Benefits of a space transportation industry	425
115	Space commerce model	425
116	Private lunar enterprise: ISELA	430
117	Innovative designs of launch vehicles	432
118	Current ESA spacecraft	435
119	Space business through automation and robotics	440
120	Interagency space agreements: SOHO	450
121	Multinational space synergy	452
122	Providing lunar infrastructure	458
123	*The Space Show*	461
124	Educating tomorrow's spacefarers	465
125	Why the Moon?	470
126	Characteristics of the Moon	473
127	Lunar world	476
128	Japanese lunar macroplanning	480
129	Lunar science opportunities	483
130	Lunar Prospector Mission Profile	487
131	Planned International Lunar Observatory	492
132	Lunar enterprise forecasts	493
133	Lunar industrialization	495
134	Lunar shelters	509
135	Possible lunar lander	510
136	The consequences of growth and no growth: Krafft Ehricke	517
137	Future possibilities	518
138	Space light: The Moon as seen from Earth beckons us!*	524
139	Orbital envoys of humankind	531
140	Human emergence in space	532
141	Living and working offworld	534
142	Bridge between two worlds: Earth/Moon system	536
143	Demonstration lunar power base	544
144	Comparison of 21st-century energy systems	548
145	21st-century power crisis	550
146	Lunar power system	551
147	Space energy commentary	554
148	Concept: lunar power system	556
149	LSPS: details and challenges	558
150	Lunar industrialization possibilities	560
151	Privatizing space transportation	575
152	Lunar medics	578
153	Predicted effects of weightlessness	581
154	Space psychology	588
155	Personal spaceflight health issues	589
156	Space Nursing Society symbol	590

157	The future of space nursing	593
158	ILO 2	598
159	ILO 3	603
160	Concept of a lunar rover	606
161	Starry nights	612

We'll continue our quest in space. There will be more shuttle flights ... More teachers in space. Nothing ends here. Our hopes and journeys continue.

U.S. President Ronald Reagan, 1986

Abbreviations and acronyms

3SP	Space Settlement Studies Project
A&R	Automation and Robotics
AAS	American Astronautical Society
AHNA	American Holistic Nurses Association
AI	Artificial Intelligence
AIAA	American Institute of Aeronautics and Astronautics
AMA	American Medical Association
APSU	Auxiliary Propulsion System Unit
ARC	Alliance to Rescue Civilization
ASCONT	Association for the Advancement of Science and Technology
ASI	Italian Space Agency
ASLV	Automated Satellite Launch Vehicle (India)
ASTD	American Society for Training and Development
ATM	Automated Teller Machine
BIS	British Interplanetary Society
BNSC	British National Space Centre
CASC	China Aerospace Science and Technology Corporation; China Aerospace Corporation
CASE	Competitive Alliance for Space Enterprise
CAST	Chinese Academy of Space Technology
CATS	Cheap Access To Space
CBM	Condition Based Maintenance
CCSSE	Challenger Center for Space Science Education
CELSS	Controlled Ecology Life Support System
CETEC	Center for Extraterrestrial Engineering and Construction
CEV	Crew Exploration Vehicle
CFHT	Canada France Hawaii Telescope

CHOM	Common Heritage Of Humankind
CIS	Commonwealth of Independent States
CLEP	China Lunar Exploration Program
CNES	Centre National d'Etudes Spatiales (National Center for Space Studies) (France)
CNSA	China National Space Administration
COMET	Commercial Experiment Transporter
COMSTAC	Commercial Space Transportation Advisory Committee
COSPAR	Committee on Space Research
COTS	Commercial Orbital Transportation Services
CPM	Critical Path Method
CASA	California Spaceport Authority
CSA	Canadian Space Agency
CSG	Guiana Space Center
CSIRO	Commonwealth Scientific and Industrial Research Organization
CZ-1	Chang-Zheng-1
DARPA	Defense Advanced Research Project Agency
DECU	Development and Educational Communications Unit
DLR	German Aerospace Centre; German Space Agency
DOS	Department of Space
DS	Descent Stack
EAC	European Astronaut Centre
ELDA	European Lunar Demonstration Approach
ELDO	European Launcher Development Organization
ELV	Expendable Launch Vehicle
EOSAT	Earth Observation Satellite
ERS	European Remote Sensing Satellite
ESOC	European Space Operations Centre
ESR	Earth–Space Review
ESRIN	European Space Research Institute
ESRO	European Space Research Organization
ESTEC	European Space Research and Technology Centre
ETM	Extraterrestrial Material
EU	European Union
EURECA	European Retrievable Carrier
EVA	Extravehicular Activity
FAA	Federal Aviation Administration
FALCON	Fission Activated Laser Concept
FESTSP	Future European Space Transportation System Program
FINDS	Foundation for International Non-Govermental Development of Space
FMF	First Millennium Foundation
GEO	Geosynchronous Orbit
GES	Global Exploration Strategy

GNP	Gross National Product
GPS	Global Positioning System
GSLV	Geostationary Launch Vehicle
GST	Global Space Trust
GWIC	Great Wall Industry Corporation
HAR	Human Assets Registry
HBSSS	Human Behavior Space Simulation Studies
HEDS	Human Exploration and Development of Space
HIMES	Highly Maneuverable Experimental Space (vehicle)
HPEE	Human Performance in Extreme Environments
HRD	Human Resource Development
HTCI	High Technology and Commercialization Initiative
I-PAF	Italian Processing and Archiving Facility
IAA	International Astronautical Association
IAAA	International Association of Astronomical Artists
IAF	International Astronautical Federation
IARC	International Agreement of Recognitions and Capacity
ICAO	International Civil Aviation Organization
ICBM	Intercontinental Ballistic Missile
ICE	Isolated, Confined Environment
ICSU	International Council of Scientific Unions
IDEEA	International Design for Extreme Environments Assembly
IDPT	Integrated Design and Process Technology
IGA	Intergovernmental Agreement
IISL	International Institute of Space Law
ILC	International Lunar Conference
ILEC	International Lunar Exploration Conference
ILEWG	International Lunar Exploration Working Group
ILO	International Lunar Observatory
ILOA	International Lunar Observatory Association
Inmarsat	International Maritime Satellite Organization
IOSC	International Organization for Spacekind Cultures
IPN	Interplanetary Network; Interplanetary Internet
IPO	Initial Public Offering
IPP	Innovative Partnership Program
IPT	Integrated Product Team
ISAS	Institute of Space and Astronautical Science
ISE	International Space Enterprises
ISEMSI	Isolation Experiment for the European Manned Space Infrastructure
ISO	Infrared Space Observatory
ISRO	Indian Space Research Organization
ISRU	*In Situ* Resource Utilization
ISU	International Space University
ISY	International Space Year

ITAR	International Traffic in Arms Regulations
ITT	International Telephone & Telegraph
IUE	International Ultraviolet Explorer (satellite)
IVA	Internal Vehicular Activity
JAXA	Japanese Aerospace Exploration Agency
JEM	Japanese Experiment Module (Kibo)
JPL	Jet Propulsion Laboratory
KSC	Kennedy Space Center
LCLV	Low Cost Launch Vehicle
LDC	Lunar Development Corporation
LEC	Lunar Enterprise Corporation
LEDA	Lunar Economic Development Authority; Lunar European Demonstration Approach
LEM	Lunar Excursion Module
LEO	Low-Earth Orbit
LESA	Lunar Exploration System for Apollo
LOX	Liquid Oxygen
LPSC	Lunar Power Systems Coalition
LSPS	Lunar Solar Power System
LST	Living Systems Theory
LTPO	Long-Term Program Office
LTV	Lunar Transfer Vehicle
MDA	Material Dispersion Apparatus
MEDA	Mars Economic Development Authority
MEMS	Microelectronic Mechanical System
MFPE	Mission From Planet Earth
MMU	Manned Maneuvering Unit
MNT	Micro/NanoTechnology
MOA	Ministry Of Astronautics
MORO	Moon Orbiting Observatory
MSE	Manned Spaceflight Engineer
MSS	Mobile Servicing System
MTPE	Mission To Planet Earth
NAFTA	North American Free Trade Agreement
NAL	National Aerospace Laboratory
NASA	National Aeronautics and Space Administration
NASDA	National Space Development Agency
NASEA	North American Space Enterprise Agreement
NASTAR	National Aerospace Training And Research Center
NATO	North Atlantic Treaty Organization
NCOS	National Commission On Space
NCSR	National Committee for Space Research
NEEMO	NASA Extreme Environment Mission Operation
NEO	Near Earth Object
NGO	Non-Governmental Organization

NIAC	NASA Institute for Advanced Concepts
NOAA	National Oceanic and Atmospheric Administration
NRC	National Research Council
NRSA	National Remote Sensing Agency
NSBRI	National Space Biomedical Research Institute
NSF	National Science Foundation
NSS	National Space Society
NTL	National Training Laboratory (Institute of Applied Behavioral Science)
NTM	Non-Terrestrial Material
NUTEC	Norwegian Underwater TEchnology Center
NYPA	New York Port Authority
OCST	Office of Commercial Space Transportation
OIMS	Operations Integrity Management System
OLMSA	Office of Life and Microgravity Sciences and Applications
OPAL	Orbiting Picosatellite Automated Launcher
OSAT	Office of Space Access and Technology
OSC	Orbital Sciences Corporation
OTA	Office of Technology Assessment
PERT	Program Evaluation and Review Technique
PISCES	Pacific International Space Center for Exploration Systems
PMSS	Project Management State Space
POCKOCMOC	Russian Federal Space Agency (Rosaviakosmos or Roskosmos)
PRC	People's Republic of China
PRL	Physical Research Laboratory
PSF	Personal Spaceflight Federation
PSLV	Polar Satellite Launch Vehicle (India)
R&D	Research and Design
RA	Robotic Arm
RAR	Robotic Assets Registry
RDM	Research Double Module
RKA	Alternative name for Russian Federal Space Agency
RLEP	Robotic Lunar Exploration Program
RLV	Reusable Launch Vehicle
ROI	Return On Investment
RpK	Rocketplane Kistler
SAC	Space Applications Center
SAF	Space Agency Forum
SAR	Synthetic Aperture Radar
SRB	Solid Rocket Booster
SCN	Suprachiasmatic Nucleus
SDI	Strategic Defense Initiative
SEAC	Space Economic Advisory Council
SEDS	Students for the Exploration and Development of Space

SEI	Space Exploration Initiative
SETI	Search for ExtraTerrestrial Intelligence
SFF	Space Frontier Foundation
SHAR	Sriharikota Range (now Satish Dhawan Space Center, India)
SITE	Satellite Instructional Television Experiment (India)
SLV	Satellite Launch Vehicle (India)
SMA	Space Medicine Associates
SMART	Small Mission for Advanced Research Technology
SNS	Space Nursing Society
SOHO	Solar Heliospheric Observatory
SpaceX	Space Exploration Technologies
SPC	Space Age Publishing Co.
SPDS	Space Personnel Deployment System
SPHERES	Synchronized Position Hold, Engage, Reorient, Experimental Satellites
SPM	Strategic Planning and Management
SPN	Space Power Network
SPS	Solar Power from Space; Solar Power Satellite
SRR	Space Resources Roundtable
SSC	Swedish Space Corporation
SSI	Space Services Incorporated
SSTO	Single Stage To Orbit
STC	Scientific and Technological Center
STEP	Solar–Terrestrial Energy Program
STS	Space Tourism Society; Space Transportation System
SWOT	Strengths, Weaknesses, Opportunities, and Threats
TASME	The American Society for Macroengineering
TERLS	Thumba Equatorial Rocket Launching Station
TMIS	Technical and Management Information System
TOS	Transfer Orbit Stage
TPS	The Planetary Society
TT&C	Telemetry, Tracking, and Control network (China)
TVA	Tennessee Valley Authority
UNCOPUOS	United Nations Committee on the Peaceful Use of Outer Space
UNESCO	U.N. Educational, Scientific, and Cultural Organization
UNISC	U.N. International Space Center
USIS	United Societies in Space
UTMB	University of Texas Medical Branch
VKS	Military Space Forces (Russia)
VLF	Very Low Frequency
VSE	Vision for Space Exploration
VSSC	Vikram Sarabhai Space Center
VTOL	Vertical Take-Off and Landing
YIBS	Yale Institute for Biospheric Studies

About the contributors

The author is grateful to his colleagues listed below for sharing their expertise for the benefit of our *Space Enterprise* readers. Their biographies are arranged in the sequence in which their contributions appear in the book.

Foreword contributor:

DAVID G. SCHRUNK, M.D. is an aerospace engineer and medical doctor with board certifications in the medical specialties of nuclear medicine and diagnostic radiology. Dr. Schrunk retired from the practice of medicine in 1994 and now dedicates his time to his two passions: the future exploration and human development of the Moon, and the science of laws. He has authored many scientific papers on lunar development issues and is a co-author of the book, *The Moon: Resources, Future Development, and Settlement.* Published originally by Wiley/Praxis in 1999, the second edition of this classic "Moonbook" was released by Springer/Praxis in 2007. Dr. Schrunk founded the Quality of Laws Institute in 1995 and authored the book *The End of Chaos: Quality Laws and the Ascendancy of Democracy*, published in 2005 by the Quality of Laws Press. A graduate of Iowa State University, he is promoting an "Iowa on the Moon" project to involve his Alma Mater in the space program. Dr. Schrunk lives in Poway, California with his wife Sijia, son Erik, and daughter Brigitte. [PO Box 726, San Diego, CA 92074; tel.: 858/382-1789; email: *docscilaw@aol.com*; website: *www.qualitylaws.com*]

Spaceport Case Study, Section 7.3.1 contributor:

DEREK WEBBER is the Director of Spaceport Associates in Washington, D.C. He has directed three landmark studies in commercial space business planning: the ASCENT Study of Space Markets, for NASA's Marshall Space Flight Center; the Futron/Zogby Study of Space Tourism Demand; and the Adventurers' Survey which was completed jointly with Incredible Adventures. He has provided testimony to the President's Commission on the Future of the U.S. Aerospace Industry, and has been recognized at the Space Tourism Society's Orbit Awards for his contributions to the formation of the space tourism industry. Mr. Webber is an active member of the Reusable Launch Vehicles Working Group and the Launch Operations Support Working Group of the Federal Aviation Authority's COMSTAC Committee, and is an Affiliate and technical advisor to SpaceWorks Engineering, Inc. (SEI).

Derek's career began as a launch vehicle and satellite engineer in the U.K. in what is now EADS/Astrium Space Systems. He worked on thermal control system design for a number of satellites, including the Skynet satellite, and on kinetic heating analyses for launch vehicles, including postflight analyses of telemetry from the Europa vehicle firings from the Woomera range. He became Head of Procurement at Inmarsat (responsible for contracting for over a billion dollars worth of communications satellites, their launch vehicles and ground segment), and Managing Director of Tachyon Europe (providing satellite broadband and Internet access across the continent). Mr. Webber operated a consulting and training company based in San Diego with clients including Comsat Corporation and the then Hughes Space and Satellite Systems. Consulting assignments included PDR, CDR Design Review Board membership, international negotiation skills training, and systems engineering tasks.

He has entries in *Who's Who in the World* and *Who's Who in Science and Engineering*, and is a published author in *Beyond Earth: The Future of Humans in Space*, and *Kids to Space*. He holds a B.Sc. Honors Degree in Physics and Mathematics from Newcastle University (U.K.), and postgraduate qualifications in Space Science (from University College London) and Management (from the University of Westminster, U.K.).

He is a Senior Member of AIAA, and a Fellow of the British Interplanetary Society. Mr. Webber has been a member of the Adjunct Faculty of the International Space University and at the U.S. International University in San Diego, where he provided courses on international management. Derek lives in Bethesda, Maryland, with his wife Sarah. [Spaceport Associates, 5909 Rolston Road, Bethesda, MD 20817; tel.: 301/493-2550; email: *Dwspace@aol.com*; website: *www.SpaceportAssociates.com*]

Appendix A contributor:

GEORGE S. ROBINSON, D.C.L. is an Attorney at Law, currently in commercial/space law practice with his sons and daughter-in-law in Virginia and Maryland. Recently, he retired as associate general counsel of the Smithsonian Institution. His previous legal counsel services have included NASA as an International Relations Specialist, the Federal Aviation Administration, and the U.S. Department of Transportation (Commercial Space Advisory Committee). Although his undergraduate studies were in biology and chemistry at Bowdoin College, his LL.B. was from the University of Virginia. He continued his professional studies in Montreal, Canada, receiving an LL.M. and first Doctor of Civil Laws degrees from McGill University's Institute of Space Law. He is also a member of several professional and governmental advisory committees promoting space commerce, Dr. Robinson has taught commercial law at several universities, including as an adjunct professor at George Mason University and a 1965 lecturer at Oxford University. President of Ocean-Space Services, George was formerly Chairman, Council of Regents/United Societies in Space. A prolific writer in both space philosophy, governance, and law, he was co-author of the classic book *Envoys of Mankind: A Declaration of First Principles for the Governance of Space Societies* (1986), and *Space Law: A Case Study for the Practitioner* (1992). [Dr Robinson's law office address is 8458 Meadows Road, Warrenton, VA 22186; tel.: 540/459-1630; fax.: 540/540-2210; email: *astrolaw@aol.com*]

Appendix B contributor:

DAVID R. CRISWELL, Ph.D. is Director, Institute for Space Systems Operations at the University of Houston; associate administrator of the Texas Space Grant Consortium for universities throughout that state. With a doctorate from Rice University, he has been an astrophysicist for 30 years. His prolific research has resulted in many patents, including one as co-inventor of the Lunar Solar Power System. An AIAA Associate Fellow, Dr. Criswell has published over 150 technical papers and conducted numerous studies for the University Space Research Association and NASA, one of which resulted in *Automation & Robotics for the National Space Program* (NAGW629, 1985). His career has included positions with TRW Inc., the Lunar and Planetary Institute, and the California Space Institute. In 1995 as a recognition of his accomplishments, the World Bar Association bestowed on him its *Space Humanitarian Award*, and the National Space Society elected him to its Board of Directors. [Dr. Criswell's address is ISSO, University of Houston, Houston, TX 77204; tel.: 713/743-9135; email: *dcriswell@UH.edu*]

Appendix C contributor:

THOMAS L. MATULA, Ph.D. is a professor of management and space commerce. Dr. Matula has a Bachelors degree from the New Mexico Institute of Mining and Technology (1983) and both an MBA degree (1984) and Ph.D. in Business Administration from New Mexico State University (1994). His dissertation focused on development of a model designed to identify factors that would influence public support for a commercial spaceport. He since has published numerous articles on space policy and economic development strategies for the space industry. Dr. Matula has served on the American Society of Civil Engineers' Space Engineering and Construction Committee and its Subcommittee on Lunar Surface Operations. His academic career includes 15 years of teaching and research on business strategy and marketing. Currently he is an online consultant for numerous universities and President of T. L. Matula and Associates, a business strategy consulting firm. [Dr. Matula's address is 13354 Bavarian Drive, San Diego, CA 92129; tel.: 858/382-1208; fax.: 858/538-5100; email: *tmatula@tmatula.com*]

xxxvi **About the contributors**

Appendix D contributor:

LINDA MARIE HAIGHT PLUSH (RN, MSN, CNS/ FNP, FRSH) is a Nurse Practitoner with a post-masters certificate from Azusa Pacific Unitversity, California, which in 2004 added her to their Graduate Academic Hall of Honor. Currently, she is CEO of two health care corporations with a third business under way. Linda founded: Plush Systems Inc., a health education and consulting business in 1991; West Palm Inc., a mobile acute dialysis service in 1995; and is in the process of building a chronic outpatient dialysis unit, which opened in 2006. In addition, Linda is an advanced practice nurse–CNS in adult nursing and nephrology. She has been an adjunct instructor for junior college nursing programs, and active in nephrology nursing, serving in various positions over the years from staff nurse to Director of Nurses. She is a member of numerous professional nursing organizations including Sigma Theta Tau, ANA/c, the Rogerian Scholars, and the Space Nursing Society. In addition to the above U.S. nursing organizations, in 1998 Linda was elected to the position of Fellow in the Royal Society of Health, U.K. She also is Past President of the local chapter of Soroptimist International (Antelope Valley), which is the world's largest professional business women's service club that enjoys NGO status at the U.N. In October of 2003, Linda was recognized by The Republican National Congressional Committee with the "Republican Gold Metal" for her work on reducing taxes on small business.

Linda is a student pilot and a PADI-certified open water diver, with an avid interest in health care in extreme environments. This interest led to her becoming the driving force behind the founding of the Space Nursing Society in 1991. She served as the founding President from 1991 to 1996, and is currently the Executive Director for SNS. She has been on several panels and task forces related to her interests in health care in space and extreme environments. She has been a topic coordinator for the *Life Support & Biosphere Science Journal* since its founding, and she is on the Editorial Board of *The Journal of Human Performance in Extreme Environments*. Linda is also a Life Member of the Aerospace Medical Association, and was a student member of the American Society for Gravitational and Space Biology.

In the past she has worked on magnetic levitation, including the health and safety issues involved in its use. Currently she is working on the use of medications in the space environment, and health care issues for long-duration spaceflight. On November 30, 2000, Linda became the only advanced practice nurse to participate on the Therapeutics and Clinical Care Integrated Product Team at NASA-Johnson Space Center. This is an honorary advisory panel which reviews and advises scientists and flight surgeons on operational issues related to clinical management and drug treatment, choice of medications and pharmaco-therapeutic research goals, emerging technologies and procedures for the U.S. Space Program. In addition, Linda was asked to participate on the Advanced Projects Team at NASA-Johnson Space Center. This team is currently working with telemedicine issues for long-duration spaceflight. Recently, Linda became a co-founder and Vice President of Space Medicine Associates, a company supporting the health care of tourists/passengers in space environments. [Plush Systems Inc./West Palm Inc., 43932 15th Street West, Ste. 105, Lancaster, CA 93534; tel.: (661) 949 6780 wk; fax.: (661) 949 7292; email: *lplushsn@ ix.netcom.com*]

Appendix D contributor:

ELEANOR A. O'RANGERS, Pharm.D., is President and CEO of Space Medicine Associates, Inc. A clinical pharmacist by training, she has a sub-specialization in cardiovascular pharmacology. Eleanor maintains an active interest in microgravity pharmacokinetics/dynamics and has lectured on the subject. She has contributed to the National Space Society's publication, *Ad Astra*, writing on space medicine and pharmacology-related topics. She also edited an issue of the *Journal of Pharmacy Practice* dedicated to space medicine. Eleanor has organized five space medicine program tracks and two space medical emergency simulations for the National Space Society's International Space Development Conference, as well as being an NSS Workgroup Leader for Space Medicine Issues.

Dr. O'Rangers has served as a pharmacology member of the Nutrition and Clinical Care Integrated Projects Team at NASA-Johnson Space Center, whose mission is to provide a non-agency perspective on pharmacology and nutrition research needs for the U.S. Manned Spaceflight Program. Eleanor continues to be involved in the development of monographs for Space Shuttle and International Space Station medications. Eleanor also volunteers twice monthly at the National Air and Space Museum in Washington, D.C., where she has been a docent since 1995. In addition, Eleanor was invited to serve as the sole Civilian Pharmacist Specialist for the Curriculum and Examination Board, U.S. Special Operations Command, Department of Defense. She also has been an active supporter of the Space Nursing Society.

Prior to co-founding Space Medicine Associates, Inc., Dr. O'Rangers worked in the pharmaceutical industry as a medical director for a major pharmaceutical brand and also served as a field-based scientist. In addition to her SMA activities, Eleanor continues to work as a scientific strategist/consultant for a number of pharmaceutical clients in the cardiovascular and endocrinologic therapeutic areas. [Dr. O'Rangers address is Space Medicine Associates, 4720 Water Park Drive #M, Belport, MD 21017; tel.: (917) 816-7547 cell; email: *eorangers@spacemedicineassociates.com*; website: *www.spacemedicineassociates.com*]

Appendix E contributor:

STEVE DURST, M.S. has been Editor and Publisher at Space Age Publishing Co. since 1976, and operates its Hawaii and California offices. Space Age publishes *Lunar Enterprise Daily* and *Space Calendar* weekly, and supports pioneering ventures such as the International Lunar Observatory, Stanford on the Moon, and Ad Astra Kansas initiatives. His company pursues a business plan consistent with establishing eventually a third office on the Moon.

Steve's commitment to the lunar imperative and to see people on the Moon within the decade reflects his understanding of humanity's greatest advance, and of the quickest way to great wealth, and to the stars. A graduate of Stanford University where he received his Bachelor and Master's degrees, he has a space project under way with the alumni of his Alma Mater. [Space Age Publishing Company/International Lunar Observatory Association, 65-1230 Mamalahoa Highway D20, Kameula, HI 96753; tel.: 808/885-3474; fax.: 808/885-3475; email: *info@spaceagepub* or *iloa.org* ... 480 California Ave. #303, Palo Alto, CA 94306; tel.: 650/324-3716; fax.: 650/324-3716]

About the author

PHILIP ROBERT HARRIS, Ph.D is a management/ space psychologist, as well as a prolific author and futurist. He is president of Harris International, Ltd. in La Jolla, California, founded in 1971 as a global management consultancy for human resource and organization development. Before that, he was vice president of Copley International Corporation, Senior Associate for Leadership Development Inc., and executive director of the Human Emergence Association, Inc.

Dr. Harris received his Ph.D. and M.S. in psychology from Fordham University, and a B.B.A. in business from St. John's University. In 1959, he was licensed as a psychologist by the Education Department, University of the State of New York. He is also a GS15 Federal Consultant. His biography is listed in many directories (e.g., *Who's Who in America* and *The Writer's Directory*), as well as on his website (*www.drphilipharris.com/*).

Author: For the past 60 years, this behavioral scientist has been engaged in leadership development, focusing his research and services on change, culture, communication, management, and space human factors. Dr. Harris has edited three journals, published over 250 articles, authored or edited some 48 books. In 2008, *Space Enterprise: Living and Working Offworld in the 21st Century* (the present book) was published by Springer/Praxis, along with *Toward Human Emergence: A Human Resource Philosophy for the Future* released by Human Resource Development Press. The latter previously published his *Managing the Knowledge Culture*, and the three-volume *New Work Culture Series* (*www.hrdpress.com*). In 2007, the seventh edition of his classic *Managing Cultural Differences* was released by Elsevier Science (*www.books.elsevier.com/business*). Adopted as a text in over 200 universities and colleges worldwide, this work is the parent of some 15 supplementary titles in the *MCD Series* which he co-edited. Phil, as he is known to colleagues, also co-authored the fourth edition of *Multicultural Law Enforcement*, recently published by Prentice Hall (*www.prenhall.com/criminaljustice*). In his work as a space psychologist, he has authored a professional book, *Living and Working in Space*, as well as a science-based novel, *Launch Out* (both available from *www.univelt.com*). Besides Amazon.com, his publications are listed at *www.booksandauthors.com*. For the past ten years, Phil has been a member of the editorial advisory board for the *European Business Review* in England.

Consultant/Educator: In his multifaceted career, this international consultant has served more than 200 organizational systems, including multinational corporations, government agencies, military services, professional associations, and educational institutions. During the Apollo period, his consulting with NASA took him to their headquarters and four field centers. Dr. Harris began his career as a secondary-school teacher and guidance director. In higher education, he was a vice president at St. Francis College in Brooklyn Heights, New York, then a visiting professor at The Pennsylvania State University and Temple University, as well as lecturer at numerous universities worldwide (e.g., Michigan State University, University of California-San Diego, University of Northern Colorado, Pepperdine University, University of Northern Colorado, University of Strathclyde in Scotland, Sophia University in Japan, East-West Center in Hawaii, and currently the California School of International Management).

Awards/Honors: Dr. Harris has been recipient of numerous awards and grants, such as from the U.S. Office of Naval Research, Fulbright Professor to India, NASA Faculty Fellow at the California Space Institute, Fellow of the NTL Institute of Applied Behavioral Science, and Associate Fellow of the American Institute of Aeronautics and Astronautics. In 2005, the Foundation for the Future invited Phil to be a participant in a *Humanity 3000 Symposium*, and his remarks appear in their proceedings (*www.futurefoundation.org*). The American Society of Training and Development presented him with the *Torch Award* for outstanding contributions to the field of human resource development. The Junior Chamber of Commerce named him *New York City Young Man of the Year for 1959* for his work in the renewal of downtown Brooklyn, New York.

Media: Phil has been interviewed on many radio and television programs. He has also been a producer of major media projects (e.g., for NBC/*Sunday Today Show*; Westinghouse Learning Corporation, and the U.S. Marine Corps). The Aviation/Space Writers Association bestowed several Journalism Awards for Excellence for his published space articles. At the apogee of his career, this 82-year-old scholar is hoping that *Space Enterprise* may become a classic in the global space community, as well as a university textbook, and basis for an educational media project! [Email: *philharris @aol.com*]

January 22, 2008

Prologue

Since the time our ancestors climbed down from trees and walked upright, the human species has always probed new frontiers! Over three million years, we manifested this characteristic for exploration repeatedlyas when our early ancestors migrated from Africa across the planet, or when Europeans opened up the Pan American and African continents, or when Americans pushed across the Western frontier. With each terrestrial expansion, we created more sophisticated mental models of our world, more complex social relations, while increasing our control over matter. Through these achievements, civilization has advanced, vast resources developed, and settlements established. In the last part of the 20th century, humans took their first faltering steps off their earthly homeland, moving beyond the atmosphere to another frontier which Isaac Asimov describes as much vaster and incomprehensibly richer. Going aloft to explore and uilitize these resources, he envisions as a great goal for Earth's peoples, one to fill our hearts and minds with glory and satisfaction, making narrow suspicions and hatreds seem small and unimportant.

But relatively few appreciate how these are truly "giant steps for mankind" as we fly into the real *new world*, one free from gravity and population limitations, as well as atmospheric impurity and endless energy. Outer space is an ideal realm for experimentation and production that is impossible on Earth. Its multitude of assets can enrich the human family, possibly eliminating poverty on this planet. Permanent stations, outposts, bases and eventually cities in that orbital environment provide an unparalleled vantage point for scanning the cosmos and understanding the universe. Already both manned and unmanned spacecraft transmit to earthlings, information and images about other planets and galaxies within our Solar System. Space satellites have proven most persuasively their value for improving our global communication and agriculture, for predicting the weather and tracking human activities, for studying the Earth's topography and oceans, for understanding our own fragile biosphere, in terms of both problems and resources. In accomplishing such advances for the benefit of the global human family, we have taken high risks, investing large financial, material, and human resources. We have witnessed peak human performance among the thousands of space workers, both on the ground and in orbit, who make it all possible but especially among the several hundred or so of our species who actually made it aloft. We also have experienced failures and loss of life in the last fifty pioneering years which inaugurated the Space Age. But in the process, humankind has literally gone from the cradle of this Earth to its Moon and beyond. The impact of these initial space endeavors has tremendous implications and hope for our future, impelling us to change our collective image of our species. *We are no longer earthbound maybe our real home is out there!* Exhibit Pi depicts where our new world begins.

So after millions of years as terrestrial beings *Homo sapiens* in ever increasing numbers is beginning to migrate offworld. Human enterprise in space, so far, is manifested in both unmanned and manned activities. It is the creative human mind that is responsible for launching automated spacecraft to far-off planets, as much as for the building and staffing of a space shuttle or station. Extraterrestrial human activity increasingly involves artists, architects, film makers and entrepreneurs, as well as planetary scientists, engineers, technicians, and aerospace planners. The

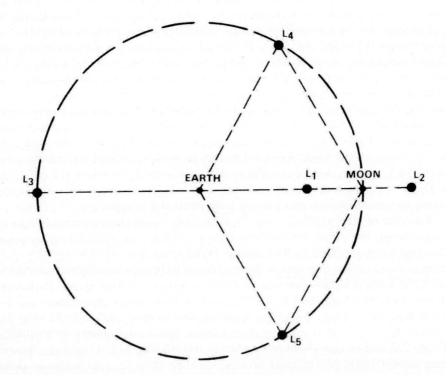

Exhibit Pi. Libration points/Zones in Outer Space. The Lagrange or Libration Point One is 35,000 miles above the Moon's center, in direct line with Earth. The five L-points pictured are ideal positions for launching humanity into other worlds. Source: National Commission on Space, *Pioneering the Space Frontier*, New York: Bantam Books, 1986, p. 132.

problems of transporting equipment and personnel safely into both lower and geosynchronous orbits necessitate the current emphasis in the space program upon the natural sciences, engineering, and technology. We need hardware and software solutions to get out of *Earth's gravity well*, especially inexpensive, reusable, and safe launch systems. However, once we gain full access to space, hopefully at lower costs, what do we do when we get there? The National Commission on Space cited in the above exhibit articulated a vision for people *Pioneering the Space Frontier*. Exhibit Pii from that momentous report illustrates the realm of space that is the focus of this book.

Space developments for the remainder of this the 21st century will particularly demand greater contributions from the behavioral, biological, information, and engineering sciences. Expanding human presences throughout the Solar System will accelerate scientific investigations by psychologists, sociologists, anthropologists, and communication specialists, as well as by biologists, physicians and nurses. As well as experts in the environment, ergonomics, and ecology. Construction of space stations and lunar bases lay the groundwork of a space infrastructure for the next fifty years, leading to further space settlements, manned missions to Mars, mining of the asteroids, and eventually to more human colonies orbiting in space, or established on other planets. Beginning with a handful of astronauts and cosmonauts, extending to space construction workers or *technauts*, human population up there is likely to escalate during this Millennium to thousands of spacefarers, whether as visiting scientists, consultants, tourists, or settlers. The politician, teacher, journalist, or tourist who may now fly the Shuttle or a rocket are forerunners of the millions of *spacekind* to come!

To study this emerging phenomenon of human behavior, culture, and potential aloft, this book was first compiled, largely from the perspective of behavioral scientists. The possibilities for such a work became evident during the exhilarating Apollo missions of the 1960s when I served as a management and psychological consultant to NASA. Involved then in an executive and management development project which took me from the Agency's headquarters to many of its field centers, I began to sense that the organizational cultures of that space agency and its aerospace contractors were centered naturally upon the technical and engineering aspects of space undertakings, while influencing broader decisions on human living and working aloft. Later in 1984 as a NASA Faculty Fellow during a summer study at the California Space Institute, I participated in analysis of space resources as the technological springboards for developments in the next century, the proceedings of which were eventually published in five volumes.[1] Fortunately, during this strategic planning for a lunar base, one of the four workshops was devoted to examination of economic and system tradeoffs, as well as management, political and societal issues related to space activities. As team leader then for a group of diverse scholars considering the human dimensions of the lunar enterprise, I was exposed to interdisciplinary input on the

---

[1] McKay, M. F.; McKay, D.; and Duke, M. (eds.) *Space Resources*. Washington, D.C.: U.S. Government Printing Office, 1992 (NASA SP-509), 5 vols. Your author's contributions are in Vol. 4, *Social Concerns* (now available from *www.univelt.com*).

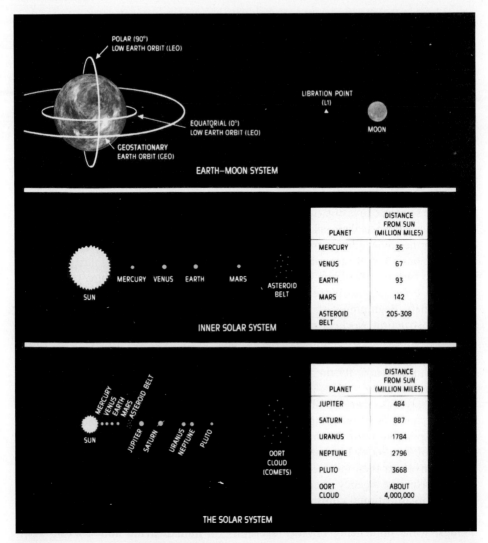

Exhibit Pii. Our realm of Outer Space. Exploration begins in the Earth-Moon System, extending out into our Solar System, and then out into the universe/s! Source: National Commission on Space, *Pioneering the Space Frontier*. New York: Bantam Books, 1986, p. 7.

subject. It came from varied perspectives of psychiatrist and psychologist, political scientist and attorney, biologist and design engineer, anthropologist and architect, in addition to an economist and management specialist. Since NASA encouraged us to publish our insights, some of these findings were summarized and shared by me not only in numerous journal articles, but in the original two editions of *Living and Working in Space*. This new version, *Space Enterprise: Living and Working Offworld in the 21st Century*, represents forty years of my behavorial science research on space.

Bur this edition, is a completely revised, updated, and expanded work, timed to enhance understanding of our new national space policy, *Vision of Space Exploration* (VSE).

Space Enterprise consists of ten chapters, which begin with an examination of emerging global space visions, ethos, and enterprises so vital to our extraterrestrial expansion. We consider some of the human dimensions involved, apart from scientific and engineering considerations. Current and potential behavioral science contributions in the transition of *Homo sapiens* beyond our home planet are analyzed, especially from the perspective roles of anthropology, sociology, psychology, and living systems theory. Then reader attention is directed to the influence and the impact of space exploration on Earth's cultures, including the cultures of space organizations, as well as in creation of a new space culture. Next a chapter is devoted to a review of human performance aloft, and what can be done to ensure higher levels of orbital team behavior. A strategy is proposed relative to creation of a space personnel deployment systems for more effectively sending and returning greater numbers of people to and from orbit. Another chapter describes the kind of macro-thinking necessary for global space planning, whether by the public or private sectors. Innovation in managing space enterprises follows to stimulate reader mind-stretching about adopting new strategies of macromanagement. In addition to several national case studies in the opening chapter, two special cases appear in other chapters on the International Space Station and spaceports. In treating the diverse challenges in space industrialization and settlement, special emphasis is placed upon the prospects of space commerce, law, and policy. Finally, the last chapter closes by looking at alternative strategies for lunar economic developments since the Moon has to be the first platform toward exploring our universe. Each chapter has a bibliography of the related references should the reader wish to probe further into the subject matter, plus a section on resources at the end of the book. Previously, all of the chapters in this exposition have been published in professional journals and subject to peer review. Four of them have received journalism awards for excellence from the Aviation/Space Writers Association.

To facilitate the creation of a spacefaring civilization, the Epilogue proposes *action* plans for translating space visions into realities. Finally, the book concludes with five additional appendices by other space experts which supplement my chapter content.

Space Enterprise is intended primarily as a text for professionals and advanced students in both the behavioral and space sciences. However, I hope it will appeal to other audiences, such as:

- thoughtful and intelligent people who are intrigued by the human adventure of advancing upward and outward into our universe. The latter includes not only the public who are generally supportive of space exploration, but the hundreds of thousands of space activists worldwide who accelerate "high ground" developments through their organizations, publications, and lobbying
- university students and researchers involved in academic courses related to space science, habitation and settlement

- social scientists and futurists wishing to advance their own knowledge and contribution to interstellar migration
- business leaders, managers, and entrepreneurs concerned about challenges in space commerce
- planetary scientists and space engineers exercising their technical capabilities in space activities who wish to understand more about the human needs aloft of those who are to benefit from their expertise
- media specialists who have the talent to communicate the significance of space development to the masses among Earth's peoples.

We remind all readers that this volume is again dedicated to those *envoys of humankind*, the world's astronauts and cosmonauts, some of whom died to advance human migration into outer space. On October 4, 2007, this book was finished as a tribute for the fiftieth anniversary of Sputnik, which inaugurated the Space Age. It also happens to be the feast day of Francis of Assisi, the peacekeeper who wrote poetry about Sister Moon in the 13th century!

Philip Robert Harris, Ph.D., Author

Management/Space Psychologist,
Associate Fellow, American Institute
of Aeronautics and Astronautics
La Jolla, California, U.S.A.
www.drphilipharris.com/

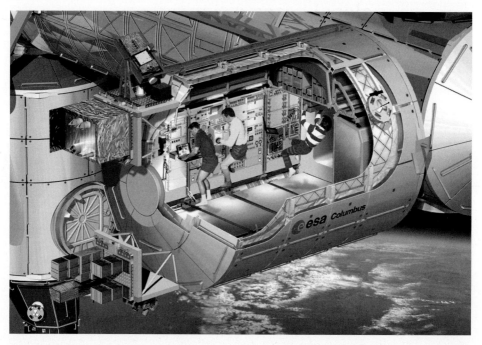

Exhibit Piii. Offworld human activity. An artist's impression (a cutaway view) of astronauts working in *Columbus*, the European laboratory module of the International Space Station. Source: ESA—D. Ducros.

1

Toward a global space vision, ethos, and enterprise

> *Yet the greatest revelation that has emerged from space research and exploration—one that lies at the heart of every scientific endeavour—is that the Universe is knowable. And we have begun to know it ... The 20th century closed with an orbiting space station, a near comprehensive tour of the solar system, and the discovery of the fifth force in nature—an inflationary Universe and awareness of more planets outside our solar system than within. We have begun to know the workings of our own planet, its oceans and its origins. We have begun to discover the secrets of life itself.*
>
> Dana Berry, *Smithsonian Intimate Guide to the Universe*, Washington, D.C., 2004

> *With the 50th anniversary of the Space Age, this book seeks to share insights from the behavioral sciences on the momentous significance of humanity going offworld: past, present, and future.*
>
> Philip R. Harris, Author

Human enterprise in this new millennium will focus on the development of a *spacefaring civilization*! The 21st century will expand space commerce and industrialization, with consequent space settlement, starting with the Moon. Such progress will be feasible because of increasing collaboration among nations and organizations in space exploration and development. The very complexity and scope of undertakings beyond Earth necessitate such an approach. The late U.S. Senator Matsunaga from Hawaii realized this when he observed:

"At a certain point, anything other than international exploration of the cosmos from our tiny planet will cease to make any sense at all ... we must develop policies that respond to the unfolding realities of the Space Age, that move out to meet it on its own uniquely promising terms. Without such policies, earthbound civilization can only wind up recoiling upon itself [1].

Organized events and institutions like the International Space Year, World Space Congresses, professional space societies, world space agencies, and the United Nations, especially through its Office of Outer Space Affairs, contribute to this process by providing the forums for such sharing and planning of activities beyond Earth. Humanity is motivated to pursue such enterprises by envisioning its future, setting goals and policies, establishing objectives and targets, while proceeding with implementation, step by step.

SPACE VISION

What is happening to humanity off this planet should be viewed within the context of "Cosmic Evolution". In a seminal book on this subject, Eric Chaisson maintains that we are connected to space and time not only by our imaginations, but also through a common cosmic heritage.[1] This Tufts University research professor argues that emerging from modern science is a unified scenario of the changing cosmos. Our richly endowed universe manifests a pattern of order and structure for every known class of object. From this perspective, sentient beings like ourselves play an integral role in the origin and development of matter. Our movement as a species offworld is part of this cosmic process.

In formulating a space vision, the basic question to be answered is *why go into space at all?* Jesco von Puttkamer, a former NASA strategic planner, sees this evolutionary opening of cosmic space as a *cultural process*. For him, going into orbit is a social phenomenon, a force for change and growth in the human family. Having worked with the futurist, Wernher von Braun on the Saturn/Apollo and Skylab programs, this aerospace engineer views spaceflight as a source of innovation bringing immense returns. Von Puttkamer believes that we have an ethical obligation to invest in space on behalf of future generations. Accessing space requires a *consciousness change*, so we see humans and Earth–Space as one creative system, as illustrated in Exhibit 1. We go into space for the same reasons that prehistoric ancestors left the seas to breathe on land. Planetary scientist, Carl Sagan, also assured us that human expansion into space is genetically driven, an inevitable part of the natural evolution of the universe. Princeton physicist, the late Gerard K. O'Neill, thought: "the opening of the high frontier represents opportunity that challenges the best in us—as humanity lives and works off this planet, we are given new freedoms to search for better ways of life."

---

[1] Chaisson, E. J. *Cosmic Evolution: The Rise of Complexity in Nature.* Boston, MA: Harvard University Press, 2001 (*www.hup.Harvard.edu/catalog/CHACOS.html*).

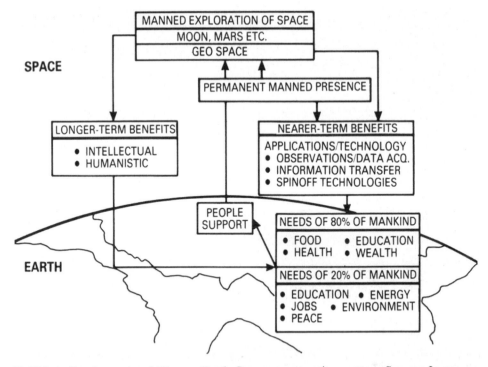

Exhibit 1. Development of Human–Earth–Space, one creative system. Source: Jesco von Puttkamer, from his "Foreword" to the first edition, *Living and Working in Space*, by P. R. Harris. Chichester, U.K.: Ellis Horwood, 1992, p. 18 ... See also the biography of America's pioneering space visionary, *Wernher Von Braun: Crusader for Space* by Ernst Stuhlinger and Frederick Ordway III (Melbourne, FL: Krieger Publishing Company, 1996; info@krieger-pub.com).

1.1 SPACE VISIONS OF HUMANKIND

For eons, humans thought they were Earthbound, but dreamed of leaving the planet and even of going to the Moon. The Man in or on the Moon was a recurrent theme in our mythology, so philosophers, poets, and novelists wrote about the possibility of our species moving beyond this terrestrial home of origin. Such dreaming became reality when a 27-year-old Russian cosmonaut, Yuri Gagarin, orbited the Earth in the "cosmic ship" Vostok on April 12, 1961. Then, on July 20, 1969 the Apollo 11 LEM landed on the lunar surface, allowing astronauts Neil Armstrong and Buzz Aldrin actually to walk on the Moon! [2] (see Exhibit 2). As longshoreman–philosopher Eric Hoffer reminded us, we are now forced to change our image of humankind: we are *no longer Earthbound*; maybe we are only transients on this planet, perhaps our real home is out there! Space achievements challenge us to remove the psychological binders and blinders on our collective self-concept, fostering the actualization of human potential. But as we move ahead on our joint space

Exhibit 2. First lunar landing of humans. Apollo 11 astronaut "Buzz" Aldrin is photographed on the Moon, July 21, 1969, by crewmate Neil Armstrong reflected in the visor; he is the one who made the first giant step for mankind, saying "That's one small step for a man, one giant leap for mankind." Source: NASA Headquarters [2].

odyssey, we need to be pragmatic. While nothing worthwhile is achieved without vision, reality has that nasty habit of undermining or even obliterating that vision! This will become evident in our attempts at fulfilling the latest *Vision of Space Exploration (VSE)*, a national policy that will be discussed latter in this chapter. For now, we examine the evolution of that particular space vision.

As humans change mindsets about who and what we are, so will we alter our vision of outer space and our purposes in its exploration, development, and settlement. In a Los Alamos conference and proceedings on interstellar migration and the human experience, the assembled scientists set forth this vision [3]:

- For the past five million years of human physical and cultural evolution, the learned use of technology—both material and social—has enabled this exploring species to expand and cope with all aspects of this planet's environment and beyond ...
- Technological prowess, revolutions, and contributions have not only driven human culture and procreation for the four billion inhabitants of this Earth, but are creating an entirely new space culture and being ...
- Just as the ancient Polynesians navigated through images in their heads to create a new culture by colonization of the Pacific islands, so our descendants with emerging space technologies and explorers' bent may use the asteroids, comets, and planets to populate, settle, and civilize our Solar System ...
- Since *Homo sapiens* adapts and evolves both biologically and culturally, the conquest and colonizing of space will facilitate a quantum leap in human evolution and diversity, eventually leading to crossing the gulf of light years to other star systems.

1.1.1 Evolution and space development

In an anatomy of new realities, the late Jonas Salk brilliantly philosophized on man unfolding and the "survival of the wisest" [4] With these ponderings, the polio pioneering physician speculated on a "metabiological evolution" that results in an increase of consciousness, including of our evolution. Just as biological evolution depends on genes in a cell, the metabiological equivalent is *ideas* generated in the human mind. Such ideas affect the nature, characteristics, and behavior of the metabiological organism, whether as an individual or collectively in society. My friend, Jonas the scientist, was intrigued by the way the human mind deals with the challenges posed by our environment, developing *metabiological* traits that are transmitted, like genes, to succeeding generations. Dr. Salk envisioned the human family as a single organism requiring new responses and relationships, especially between our intuition and reason. He described evolution as a process of changing relationships. If Salk's thesis is correct, that human thought and creativity develop in the response to our environment, then it is your author's contention that the space environment will produce the real metabiological evolution with immense cultural diversity. There, *beyond the Earth's atmosphere*, humankind may develop the wisdom to survive and diversify as part of the cosmos, thereby ensuring our continuance as a species!

When our primordial ancestors made the transition from ape to human, they moved from jungle to savanna, employing tools in their quest for survival. The initial stages of human development were first tribal, centered around hunting and food gathering; followed next by the agricultural period focused on farming, which stimulated the beginnings of market places or cities. In both previous stages, our ancestors increasingly utilized technology that produced personalistic, flexible, and small-scale organization, as well as conflict. Subsequently, the third industrial stage employed mechanized technology and systems, laying the groundwork for going off the Earth. In our present fourth developmental stage, the metaindustrial, our energies are directed toward using information and communication technology, leading to the phenomenon of globalization [5]. One outcome today is that the management of complex, large-scale systems requires cooperation, all factors contributing to the movement from terrestrial to extraterrestrial living. To succeed aloft requires innovative research and synthesis in both space engineering and behavioral science technologies. In this century, the emphasis in on creating a *knowledge culture*, which will be best accomplished offworld [6]. As humans expand outward into the universe, anthropologist N. P. Tanner from the University of California-Santa Cruz comments that this migration into solar and star systems will necessitate cross-cultural skills and genetic diversity, whether we meet extraterrestrials out there or whether *we become* the differing extraterrestrials! (see Exhibit 3).

1.1.2 Multiple visions of space

Multiple visions have been put forth about humankind's diaspora in space by cosmic philosophers and prophets. Perhaps science fiction writers best bestirred our imaginations about space, from authors like the 17th-century Jules Vernes to the 20th-century Robert Heinlein (*www.Heinleincentennial.com*). The professional visionaries range from 19th-century Russian science teacher Konstantin Tsiolkovski to 20th-century American engineering professor Robert Goddard and Romanian mathematician Hermann Oberth. Their utopian dreams were translated to realities by teams of German, Russian, or American rocket engineers led by men like Sergei Korolev, Wernher von Braun, and Krafft Ehricke [7]. Their forecasts of yesterday are today's realities, such as satellite telecommunications, remote sensing, Earth observation, manned and unmanned missions aloft, and more. During the past several decades, visions of orbital space colonies and artificial worlds have been set forth in classic books in many languages; some of these authors have even suggested planetary engineering or *terraforming* for the restructuring of other planets into New Earths [8].

Since the 16th century, artists and now a variety of media specialists have illustrated and interpreted such space visions for the masses, from artists such as Michelangelo, Lucian Rudaux, Chesley Bonestell, R. A. Smith, and David A. Hardy, to film producers such as George Pal, Stanley Kubrick, Arthur C. Clarke, Gene Roddenberry, Steven Spielberg, George Lucas, and Tom Hanks. Further, the International Association of Astronomical Artists (IAAA) has drawn together artists from 14 countries to visualize our future in space; some of their artistic renderings are reproduced in the chapters which follow. Using all mediums and even computer graphics, these creative professionals help us by imagination to journey where we cannot yet go, even to envision space settlement and interplanetary travel. Their Association members also include those who have been in orbit, such as cosmonaut Alexei Leonov and astronaut Alan Bean [9]. In addition, creative professionals in fields such as simulation, animation, and virtual reality, have produced artificial space experiences ranging from videos to games. Currently, the differences in visions of human futures in space center around whether this frontier is to be used to promote peaceful scientific exploration, commerce, and settlement, or for military defence and eventually war [10].

Fortunately, with the demise in the 1990s of the former Soviet Union and the Cold War, the former view is in ascendance, while the approach of "Star Wars" and the Strategic Defence Initiative diminishes somewhat. In the U.S.A., as a *case in point*, the National Commission on Space (1986) set forth an American vision of civilian space goals for the next 50 years. The Commission's report, *Pioneering the Space Frontier*, offered a scenario for a multitrillion-dollar, space-based economy. These distinguished proponents of space commercialization called for private–public sector partnerships in a threefold increase in space investments, research, technologies, and transportation. The Commission advocated living and working in space on a permanent basis, and offered a roadmap for goal achievement. This American dream of a space renaissance, unfortunately, was overshadowed by other realities, such as a failing economy, so that the nation's plans for space were curtailed [11].

1.1.3 Articulating a space vision

The space activist movement endorsed the direction of that Commission's vision. Among them, one of the largest organizations, the National Space Society, articulated specific goals and a timetable for creating a *spacefaring civilization*, such as [12]:

- immediate inauguration of a lunar polar orbiter, lunar transport system, Mars geothermal orbiter;
- completion of a Space Station design aimed at a manned return to the Moon;
- provision of adequate Earth-launch capability to support build-up of a permanent presence on the Moon by the processing/utilization of lunar resources by 2010;
- development of nuclear space power systems both for transportation systems to support larger space settlements, while reducing launch payload costs and limitations;
- public policies and subsidies for a U.S.A. commercial space transportation industry.

In an attempt to sharpen its own planning and goal setting, as well as to enlist public support, the National Aeronautics and Space Administration has promoted a number of in-house studies to present its near-term views of the U.S.A.'s space future. Some of these reports and special publications were completed in association with the NASA Advisory Council, the National Academy of Science, or with the National Academy of Public Administration [13]. Under the aegis of astronaut Sally Ride, the Agency expressed its most focused vision in *Leadership and America's Future in Space*. This report discussed the immediate future in terms of proposals for

- examination of global phenomena on this planet from the international perspective of geostationary and polar-orbiting platforms: the "Mission to Planet Earth';

- facilitation of a planetary science program of the outer planets and smaller celestial bodies (e.g., asteroids), possibly including an unmanned Mars sample return: the "Exploration of the Solar System";
- establishment of a scientific outpost on the Moon, beginning with robotic exploration and Lunar Geoscience Observer until several operational phases result in a base for 30 people engaged in scientific and industrial pursuits (see Exhibit 5);
- inauguration of a Mars manned mission program culminating in a human landing on that planet or its moons, and eventual establishment of a permanent base.

Alas, that country's sociopolitical, military, and economic situation again limited the implementation of most of these visionary plans. But in the 21st century, the National Space Society is energetically supporting efforts to implement the new VSE policy. However, government actions alone are insufficient if a nation's space vision is to be communicated to and confirmed by a majority of citizens. Too often technologists drive the space program, neglecting to involve the public in the process [14].

When in 1992 the United Nations endorsed the International Space Year (ISY), it called for activities that would improve [15]

- the management of the Earth and its resources;
- long-term education in space science, technology, and applications;
- public education on the role of space science and technology, with reference to the Earth's environment and the rest of the universe.

But, to ensure success in permanent space habitation requires broad interdisciplinary and institutional participation, as well as interagency and international collaboration. Professional associations in the space field have long promoted conferences and publications to express their vision of the future (such as the International Astronautical Federation, the British Interplanetary Society, the American Institute of Aeronautics and Astronautics, the American Astronautical Society, and comparable organizations). Invariably, their published proceedings contribute to a body of literature making the case for space investment. Other clarifications of space vision may come from the published reports of research and non-profit organizations (e.g., the United Nations Association of the U.S.A. in their 1986 report, *The Next Giant Leap in Space: An Agenda for International Cooperation*). Annually, for instance, voluntary organizations gather to discuss our space futures, and again publish the proceedings which express their vision: two examples are the *International Lunar Exploration Conference* (www.spaceagepub.com), and the *National Space Society's International Space Development Conference* (http://www.nss.org). Sometimes informal networks, often gathered through the Internet, best express the space aspirations of inhabitants via the information superhighway (e.g., www.space.com and www.thespacereview.com). Now podcasts are being used to permit space experts to share their thinking with worldwide audiences (www.thespaceshow.com).

The visions of planetary science, in particular, inspire and fascinate us, while shedding new light on the workings of our own planet, thus facilitating the creation of the field of Earth Science. In summary, the various visions of humankind's future on the high frontier generally conclude that *space is a place* for

- advancing human evolution and culture through space technology and the practice of synergy in the application of information and knowledge;
- improving the quality of human life both on Earth and offworld by utilizing the unlimited space resources;
- engaging in peaceful human pursuits, such as scientific exploration and experimentation, industrialization, and settlement.

What is needed in this century is *global expression* of other space powers, beyond the U.S.A. and Russia, that includes the People's Republic of China, Japan, India, and the European Space Agency. Eventually, possibly through the United Nations, humanity will articulate its future plans for outer space.

To conclude this section, consider these varied statements relative to human evolution in Exhibit 3.

As the Renaissance defined European identity in the past by liberating individual creative power, so the Space Age gives meaning to today's generations, freeing us to move beyond terrestrial thinking. Migrating off this planet opens our minds to new discoveries and knowledge, challenging us to formulate new visions of humanity's future. The nature of our species is to explore the unknown, to pursue the far horizon, to forge new frontiers [18]. Driven by destiny to extend human civilization beyond our own Solar System, the bold journey through space satisfies our spirit to know, increases our coping skills, and enriches human culture. Perhaps my esteemed colleague and space psychologist Dr. Albert Harrison best sums up the case for moving beyond planet Earth in Exhibit 4.

SPACE ETHOS

Perhaps the underlying need for humankind is to articulate and support this vision with a new space ethos. *Ethos* is defined as the *fundamental character* or *spirit of a culture*. It is the underlying sentiment that informs the beliefs, customs, practices of a society. Moving beyond Earth causes a redefinition of the American, Russian, European, or Asian ethos. As will be seen from the following four mini regional case studies, nations are struggling to redevelop their purpose, policies, and priorities as to their roles in the development of outer space. Currently, the majority of global inhabitants do not perceive space, its exploration and utilization, as central to their wellbeing. In general, humanity is still terrestrially oriented. People have yet to fully grasp the deep significance of migrating aloft, and its importance to this planet and its peoples! Each of these cases describes the emerging *space ethos* in different world regions that is taking us beyond Earth [19] Humanity is beginning to appreciate that we explorers are required to undertake bold endeavors [20].

Exhibit 3. HUMAN EVOLUTION AND SPACE.

- **John Desmond Bernal**, British physicist, in his 1929 classic book, *The World, the Flesh, and the Devil*: "However, once acclimatized to space living, it is unlikely that Man will stop until he has roamed over and colonized most of the sidereal Universe, or that this will be the end. Man will not ultimately be content to be parasitic on the stars, but will invade and organize them for his own purposes."
- **Daniel Boorstin**, Librarian of Congress Emeritus and author of *The Discoverers*: "The most promising words ever written on the maps of knowledge are terra incognita—unknown territory ... If the mere displacing of the Earth from the center of the solar system was so disturbing to thoughtful laymen then, what must be the consequences in our time of the discovery that our whole solar system, our whole Milky Way, our whole galaxy, our whole Universe, is only a negligible peripheral one of countless billions ... Perhaps we are no longer *Homo sapiens*, but rather *Homo ludens*—at play in the fields of stars" [16].
- **Alvin Toffler**, futurist on *The Space Program's Impact on Society*: "Space will not only shape our descendants view of our time; it also shapes our view of the future ... There is at the same time a growing post-national consciousness around the world, a globalist ideology with a planetary view, and space activity has clearly been a major contributor to its spread. The very image of Spaceship Earth has been a potent diffuser of post-national globalism" [14].
- **Patricia Daniels** and **Stephen Hyslop**, authors and editors of *Away from Earth*: "The Space Age began with a 184 pound ball hurled into Earth's orbit by a Soviet rocket on October 4, 1957. Fewer than 20 years later, two Americans strolled the surface of the Moon. These events represent the two options for exploring space—with or without humans ... Earth's population is forecast to exceed nine billion by 2050 ... Past predictions that human beings will overwhelm Earth's capacity to support them have all proved false. Technological advances have postponed that day of reckoning. There is reason to be hopeful that humankind will continue to escape such fate, but there is no guarantee of success" [17].
- **Carl Sagan**, astronomer and winner of the NASA Apollo Achievement Award: "Humans had lived all their lives, back to the first person a few million years ago, confined to one small world among a vast number in the cosmos. Suddenly, we stepped out and set foot on another. In a way, Apollo 11 was merely the continuation of human exploration and discovery ... a turning point in history ... It taught us that the future can be greater than the past. There was a time, not long ago when we soared. Let us soar again ... It may be too expensive to go it alone, But together then human species can do it. Why don't we?" [18].
- **Michael Griffin**, current NASA Administrator: "Humans will colonize the solar system and one day go beyond ... I think that we will be poised by 2020 for a rational, national decision to aim resources at going to Mars" (NSS *Ad Astra*, Fall 2007, pp. 20, 24).

Exhibit 4. WHY GO OFFWORLD?

We seek to expand our presence in space in order to conduct science and pursue knowledge, tap the fabulous resources of the universe, grow psychologically, and help resolve international tensions ... Utilitarian motives include science and industry, national defence, generation of wealth, and other practical activities that have preoccupied us in the past ... Transformational motives include developing human potential, eliminating conflict and war, and promoting social renewal by offering people a fresh start. These represent the next step forward, evolving beyond creatures of the world to become creatures of the universe. Transformational motives are qualitatively different from business as usual: they involve innovation and will change the *status quo* ... As a species, we humans have always wondered about the planets and stars, but only recently have we been able to venture off our planet's surface. This biological foundation is strengthened by several cultural themes, including our interest in technology, our fascination with exploration and conquest, and our hope for new opportunities. (Source: Albert A. Harrison, *Spacefaring: The Human Dimension*. Berkeley, CA: University of California Press, 2001, pp. 17–18 ... See also John S. Lewis, *Mining the Sky: Untold Riches from the Asteroids, Comets, and Planets*. New York: Helix Books/Addison-Wesley, 1996.)

WHY BOTHER WITH HUMAN EXPLORATION

... When robots do such a good job and do it much more cheaply? Proponents of human spaceflight argue that only humans have the supple physical coordination and mental agility to get the most from an expedition. But the most compelling argument for human spaceflight may remain the one that worked at the beginning. Space exploration is ultimately about human dreams ... As Michael Griffin, present NASA administrator, says, "going offworld is about the drive to extend our reach—human destiny is reason enough!" (Source: Guy Gugkiotta, "Space the Next Generation," *National Geographic*, October 2007, p. 124.)

1.2 NORTH AMERICAN SPACE CASE STUDY: U.S.A., CANADA, MEXICO

Over the past few hundred years, the New World experience has generated a unique national ethos: attitudes and convictions inherent in the heterogeneous cultures of both the United States of America and Canada. As Robert Reich noted, when U.S. Secretary of Labor, "One of the saving graces of Americans has been a willingness and ability to roll up their sleeves and get on with the task, putting ideology aside" (*Los Angeles Times*, May 3, 1987, II:1). This has been demonstrated in the way that country mobilized behind a national goal in the 1960s to put a man on the Moon: some 400,000 people were involved in those Apollo lunar missions. In this decade, such national purpose has again been manifested to ensure permanent human presence on the space frontier by year 2020 [21]. When concerns for space ventures truly become part of North America's ethos, they will dominate the public's assumptions, values, and dispositions.

If the United States had such an ethos, the issuance of the watershed report of the National Commission on Space (1986) would have been really big news; implementation today of its proposals would now be a priority. That work was followed by NASA's own "Ride Report" (1987) [13]. If the recommended strategies of these and many other such studies had been instigated, such as envisioned in Exhibit 4, humans would have been working on the Moon for decades, instead of only planning for that return. Unfortunately, lacking an adequate space ethos in both the U.S. and Canada, their space visions never get fully translated into actual policy, backed by both government and public commitment.

Had a national space ethos been cultivated, then both countries would have in place meaningful space policies and strategies that gain both citizen and investor support. Thus, North American leadership in space would be assured through the 21st century by adequate financing and implementation of bold plans. Although the American people have been sporadically enthusiastic about space endeavors, particularly as related to space movies and television programs, their interest has not been translated into sustained resolution behind long-term space exploration. About 2 million people in the aerospace industry, government and military, apparently may be counted among their space community. According to various polls, the space supporters in the U.S.A. and Canada represent only a small fraction of the general population. Some 400,000 activists who do possess a space ethos have joined about 50 space advocacy groups to influence policy and legislation, to inform themselves on space matters and participate in space progress. In the nation's capital, *Spacecause*, affiliated with the National Space Society, has effectively represented and delivered the pro-space message through lobbying, conferences, and public affairs efforts at the grass-root level.

The largest of these space groups, the Planetary Society and the National Space Society (NSS), conduct and publish an annual survey of their membership on space policy and goals. Their members manifest their *space ethos* by overwhelmingly voting for cheaper access to space and greater involvement of private enterprise in space exploration. When asked why explore and develop outer space, their principal reasons given were

- furthering technological development
- expansion of settlement frontiers
- improving the Earth's environment.

In reply to the question on how the U.S. Congress should support commercial space activities, the vast majority of activists urged

- eliminating adverse government regulations of space industry;
- increasing government R&D contracts to private enterprise;
- expanding privatization of space activities;

- eliminating direct competition between NASA and private firms;
- providing tax exemptions on space-developed products and investments.

Yet, a groundswell of popular mass conviction for space enterprise has yet to energize the average person in North America, especially in favor of synergistic support for a new space transportation system, the International Space Station, lunar base, and manned Mars mission. The situation today has slightly improved from 20 years ago when Thomas Paine, former NASA administrator and Chairman of the National Space Commission, succinctly expressed the frustration of the scientific and commercial communities:

> "The biggest problem is the lack of direction of the U.S. space program which adds to the difficulty in mounting major international programs that could enhance the exploration of space" (*Los Angeles Times*, May 25, 1987, I:3).

Since that astute observation, there is still no national unanimity in 2008 as to what America should do next in space and how much should be invested. There is no national *space ethos*, and only little evidence of *space synergy*. The space agency signs international agreements, such as to complete the International Space Station, then the U.S. Congress squabbles over the funding to implement such. The nation's resources are wasted on defence and war, while the space agency budget is further trimmed. The old arguments of manned or unmanned spaceflights continue, so planetary scientists with vested interests support only automated missions of exploration, while opposing advocates of human space discovery and settlement. However, there are two hopeful developments indicating a possible resurgence in U.S. space endeavors. First, in the opening decade of the 21st century, the President did task NASA to give priority to returning to the Moon permanently by 2020 (see Exhibits 5 and 125). Second, private enterprise has scored some successes in the fields of space communication, transportation, and tourism.

1.2.1 United States space case study

To appreciate the present space vision and ethos in this nation, it is important to understand its recent past. Eight decades ago, American scientist Robert Hutchings Goddard fired that first successful liquid-fuel rocket from a Massachusetts' pasture, providing his country with the means to enter the space frontier. Since then, Dale Meyers reminds us that the U.S.A. has used that key to good purpose: sending inquisitive extensions of human intelligence to eight planets in our Solar System; examining our own star, the Sun, as well as the cosmos from outside the obscuring atmosphere of Earth. This country's ultimate feat was placing a dozen astronauts on the surface of the Moon [22]. Meyers, a former NASA deputy administrator, noted that these accomplishments marked the first time in the 4.5-billion-year history of our planet that creatures, evolved from the chemicals in its crust, had achieved the ability to leave its surface and visit other celestial bodies in the Solar System.

Exhibit 5. Outposts on the Moon. *Selenians* or lunar dwellers, living and working with automated helpers on the Moon! If the goals of current policy of the *Vision for Space Exploration* are realized, humans will be back on the lunar surface permanently by 2020, and then move on to Mars and beyond. Source: Original painting by Dennis M. Davidson for *Space Resources* (NASA SP-509). Washington, D.C.: U.S. Government Printing Office, 1992.

The American *space ethos* first began to emerge in response to the former Soviet Union's challenge with the launching of its first Sputnik in 1957. Fifty years ago, President Dwight Eisenhower signed on July 29, 1958 the National Space Act, thereby creating the space agency popularly known now as NASA. Several years later, on May 25, 1961, President John F. Kennedy called for national leadership in space achievement, and committed the country to a lunar conquest within a decade. Subsequent presidents gave mediocre support and declining resources to the space program up until the 21st century. One way to evaluate a country's space ethos is to examine what it invests in space development: the U.S., for example, on average has been spending as much on space as the rest of the world put together. The annual battle of the Federal budget allocations to the space program is up to approximately

$18 billion annually, a fraction of what the nation spends on defence. Despite 50 years of remarkable and comparative progress in space technology and operations, the country's orbital efforts were sidetracked by the loss of two Space Shuttles, *Challenger* and *Columbia*, along with their crews of 14 astronauts (see the dedication on p. xii for their pictures and names).

As far back as 1984, the Office of Technology Assessment (OTA) reported that the U.S. government had not adapted to the changing circumstances regarding its space program, and the lack of a national consensus about long-term goals and objectives for this country in space. To remain competitive, even then the OTA urged provisions for greater private sector investment in civilian space activities, as well as more international cooperation in space endeavours. The OTA study advocated involvement with Japanese and European partners for this purpose, calling for a more synergistic national policy and direction for space development. Only now has such recommendations begun to be realized as in the multinational agreements relative to the International Space Station. Hopefully, the same type of partnerships will soon emerge relative to building a lunar base!

Normally, we expect a head of state to articulate a country's vision and ethos. Among the Presidents of the United States, only John Kennedy has succeeded in real national leadership with regard to space. From the beginning of Ronald Reagan's administration, that President also consistently voiced support for space enterprise. This quotation on July 4, 1982, Independence Day, is typical of his vision communicated to the citizenry over the course of two terms:

> "Today we celebrate the 206th anniversary of our independence ... The fourth landing of the *Columbia* is the historical equivalent of driving the golden spike which completed the transcontinental railroad. It marks our entrance into a new era ... And now we must move forward to capitalize upon the tremendous potential offered by the ultimate frontier of space ... Simultaneously, we must look aggressively to the future by demonstrating the potential of the Shuttle and establishing a more permanent presence in space."

With that speech, the President issued a directive on the same day detailing the basic goals of his space policy. It was to ensure the country's security and space leadership; obtain economic and scientific benefits through space exploitation; expand private sector space investments and involvement; promote cooperation with other nations in space for the enhancement of humanity.

On March 23, 1983, in his State-of-the-Union Address, the "Great Communicator" endorsed the free enterprise philosophy:

> "Our second great goal is to build on America's pioneer spirit and develop our next frontier: space. Nowhere do we so effectively demonstrate our technological leadership and ability to make life better on Earth ... America has always been greatest when we dared to be great. We can reach for greatness again. We can follow our dreams to distant stars, living and working in space ..."

Then Reagan instructed NASA to develop a permanently manned space station within a decade, and to invite other countries to participate in the enterprise as a means of strengthening world peace. To ensure that companies interested in putting payloads into space have access, this President also directed the Department of Transportation, through its new Office of Commercial Space Transportation, to assist in developing a private sector launch service industry, and to remove regulatory barriers to space industrialization. Within six months, that President was to approve a National Space Strategy promoting initiatives for long-range studies of new launch vehicles, military operations in space, identification of major space goals, Shuttle cost recovery, and reaffirmation of an $8 billion space station. (*Author's note*: Over two decades later and with expenditures of billions of taxpayer's dollars, that International Space Station is now in orbit, and soon to be completed with the help of global partners, especially Russia. See the station case study in Chapter 8.)

As a national leader, President Ronald Reagan's many speeches apparently expressed a tentative *space ethos* for his country's citizenry. In 1988, toward the end of his Presidency, Reagan issued a definitive National Space Policy that endorsed commercial projects and development of Mars technology. But many citizens perceived a gap between Presidential rhetoric and decisions that advance national space consensus and policy. The President's words never reached down and moved the masses within the country to get behind space undertakings. Furthermore, most of the nation's financial resources devoted to space during the administration of that 40th President did not go to civilian space programs, but to military space programs such as "Star Wars" or the Strategic Defense Initiative (SDI).

A people's ethos is forged through both success and tragedy. The Apollo Moon landing in 1969 represented a peak experience for America, as well as for humanity in general. But, in this nascent period of the American space program, the death of astronauts both on the ground and in orbit tested public willingness to accept the high risk inherent in space exploration. Until 1986, the Shuttle transportation system seemed to be another model of achievement until both technological and human system failure doomed the *Challenger* and later *Columbia*. But even investigations of both failures contributed to the refinement of a *space ethos*, leading to new advancements in space transportation and safety. Unfortunately, these setbacks also resulted in Presidential directives giving mission priorities on the Shuttle to defense, not its commercial uses. While praising NASA reform efforts, U.S. Senator Donald Riegle reminded the nation in 1987 that NASA cannot craft space policy; that must be done by the Executive Branch with the ratification of Congress and the people. When he was chairman of the Senate Subcommittee on Science, Technology, and Space, he urged a national debate on how much of the American space effort should be military in scope, in contrast to commercial emphasis. Such widespread public discussion has yet to occur.

The closest such discussions happened was the National Commission on Space, when 15 experts appointed by the President, conducted 15 public forums held across the country, as well as utilizing citizen input from communications received by mail, computer network, or testimony at Commission meetings. Their farsighted report in 1986 entitled *Pioneering the Space Frontier* was the closest expression to date of developing American *space ethos*. It represented a cross-section of American thought on space planning up to the year 2035, recommending programs to [11]

(1) advance understanding of our Earth, its Solar System, and the universe;
(2) explore, prospect, and settle the inner Solar System;
(3) stimulate space enterprise for the direct benefit of Earth's people.

To accomplish these goals over the next half century, the Commission estimated an investment of $700 billion by America and its international partners. However, the Commissioners also envisioned big returns on that investment, as the first 50 years of space enterprise has already proven. They were aware of a Government Accounting Office estimate in 1982 that the payback is approximately $7 for every $1 invested in space. Therefore, NCOS expected returns on space investment which would be primarily in

- advancement in national science and technology capabilities, so critical to economic strength and security;
- economic returns from space-based enterprises and capitalize on broad, low-cost space access;
- opening up the vast resources of the space frontier to supplement the limitations of our own planet.

Over 20 years ago, this National Commission not only set forth a vigorous vision for the American civilian space program, but the rationale, methods, and aspirations for its implementation. Its "Declaration for Space" communicated a possible *space ethos* with statements such as this:

"With America's pioneer heritage, technological pre-eminence, and economic strength, it is fitting we should lead the people of this planet into space. Our leadership role should challenge the vision, talents, and energies of young and old alike, and inspire other nations to contribute their best talents to expand humanity's frontier .... As we move outward into the Solar System, we must remain true to our values as Americans: to go forward peacefully and to respect the integrity of planetary bodies and alien life forms, with equality of opportunity for all."

Then, the non-governmental National Space Society with some 50,000 members endorsed this vision, adding:

"We believe the technologies and industries born on the space frontier in the next few decades will drive the world's leading economies in the next century. Our role is to educate the public on the benefits of space developments and work with allied organizations to create the cultural and political context for an open frontier in space. We believe the United States must be a leader of that frontier, or it will cease to be the greatest hope for human liberty and freedom" (*Space World*, January 1988, p. 32).

Unfortunately, because of political conditions at the time, such as an Administration crisis over the "Iran Contra Affair", the excellent report and recommendations of the National Commission on Space were largely ignored at the White House. It was not until September 29, 1988, that the U.S. manned space program was reborn! This watershed event may have contributed to the re-emergence of a space ethos. After three years of launch failures and frustrations as a result of the *Challenger* accident, a NASA spokesman announced, "Americans have returned to space." The successful launch and subsequent return of the Shuttle orbiter, *Discovery*, brought over a million well-wishers to the Kennedy and Vandenberg space centers (see Exhibit 6 for launch photograph). Excitement spread across the nation: the masses were again inspired by the feats made possible through the high performance of space scientists, engineers, and the astronaut team. The world again watched and cheered, along with the average U.S. citizen. The media joined in celebrating the new beginning, but commentators cautioned that there is a need for the country to set new directions in its space endeavors. A tentative ethos was manifested after being in a hiatus since the Apollo Moon mission days. Broadcasters, such as CBS's Dan Rather, described the *Discovery* launch on September 29, 1988, "As the day Americans paused to dream again." Media professionals joined their countrymen by urging them to forge ahead aloft, but wisely asked, "What does this nation want to do in space?" Response to that key question requires public participation if a critical mass of citizen support is to result: that has not yet sufficiently occurred!

The Administration of the first President George Bush attempted to set that new direction for the nation's future in space. This was done by reinstalling the National Space Council under Vice President Dan Quayle, appointing a former astronaut as the new NASA Administrator (Admiral Richard Truly), and designating a blue-ribbon panel to untangle the fiscal and managerial knots that bond the space agency. Called the Advisory Committee on the Future of the U.S. Space Program, it produced in 1990 another significant report on American space agency problems, expressing these nine concerns:

(1) lack of a national consensus as to what should be the goals of the civil space program and how they should be accomplished;
(2) NASA's over-commitment in terms of program obligations relative to too little resources and allowance for the unexpected;
(3) management inefficiencies caused by constant changes in project budgets of the space agency, exacerbated by actions sometimes necessary to extricate projects from technical difficulties;
(4) institutional aging and insufficient planned organizational change with NASA;
(5) incompatible personnel policies between civil service requirements and space agency needs for leading-edge technical specialists and managers;
(6) natural tendency for space projects to grow in scope, complexity, and cost, and the need to curb such unplanned expansion;

Exhibit 6. Launch out: space exploration. From Cape Canaveral, the Shuttle Space Transportation System lifts off from the Kennedy Space Center in Florida. Here as crowds watch nearby on the banks of the Indian River, more than 7 million pounds of thrust in a controlled explosion hurls a 4.4-million-pound object into the heavens at 25 times the speed of sound. Source: NASA. A second U.S. launch pad is at the Vandenberg Air Force Base on the West Coast in Southern California.

(7) deterioration of NASA technology base which must be quickly rebuilt;
(8) need for engineering systems to continuously monitor flaws in spacecraft technology and correct problems before they escalate into disasters;
(9) overdependence of the civil space program on the Space Shuttle, and the need to use other heavy-lift launch vehicles for all but missions requiring human presence.

Under its chairman, Norman R. Augustine of Martin Marietta Corporation, this prestigious expert panel made four principal recommendations concerning space goals, programs, affordability, and management. Relative to the themes of this book, these included

- Not only "Mission to Planet Earth" (MTPE) focusing on environmental measuring from space, but a "Mission from Planet Earth" (MFPE) with a long-term

goal of human exploration of Mars, preceded by a modified space station which emphasizes life sciences, an exploration base on the Moon, and robotics. MFPE would operate on a "go as you pay" basis with missions scheduled as adequate funds are available ...
- Major reforms in management and personnel operations of NASA, including appointments of an Associate Administrator of Human Resources responsible for acquiring and retaining the highest quality employees; an Associate Administrator for Exploration responsible for robotic manned missions to the Moon and Mars; establishment of a systems concept and analysis group in a Federal Research and Development Center.

In the long term, the senior President Bush may have contributed much to articulating the U.S. space ethos principally through his Space Exploration Initiative (SEI). In a speech on July 20, 1989, commemorating then the 20th anniversary of the Apollo 11 Moon landing, George Bush tried to lay out a framework for American space developments for the decade of the 1990s and the 21st century. SEI was to be a peaceful civilian endeavor in contrast with Reagan's Space Defense Initiative (SDI). The Space Exploration Initiative expressed noble objectives, including completion of the Space Station in the 1990s; a return to the Moon to stay at the beginning of the new century; and a journey into tomorrow with a manned mission to Mars. The President called this vast endeavor a "New Age of Exploration" and NASA estimated SEI would cost $400 billion. Unfortunately, that Chief Executive was unable to get Congress to fund his Initiative beyond preliminary studies of this ambitious program.

To help in analysis of the SEI prospects and priorities, the National Space Council also tried to reach beyond the space agency for innovative input about what directions and actions should take place in the near future. Thus, the Council encouraged NASA to create the "Outreach Program", a nationwide search for creative ideas and technologies to further the exploration of the Moon and Mars. To move toward new levels of ingenuity, this 1989 effort tapped diverse sources from the National Research Council, and national laboratories to the strategic defense specialists, to even the public at large by means of an "800" free telephone connection. To assist in the surveying and reporting of the solicited input throughout the country, NASA contracted with two organizations, the Rand Corporation (a non-profit research group and think tank in Santa Monica, California) and the American Institute of Aeronautics and Astronautics (a professional society of the aerospace industry in Washington, D.C.). In addition, the various federal agencies were invited to participate, and the many respondents were U.S. Departments of Energy and Defense, as well as NASA personnel. The Rand "toll-free" number produced 19,048 calls, while the deadline resulted in 1,697 ideas to that collector, and another 530 ideas through the AIAA conduit (1991). In effect. this was another major opportunity for the country to articulate its space ethos. After a preliminary screening by Rand and AIAA, the best and the brightest ideas went to an independent advisory panel in Crystal City, Virginia, headed by ex-astronaut and former air force Lieutenant General Thomas Stafford. Called the "Synthesis Group", it reported the results to the National Space Council in Spring 1991. The group organized into teams to review the ideas in terms of how a mission would leave the Earth, then travel to and land on the surface of the Moon and Mars. Both Rand and AIAA held public conferences around their findings. While limited in its outreach, the whole undertaking provided more than a coherent architecture for the Space Exploration Initiative. Indeed, it was a democratic and synergistic process that broadened the base of public involvement in the U.S. space program. As the director of program architecture for this Synthesis Group, Lt. Gen. Sam Armstrong (USAF Retd.) observed:

> "A pioneering society is a growing society. We've always been a nation of pioneers. But the people have to want to do it" (*Final Frontier*, January/February 1991, p. 41).

While this SEI vision statement has yet to be implemented, there was another initiative by the first Bush Administration in the Fall of 1990. Namely, synergistic talks were begun not only with Canada, Japan, and the European community, but also with the then Soviet Union on the possibilities for an international lunar base and human expedition to Mars. American leadership in the past has inspired international cooperation in space for the world community. Globally, space agencies now have agreements with one another, not only in developed free enterprise nations of the First World, but with the former or present socialist economies of the Second World, as well as in many developing countries of the Third World.

Another indicator of the lack of a vigorous space vision and ethos has been evident in recent American Presidential campaigns. The politicians hoping to be elected to the country's highest office rarely discussed space development as a central issue for public concern. One exception came in 2004 when the Democratic Party's candidate, John Kerry, criticized reigning President George W. Bush for his lack of space vision. Despite a bold plan for NASA which the second President Bush laid out for NASA in January of that same year, Senator Kerry commented it was "big on goals, but short on resources to reach them" (Space.com/news, 6/19/04). Although that candidate lost that election by a close margin, he had at least issued a favorable space policy statement during the election period. Previously that Senator had co-sponsored a bill (S.1821) to re-establish a National Space Commission at the White House to coordinate all space activities; it never was passed as law. So far in the 2008 President election drive, the leading candidates again have had little to say about the country's future in space, except Senator Hillary Clinton!

Thus, many within the American space community feared that the nation's space effort and ethos might end up like those of China's overseas exploration initiative some 600 years ago (see Exhibit 7).

With a hiatus of 35 years since Americans landed astronauts on the Moon, those in this space enterprise wondered if the U.S.A. was losing its leadership in space, just like the Chinese did with the oceans six centuries ago. Perhaps U.S. space fortunes will soar again with the new VSE space policy discussed below ... Further, China seems to have also learned a lesson from Admiral Zheng He's experience as it

> **Exhibit 7. CHINESE OVERSEAS EXPLORATION.**
>
> A Muslim eunuch from Central Asia, Zheng He, persuaded the Chinese Ming emperor, Zhu Di, to finance the most extraordinary expeditions the world had ever seen. Between 1405 and 1433, Zheng and his men then sailed through all of Southeast Asia, to both coasts of India, and finally to Africa's eastern coast as far as Ethiopia! The mightiest ship of this Chinese fleet was 450 feet long by 185 feet wide, with nine masts, about five times the size of what the Europeans could then build. Admiral Zheng's 317 ships carried 28,000 men. Among them were diplomats who established Chinese embassies along the route, plus traders who brought back vast supplies of goods from the places visited.
>
> Equipped with such technology and organization, China might have discovered Europe long before the Europeans sailed into China and dominated it. But Zheng's seventh such voyage was to be his last, and the sea-going eunuch fell from favor. By year 1500, it was a capital offence in China to go to sea in a two-masted ship without the Emperor's permission. With the scuttling of the ships of its visionaries, China began a long period of isolation like that imposed upon Japan in the 17th century. Also with the mothballing of Zheng's fleet, came the beginning of the end of Chinese superiority. Now 21st century Chinese celebrate the accomplishments of the Admiral's fleet and naval technology, recalling how it had once turned the western Pacific and Indian Ocean into a "Chinese lake".

Source: Adapted from "Chinese History: The Admiral of What Might Have Been," *The Economist*, July 16, 2005, p. 15.

pushes ahead today with its own vigorous space program that is a real challenge to the West.

21st-century U.S. space vision and ethos

Despite involvement in wars on terrorism in Afghanistan, Iraq, and elsewhere, President George W. Bush has tried to implement the unsuccessful *Space Exploration Initiative*. Fourteen years after his father's speech in 1992 about returning to the Moon and going on to Mars, the son authorized his version of a national space policy on August 31, 2006. His *Vision for Space Exploration* outlined an agenda again for the Moon, Mars, and beyond! With the seeming support of the U.S. Congress, industry, and the space community, this Vision or VSE has begun to make progress in its goal to return permanently to the Moon by year 2020. Under the leadership of its new Administrator, Dr. Michael Griffin, NASA is ramping up the program, despite the prospect that its FY2007 budget may provide the Agency with half a billion dollars less. That same year, the newly elected Democratic majority in Congress has space supporter Senator Barbara Mikulski (D-MD) chairing the subcommittee of the Senate Appropriations Committee that oversees NASA funding. A 2007 study released by the University of Chicago revealed that in terms of government-spending priorities, space exploration ranked 21st out of 22 funding areas! Yet, as NASA's budget is being redirected again to human space travel, the Agency science budget is likely to remain the same through 2011, $5 billion a year, while its current spending on the Moon–Mars initiative doubles to $8.7 billion!

Two Gallup polls on this subject show mixed public reactions: in 2005, 58% of those polled were opposed to setting aside money to land humans on Mars, but in 2006, two-thirds of the respondents were in support of the new Vision program. More worrying is that younger Americans, dubbed Generation X, have expressed high levels of disinterest or opposition to the VSE. Obviously, for a strong space ethos to emerge, NASA and the whole space community have to explain more clearly and simply to the general public why humans should return to the Moon and journey eventually to Mars!

As space policy expert John Logsdon ruefully observed: "the pathway to the Moon and Mars leads straight through Capitol Hill," meaning the U.S. Congress. In the current climate of budget reductions and deficits, exacerbated by serious domestic problems and involvement twice in Middle Eastern wars, public funding of space missions is constrained. Yet, the Vision plan has bipartisan support in both houses of the legislature, and the NASA authorization bill of 2005 explicitly endorsed VSE!

Within the Agency and its aerospace contractors, there has been a struggle to find the right combination of spacecraft to get us safely back to the Moon. Design debates go on as to which solid rocket booster to use for upper stages of flight, such as Ares 1 or Orion. The political winds could also affect the decision to continue moving ahead on the Vision for Space Exploration, much depends on the new U.S. President and Administration in 2009 (see Jeff Foust, editor of *The Space Review*, 1/26/07 at *www.thespacereview.com*).

The White House Office of Space Technology Policy document of 2006 calls for a sustained, innovative, affordable *human and robotic* exploration program across the Solar System. It supports space nuclear power systems, urges both commercial space undertakings and international cooperation, as well as the minimization of space debris. Seeking to enhance national security after the 9/11 terrorist attacks, there is also a section calling for the Secretary of Defense "to ensure free space", charging the Director of National Intelligence to provide "a robust foreign space intelligence and analysis capability". This Presidential order would deny adversaries access to space for hostile purpose. However, the new policy did not call for the deployment of weapons in space, but emphasized instead U.S. commitment to the "peaceful uses of space by all nations and that space systems enjoy the right of free passage." National Security Council spokesman Frederick Jones observed:

> "Technology advances have increased the importance of and use of space. Now we depend on space capabilities for things like ATMs, personal navigation, package tracking, radio services and cell phone use" (*www.space.com/news*, 10/9/06 ... *San Diego Union*, October 19, 2006, p. A12).

The noted space lawyer George S. Robinson speculates that the new U.S. national space policy may be pushing the limits of the United Nations space treaties. Dr. Robinson maintains that legitimate concerns about national security cannot be allowed to set aside international treaty law. For example, the "1967 Outer Space Treaty" and other related agreements recognized by the United States, establish the law relative to the global exploration of space and celestial bodies, as well as the occupation, settlement, and use of its resources for the benefit of all humankind. In 2002, attempts were made in Congress to pass a "Space Preservation Act" that would authorize the President to start negotiations toward "a treaty that would ban space-based weapons and the use of weapons to destroy or damage objects in space that are in orbit." Robinson argues that such treaties are important so as to curb "Star Wars" or the *unlawful use* of space for military purposes. This expert believes that no national space policy should undermine the spirit and intent of previous U.N. space treaties (see Appendix A).

The release of the 2006 National Space Policy has led to vigorous debate in the American space community that contributes to the refinement of U.S. space vision and ethos. For example, the Space Frontier Foundation, a U.S. space advocacy group, issued a study criticizing the current space transportation system planning, while favoring reconsideration of the Atlas 5 and Delta 4 families of launchers. Further, the Foundation argued that NASA planning for VSE should make greater us of the "new space industry, energized by free enterprise and entrepreneurship." Their 18-page white paper, "Unaffordable and Unsustainable: NASA's Failing Earth-to-orbit Transportation Strategy," calls for the Agency to transform its relationship with the private business sector. Rather than develop, build, or operate vehicles for orbital crew or cargo missions, the position paper maintains NASA should be buying these from private service companies (as reported by Leonard David in *www.space. com/news*, 7/8/06).

Another illustration is the panel discussion in October 2006, *21st Century Exploration Goals* at the Lunar Planetary Institute. Co-sponsored by the Association of Space Explorers and the Planetary Society, this event brought together distinguished astronauts and planetary scientists to consider the latest scientific results from space missions, the interactions between humans and robots in future space endeavors, as well as potential discoveries and destinations. The panelists focused on international exploration of our Solar System within the next few decades (*The Lunar Enterprise Daily* 9/14/06 ... www.spaceagepub.com). A further helpful effort for involving the American people in VSE was the launching in 2006 by the National Space Society of a public Internet portal offering unlimited access to a comprehensive collection of documents, studies, and other resources concerning space settlement and other related issues (*www.nss.org/settlement*).

A final example of the VSE exchanges is among the Coalition of Space Exploration, experts from the American space community utilizing the Internet through electronic mail, chat rooms, or blogs. These candid and spirited communications offer different viewpoints and responses relative to

- the infinite resources of space;
- development of a space-based economy;
- lunar exploration strategy and architecture;
- returning to the Moon for human settlement there;

- creation of a Lunar Economic Development Corporation or Authority;
- manned Mars missions as the next step after returning to the Moon;
- space policy based on historical fact, economic realities, and not fiction;
- opening up the space frontier to the American people;
- needed life science research in returning to the Moon and then to go beyond;
- development of new multi-legged space paradigms, including support for space entrepreneurs;
- COMSAT as a different funding model;
- governance structure for space enterprise;
- involvement of other nations in joint VSE lunar expeditions.

For further information contact electronically *Feng.HSU@NASA.Gov*.

In 2007 as this book was being finished, the House of Representatives in the U.S. Congress approved a funding bill providing NASA with a budget of $1.2 billion, an increase over the previous year, but with the provision that specially forbids NASA from investing in *human exploration of Mars*. The National Space Society opposes this restriction. Another view is that with limited financial resources, the U.S.A. should focus on lunar development for now, and eventually, possibly within 50 years, there will be human missions to the Red Planet.

Finally, as the United States plans to retire its Shuttle fleet in 2010 and its emphasis on low-Earth orbit, a new generation are in the making of more powerful spacecraft by new manufacturers, so as to achieve the VSE goals [23]. NASA has contracted $500 million to Space Exploration Technologies and Rocketplane-Kistler to produce a new vehicle named Orion. While it draws on Apollo–Saturn technologies and design, it will be larger. It will sit atop a two-stage rocked named Ares I or Ares V, assisted by external tanks and five engines capable of lunar return. This Crew Exploration Vehicle (CEV) may come in many configurations and will offer a challenge to human factor researchers.

1.2.2 Canadian space case study

In Canada an ethos toward space is gradually arising. It was born with the launching of Alouette 1 in September 1962 on the U.S. Thor-Agena vehicle, a cooperative program among Canada, the U.K., and the U.S.A. In 1972, Canada launched its first domestic geostationary communications satellite system. Under the leadership of the Canadian Space Agency, created in 1989 and headquartered on a new campus in St. Hubert near Montreal, this nation's diversified space efforts coalesced. The CSA president oversees five core functions: space technologies, space science, astronaut office, space operations, and executive functions. When visiting the CSA homepage on the Internet, note the emphasis on four major activities: Earth observation, satellites, science, and exploration (*www.space.gc.ca*). For example, in 1995 RADARSAT 1 was this country's new age satellite for Earth observation, and has been used to provide a mosaic from outer space of Australia and Central Africa. MOPITT is the CSA instrument to measure pollution in the troposphere of the Earth's atmosphere.

Until recently that organization and the indigenous aerospace industry were principally aligned with NASA and its space projects, having developed key components in both hardware and software for spacecraft and satellites. Its principal activity has been creating a Mobile Servicing System (MSS), the *Canadarm* for use on the International Space Station, based on Canadian astronauts' experience with a Shuttle robotic arm. With a budget of $300 million, space expenditures are modest compared with their NASA counterpart.

There is a growing involvement of CSA and Canadian researchers not just with U.S. space missions, but with the space agencies of Europe, Russia, and China. After 20 years of collaboration as an observer, on March 21, 1991 five agreements were signed between CSA and ESA for Canada's participation in these European programs: Hermes spaceplane development, Earth observation, and telecommunications. Other space ethos indicators evident in Canada are

- a growing space community with many outreach programs, publications, and special events;
- a steady increase in funding allocation of the national budget for space activities;
- a THEMIS constellation of satellites launched in 2007 to investigate the Earth's atmosphere and learn how our planet's magnetosphere works, this project will have 16 ground observatories installed in the Canadian North.

When CSA's former director, Dr Roland Doré, became president of the International Space University with its main campus located now in France, Canadians began to identify more with the global space community. The signs of this first became evident in the ISY '92 7th Conference on Astronautics by the Canadian Aeronautics and Space Institute with its theme "Canada in Space: A Coming of Age and New Horizons". The discussions covered a wide range of space topics from research, technology, and solar power to astronauts, education, and business prospects. Perhaps the Canadian space vision and ethos is best summarized in his prophetic words of Exhibit 8. However, Canada's space community is still a small subculture that has yet to penetrate mainstream thinking [24]. For more information on CSA activities, publications, and workshops, visit *www.vmware.com/Canadian_space_agency.pdf* and *en.wikipedia.org/wiki/Canadian_Space_Agency*.

1.2.3 Mexico's space vision and ethos

Although culturally aligned with Latin America, Mexico is geographically part of North America and participates in the North American Free Trade Agreement [25]. While Central and South America have signed a few space agreements with NASA and provided one talented astronaut, Mexico's government, with its developing economy, has yet to be involved formally in any space undertakings of its own. However, some Mexicans are active in the space community, as can be witnessed in chapters there of the National Space Society and the Mars Society. Jesús Raygoza Berrelleza is president of the Mexican Space Society and a leading proponent of the

Exhibit 8. CANADIAN VISION AND ETHOS.

> With the new millennium, we are entering a new space age in which space is put to the service of humanity, bridging ideological and geographical frontiers. Space will be valued for its ability to serve humanity in telecommunications, earth observation, space science and human exploration. We will judge the value of space technologies to serve human needs such as food production, medical and health services, rural education, disaster warning and mitigation, environmental protection, navigation services, communication, understanding of the universe while extending humanity beyond the Earth frontier. Fulfilling user needs will be of prime importance in coming years. This will trigger a shift from government-led programs to activities initiated and fully developed by the private sector.

Source: Roland Doré, Foreword, *Living and Working in Space* by Philip R. Harris. Chichester, U.K.: Praxis/Wiley, 1996, Second Edition, pp. xxi–ii.

proposed *Mex-Lunarhab* [26]. This is a plan to build a Lunar Habitat in Mexico, co-sponsored by the Institute of Advanced Sciences, Mexican Space Society, Mars Society España, Pryoyechtos and Construcciones MV, and University of Xalapa.

The Mex-Lunarhab would be a Hispanic-Mexican terrestrial analog of a settlement on the Moon, complete with robots. Its purposes are to (1) generate Hispanic-Mexican interest in space activities; (2) provide simulated scientific lunar exploration based on real scientific and technological research; (3) assist the Lunar Economic Development Authority, Inc. in planning for twin-planet economy; (4) design a simulated spherical-shaped habitat that eventually could be built on the Moon; (5) create a space theme park and tourist attraction that would help the local Mexican people both financially and educationally. The proposal calls for a site at the base of the Pico de Orizaba Mountain, which is 5,747 meters above sea level on the leeward side of the State of Puebla, near the Gulf of Mexico. Though ambitious, it is an illustration of emerging economies searching for a place in space development. With the widespread use of space resources and its spinoffs, such as television satellites and mobile phones, we can expect many Third World countries like Mexico to benefit economically.

1.2.4 Conclusions about North American space vision and ethos

The United States, like its neighbors Canada and Mexico, is still struggling to articulate its space vision and ethos in such a way as to gain *mass public support* for space exploration and development. To this end, government has established numerous studies, commissions, and even passed legislation to further space activities. Through its space agency, NASA, the United States has even undertaken some daring and technologically sophisticated space missions. By using mass communications and the Internet, the U.S.A. needs to engage in serious national discussions on implementing recommendations from the National Commission on Space. Since the

Administration has already launched its *Vision for Space Exploration* without public or international input on this policy and program, the challenge will be to obtain the funding and global participation enabling humanity to return to the Moon permanently and then go beyond.

Currently, private enterprise is demonstrating leadership in designing new spacecraft and companies that will facilitate both space exploration and tourism (see Chapter 9 and Appendix C). To foster a space ethos, more innovative methods have to be undertaken to involve citizens in supporting the space program beyond its present constituency. Perhaps the new Presidential Administration in 2009 might carry out two recommendations of the NASA study group report (SP-509), namely to sponsor (1) a *White House Conference on Space Enterprise* (see Epilogue); (2) a *national space lottery* to raise funds for high-frontier ventures [27].

To underwrite human enterprise in space involves hard national choices on allocation of scarce resources to planetary science and astrobusiness enterprise. The decision-making process for long-term, large-scale space projects requires a sociopolitical environment that will ensure public support and participation (see Chapters 7 and 8). Translating today's space goals and policies into realities demands both innovation and global cooperation. More mass media productions like *2001*, *Star Trek*, and *Star Wars* do help to create a subculture in favor of space. But the mainstream culture in North America has yet to be "turned on" to the opportunities ahead on the final frontier offworld. Educational institutions have yet to prepare youth for their new millennial prospects of colonizing the galaxy!

For Pan America to create a vibrant space vision and ethos, U.S. space leadership needs to start with its closest neighbors, Canada and Mexico. Like NAFTA, perhaps consideration should be given to formulating a North American Space Enterprise Agreement (NASEA)? Then perhaps these three nations might turn south to Latin America, helping students, scientists, and entrepreneurs there to meet the quintessential challenges of space science, technology, commerce, and settlement. Brazil may be the first collaborator because of its existing space interests and investments.

1.3 RUSSIAN SPACE CASE STUDY

The International Space Year proved in 1992 to be a turning point in another way: the Union of Soviet Socialist Republics imploded and collapsed as a political entity. It ended the Cold War and the Space Race between the superpowers (namely, the U.S.A. and the U.S.S.R). Certainly, both shared a position of space leadership, for the Soviets put the first satellite in orbit, as well as the first man and woman in space, amassing during their regime three times the "manned" time in space in contrast to Americans aloft. Russians not only were far ahead in the number of rocket launches, but have had a functioning space station Mir in orbit for 13 years of extended missions. Premier Nikita Khrushchev once boasted, "the launch pad for our cosmonautics is socialism." Under the Soviets, it was always difficult to estimate how much was expended on both military and civilian space endeavors, but in 1989

Sec. 1.3] Russian space case study 29

Exhibit 9. Energia space launchpad: Kazakhstan. The most powerful rocket in the world today is Energia, which is launched outside Russia in Tyuratam/Baikonour, spaceport in the CIS republic of Kazakhstan. Source: C. M. van den Berg, *Earth Space Review*, **1**(4), 1992, 14 (Gordon & Breach, 820 Town Center Dr., Langhorne, PA 29047).

the leadership stated that $44.8 billion was being spent on civil space, though Western analysts considered the figure somewhat high. After the demise of the U.S.S.R., the Russian space program declined because of sociopolitical–socioeconomic chaos in the country (particularly lack of funding). For example, a hapless cosmonaut, Sergei Krikalev, had to stay in orbit on Mir an extra five months because the government that sent him was no longer in power and the new one did not have money allocated for his retrieval rocket! This contributed to his holding the record for the most time spent in space: 804 days in six flights on Soyuz, Mir, the Space Shuttle, and ISS. The Soviet system had been replaced by a shaky political entity known as the Commonwealth of Independent States, which in 1992 formed its own CIS Space Agency. This new authority deveoped an inter-republic space agreement such as indicated in Exhibit 9. The major spacefaring republics are Russia, Kazakhstan, and the Ukraine. The reality is competing power centers with questionable goals! Yet, today because of its engineering and technology competence, as well as participation in international space activities, the country's space enterprise is rebounding.

Soviet space program: the way it was [28]

When my book *Living and Working in Space* was first published (1992), it appeared the Russians were well ahead in creating their own unique *space vision and ethos*. The foundation for this had been laid by the 19th-century scientist-teacher Konstantin Tsiolkovski. This space philosopher (1837–1935) was called "the father of Cosmonautics", the inspiration for Russia's leading rocket men of the 20th century [29]. The most outstanding among the latter was Sergei Korolev, once a Stalin concentration camp victim who was always held in suspicion by Soviet authorities. Yet, he was a true space visionary of incomparable competence in implementation. He was responsible for putting Sputnik and Gagarin in orbit; for robotic landings on the Moon; for successful design of Russian rockets, spacecraft, and satellites; for his country's early planetary probes.

Recall in the "way it was" that the former Soviet Union was a closed system with centralized, totalitarian control. Back then, an interview with Dr. John Logsdon, a leading space policy analyst, brought forth this interesting comment reported in *Space World* (August 1986, p. 16):

> "What is striking about the Soviet Union is the country's commitment to space, which enables the government to carry out the program. That commitment comes from the Soviet leadership—particularly Khrushchev, but also Brezhnev—*defining space in a way that society internalized the belief that this is part of their future.*"

That internalized conviction was an expression of the Russian ethos, which some observers maintain was a substitute for the religious beliefs and practices that officially had been frowned on for almost 70 years of Communism's reign.

Yuri Gagarin, the first human in space, and Alexei Leonov, the first person to walk in space, were revered like demigods, and still today are held in high regard by the masses. Then, the public adulation of Soviet spacefarers was evident in the busts of gallant space giants which adorned buildings and parks. Museums, monuments, and murals paid homage to men, women, and machines of the Space Age (*Air & Space*, August/September 1987, p. 115). In their pantheon of heroes, the cosmonauts were of mythical proportions: cosmodromes and space cities were built for their care and activities; space missions were eagerly followed by the masses through the media. Soviet space museums abound and were then well attended by the populace. The extent of this prior "worship" can be gleaned from these extracts of another Smithsonian article, "Commemorating Cosmonautics" by J. Kelly Beatty, *Air & Space*, June/July 1988, pp. 96–101:

- "Crops may fail and policies flounder, but the Soviet Union's ongoing space triumphs remain an obvious source of pride among its people ... The space heroes are not merely respected, they are adored ... Atop a 25-meter spire besides Lenin Avenue in Moscow stands a 6-meter-tall likeness of Yuri Alekseyevich Gagarin, well muscled and angular ... The heroic stature attributed to Soviet space explorers is reflected in the frieze on the Cosmonaut Memorial, another gleaming spire that arches 90 meters into the sky. A Museum is situated beneath the memorial ... No word describes it better than 'shrine' ..."

- "Near the Cosmonaut Memorial, a collection of rockets and space artifacts serve as a large 'theme park' among the Exhibition of Achievement of the National Economy ... A space exhibit, the Cosmos Pavilion was added in 1966 and today it boasts more than 9,500 m of floor space and draws eight to nine million visitors a year ... A huge inscription 'Cosmonautics, The Way to Peace' wraps around the wall. The theme is repeated often: in a huge portrait that dominates the rotunda, Gagarin holds a white dove ... The first human in space serves as an enduring icon of the achievements reflected in Moscow's space museums."

Until recently, the former U.S.S.R. commitment could be perceived in the size of its space budget and personnel, supposedly double that of the U.S.A. (their ruble expenditures were said to be around $12–$13 billion annually, a questionable figure). The extent of that nation's unflagging pursuit of space was also evident in the then world's busiest spaceport, Plesetsk, northeast of St. Petersburg near the Arctic Circle; it had orbited more than 1,000 payloads and tested 40 times more missiles annually than the U.S. In a *National Geographic* review (Oct. 1986, p. 455) of "Soviets in Space" its editor, Thomas Canby, concluded that space exploits stir the Soviet soul like religion and these stirrings are fanned by the government. Some believe that space successes helped the nation to cope with its inferiority complex and losses in the Cold War.

The Russians, both past and present, also sought international participation in space. They have actively promoted scientific exchange among space scientists worldwide, and gradually opened their space facilities and programs to public view. They have included representatives on their spaceflights from both the former Eastern bloc and Western nations. In their 1988 flight plan for unmanned probes of the moons of Mars, they announced with pride that instruments from scientists of 14 nations, including the U.S.A., were on their spacecraft. At the onset of President Mikhail Gorbachev's administration, policies were implemented promoting more openness and economic restructuring, welcome forces of change that were eventually to destabilize not only Soviet society—but its space program and ethos. But by the mid-1990s, the inhabitants of the former U.S.S.R. were more concerned about survival amidst profound political and economic reform. This vast empire had broken apart and its peoples today still are in the midst of profound transition, contending with opposing forces. There are traditionalists who still favor a centralized, socialist planned economy vs. democratic reformers pushing for a market-oriented, free enterprise system. Like its American counterpart, space planning and missions in the nascent Commonwealth of Independent States is at a crossroads, grappling with rising costs, shrinking budgets, and weakening government support. Space programs and hardware that were once the showcase of the old Communist system were offered to Western customers for sale. The pride of Soviet space technology has been "mothballed" (the shuttle Buran and the heavy-lift expendable rocket, Energia) and unmanned missions to Mars had failed. Conditions at the beginning of the last decade in the 20th century were well described by two former directors of the Moscow Institute of Space Research:

- **Roald Sagdeev**, when a professor in the University of Maryland, addressed the George Washington University Space Policy Institute, and confessed that the Russian space program was caught up in the general collapse of the Soviet economy and central authority, being subject to claims of governments in the union's 15 ethnic republics (*Space Business News*, December 24, 1990, p. 3).
- **Albert Galeev**, once a leading Soviet space scientist, speaking at the Institute on Global Conflict and Cooperation in the University of California-San Diego, admitted a year before the downfall of the U.S.S.R. that dramatic scaling back of their space missions was under way. He revealed that economic and political turmoil had brought an end to discussion of manned expeditions to Mars, and the new focus instead would be on unmanned missions (*Los Angeles Times*, January 13, 1991, p. B1).

Russian Federation space program: the way it is

And what is the situation in the space program of the newly constituted Russian Federation? At the beginning of his Presidency, Boris Yeltsin was unconvinced that space exploration could immediately benefit society and remedy its ills, so he had curtailed space spending. Yet, to obtain foreign currency, he has been supportive to provide jobs to their scientists and engineers. By 1993, Yeltsin's advisors were realizing that space was a good way to further Russia's international prestige and business.

Although the bulk of its spaceports are within Russian territory, some key launch pads, such as Baikonour, are in other CIS republics. Within the Russian Federation, control of the space program fragmented as ministries and the Academy of Science were reorganized, and the Russian Space Agency (RKA) came into being. In 1995, its budget was down 40% when set at 1.187 trillion rubles, about $270 million in foreign exchange markets. Renamed the Russian Federal Space Agency (РОСКОСМОС or Rosaviakosmos), this civilian entity shares facilities and spaceflights with the Military Space Force (VKS) that controls the Plesetsk Cosmodrome. By Federation presidential decree, a Russian Center for the Conversion of the Aerospace Sector was set up to unify efforts both within Russia and the CIS. That Center not only engages directly in several large aerospace programs, but is intent on improving telecommunications, meteorological, ecological, and other space-based systems. Furthermore, the Russian government agreed to increase financial support to the country's aerospace industry via tax breaks and grants (*Space News*, November 8, 1993). By 2006, the Duma authorized a ten-year increased space budget, approximately $11 billion from year 2006–2015. Additional funding for Roskosmos comes from industry investments and commercial space launches, as well as contracts with NASA and ESA.

With a severe funding crunch and increased privatization, space organizations pursue innovative strategies to survive and become more entrepreneurial, such as

- selling or leasing space hardware and software, as well as facilities to other nations and foreign corporations ...;
- selling satellite data, especially about weather and resources ...;

- selling or exchanging Russian space expertise, particularly by consulting ...;
- converting defense and space activities to civil or commercial operations (e.g., part of the Plesetsk Cosmodrome to the local Mirny community) ...;
- training personnel from other countries in the Yuri Gagarin Cosmonaut Training Center and Star City, even sending them in orbit for a fee when the Mir space station was still in orbit ...;
- transforming the Gagarin Center, in part, into a research institute which rents facilities and services to international customers ...;
- increasing space tourism under Cosmos Tours with fees to visit former Soviet space facilities and museums, attend aerospace shows, as well as to experience cosmonaut training simulations ...;
- founding of a space consulting firm, the Space and We Co. Ltd., by ten cosmonauts to facilitate business relations and ventures in their country ...;
- contracting for services of space and nuclear scientists/engineers who are underfunded or unemployed.

With regard to the latter need, Nobel laureate Lean Max Lederman commented in *Popular Science* (August 1994, pp. 214–227):

"Civilization would be much poorer without Russian Science. Apart from their outstanding discoveries in disciplines ranging form space to biology to particle physics to mathematics, Russian scientists provide the kind of intellectual stimulation and fertilization that makes science a global community."

The painful transformation of Russian space science and technology is still under way, as these developments in the post-Soviet era indicate:

- Publication in English of *Space Bulletin*, a magazine by Moscow's Association for the Advances of Space Science and Technology; *Russian Aerospace Market*, an annual by St. Petersburg's Technoex-Conversion; *Samolynot*, a Russian/English-language aerospace magazine published in conjunction with the Aviation Maintenance Foundation International in the U.S.A ...
- Increasing attendance of Russians at International Space Conferences, and collaboration of their scientists with their peers from other nations ...
- Cooperation Agreements with other national space agencies, such as between RKA and NASA or ESA, on manned space infrastructure construction and other activities ...
- Tashkent (Uzbekistan) Agreements with ten former Soviet states, now CIS republics, provide for joint funding with Russia of civilian space activities, plus a separate accord with Kazakhstan on the future of the Baikonour Cosmodrome, located in that country (so that the Russian space program has full access without restrictions) ...
- Entrepreneurial efforts to privatize state industry, such as the founding of the Defense Industrial Investment Company of Kaliningrad as an investment bank, using funds to underwrite promising ventures within Russia to produce for the

civilian market (e.g., commercial applications of space technology); the company is made up of industrial activities that were formerly undertaken by government agencies ...
- Formation of joint business ventures between Russian space organizations and foreign business corporations, such as between the Russian Academy of Sciences and the Russian–American Science Inc. of Falls Church, Virginia (RAS); the Krunichev Strategic Missile Plant's agreement with Motorola Inc. to launch communication satellites for the American firm; and the same factory's venture with both Motorola and Lockheed Corporation on the Iridium project.
- Establishing Western marketing outlets for Russian space talent, material, facilities, and resources ...
- International Maritime Satellite Organization (Inmarsat), a London-based consortium of 64 nations, uses a Russian Proton rocket to launch one or more Inmarsat 3 communications satellites.

But the greatest feat of the Russian space program may have been continuing the Mir space station in orbit as a laboratory for space living. It hosted 16 crews on long-duration missions, which included 52 space walks, and 8 American Shuttle missions; and received 32 Progress supply ships and numerous international visitors, including the first civilian space tourists. Currently, in conjunction with an American Space Adventures company, Roskosmos sells NASA spots on its spacecraft to the ISS at a charge of $42 million per person roundtrip, including pre-departure training and re-entry services. All of the above confirm Russian ingenuity and ethos in keeping their space activities going!

Russia still conducts more launches than any other single country, more than all other spacefaring nations combined with respect to failures. Furthermore, the nation has moved ahead with new rockets, such as Proton with a lift capacity of 20 tons to LEO, and Rokot, for either crews or cargoes; Soyuz 2b will have a lift capacity of 8.5 tons in 2007. In the last 15 years, Russia entered into cooperative agreements with NASA which brought the latter's astronauts on Mir before that space station was discontinued in 1999 [30]. Additional similar agreements enabled the American space agency to complete the International Space Station because of Russian technology, equipment, and cosmonauts. The Russians built two core ISS modules: the Zarya Functional Control Block and Zvezda service module. With the loss of NASA's Shuttle *Columbia* and temporary grounding of the Shuttle fleet by the Americans, only Russia's Proton and Soyuz rockets made it possible to man and restock the ISS. Presently, the Russians plan to increase their ISS manned Soyuz flights to four a year, and Progress transport flights to eight per year. Russia has played a pivotal role in the construction and maintenance of the International Space Station!

In this century, the Russians operate a number of programs for Earth science, communication, and scientific research. By 2011, they expect to fly an upgraded Soyuz capable of carrying four passengers and to replace Progress cargo craft by Parom (Ferry), a re-usable orbital tug with expendable 12-metric-ton freight containers. By 2019, they expect to test Kliper, a winged craft with a reusable small lifting body. They hope to use it for robotic missions to the moons of Mars.

Sec. 1.3] Russian space case study 35

Perhaps the most unexpected trend of the 21st century has been Russia's 87 joint ventures with private foreign enterprises. Consider these innovative examples:

- Boeing, for example, has a major aeronautical design center in Moscow employing that country's aerospace engineers and technologists in large numbers, especially to solve problems with aerodynamics and alloys. Further, agreements with American space travel corporations has made it possible for civilian tourists to travel to the International Space Station, as they once did on Mir.
- The Soros Foundation in combination with the Russian government has opened an Internet University Center in the nation's provinces, with 30 more set up in 2002 at a cost up to $100 million. The aim is to create humanitarian universities in old Russian towns, while strengthening local educational and cultural resources. Similar centers have also been operating in Moscow and St. Petersburg for the past ten years.
- The Russian cabinet in 2005 approved a nine-year government program to expand the nation's space program, continuing the building of its segment of the International Space Station, as well as development of a new 16-ton Klipper (or Clipper). Russia also plans to increase its launches of commercial satellites, send an automated mission to the Martian moon, Phobos, to collect soil samples, as a preliminary step to manned missions to Mars.
- In 2007, Russia officially celebrated the 100th anniversary of the birth of their space legend, Sergei Korolev. After suffering years of hard labor, starvation, and torture under Stalin, thanks to the famed aircraft designer Andrei Tupolev, Korolev became chief of the Soviet rocket program, the brilliant designer of a series of Russian space achievements!

(for further information on the Russian space program, use these Internet web sites: *www.federalspace.ru*, but for English add */index.asp?Lang = English*, or *en.wikipedia.org/wiki/Russian_Space_Agency* ... *www.russianspaceweb.com*).

1.3.1 Conclusions about Russian space vision and ethos

Seemingly, the old Soviet ethos regarding space eroded in the 1990s during the nation's transition to more democratic ways. In the beginnings of the more market-oriented Russia, if the price was right, everything was up for sale, including space artifacts. Mementos of space heroes were posthumously pressed into duty to sell cars: the famous Cosmos Pavilion, a shrine of Soviet space feats, had been turned into a used-car lot in 1992, while the exhibit was closed for repairs. The Buran space shuttle became an amusement for children in Gorky Park. Cosmodromes deteriorated and fell into disrepair.

Yet, great leaders of the old Soviet military–space complex had motivations other than pleasing the state. In a 1992 lecture to the Planetary Society in California, General Gherman Titov, the first man to orbit Earth for more than 24 hours and once chief of the Soviet Defence Command, explained that he and many of his colleagues were and are concerned not only with humanity's exploration of the Solar

System—but with gaining a deeper understanding of our own planet! Again the Russian vision and ethos was revealed that same year when Vladmir L. Ivanov, flight-testing director for the Mir space station, described plans for converting the Plesetsk Cosmodrome into a fully fledged space center to launch a variety of rockets—even into geostationary orbit. Then he asserted:

> "Russia is and will be a space power. Breathtaking ambitions are necessary to reach the spiritual heights, the heights of the Russian spirit. We began with the first Sputnik and we will go far beyond" (*Space Fax Daily*, October 6, 1992).

This would seem to be confirmed by a French government report in 2007 that Russia has multiplied space spending by ten times in this decade! Currently, their annual space budget is about $800 million.

Despite continuing chaos which has taken a toll on the Russian space program and caused numerous setbacks, Nicholas Johnson, an American who publishes annual reports on their space activities, believes the Russian space program still operates over 150 satellites, and will rebound [31]. For Russians, space exploration has always had a deeper *cosmic* meaning. *Russian Cosmism*, best expressed by the visionary Tsiolkovski, is a social movement dealing with the history and philosophy of the origin, the evolutionary future of the universe, and humankind in our genetic unity and mutual influence. This "father of space travel" philosophized before his death in 1935:

> "The biggest step in the history of mankind will be to leave the atmosphere of Earth to join the other planets" [32].

The future of space programs within Russia and the Commonwealth of Independent States seems to be in joint international ventures with other nations. These developments within the global space community will be discussed in the Chapter 2. The best summary of this nation's space contributions was provided by the astute Dublin space writer Brian Harvey when he concluded [33]:

> "An enduring feature of the Soviet and Russian space programs is its sense of history. It is not one universally shared in a country which has endured much hardship and where people have more immediate and pressing concerns on their mind, but it is one held by enough people to matter ... It was a space program in which its participants and admirers could immensely take pride—a program built on a potent mixture of courage, endurance, daring, engineering genius, quality and imagination."

Yes, this was and is a major space program with historical roots and conviction that has endured for over a century, and will contribute more in the future!

1.4 EUROPEAN SPACE CASE STUDY

The multicultures of modern Europe have been developing a *space ethos* for centuries in their art, science, and literature. The early dreamers of space travel came from Europe (e.g., Italy's inventive artist, Michelangelo; France's science fiction writer, Jules Verne; and Romania's mathematician, Hermann Oberth). The drawings and writings of such visionaries inspired generations of future space scientists and engineers worldwide, but especially in Germany where rocket pioneering abounded in the first half of the 20th century. Europe's entry into the Space Age occurred in World War II, with the negative impact of missiles from Peenemünde on a devastated England. However, with their minds on the Moon, many of the German rocket scientists there under Wernher von Braun were transformed from war to peace by being absorbed into either the American or Russian space programs [34]. Since 1958, Western European nations have collaborated, originally through the European Launch Development Organization and the European Space Research Organization. The first non-Soviet Europeans to fly as cosmonauts on Russian spacecraft were in 1978: Vladimir Remek of Czechoslovakia, Miroslaw Hermaszewski of Poland, Sigmund Jahn of East Germany, and in 1982 Jean-Loup Chrétien of France.

In the 21st century, this continent and its neighbors endeavor to create cooperation through the European Union, now with 26 national members. Space synergy is evident through the workings of the multinational *European Space Agency*, founded in 1974, to coordinate the space investments and missions of its 17 member states and their national space agencies (ESA Headquarters, 8/10 rue Mario Nikis, 73738 Paris Cédex, France). Very slowly the expression of European space vision and ethos emerges within the EU nations though ESA. The Agency's space efforts are seemingly mission oriented and budget driven annually without much long-range vision, with only short-term strategic plans, especially with reference to human spaceflight and space settlement. Yet, its collaborative spirit extends to joint efforts with North American and Russian space agencies. It has provided projects and astronauts for both Mir, and now the International Space Station, with Europeans flying on NASA's Shuttle and Russian spacecraft. Presently, ESA centers its programs around space science, astronomy, and related technologies, as well as participation in larger NASA undertakings (*www.esa.eu*).

In addition to the four countries listed below, ESA comprises space agencies and other entities from Austria, Belgium, Denmark, Finland, Greece, Ireland, Luxembourg, the Netherlands, Norway, Portugal, Spain, Sweden, and Switzerland. ESA has partnership agreements with Hungary, the Czech Republic, Romania, and Poland to prepare them for full membership. ESA leaders are listed alphabetically below with a brief historical sketch.

1.4.1 France

The Committee for Space Research was organized in 1959, followed by the formation of the National Center for Space Studies (CNES) in 1962. Twenty-three years later the nation's space agency got a modern headquarters in the Ranguil suburbs of the medieval city of Toulouse where French aerospace industry and schools were also located. Finally, a Directorate General for Space was established in 1988 within the Ministry for Post, Telecommunications, and Space ... In 1965, France became the third country in orbit with its rocket Diamant and its Asterix 1 satellite; by 1990, it had launched a series of satellites beginning with Spot 2 for remote sensing. The country's principal launch vehicle is now Ariane 5 which in 2007 was readied to launch astronauts in five years. Arianespace became the private sector enterprise incorporated under French law to manage and market the Ariane launch services. In 1964, CNES chose Kourou near the equator to take advantage of the Earth's rotation for a geostationary orbit. There evolved the Guiana Space Center (CSG) which was to become Europe's official spaceport (see Exhibit 10). France has also developed the imaging system for the U.S. Clementine lunar survey, and launched numerous satellites, such as Helios 1 and Telcom 2C. Since 1977, France has worked closely with COSPAR (Committee on Space Research), reporting on its technological progress through its space agency, but also relative to aerospace medicine, astronomy, Earth resources, geodesy, meteorology, oceanography, solar physics, and other systems.

Today, CNES relies on a dozen major research laboratories as space contractors. French scientists, engineers, and technology have made outstanding contributions to global space progress, including ongoing research on cryogenic engines to power Ariane 5, and nuclear-powered propulsion of spacecraft and satellites for deep-space exploration. One ethos measure is that France spends billions on its civil space programs and millions on military space. French citizens have flown on both the American Shuttle and the Russian space station. France is a prime mover in pushing Europe toward space independence, and its Parliamentary Office for Evaluation of Scientific and Technological proposes to reinvigorate Europe's civilian and military space policy (*Space News*, February 12, 2007). Its report offers 50 strategies for this purpose, giving EU preference to European launch systems; forming an alliance with NATO for encrypted signals on the Galileo satellite navigation system; assistance to companies engaged in suborbital flights designed to create a space tourism industry; recommended purchase of nuclear-heating technology for Europe's Mars explorations (for further information, visit *www.fas.org/spp/guide/france/agency/index.html ... www.cnes.fr ... en.wikipedia.org/wiki/Category:French_space_ prog*).

But as the country's immigrant population increases, its fledgling space vision/ethos have not moved down from the intellectuals and scientists to the masses. Yet, the catalyst for Europe's space future may come from many prestigious global space organizations headquartered in France, such as the European Space Agency and the International Astronautical Federation in Paris, as well as the International Space University in Strasbourg.

1.4.2 Germany

As previously indicated, Germans were the principal rocket innovators in Europe during the 20th century., despite the setbacks of two World Wars. The German Aerospace Research Center or establishment was formed in 1969 within the Federal

Exhibit 10. Europe's spaceport in South America. An Ariane 4 launch vehicle reaches the launch zone at Europe's launch base in Kourou, French Guiana. Source: ESA/CNES/Arianespace.

Republic by a merger of three organizations, one of which dated from 1907. By 1974, responsibilities for atmospheric research were absorbed under the Ministry for Research and Technology. The German Space Agency (DLR) was founded in 1989 as a limited liability company. While Germany has participated in both NASA and ESA space projects, its own unique efforts have been directed toward the Rosat X-ray astronomical research satellite, DFS and TUBsat telecommunications satellites, the Sanger aerospace plane, and the Columbus module for the International Space Station. Germans and their experiments have also flown on both the American Shuttle and the Russian space station. The first ESA astronaut to fly in space was Ulf Merbold from Germany. With a civil space budget over $900 million, Germany is the second largest contributor to ESA funding.

Besides their participation in ESA projects, German aerospace research has been active in computer system design; data/communication systems, networks, and transmissions; technological forecasting, gravitational biology, as well as joint programs, such as Skylab, Symphonie, and Helios.The German Aerospace Center manages the nation's space life science programs. German scientists focus their space efforts on integrative human physiology; biotechnical applications of the microgravity environment; and radiation biology and response. Yet the German people at large do not seem enthused by a space vision and ethos, despite the pioneering accomplishments of their early rocket engineers (for further information on the Internet, visit *www.areospaceguide.net/ ... worldspace/germanspacepro ... www.ncbi.nlm.nih.gov/*).

1.4.3 Great Britain

At the start of the Space Age, the United Kingdom showed great promise with talented space scientists and engineers, a national launch vehicle program, and vigorous activities by its space activists and their publications. The British Interplanetary Society and BBC space specials have done much to raise citizen awareness. In 1978, the nation participated with both NASA and ESA in launching IUE, the world's first high-orbit telescope, used for ultraviolet astronomy. Successful Earth orbit projects followed with ESA, including Giotto, launched in 1985, the first deep-space mission to study comets. A positive development was the creation in 1985 of the British National Space Centre under the U.K. Department of Trade and Industry. A second was participation in 1990 by the U.K. Science and Engineering Research Council to fly the Along-Track Scanning Radiometer as payload on the European Remote Sensing Satellite (ERS 1). Another indicator of their space ethos is the publication by Elsevier/Butterworth-Heinemann in Oxford of the prestigious international journal, *Space Policy* (*www.elsevier.com/locate/spacepol*), as well as the *Space Exploration series of books by Springer-Praxis, Chichester (www.praxis-publishing.co.uk*). Other U.K. space contributions are in instrumentations for Earth observation, and ESA's Infrared Space Observatory, as well as tracking and observation systems for orbiting spacecraft.

But, the nation's ethos for space seems to be diminishing temporarily with government cutbacks of its support for the aerospace industry in general, and space research in particular. Britain takes a token role in ESA projects such as Ariane and Columbus. Its space budget is well below its European counterparts, while Britain's ambivalence about fully participating in the European Union somewhat undermines its activities within ESA (this problem with its continental neighbors goes back to the 1970s when its advocacy of Europa as a civil launch vehicle failed to gain their support). However, British spacefarers have flown on the American Shuttle, as well as on Mir and ISS. But, the average Briton and their politicians fail to realize that the way back to the glory days of "empire" lie on the high frontier (for further information on the Internet, consult *www.aerospaceguide.net/britishspaceprogram.html* ... *www.videojug.com/film/the-british-space-program* ... *www.spaceuk.org/*).

1.4.4 Italy

A country that moved early and actively by establishing a National Commission on Space Research in 1959, followed by spacecraft development. By 1964, it was cooperating with NASA to launch the San Marco 1 satellite from a mobile platform off the coast of Kenya; this has evolved today into the San Marco Scout spacecraft. By 1979, Italy had a National Space Plan that led to the founding in 1988 of the Italian Space Agency (ASI) under its Ministry of Science and Technology. The nation has been a key participant in ESA, providing the use of the Italian Processing and Archiving Facility (I-PAF) located at ASI/Centro di Geodesia Spaziale in Matera. It also contributed 65% of the costs for the ESA Vega launcher. Italy contributes about $500 million to its ESA space investment, less than previous years. Innovative Italian space research has ranged from Italsat II to programs with the U.S.A. on *Tethered Satellite Systems* and Lageos II to study tectonic motions with lasers, as well as the BeppoSAX X-ray astronomy satellite and SAR-X, a synthetic aperture radar satellite. At present, the Italian ethos for space is enthusiastic and very active, but delimited to a subculture of scientists, engineers, and students (for further Internet information, go to *www.aerospaceguide.net/italianspaceprogram*).

1.4.5 ESA emergence

After decades, the efforts by Western Europe to coordinate its space activities climaxed with the formation of the European Space Agency by a merger of two prior organizations. The growth in the space science and launch program within the original 13 member states parallels the progress of the larger entity, the European Common Market. By ESA's 20th anniversary, the latter has evolved into the expanding European Union, which is providing space with an ESA budget of over $3 billion annually. Under *Horizon 2000*, ESA's long-term science program, the Council of member states has laid a realistic foundation for European space activities related to environmental sciences, telecommunications, launch systems, and manned spaceflight. ESA's synergistic strategies have been implemented on the basis of worldwide partnerships with the U.S.A., Russia, Japan, and Canada, as will be explained in Chapter 2.

In addition to its Paris headquarters, ESA facilities include the European Space and Research Center in the Netherlands (ESTEC, Noordwijk); the European Space Mission Control in Germany (ESOC, Darmstadt); and the European Space Research

Institute in Italy (ESRIN, Frascati). In all, the European Space Agency has ground stations on the continent and Canada, plus four processing and archiving facilities in France, Germany, Italy, and the United Kingdom, plus a network of a dozen ground stations to receive radar imagery. As previously noted, its official launch center is Kourou, French Guiana, with management there delegated to Arianespace, while the European Astronaut Center is located in Cologne. The Agency has done much to encourage enterprise in space science by EU members, especially Earth observation, and satellite communications. For example, in 1995, Finland became the 14th member state of ESA to participate in its programs, expending $4 million in space R&D ... ESA also has an active program with the Swedish Space Corporation (SSC) at its Esrange center to track and control ESA satellites (Sweden spends over $100 million annually on space-related ventures) ... The Netherland's ESTEC establishment is where the Dutch are spending over $125 million on its largest ESA program: building a robotic arm for the Russian part of the International Space Station. The remaining space investment of other ESA members not mentioned above are from Austria, Belgium, Denmark , Ireland, Norway, Spain, and Switzerland, plus some private R&D spending. After the collapse of the U.S.S.R., former eastern bloc countries wanted association with the agency, so cooperation agreements were signed with the nations that are new EU members. There are many indications that the *space ethos* is increasing in Central and Eastern Europe through information exchanges and Pan-European conferences. To assist Africa with remote sensing, mapping, and surveying, an ESA workstation was set up in Kenya (Nairobi). Finally, ESA has a long history of cooperation in both unmanned and manned missions with the United States, Russia, Canada, and more recently China. It is developing a Double-Star Mission with its most important partner, the Russian Space Agency.

Despite some setbacks and launch failures, the European Space Agency has manifested steady successes, such as:

- the Ulysses spacecraft which, after an encounter with Jupiter in 1992, went into an orbit taking it over the poles of the Sun;
- ERS 1 and 2 satellites, establishing an ESA information network called Earthnet ...;
- Giotto (GEM) instruments from the comet Halley encounter to the comet Grigg–Skjellerup encounter ...;
- mapping of 120,000 stars by the satellite Hipparcos and the scientific experiments and data gathering of Eureca ...;
- participation in NASA's Hubble Space Telescope project ...;
- Human Behavior in Space Simulation Studies and the Isolation Study for European Manned Space Infrastructure, which will be discussed in Chapter 2 ...;
- Spacelab manned orbiting laboratory developed for use on the American Space Shuttle by ESA astronauts ...;
- free-flying laboratory called Columbus to be an orbiting module on the International Space Station (discussed in Chapter 8) ...;
- SCOS, a satellite control center which any European firm may use ...;
- SMART 1 probe testing leading-edge space propulsion technology ...;
- Venus Express and Mars Express unmanned missions in the 21st century.

The Agency does not invent missions, but asks for proposals from scientists, both in Europe and elsewhere, whose peers then select the missions to be pursued. Such programs have included the Solar Heliospheric Observatory (SOHO); Cluster, four identical spacecraft, sent into orbit by two Russian Soyuz launchers, and flying in tetrahedron configuration for polar orbits around Earth, observing the Sun's interaction with its magnetic field; the Infrared Space Observatory (ISO); the ESA/NASA joint venture with SOHO sending back results of its sungazing; the combined Herschel/Planck space observatory missions planned for 2009; Corot, a space telescope to search for extrasolar planets. Currently, its Aurora Progam lays out a timetable for Mars missions until year 2030. ExoMars will be launched in 2013 for a Mars rover mission, followed by another to bring back samples from that planet.

1.4.6 Conclusion about European space vision and ethos

In the European Union, a healthy vision of space potential and activities develops, but not among the majority of its people. The European Space Agency provides the leadership in competent space science projects, endorsed by politicians, academics, aerospace professionals, and students, especially within the International Space University. Europe's financial commitment to space is reflected in the ESA budget in 2006 which was approximately $3.5 billion, but when combined with the spending of member state space agencies, this investment is about $7 billion. The focus of that spending is on launch vehicles, and only 16% of the budget goes on human spaceflight.

Impressive advance planning and reports occur under ESA's Office of Scientific Programs (e.g., *International Lunar Workshop: Towards a World Strategy for the Exploration and Utilization of Our Natural Satellite*). Long before the current NASA/VSE program, that seminal report in 1994 provided a rationale for permanently returning to the Moon, including the philosophical, scientific, political, and socioeconomic reasons for doing so. ESA Administrator Roger M. Bonnet then explained their vision of the Moon as "a vast space station offering many areas of cultural, as well as industrial resources—the closest space harbor beyond Low Earth Orbit." Yet, most ESA proposals for lunar exploration and development have yet to be funded and implemented, despite these noble observations of some 14 years ago in its publication, *The European Moon Programme* (ESA/SPC 94-43, Annex 1):

> "At the turn of the second millennium, our planet is at a major turning point. Politically, economically, and socially, humanity is in search of a new equilibrium. Such an equilibrium is necessary if we are to stop the continuous increase in unemployment which affects several hundreds of millions of the Earth's inhabitants. The need for large and global programmes ... is necessary for a new start in global economic development ... through peaceful initiatives which would offer a means to occupy the talents of everyone for the common good and the future of civilization. ...
>
> Space programs are among such large programs, encompassing science and high-technology activities on a planetary scale, serving mankind, offering a mirror

with which it can survey its home planet, while exploring further the limits of habitability and its visibility of the Universe."

More recently, ESA Director General Jean-Jacques Dordain summarized the European vision and ethos in these words:

"Today space activities are pursued for the benefit of citizens, and citizens are asking for a better quality of life. They want greater security and economic wealth, but they also want to pursue their dreams, to increase knowledge, and they want younger people to be attracted to the pursuit of science and technology ... I think space can do all this ... the reason for space exploration."

(For more information, read Brian Harvey's book, *Europe's Space Programme* [35].)

1.5 ASIAN SPACE CASE STUDY

Since the dawn of the Space Age many countries in the Middle and Far East have demonstrated interest in space, particularly with reference to communication satellites. The United Nations, particularly through its Office of Outer Space Affairs, has sponsored space conferences for participants from developing countries [36]. These have emphasized space technology and resources that might contribute to sustainable development and food production, communications, and safety, in addition to modernization and infrastructure. Let us take Asia as a case in point. Japan with its First World economy once took the regional leadership position relative to its space activities, but in the 21st century it has been surpassed by China and India whose rapidly developing economies have also advanced their space technologies and programs. All three countries are Asia's catalysts with regard to space vision and ethos, stimulating the space endeavors of other Pacific Rim countries, such as Australia, Indonesia, Pakistan, South Korea, and Taiwan. Thus, space programs and budgets are increasing throughout the region, largely devoted to productive, nonmilitary use. This geographic area has seen a marked increase in launching orbital geostationary communication and broadcast satellites, thus advancing their telecommunication and entertainment industries.

Again, one ethos manifestation is what a country invests in space technology. Australia, for instance, spends approximately $7 million for that purpose, plus another $600 million for commercial purchases of space goods. The Australian Space Council's plans include development of a small satellite-manufacturing industry; promoting launch sites on that island continent, such as the Woomera rocket range and a proposed Cape York spaceport; accessing remote-sensing data, then extending its use within the Asia–Pacific market; performing space-related research through universities and high-tech aerospace corporations; and participating in international space projects ... Despite internal socio-economic and terrorist problems, Pakistan expends about $8 million annually on space. It has a national space agency, SUPARCO, which launched Badr B, its second LEO satellite, in 1995. At its

Islamabad receiving stations, remote-sensing imaging data are obtained and processed from many orbiting satellites of other countries. Perhaps the most unique of Pacific Basin space endeavors is exemplified by the small island of Tanegashima with its $1 billion, modern, advanced Japanese launch center ...

In 1994, ministers for space programs in the area first gathered in Beijing to formulate a Regional Space Technology Program under the aegis of the U.N.'s Economic and Social Commission for Asia and the Pacific. Regional cooperation or synergy may be the best way for these diverse peoples and countries to cultivate their space vision and ethos, until they truly become spacefaring nations! Ideally, Asians can advance their offworld dreams by cooperation agreements with larger space agencies in the world for sharing data, technology, and systems. The three largest space undertakings in the region are described in the following sections.

1.5.1 China

An ancient nation whose lore of Moon travel and experiments with rockets go back many centuries. In the 16th century, the Chinese inventor Wan Hu built and tested a rocket-propelled chair which launched him into fiery oblivion! As China becomes a major economic superpower in the 21st century, its space ambitions and activities expand [37]. Exhibit 11 is one indication of the People's Republic of China's sophistication in space technology.

In the 20th century, despite a civil war and cultural revolution, the country's space activities began in the late 1950s when a strategic space plan was formulated. Its space program was unwittingly spurred on by the Americans who foolishly expelled nearly a 100 scientists of Chinese ancestry. One of them was Dr. Tsien Hsueshen, an aeronautical engineer and co-founder of CalTech's Jet Propulsion Laboratory. In 1956, he became head of the PRC's new missile program. Soviet scientists helped his team until 1959. In 1964, China developed and launched its first space vehicle. In 1970, it built the first home-grown rocket known as Dong Feng (East Wind), launching it from its Jiuquan base. Then, its first satellite Mao 1 was sent into orbit via the Chang-Zheng 1 (CZ-1) launch vehicle, thus inaugurating a series of recoverable communication satellites. By April 1990, the CZ-3 was engaged in the first commercial launch for a foreign customer: the satellite Asiasat 1. By 1995, China had successfully launched more than 40 satellites with self-developed launch vehicles, particularly the Chang-Zheng (Long March) series. The country now has the Beijing Rocket Test Center and two launch facilities (Xichang Space Center and Jiuquan Space Center), as well as a TT&C satellite-tracking network. The Beijing Space Technology and Test Center was constructed to provide a clean operational environment from delivery of dispatched products of spacecraft to assembly and testing. There, the National Spacecraft Trial Base has a 100,000 m^2 floor area for space-related laboratories.

In China, the Ministry of Astronautics (MOA) oversees space research and industry. Its technical consulting unit is the Science and Technology Committee, while its R&D entity is the Chinese Academy of Space Technology (CAST). The nation and its universities attach great importance to space technologies and their applications for improving economic and social conditions within this huge, impoverished country. Its satellites are being utilized for television communication, disaster monitoring and forecasting, identifying natural resources, and even for mass education via the Satellite Distance Education Network which links 53,000 learning centers. Because China is a closed system and there is an emphasis on military satellites, it is difficult to obtain information on how much the country is spending on space. Since 1985 the state has been moving into the commercial market by the decision to make its three-stage CZ-3 and CZ-4 rockets available to foreign clients. The China Great Wall Industry Corporation is leading in the establishment of an independent space industrial system with international marketing capability. In 1993, the China National Space Administration (CNSA) was designated responsible for national space policy, while the China Aerospace Corporation (CASC) executes that policy by means of launch services and coordination of subcontractor partners. In 1998, CASC was split into a number of smaller, state-owned companies not directly owned by the government. China has agents throughout the world to represent its space interest. In the U.S.A., for instance, it is Becker & Associates of McLean, Virginia. Also under MOA, China has two other companies in space-related business: Wanyuan and Lishen Microelectronics. Presently, China is spending about $1.2 billion yearly on its space efforts.

In addition to launch and research services, China is selling re-entry satellites to customers (SETE and FSW 1/2). China is also slowly joining the international space community. For example, in 1990 it began exchanges with the European Space Agency on launch services, and in 1995 it signed a bilateral cooperation agreement with the U.S.A. In 1997, the Asia Telecommunications Co., partially owned by the Chinese CITIC Group, flew Asiasat 3 on a Russian Proton rocket. The government has always been actively engaged in U.N. programs on space applications, particularly with reference to environmental concerns such as remote sensing. In 2002, the government issued a white paper 50 years after the nation embarked on its space activities; in 2006, the government issued a full annual report on its space development and future. This included its current Fengyun and Ziyuan satellites; Long March with 46 consecutive flights; new construction at its three launch sites (Jiuquan, Xichang, and Taiyuan); telemetry, tracking, and command. In 2003, China became the first nation in Asia and third country in the world to develop manned spaceflight (Shenzhou 4 on October 15, 2003); then, its Shenzhou 7 carried three astronauts to engage in its first space walk ... That same year, China hosted the Fifth ILEWG Conference in Beijing with participants from global space agencies, and the International Lunar Conference which attracted worldwide participation.

China is now one of the world's leaders in important areas of space technology, such as satellite recovery, multi-satellite launch on a single rocket, rockets with cryogenic fuel, strap-on rockets, launch of geostationary satellites, and telemetry, tracking and control. Now China has joined the previously exclusive human spaceflight club of the U.S.A. and Russia, planning its own unmanned and manned shuttles, as well as a satellite to orbit the Moon in preparation for a lunar landing by 2020. The challenge for this emerging economic and space power is to use its space capabilities for economic development and global cooperation, instead of wasting

Sec. 1.5]   Asian space case study   47

resources on the pursuit of military dominance of the high ground. A hopeful sign is that Chinese scientists are already planning outposts and manufacturing on the lunar surface. That may be feasible with its Long March 5 vehicle capable of launching 70 tons into orbit (for further information, visit *www.csna.gov.cn* ... *www.xinhuanet. com/enligh/space/index/* ... *www.space.com/businesstechnology/technology/shen* ... *org/wiki/en. wikipediaChina_National_Space_Administration*; and for a list of Chinese space milestones, visit *www.astronautix.com*).

1.5.2 Conclusion on Chinese space vision and ethos

Among this ancient people, presently ruled by a Communist government, the possibilities of space exploration and development are appreciated largely among the educated elite. They have increasing interest in human spaceflight, life sciences and space medicine, planetary exploration, and utilization of space resources. Yet, the country's launch of its own *taikonaut*, was shared by the masses via television, and so a greater involvement of citizen interest is occurring. Col. Yang Liwei, the first Chinese in orbit, was a source of national pride, although his mission cost about $2 billion. Despite the PRC space program advancing slowly, it expects to launch 100 satellites within ten years. Under a military-run administration, its space planners are also discussing space shuttles and stations, as well as deep-space missions. There is a Chinese myth about a woman Chang who took a potion, flew to the lunar surface, and remained there; now CSNA has a three-step Moon exploration program under way. Hopefully, they will do all this in an international context by collaborating with other spacefaring nations. A step forward occurred in 2003 when China became involved with the Europeans in developing a global-positioning system. Then, again in 2006, China's president Hu Jinto invited NASA Administrator Dr. Michael Griffin to tour their space facilities, beginning an annual dialog on joint space ventures, such as China's participation in ISS and VSE programs. On the positive side, China has space partnership arrangements with ESA, Russia, Brazil, and some Asian–Pacific countries. A step backward occurred in 2007 when the PRC unilaterally conducted an anti-satellite weapons test by using a ballistic missile to knock out its aging weather satellite orbiting the Earth, some 537 miles aloft, thus adding to the problem of space debris.

Perhaps the best expression yet of China's space vision and ethos was expressed in this statement from the white paper "China Space Activities in 2006" issued by the PRC's State Council:

> "The aims of China's space activities are to explore outer space and enhance understanding of the Earth cosmos; to use outer space for peaceful purposes, promoting human civilization and social progress and benefit the whole of mankind; to meet the demands of the Chinese people for national security, and social progress, economic construction, scientific and technological development ..."

Exhibit 11. A Chinese mission control room. Source: China Great Wall Industry Corporation (GWIC) and published in Deyong Kong, "A Chinese Perspective on Space Development," *Space Governance*, **2**(1), June 1995, 20.

1.5.3 Japan

The origins of this nation's rocket program go back to the 1920s. "The father of its space program" was Hideo Itokawa, a professor at the Institute of Industrial Science, Tokyo University, who took the lead in development of its first rocket plane, and its first Pencil rocket in 1955. Since then Japan has become a leading spacefaring nation, having flown probes to the Moon, Mars, and comet Halley. Today, Japan has a single Japan Aerospace Exploration Agency formed in 2003. JAXA is responsible now for space research and development, satellite launches into orbit, and operation of future missions, both unmanned and manned. It was established through the merger of three previously independent space organizations. In 2007, its budget was around $1.6 billion.

Before JAXA, this fourth nation to go into orbit had three space organizations, each funded separately. The oldest and smallest, the Institute of Space and Astronautical Science was founded in the mid-1950s, and focused on space and planetary research. ISAS has launched scientific satellites from its Kagoshima Space Center on the southernmost main island of Kyushu. In collaboration with the

University of Tokyo and Japanese electronic/motor companies, ISAS developed large solid rocket launchers (e.g., the Mu-Series, which carries the science satellites Muses A, Solar A, and Astro D, an X-ray satellite). As a multi-university institute, professors from many institutions come to conduct research, such as on HIMES (highly maneuverable engineering space vehicle) ... The second major entity founded in 1964, was the National Space Development Agency which had a big budget, projects, and facilities for developing rockets, satellites, and the Japanese Experimental Module for ISS. In 1972, NASDA established its Tsukuba Space Center, now Tanegashima Space Center in southeastern Tanegashima (an island in Kagoshima prefecture). From this site, a series of satellites have been launched in both LEO and GEO (approximately 50 on its own spacecraft, 25 on vehicles of other countries). Their Earth Observation Center is in Hatoyama-machi, Saitama prefecture, while their Kakuda Propulsion Center is located in northern Kakuda City, Miyagi prefecture ... The third entity was the National Aerospace Laboratory of Japan (NAL). Japan's space scientists have worked on the Orbital Re-entry Experiment Vehicle, the Vehicle Evaluation Payload, and Space Flyer Unit for launch on H-II in geostationary orbits.

Japan's Space Activities Committee has attempted to coordinate the nation's space program among four government departments, facilitating joint ventures with the above agencies, universities, and private sector aerospace companies, as well as with other international agencies. Such collaborative projects include the 87-tonne J-1 rocket launched in 1995, joint studies with the National Aerospace Laboratory, and vigorous lunar development research such as on a proposed Lunar Energy Park, supposed to extend until year 2030. Among its successes aloft, Japan counts X-ray astronomy, its HALCA mission featuring Very Long Baseline Interferometry; solar observation and magnetosphere research; Earth climate observation. The critical launch of a Multi-Functional Transport Satellite 1R was important for weather forecasting, while MTSAT2 in 2006 facilitated the nation's air traffic and the communication satellite business. Though the Japanese have been innovative in their space planning, their actual accomplishments have been somewhat disappointing. With NASA as it closest partner, Japan showed much promise in space exploration and development, but has been surpassed by China and India. In this decade, the country has suffered setbacks not just in its general economy, but in its space efforts, such as loss of a rocket in 2003, a blow to its commercial launch industry; scrapping an automated mission to Mars; problems with its Hayabusa mission which finally landed on an asteroid, but whether it will collect samples and return in 2010 is uncertain; only one out of three attempts managed to launch solar sail prototypes. *The Economist* had a feature "Sayonara, Spaceflight" which stated that bureaucratic rivalries, incompetence, cost overruns, and technical difficulties undermined the nation's space efforts (August 8, 1998, pp. 65–66) [38].

Currently, JAXA is developing the Galaxy Explorer (GX), the first rocket to use a liquid natural-gas fuelled second stage, although its debut will be no earlier than 2011. Despite earlier problems, the Agency has enjoyed success with its Akari (Astro F) infrared space telescope, its Hinode (Solar B) spacecraft to explore the Sun's magnetic field, and the Kaguya (SELENE) probe to the Moon. Since 2006, JAXA has also been involved in a programme to develop and demonstrate the technologies for small satellites. The emphasis is now on further missions to the Moon, the advanced H-IIB and M-V rockets, and new Earth observation satellites.

To date, seven Japanese have been in orbit, starting with Toyohiro Akiyama, a journalist who paid $14 million for seven days on the Russian Mir space station in 1990. The other six professionally trained astronauts all flew on NASA Shuttles, except Toyohiro Akiyama who flew on Russia's Soyuz TM-11 in 1990. Those orbited on American spacecraft include Mamoru Mohri on STS-47 in 1992, STS-99 in 2000; Chiaki Mukasi on STS-65 in 1994, STS-95 in 1998; Koichi Wakata, STS-72 in 1996; and Takao Doi, his country's first space walker on STS-87 in 1997, and Akihiko Hoshide, mission specialist on STS-124 in 2008. Three more Japanese astronauts have also been training for service on the International Space Station.

Next, JAXA has set a goal of sending its own astronauts to the Moon by 2020, and constructing a lunar base by 2030. (In a science-based novel on lunar industrialization entitled *Launch Out*, your author conceived a similar scenario but as a joint venture by private enterprise in Japan and the U.S.A. [39].) JAXA is also developing technology for the next generation of supersonic aircraft. In addition to seven research, testing, and propulsion centers, the Japanese Aerospace Exploration has headquarters in Chofu, Tokyo (*www.jaxa.jp/index.html*). The best book resource on this and the next case study is Brian Harvey's *The Japanese and Indian Space Programmes: Two Roads into Space*, published in 2000 by Springer/Praxis (*www.praxis-publishing.co.uk*).

1.5.4 Conclusion on Japan's space vision and ethos

On the crowded islands of Japan, there is need to expand, not to Greater East Asia—but to outer space. There seems to be a strong and growing space vision and ethos in Japan among the educated Japanese, especially in academic, scientific, and corporate communities. Unfortunately, the younger generation has yet to buy into space enterprise. For more than 40 years, Japan has been expanding its international space role, with emphasis on satellite launches, material processing, life science experiments, preparation of payload specialists for the U.S. Shuttle, and developing the JEM (Japanese Experimental Module) for use on the International Space Station. But, divisions and delays within this nation's space community have curtailed its progress. The country has yet to launch its own manned spaceflight. Another disappointing example came in 1995. Japan announced a plan to plant two seismic sensors on the lunar surface, so as to learn more about the origins of Earth's only natural satellite, but the mission has taken so long to implement, that in 2007 the revamped Japanese Aerospace Exploration Agency decided to scrap Lunar A, and offer the penetrator probe technology to other space agencies, possibly Russia. Fortunately, Japanese space scientists have always collaborated with NASA, ESA, and RKA, so its future may lie in global joint space ventures. Finally, JAXA is pursuing international lunar initiatives with both a new lunar orbiter, a lander, and possibly some kind of roving vehicle.

1.5.5 India

The people of India have been fascinated with spaceflight for centuries, for it is part of Hindu mythology: ancient Sanskrit texts tell of a rocket ship, the *Vamana* which ascended to the heavens! In the 18th century the reality of this emerged when Tippu Sultan of Mysore bombarded the British occupiers with rockets! By 1957, a corps of Indian scientists under Vikram Sarabhai were engaged in studies of cosmic ray physics at Ahmedabad's Physical Research Laboratory (PRL). In 1962, Prime Minister Jawaharlal Nehru created a National Committee for Space Research (NCSR) with Dr. Sarabhai as chairman. Although India was a Third World nation, Nehru encouraged space investment to solve socio-economic problems, and to make the country self-sufficient. For over four decades, India has been seeking answers to its development dilemmas through space science and technology [38]. With India's economic progress in the 21st century, its space investments have increased, aided by competent scientists and technologists. Fortunately, the headquarters of the Indian Space Research Organization is in Bangalore, the center of the country's high-tech revolution. Today, India, like its economic rival China, has a robust space program and plans for lunar missions!

Sarabhai, founder of the Indian Institute of Management and "father of the Indian space program", began by setting up a launching range for the United Nations at Thumba. This first space base near Trivandrum, Kerala, was initiated for the manufacturing of French Centaur rockets. In 1963, they launched their first sounding rocket, an American Nike-Apache. Nearby, the Vikram Sarabhai Space Center (VSSC) was to evolve under the aegis of the Indian Space Research Organization (ISRO), created in 1969 and now part of the government's Department of Space (DOS). India has a history of joint ventures with the major spacefaring leaders: in 1972, the former Soviet Union launched India's first satellite Aryabhata; in 1984, India's first cosmonaut Rakesh Sharma flew aboard Salyut 7 for a week; in 1988, the U.S.S.R. entered into the commercial satellite market by launching India's IRS IA for remote sensing of its natural resources ... In 1975, ISRO undertook the SITE satellite television project via NASA's ATS6 satellite, thus demonstrating the social benefit of space-based broadcasting (farmers and students in 2,400 villages received the communications in six states for a year). Because of Indian expertise in cosmic particle research, its payload experiments were part of an ESA 11-nation consortium which flew on NASA's Spacelab in the 1980s. At Cape Canaveral in 1982, INSAT IA was lifted atop a Delta rocket, thus launching a communications revolution for the huge country and its teeming population. Built under contract by Ford Aerospace for the Indian Space Department, this was the world's first geostationary civilian satellite to combine telecommunications, TV broadcasting, and weather forecasting. INSAT IB was launched from a U.S. Space Shuttle, helping to establish an INSAT National Satellite System for disaster warnings, as well as broadcast communications; this spacecraft extended TV coverage to 70% of India's population.

Despite widespread poverty and inadequate infrastructure, India has invested over $2 billion in space research and facilities that is having a payoff today. In 2007, the nation spent about $1 billion on space programs which employ some 20,000 citizens. Further, it plans to spend up to $3 billion on its new human spaceflights. The ROI has been largely in multipurpose domestic satellite remote sensing, and launch vehicles. With a total of nine space facilities, including three launch sites, the heart of the Indian space program is Bangalore in the south of the subcontinent, headquarters for both DOS and ISRO, plus the latter's Satellite Center for design and fabrication ... At the southern tip of India in Trivandrum, VSSC has become the center for space R&D technology, especially the SLV rocket (used to launch India's first space satellite Rohini), and the later ASLV and PSLV series of launch vehicles, as well as operation of the U.N.'s TERLS station ... On Sriharikota Island near Madras is the SHAR facilities for integration, testing, and launching of space vehicles, such as the SLV 3, in addition to its ground-tracking system for orbiting spacecraft ... In the northeast at Ahmenabad, the Space Applications Center (SAC) engages in space science and technology applications, such as telecommunications, space-based surveys, space meteorology, and satellite geodesy ... The Auxiliary Propulsion System Unit (APSU) has laboratories at both Bangalore and Trivandrum which design, develop, and supply propulsion control packages. Apart from the entities already cited, the government of India also sponsors the Development and Educational Communications Unit (DECU) and the National Remote Sensing Agency (NRSA).

By 1996, India (now one of the six leading spacefaring powers) successfully deployed the Polar Satellite Launch Vehicle, placing IRS P3 into orbit. In 2007, ISRO announced plans to spend up to $3 billion on its first manned spaceflight. ISRO is also developing its first interplanetary space mission Chandrayaan 1 for launch in 2008. This is the beginning of science series missions to the Moon and Mars, including a manned lunar mission by 2019 as part of the American Vision for Space Exploration. Further, India is seeking more space synergy, by increased collaboration with NASA, ESA, and RKA. Cooperative programs are under way on satellite navigation systems: Raytheon is working with the Indians for a terrestrial global-positioning system that will supplement Landsat data with India's Resourcesat information. Its Geostationary Launch Vehicle System with an indigenous cryogenic engine is India's most powerful launch vehicle. In 2007, GSLV's upper stage will be orbited. In 2008, GSLV3, a heavy launch spacecraft with liquid main stage and two boosters, will be flown in preparation for lunar orbit probes. Other countries may attach payloads in this survey of the Moon's resources. ISRO is continuing research on scramjet air-breathing engines, especially for use in RLVs (reusable launch vehicles). ISRO is also seeking small-satellite payload business up to 10 kilograms from other countries, such as Israel and Indonesia (electronic inquiries are to be addressed to *scc@isro.org*).

In summary, India is making slow, steady, and innovative progress in a self-reliant space program mainly devoted to satellite development, including remote sensing and communication spacecraft like INSAT 2. ISRO's space science receives less budget than comparable programs in other countries (see Exhibit 13). Instead, the organization set up its own commercial company to bolster high-technology exports; its profits are allocated to more space R&D. The aim is to promote Indian commercial space enterprise!

1.5.6 Conclusion on India's space vision and ethos

India's space program has always demonstrated unique vision, for it is expected to contribute pragmatically to the nation's economic, technological, and social development. Its huge population has directly benefited from space technology. Because of this heritage and improvements in quality of life that satellite communications has brought to the masses, it would seem that the space ethos in India is more widely diversified. It is evident in the fierce protection of village television by peasants, and thousands of Indians who flock to the Thumba space museum when it is open. The people of India seem proud of their country's space achievements, so now accept and even enthusiastically support space spending: now a modest budget of $700 million annually, a small percentage of total government expenditure. The public expectation is that ISRO provide basic communications, education, and environmental information via its satellites. In the long run, its space ethos will spread among this huge population only if the late Vikram Sarabhai vision is fulfilled; namely, using India's rockets and satellites to deliver greater literacy, better living conditions, and less poverty! [40] (for further information, consult www.isro.org/ ... en.wikipedia.org/wiki/Indian_Space_Research_Organization ... www.fas.org/spp./guide/India/agency/index.html ... www.bharat-raksgak.com/Space/space-history1.html).

1.5.7 Australia

With a Pacific island continent of almost 3 million square miles, a population over 19,731,000, including the Aboriginals, and a per capita income of $23,200 annually, you would expect that Australia would be serious about entering the Space Age. But, the government has yet to establish a space agency, depending instead on the Australian Space Research Institute to advance the nation's space science and technology (www.asri.org.au/). For the first time in three decades, the nation will launch a scientific satellite hopefully in 2008 (www.spacdecom/mission/launches/australian_satellite_000509.html). But even that microsatellite mission resulted from an agreement between NASA and CSIRO (Commonwealth Scientific and Research Organization). With their support, FEDSAT represents a joint venture of these two entities with Australian universities (6) and companies (5). NASA has a long and successful history of cooperation with Australia, especially its Center for Cooperative Research (www.crcss.ciro.au) ... At the same time, agreements have been reached for the United States to build a military satellite communication base at Geraldton, 250 miles north of Perth. There exists now a smaller facility at that site to intercept satellite communications from Antarctica to Siberia. The Americans already have a space base at Pine Gap near Alice Springs which provides early missile launch warnings for Asia and the Middle East, as well as another at the Northwest Cape in Western Australia for sending underwater signals by satellite to their nuclear submarines.

The time has come for Australia to again exercise space leadership, like it did many years ago when it was the third nation to launch its own satellite (WRESAT)! Decades ago, Woomera, 350 miles northwest of Adelaide, was the site between the

1950s and 1970s for secret American rocket tests at the nation's first launch pad. Recently, an American company Rocketplane Kistler announced plans for a $100,000 spaceport there by 2008, so as to launch spacecraft to the International Space Station ... At the same time, the South Australian government in Adelaide, a hub for space-related technologies, is backing a space initiative for that region to establish a billion dollar space industry. Perhaps the Australian NASA astronaut Andy Thomas summarized best his country's challenge:

> "Australia can certainly have a role in space exploration if Australia wishes. It is a question of *national will*, not of technical or intellectual capability, or education or manufacturing resources. And there are more than enough young people who would love to be involved in something like this" (*The Advertiser*, an Adelaide newspaper, August 21, 2006, p. 1).

Space vision among most Australian politicians has been lacking; perhaps private space entrepreneurs will take the lead? South Australian Senator Grant Chapman highlighted the matter:

> "We are 50 years into the Space Age but still have no government endorsed national space policy, space program or space agency."

A governmental review is under way on the Senator's proposal, *Space a Priority for Australia* (*www.industry.gov.au/space*). Apart from its aerospace and high-tech industries, there is support for greater space enterprise among the Australian public. For example, the National Space Society, based in Washington, D.C., has five local Australian chapters mainly in New South Wales (Sydney, Wyoming, Newcastle), as well as in Queensland (Nundah). Down in remote Tasmania lives space advocate Kim Peart, who publishes an electronic newsletter, *Star News*, as well as articles on his space vision (email: *kimpeart@keypoint.com.au*). Exhibit 12 provides an excerpt of his futuristic thinking.

1.5.8 Conclusion on Asian space vision and ethos

The scope of interest in space technology has risen dramatically among Oriental peoples in the last decade. Given Japan's past economic and technical leadership, one would expect it to be pre-eminent in this regard in the region. But, as far back as 1976, Indonesia became the first Asian country to enter the domestic satellite era when it acquired the U.S.-built Palapa system: today with over 40 ground stations, it provides communication links for 3,000 islands over 5,000 km of ocean. The current generation of this Indonesian satellite uses surplus capacity on behalf of Malaysia, Singapore, Thailand, and the Philippines. But, what is remarkable is the vigor and progress in developing economies, such as China, India, and even Pakistan. Through its Islamic Space Institute, the last-mentioned republic has formed partnerships with Bangladesh, Egypt, Indonesia, Iran, Saudi Arabia, and Turkey to pool their resources in satellite construction, instrumentation, control systems, launch vehicles,

> **Exhibit 12. CREATING A SOLAR CIVILIZATION.**
>
> An Earth citizen's vision for space may begin with a few good-hearted folk, but to have an impact, participation and support would need to grow to a million strong and beyond. In this way, the vision for space could become a mainstream activity with a view to saving human society as a whole by working toward our survival and prosperity in space, as well as peace on Earth and in our Solar System. The benefits of space must reach the whole human family for that vision to be realized. With a large involvement of people working toward practical outcomes on Earth and in space, it will be possible to consider serious fund-raising to drive a citizen's space initiative from Earth to the stars! If our survival matters, then the challenge ahead is to spread this story that inspires participation and action.
>
> We now have a golden opportunity to consider our future on Earth and in space. Our fate is in our hands, for should we miss this chance to fly from the Earth and live among the stars, we have only ourselves to blame. It will be a huge effort to succeed, but succeed we must, if we would like to ensure our survival on the home planet, as well as in space. It may be by flying from our earthly nest that we can begin to realize our full potential as children of the stars. Are you ready for the celestial challenge in the Solar System and beyond?

Source: Adapted from Kim Peart's *Creating a Solar Civilization* (August 2008, p. 19). Contact *Star News* (33 South Terrace, Lauderdale, 7021 Tasmania, Australia; tel.: 03 6248 1373).

and facilities. Pakistan is also an example of how a Third World country can benefit from international cooperation, as exemplified by the hundreds of sounding rockets launched from Sonmiani beach near Karachi carrying payloads from the U.S., France, and the U.K. Regional satellite systems are another example of *space synergy*, the subject matter of Chapter 2. This is exemplified by the communication satellite Arabsat being financed by 22 countries in West Asia.

The last case study above confirms the existence of space vision in Asia, but the issue is its transmission among the masses of inhabitants. Obviously, the *space ethos* is growing throughout the Middle and Far East, but India seems to have taken the lead in terms of its impact on its inhabitants.

1.6 OTHER NATIONS IN SPACE

Beyond the four regional case studies provided in this chapter, other nations continue to become members of the world space community. The European Space Agency has sponsored cooperation conferences with 12 Mediterranean countries (Algeria, Cyprus, Egypt, Israel, Jordan, Malta, Morocco, Syria, Tunisia, Turkey, Lebanon, and even Palestine). In its EGNOSS program within the North African region, global satellite navigation leaders envisage using this capability for telemedicine, Earth observation, and coastal zone and water resource management. Currently, Turkey, for example, wants to have its own military television satellite in orbit by 2011, and has invested $200 million in the project. In addition, space undertakings, principally with orbiting satellites, occur within Argentina, Brazil, Indonesia, Iran, Israel, South Korea, Malaysia, Pakistan, Taiwan, and the Ukraine. More and more countries are exercizing their rights under U.N. space treaties which affirm celestial bodies as the heritage of all humanity! No wonder Stephen Pyne, a historian at Arizona State University, sees a correlation between geographic exploration and general cultural vitality. Thus, Tom Harris, a Canadian lecturer, noted that:

> "we explore for our own good. Just as the discovery of the new world changed Western civilization, so our ventures into the galaxy will change who we are" (Commentary, *Los Angeles Times*, December 7, 1995).

SPACE ENTERPRISE

Dictionary definitions of the word *enterprise* refer to undertakings or ventures that are of great scope, complexity, and risk. Those who would be enterprising need to be imaginative and innovative, able to face challenges with initiative, persistence, and coping skills to deal with uncertainty, as well as the unknown and untried. All such descriptions readily apply to the arena of *space enterprise*. It is not enough to have a bold space vision and a strong space ethos—we need to combine these qualities with an enterprising spirit and competence to enable us to carry out what is envisioned. Human enterprise is part of our nature which motivates us to look beyond horizons, overcome constraints, and explore or experiment with the unknown. In the long history of our species, human beings have amply demonstrated enterprise worldwide. Now we are doing it offworld, beginning with our Solar System! It is no accident that in the *Star Trek* television series, author Gene Roddenberry named his interplanetary spacecraft, *Enterprise*, the same name adopted by NASA for its first Space Shuttle.

Space enterprise is to be found both on the ground and in orbit, in unmanned or manned missions. It may occur in both the public or private sector, as well as in all disciplines, professions, and industries. For the past 50 years, the public sector, in partnership with the aerospace industry, has provided the leadership in this enterprise beyond Earth. Achievements are evident in mobile satellite communications, navigation with Global Positioning Systems (GPS), and in space science with orbiting telescopes, like Hubble. One measure of the scope of this enterprise is the investment by governments of public funds in myriad undertakings aloft. Exhibit 13 illustrates how much is presently spent on space activities by geographic region and nation: a total of over $28 billion. Yet, such expenditures produce payback in ROI, such as spinoffs that benefit humanity on the ground.

But, expect to see an expansion of space enterprise in this century as the private sector becomes more dominant in commercial, industrial, and settlement ventures beyond our planet. One indicator is the first industry to derive from space technology:

Exhibit 13. GLOBAL SPACE SPENDING.

Approximate annual space expenditures by nations in U.S. dollars, public sector only (2007):

- North America   U.S.A. $18 billion; Canada $300 million.
- Russia $800 million.
- European Union   ESA $3.5 billion (or $7.5 billion when combined with its space agencies).
- Asia   China $1.2 billion; Japan $1.6 billion; India $700 million.
- Other   Brazil $35 million.

Source: *en.wikipedia.org/wiki/Indian_Space_Research_Organization*, 2007, p. 10.

orbital satellites. One financial projection for this business is that by 2008, it will produce $171 billion in operating revenue. And what will be the next big industry off the planet: space tourism, mining, energy? Emerging space commerce and industrialization will be called *astrobusiness*, and this subject will be discussed further in Chapter 9 and Appendix C. Literally, we are in the process of building an interplanetary economy, starting with the Earth–Moon twins.

But, there will be obstacles to be overcome if space enterprise is to become viable in the decades ahead. Dr. David M. Livingston, the founder of *The Space Show* podcast, has identified these principal barriers (*www.thespaceshow.com*) [41]:

- government policy, bureaucracy, inertia, and resistance toward private sector entry into their domain;
- obsolete laws, regulations, and traditions opposing or hindering private access to space;
- conflicts and inter-agency rivalries between various branches of government that hamper space commerce;
- potential lack of business ethics in commercial space industry requiring more external regulation;
- misperceptions in the marketplace and among financial investors about space enterprises;
- conflict in free-market enterprises with United Nations space treaties regarding outer space.

In an article on the "Future of Private Commercialization of Space", prepared for the 2002 *Annals of Air and Space Law*, George Robinson, the first person to receive a doctoral degree in space law, offered this sage advice:

"What private space enterprise has to be 'smart' about is not just traditional concerns of securing risk capital, protecting products from private competitors, and ensuring that the multitude of financial/investment technology transfer regulations, domestic and international, are satisfied. What space entrepreneurs have to be smart about is unscrupulous competition offered by government regulators, domestic and foreign, as well as from politicians and major international corporations ... Space commerce is both a concept and a practice that has created tension among constituents in the business community, politicians, military strategists and tacticians, statesmen, and ideologues."

To advance offworld enterprises will not only take an inspiring space vision and ethos—but courage and competence—in addition to outstanding marketing and management skills. There are many already demonstrating these qualities such as SpaceShipOne manufacturer Burt Rutan; Virgin Galactic entrepreneur Richard Branson; Bigelow Aerospace inflatable habitat entrepreneur Robert Bigelow; SpaceDev founder Jim Benson; K-1 reusable rocket investor Walter Kistler; International Space University and X-Prize founder Peter Diamandis; Space Age publisher and lunar visionary Steve Durst; and many, many other space business leaders worldwide [44] (for further discussion, see Appendices C–E).

1.7 CONCLUSIONS ON GLOBAL SPACE VISION, ETHOS, AND ENTERPRISE

Humankind, whether in the East or in the West, senses that our species seems to be in an epochal transition to space-based living and creation of an entirely new space culture. Whether by manned or unmanned spacecraft, current generations are pushing back the high frontier, expanding ever farther into outer space for our descendants to inhabit. With the help of mass media, computers, and the Internet, more and more of this planet's peoples are beginning to appreciate its contributions to our quality of life on Earth. In the 21st century, only a small percentage of the world's population envision space prospects and possibilities!

If ethos defines the character of a people, then it is evident that there is a microculture throughout the world which already possesses a *space ethos*! This subculture exchanges electronically on the Internet, or with CDs and DVDs; through professional organizations, conferences and publications; and through joint space ventures. Even within countries that lack a strong space program, one will find space advocacy movements and proponents of national space satellites. As more spacefarers go aloft and experience the "overview effect", they gain a new perspective from that vantage point in orbit, so that even our planet is viewed as an interconnected ecosystem.

With this enhancement of our collective consciousness, "Earthbound Everyman" may be ready to cultivate not just a space ethos—but more—as this quotation from attorneys Robinson & White so aptly implies [42]:

"Thus we stand in the late twentieth century on the threshold of extending old civilizations into space, perhaps even creating new ones in which our own sons and daughters may be extraterrestrials from every point of view."

One space visionary, Konstantin Tsiolkovski, Russian space philosopher, expressed such foresight in 1911:

> "Mankind will not remain forever on Earth. In pursuit of light and space he will timidly at first probe the limits of the atmosphere and later extend his control throughout the Solar System."

Astronaut Virgil "Gus" Grissom described his orbital experience in 1965:

> "There is a clarity, a brilliance to space that simply does not exist on Earth, even on a cloudless summer's day in the Rockies. Nowhere else can you realize so fully the majesty of our Earth and be so awed at the thought that it's only one of untold thousands of planets."

Speaker of the U.S. House of Representatives, Congressman Newt Gingrich observed in 1995:

> "Bureaucracies are designed to minimize risk and to create systematic procedures ... The challenge for us is to get government and bureaucracy out of the way and put scientists, engineers, entrepreneurs, and adventurers back into the business of exploration and discovery. The 21st century should be a great century of exploration for humanity."

Space enterprise will flourish when we can reduce the cost of access to outer space. Currently, the public sector spends approximately $10,000 per pound to move beyond the Earth's gravity well. Private sector entrepreneurs are seeking to reduce the flight into orbit to between $3,000–$5,000 a pound! Elliot Putnam, senior vice president of the U.S. Space Foundation, maintains that the current high expense of getting offworld affects the cost of all space enterprise, from global telecommunications, to microgravity research, to national security systems. He states,

> "more than half the cost of orbiting a commercial satellite goes to buying and insuring the launch vehicle—costs that show up in your phone, cable, and TV bills, or direct in-home or office satellite service."

Whatever the costs and challenges, in this millennium humanity will lay the foundations for a *spacefaring civilization*. Since our ancestors climbed down from trees and walked upright, our species has explored and probed new frontiers. After 3 million years as terrestrial beings, we are now slowly migrating beyond our home planet and fulfilling our destiny. Human enterprise will take our race to the far corners of the universe [43]. *Ad astra*, then, to the stars!

> **SUMMARIZING THOUGHTS: THE AUTHOR'S SPACE VISION**
>
> One of the ways we could maximize human potential is in the education of youth for their future. And their future, of course, is in the knowledge culture, in knowledge work, but it's also offworld ... Space exploration and development requires knowledge workers. Space will require a new kind of human being. We're going to emerge into spacekind that will be different from earthkind because of the former's microgravity environment ... I am convinced that one of the major trends of the future impacting us in the next thousand years is the creation of a spacefaring civilization!
>
> Source: Philip R. Harris, Ph.D. as quoted (p. 161) in the *Humanity 3000: 2005 Symposium Proceedings* published by the Foundation for the Future (123, 105th Avenue Southwest, Bellevue, Washington, D.C. 98004 (*www.futurefoundation.org*).

Exhibit 14. The Vision for Space Exploration. VSE is current U.S. space policy. Source: Artistic rendering provided by NASA Public Affairs Office, 2005 (*www.space flight1.nasa.gov/gallery/images/vision*).

1.8 REFERENCES

[1] Matsunaga, S. "An Introduction Statement for the U.S. Senate Joint Resolution Relating to NASA and Cooperative Mars Exploration," Speech before Congress, January 21, 1985.
[2] Reynolds, D. W. *Apollo: The Epic Journey to the Moon.* New York: Harcourt/Tehabi, 2002 ... Hansen, J. R. *First Man: The Life of Neil A. Armstrong.* New York: Simon &

References

Shuster, 2005 ... Thimmesh, C. *Team Moon: How 400,000 People Landed Apollo 11 on the Moon*. New York: Houghton Mifflin, 2008 ... Lattimer, D. *All We Did Was Fly to the Moon*. Gainesville, FL: The Whispering Eagle Press, 1988.

[3] Finney, B. K.; and Jones, E. M. (eds.) *Instellar Migration and the Human Experience*. Berkeley, CA: University of California Press, 1985 ... Krone, B.; Mitchell, E.; Morris, L.; and Cox, K. (eds.) *Beyond Earth: The Future of Humans in Space*. Ontario, Canada: Apogee Space Books, 2006 ... Burrows, W. E. *This New Ocean: The Story of the First Space Age*. Modern Library Paperbacks, New York, 1999.

[4] Salk, J. *Man Unfolding*. New York: Harper & Row, 1972 ... *Survival of the Wisest*. Harper & Row, 1973.

[5] Harris, P. R. *The New Work Culture: HRD Strategies for Transformational Management*. Amherst, MA: Human Resource Development Press, 1998.

[6] Harris, P. R. *Managing the Knowledge Culture*. Amherst, MA: Human Resource Development Press, 2005.

[7] Neufeld, M. J. *Von Braun: Dreamer of Space, Engineer of War*. New York: Alfred Knopf, 2007 ... Von Braun, W.; Ordway, F. I.; and Dooling, D. *Space Travel: A History*. New York: Harper & Row, 1985 ... Burdett, G. L.; Hearth, D.; and Soffen, G. A. (eds.) *The Human Quest in Space*. San Diego, CA: UNIVELT, 1987, AAS Vol. #65 ... Antonio, M. *A Ball, A Dog, and a Monkey, 1957: The Space Race Begins*. New York: Simon & Shuster, 2007 ... Dudenkov, V. N. *Russian Cosmicism: Philosophy of Hope and Salvation* [in Russian]. St. Petersburg, RF: Sintez Publishing, 1992.

[8] O'Neill, G. K. *The High Frontier*. Princeton, NJ: The Space Studies Institute, 1989 ... Oberg, J. E.; and Oberg, A. R. *New Earths—Living on the Next Frontier: Pioneering in Space*. New York: McGraw-Hill, 1986 ... Savage, M. T. *The Millennial Project: Colonizing the Galaxy in 8 Easy Steps*. New York: Little Brown Publishers, 1994, Second Edition ... Stine, G. H. *Halfway to Anywhere: Achieving America's Destiny in Space*. New York: M. Evans & Co., 1996.

[9] Hardy, D. A. *Visions of Space: Artist's Journey through the Cosmos*. New York: Gallery Press, 1990. (For information on the International Association of Astronomical Artists, contact David Hardy, 99 Southam Rd., Hall Green, Birmingham B28 OAB, U.K.)

[10] Zubrin, R. T. *Entering Space: Creating a Spacefaring Civilization*. J. P. Tarcher/Putnam Books, 1999 ... Ordway, F. I.; and Libermann, R. *Blueprint for Space: From Science Fiction to Science Fact*. Washington, D.C.: Smithsonian Press, 1992 ... Lakeoff, S.; and York, H. F. *A Shield in Space: Technology, Politics, and the Strategic Defense Initiative*. Berkeley, CA: University of California Press, 1989.

[11] National Commission on Space. *Pioneering the Space Frontier*. New York: Bantam Books, 1986.

[12] National Space Society, 1620 I Street, Washington, D.C. 20006 (tel.: 1/202/429-1600; *www.nss.org*). Contact this space advocacy group regarding membership, chapters, *Ad Astra* magazine, position papers, and other publications.

[13] Ride, S. *Leadership and America's Future in Space*. Washington, D.C.: U.S. Government Printing Office/NASA, 1987 ... NASA Advisory Council publications, such as America's Plans for Space, Planetary Exploration through Year, 2000, etc. For a list of special publications and reports, contact NASA Center for Aerospace Information, Linthicum, MD (tel.: 1/301/621-0390, Ext. 391). For a free subscription to *Technology Innovation: Magazine for Business & Technology*, contact NASA Innovative Partnership Programs, Washington, D.C. 20546 (*www.ipp.nasa.gov*).

[14] Toffler, A. "The Space Program's Impact on Society," in Korn, P. (ed.) *Humans and Machines in Space: The Payoff*, 1992, pp. 77–106. Published for the American Astronautical Society by UNIVELT, PO Box 28130. San Diego, CA 92198 (*www.univelt.com*).

[15] *ISY World Space Congress Abstracts, 1992* is a 716-page volume of summaries for the 2,700 technical papers presented at this event attended by your author. It is available from the American Institute of Aeronautics and Astronautics, 1801 Alexander Bell Ave., Ste. 500, Reston, VA 22092 (*www.aiaa.org*).

[16] Boorstein, D. "Realms of Discovery Old and New," August 31, 1992. Inaugural address at the *World Space Congress* (IAF-92-1000). Available from the AIAA at the above address ... Burrows, W. E. *This New Ocean of Space: The Story of the First Space Age*. New York: Random House, 1998.

[17] Daniels, P. S.; and Hyslop, S. G. "The Frontiers of Science: Away from Earth," *Almanac of World History*. Washington, D.C.: National Geographic, 2007, pp. 324–325 ... National Geographic. *Encyclopedia and Space Odyssey* (*http://shop.nationalgeographic.com*).

[18] Sagan, C. "Are We Ready to Go Exploring Again?" *Parade Magazine*, July 17, 1994, p. 16.

[19] Harrison, A. A. *Spacefaring: The Human Dimension*. Berkeley, CA: University of California Press, 2001.

[20] Stuster, J. *Bold Endeavors: Lessons from Polar and Space Exploration*. Annapolis, MD: Naval Institute Press, 1996.

[21] Schmitt, H.H. *Return to the Moon: Exploration Enterprise, and Energy in the Human Settlement of Space*. New York: Springer/Copernicus/Praxis, 2006.

[22] Light, M. *Full Moon*. New York: Alfred A. Knopf, 2002.

[23] *Presidential Commission on Space Shuttle Challenger Report*. Washington, D.C.: U.S. Government Printing Office, 1986 ... Harland, D.M. *The Space Shuttle: Roles, Missions, Accomplishments*. Chichester, U.K.: Praxis/Wiley, 1998.

[24] For information on the organization and publications, contact Canadian Aeronautics and Space Institute, 130 Slater St., Suite 818, Ottawa, ON KIP 6E2, Canada (tel.: 613/234-00191).

[25] Abbott, J. D.; and Moran, R. T. *Uniting North American Business: NAFTA Best Practices*. Burlington, MA: Elsevier/Butterworth-Heinemann, 2002.

[26] Raygoza, B. J. "Mexican Lunar Habitat: A Hispanic-Mexican Habitat for Settlement of the Moon," *Space Governance*, 2002, **18**, 94–99. Available from United Societies in Space, Inc., 499 S. Larkspur Road, Castle Rock, CO. 80104 (email: *mailto:djopc@quest.net*).

[27] McKay, M. F.; McKay, D. S.; and Duke, M. (eds.) *Space Resources*. Washington, D.C.: U.S. Government Printing Office, 1992, NASA SP-509, 5 vols. Now available from *www.univelt.com* (see Vol. 4, *Social Concerns*) ... Lewis, J. S. *Mining the Sky*. Reading, MA: Addison-Wesley, 1996.

[28] Harvey, B. *Race into Space: Soviet Space Programme*. Chichester, U.K.: Ellis Horwood.

[29] Hall, R.; and Shayler, D. *The Rocket Men: Vostok and Voskhod, The First Soviet Manned Spaceflights*. Chichester, U.K.: Springer/Praxis, 2001.

[30] Harland, D. M. *The Mir Space Station: A Precursor to Space Colonization*. Chichester, U.K.: Wiley/Praxis, 1997 ... Burrough, B. *Dragonfly: NASA and the Crisis aboard Mir*. New York: HarperCollins, 1998.

[31] To remain current on the space program of the Russian Federation, consult *Space Directory of Russia*, as well as the ESA's *European Space Directory*. Both published by Sevic Press, 6 rue Bellart, F-75015 Paris.

[32] Lytkin, V.; Finney, B.; and Aleptko, L. "The Planets Are Occupied by Living Beings: Tsiolkovski, Russian Cosmism, and ETI," *Quarterly Journal of the Royal Astronomical Society*, **36**, 369, 1995.
[33] Harvey, B. *Russia in Space: The Failed Frontier?* Chichester, U.K.: Springer/Praxis, 1997.
[34] Freeman, M. *Challenges of Human Space Exploration*. Chichester, U.K.: Springer/Praxis, 2000 ... *How We Got to the Moon: The Story of German Space Pioneers*. Washington, D.C.: 21st Century Science & Technology, PO Box 16285, zip code 20041, 1994.
[35] Harvey, B. *Europe's Space Programme: To Ariane and Beyond*. Chichester, U.K.: Springer/Praxis, 2001.
[36] U.N. Office for Outer Space Affairs, PO Box 500, A 1400, Vienna, Austria. Ask for a list of publications, such as their *U.N. Report Committee on the Peaceful Uses of Outer Space* (Supplement No. 20, A/49/20).
[37] Harvey, B. *The Chinese Space Programme: From Conception to Future Capabilities*. Chichester, U.K.: Wiley/Praxis, 1998.
[38] Harvey, B. *The Japanese and Indian Space Programmes: Two Roads into Space*. Chichester, U.K.: Springer/Praxis, 2000.
[39] Harris, P. R. *Launch Out: A Science-Based Novel about Space Enterprise, Lunar Industrialization, and Settlement*. West Conshohocken, PA: Infinity Publishing (*www.buybooks ontheweb.com*), 2003.
[40] Wikipedia on-line encyclopedia, January, 2007 web search of *http://en.wikipedia.org/ wiki/Russian_Space _Agency ... European_Space_Agency ... China_National_Space_ Administration ... Japan_Space_Agency ... Indian_Space_Research_Organization*
[41] Livingston, D. M. *Barriers to Space Enterprise*, June 7, 1999, 10 pp. Published by the Cato Institute (PO Box 95, Tiburon, CA 94920, *http://www.spacefuture.com/archive/barriers_ to_space_enterprise; www.thespaceshow.org; www.space-frontier.org; www.spaceagepub. com*) ... Johnson-Freese, J.; and Handberg, R. *Space: The Dominant Frontier: Changing the Paradigm for the 21st Century*. Westport, CT: Praeger, 1997, p. 154.
[42] Robinson, G. S.; and White, J. M. *Envoys of Mankind: Declaration of First Principles for Governance of Space Societies*. Washington, D.C.: Smithsonian Institution Press, 1986.
[43] Harris, P. R., "New Millennium Challenges for Living and Working in Space," *Earth Space Review: The Magazine for the International Space Community*, **10**(1), pp. 57–68, 2001 (Gordon & Breach Publishing Group, PO Box, 433, St. Helier, Jersey JE4 8QL, Channel Islands, U.K.).
[44] Belfore, M. *Rocketeers: How a Visionary Band of Business Leaders, Engineers, and Pilots Is Boldly Privatizing Space*. Washington, D.C.: Smithsonian Books, 2007.

SPACE ENTERPRISE UPDATES

- CANADA: a 2008 book celebrates *Canada's Fifty Years in Space* by Gordon Shepherd and Agnes Kruchio—ISBN 1894959167 (*www.apogeespacebooks.com/Books/50years.html*).
- CHINA announces plans to shoot for the Moon by launching a lunar orbiter in 2008 and followed that year by a space walk (EVA); in 2012 the PRC hopes to launch a lunar rover, and in 2020 put its own space station in orbit.
- EUROPE: see Brian Harvey's *Europe's Space Programme: To Ariane and Beyond* published by Springer/Praxis, Chichester, U.K in 2003 (*www.praxis-publishing.co.uk*) for the latest European situation.
- JAPAN: in June 2008, JAXA delivered by Shuttle *Discovery* to the International Space Station, the second component of their $1 billion space laboratory which took 20 years to build. The Kibo lab is 37 feet long and weighs more than 32,000 pounds. The third and final section will be delivered next year.

2

Human space exploration and settlement

Mankind will not remain forever on Earth. In the pursuit of light and space, we will first probe the limits of the atmosphere, and later extend our control throughout the solar system. Humans will ascend into the expanse of the heavens and found a settlement there. The impossible of today will become the possible of tomorrow.

Konstantin Tsiolkovski

The human species has demonstrated its capacity to dream and envision the future, as well as to contemplate its past. Through the power of imagination, we can conceive and often execute grand plans. This conceptual ability influences our behavior, whether in terms of individuals or institutions, nations, or humanity. It results in the setting of goals which energize people to impressive achievements. And so it is with outer space and human expansion into the universe. The goal of permanently returning to the Moon by year 2020 will produce a lunar launch pad to Mars and beyond! Some astrobiologists, for instance, recommend Titan as a possible target for human space exploration because it is the most diverse place in the Solar System beyond Earth [1].

EXPLORATION

To explore has been defined as travel to unknown or unfamiliar places or regions for purposes of discovery. It is an investigation of uncertain possibilities—the systematic search for new ways, methods, sites, locations. It means crossing over borders, constraints, and limitations so as to ascertain our potential. It requires sailing into uncharted waters or flying into outer space. The act of exploring or exploration may occur within the human mind, as with scientists experimenting in a laboratory; or it may mean actually moving physically beyond the here and now.

Our ancestors did it some 100,000 years ago when they began to migrate across Africa and then across the planet. Today, contemporary generations explore beneath the seas and off the Earth. With the latter, we explore by both human spaceflight, or by extending our minds through unmanned, automated missions to the far corners of the universe.

The person who explores may be forgotten or honored in time. Those early men and women who first explored Australia some 60,000 years ago are anonymous to us moderns. But, history helps us to immortalize those who, like Tennyson's *Ulysses*, strive and seek without yielding. Some we even manage to remember their names and achievements, such as Chang Chien who opened up Asia's Silk Road to East–West trade, while exploring China; Christopher Columbus who "discovered" the New World while searching for a western route to Asia; James Cook who surveyed the coasts of Newfoundland and Labrador, as well as charting the coasts of Australia, New Zealand, North America's Pacific Coast and the Bering Straits; Vasco da Gama, who had the first recorded voyage from Europe to India; Ferdinand Magellan, who lead the first expedition to circumnavigate the globe; Roald Amundsen, first person to sail the Northwest passage and to see the South Pole; Robert Peary, first person to reach the North Pole; Yuri Gagarin, first human in space; Neil Armstrong and Buzz Aldrin, first men on the Moon; and many more too numerous to mention here [2]. Of course, there are those explorers of the mind who never get physically to travel aloft. For example, among space pioneers who never went into orbit, but in a sense were explorers: Wernher von Braun, designer of Apollo lunar missions and Sergei Korolev, chief designer of all the early Russian space triumphs. There are many who "explore" space from the ground. One was Max Faget, director of NASA's engineering and development in the 1960s, who contributed many of the design concepts for the Project Mercury manned spacecraft, and other missions to follow, including the Space Shuttle. Today another such "explorer" is Dr. Krishnaswamy Kasturirangan who is not only director of the National Institute for Advanced Studies in Bangalore, but a member of the Parliament of India [3].

Who will be the great offworld explorers of this century? Remember also that explorers often do not accomplish their intended goals, or like Columbus discover something unexpected—but not the route to the Indies. But, all explorers contribute to pushing back the frontiers of knowledge, expanding human understanding. Today, it is difficult for the average person to comprehend the impact of future space exploration on humankind! But, we have already learned much from the bold endeavors of both polar and space explorers [4]!

2.1 LEAVING EARTH'S CRADLE: JOINT VENTURING

The artist's rendition in Exhibit 15 captures for all the challenge in leaving our home planet to explore and settle the universe. Eventually, spacefarers will become *spacekind*, multicultural beings of both genders who reach out to *earthkind* to support them

Exhibit 15. Leaving the home planet. Source: Artistic rendering from NASA's Lyndon B. Johnson Space Center, Houston, TX 77058.

in their quest, be it on orbiting vehicles and platforms, or on new worlds, such as the Moon and Mars, or beyond.

We have only mounted the threshold of outer space. Since 1957 when Sputnik was launched, the central thrust of space endeavors for the past 50 years has been on technical—not social—development offworld. It is understandable why priority would have to be given to hardware and software which makes feasible both unmanned and manned operations in space. Success in this regard enables our species to put humans on the lunar surface, to produce orbiters and space laboratories, to probe our Solar System with automated spacecraft [5]. The *first stage of space development* was a time when rocket specialists, engineers, and technicians dominated planning and management. In this period of short-duration manned flights, the consistent, underrated elements in much of space research and operations, conferences and publications, had been the broader human factors and life sciences. The Russian space program has shown more concern for these dimensions, while NASA's attention has been directed principally toward physiological and medical studies of the astronauts, and to a lesser extent the psychological qualities for effective living in outer space. It is no wonder, then, that veteran astronaut Norman Thagard, himself a physician, experienced such acclimatization problems when he became the first American to stay aloft for 115 days. Relative to his adjustment difficulties aboard the Russian Mir station, the NASA administrator publicly apologized for the Agency's lapse in Thagard's psychological preparation and support services, promising improvements before the second U.S. astronaut flew to that space station in March 1996 (*San Diego Union Tribune*, July 8, 1995).

The initial decades of space exploration and exploitation have also been marked by debate over the advantages of unmanned vs. manned flight. Planetary scientists make the case for less cost and hazards by the use of automation and robotics in spacecraft without crews. To illustrate this continuing debate within the U.S.A., the Space Science Board of the National Academy of Sciences expressed its vision that the main goal of the nation's civilian space program should be the advancement of scientific knowledge and its application to human welfare (*Commercial Space*, Winter 1987, p. 13). Furthermore, that Board recommended the use of automated spacecraft for all missions beyond low-Earth orbit through year 2015, a position at odds with the National Commission on Space's advocacy of permanent colonization of the Moon and Mars in the first quarter of the 21st century. The Board's argument was that the focus should be on studies in astronomy and astrophysics, microgravity, solar/space physics, planetary exploration, and space medicine. The last mentioned point is that more research is needed on the effects of prolonged weightlessness before long-duration missions can be undertaken. One of that nation's pre-eminent space scientists, James Van Allen, articulated this viewpoint in a lecture at the University of California-San Diego. He maintained that the U.S.A. spent $150 billion on manned flight to get 133 different individuals aloft, and contrasts those accomplishments with magnificent scientific results obtained from unmanned flights at one-fifth the cost (*Los Angeles Times*, November 20, 1986, II:4). Professor James R. Arnold, a lunar chemist and first director of the California Space Institute, countered on the same occasion that longer manned flights go beyond science and the gathering of knowledge. He pointed to the inherent human drive to explore, to transform dreams of living and working outside the Earth's atmosphere into tomorrow's realities. This divergence in viewpoints is manifest most relative to missions to Mars.

In this 21st century, progress is being made on human spaceflight through organizations, such as the National Space Biomedical Research Institute. The NSBRI outreach program employs a coordinated, multi-institutional U.S. strategy to communicate the significance of life sciences in space exploration and related research. Furthermore, its outreach team and activities enhance NASA exploration capabilities and educational programs, so as to advance the national policy called Vision for Space Exploration. The six primary partners in this venture are prestigious research and educational institutions: Baylor College of Medicine, the Colorado Consortium for Earth and Life Science Education, Massachusetts Institute of Technology, Morehouse School of Medicine, Rice University, and the University of Texas Medical Branch (*www.usra.edu/*). Their post-doctoral fellowships, professional development for teachers, production of learning materials, and support for the Challenger Centers are opening new knowledge and career pathways that further space life sciences.

The consensus emerging among today's planners and policy makers is that both types of space complimentary enterprises (human and automated missions) are required. Space is not just for doing science and astronomy, but a place for exploration, settlement, and utilization of its vast resources. Whether manned or unmanned, both represent *human* extension into space. Experience has demonstrated the value of human inventiveness in space when computers and mechanical equipment fail and need to be repaired or replaced. An interview with former Skylab astronaut Owen Garriott, who had two months of spaceflight experience, revealed the liveability within that prototype U.S.A. space "station". While discussing the planning for the "international station", he observed that NASA had a Congressional mandate to provide a "man-tended capability" in the research to be done there, as well as in "telescience": interactive research combining a ground-based team, a spaceborne team, and automated instruments (*Space World*, December 1986, 27–31). In the current era of tight budgets, it is shortsighted for planetary scientists to battle against human spaceflight as they struggle for more funding of robotic spacecraft and unmanned science missions.. The current U.S. national space policy entitled "Vision for Space Exploration" calls for both types of exploration missions to the Moon and beyond. Australia, Europe, China, India, and Japan follow the established pattern of beginning with automated missions, to be followed by human spaceflight. These and other international partners are presently joining in the VSE lunar program with the U.S.A. and Russia. NASA's global exploration strategy is a human-centered plan to return to the Moon permanently, and then go beyond. ESA would participate in that endeavor, but focus on (1) robotic missions, such as their plans for a *Mars Sample Return* and *ExoMars*; (2) science studies like building a radio telescope on the Moon. For its VSE role, Britain's scientists propose two robotic lunar missions: Moonraker to land on the lunar surface an date geological samples; Moonlite to fire penetrators into the Moon and listen to the noise made (*The Economist*, January 13, 2007, pp. 71–72.)

But, the ultimate argument is that humanity goes into space because it is its nature and destiny, so other rationalizations are subservient to that overall purpose. Furthermore, public investment in the space program demands a human component. In a *New York Times* editorial (February 20, 2007, p. A19), planetary scientist Carolyn Porco, director of Cassini Imaging Central Laboratory for Operations, gave this realistic assessment of American prospects for again reaching the Moon, but staying there:

> "To reach that future will require two critical ingredients: adequate financing and a long-term, cross administration commitment that supports steady, uninterrupted progress ... While sustained budgets of the Apollo magnitude are now out of the question, what is not out of the question is our ability to pay to keep space goals front and center. We are now spending in Iraq, in a single month, $9 billion—more than half the annual budget NASA needs to stay on course ...
>
> Humanity's future need not be confined to mere survival on the home planet. Other worlds beckon, we know how to reach them, and we will once again be outward bound ... There could be no better way to encourage an equally optimistic belief in the future than to embark on an odyssey that presents tremendous challenges, demands discipline and vigor, requires decade-long focus, inspires international cooperation, promotes lasting peace, improves life for all, and paints a stirring vision of an expanded human presence beyond the Earth. There could be no better way to say: the future is boundless and it belongs to us!"

One practical example of this rationale is the joint venture of multiple national space agencies to erect and complete an International Space Station. This manned capability in orbit is discussed in a Chapter 8 case study on macromanagement. One reason for more extended spaceflights and life science research aloft is to prepare humans for a 30-month round trip to Mars. In the near term, space scientists worldwide are planning before 2050 to establish manned bases on Mars or its moons as a prelude toward colonization of the Solar System in the centuries to come [6]. Yes, we presently have the technical "know-how" for such accomplishments, but lack global will and consensus to undertake interplanetary explorations.

Yet, this world thrust toward an enduring human presence in orbit marks *the next stage in space development*, a time when the biological and behavioral sciences must cooperate together in major contributions. Although closed ecosystems are possible in space, long missions beyond 100 days necessitate much more research to create and maintain balanced ecological systems that sustain the quality of human life in a hostile environment. The perceptions of engineering planners tend to ignore or undervalue many factors that biologists consider critical to life and the reduction of stress on the spacefarer [7]. These factors, such as oxygen, food, water, circadian rhythms, or personal pollutants, tend to magnify on longer spaceflights. The insights in Exhibit 16 of an interview with an eminent space psychologist, Dr. Albert Harrison of the University of California-Davis, underscore the difficulties of living and working in space.

We have only begun to comprehend the near-Earth environment, no less ponder the human challenges beyond. Dr. Anatoli Grigoriev, director of Moscow's Institute of Biomedical Problems, highlighted the issue:

"The one-million-year process of human evolution occurred because of gravity. Take away gravity and man starts to rid himself of the complex skeletal structure necessary for upright locomotion in exchange for a frame better suited to floating in water, or in this case, space."

That is why spacekind will eventually be markedly different from terrestrial beings. The first 50 years offworld was short-duration flights in low-Earth orbit (LEO), while the 21st century will bring longer human missions in orbit. Since there is so little human experience for such extended spaceflight, behavioral and biological scientists are joining in combined research with space scientists and engineers regarding humans living and working in microgravity for a year or more! A round trip to Mars, for instance, with 18 months on its surface, would involve 30 months (i.e., $2\frac{1}{2}$ years off the home planet!). Thus, in January 2007, the National Space Biomedical Research Institute in the United States solicited proposals for ground-based studies for human health in space (*www.nsbri.org*). NSBRI is looking for investigations in these nine areas: bone loss; cardiovascular alterations; human performance factors, sleep, and chronobiology; muscle alteration and atrophy; neurobehavioral and psychological factors; nutrition, physical fitness, and rehabilitation; sensorimotor adaptation; smart medical systems; and medical technology development.

The social science contribution to this research arena has barely begun [8]. In addition to the analysis of terrestrial analogs, such as those on nuclear submarines, for possible applications to lengthy off-terrestrial living, there is much need for increased studies of

- life support systems in an orbital artificial atmosphere to counter the physical effects of living without gravity and ensure habitability;
- group behavior and dynamics in spaceflight, especially with heterogeneous, multidisciplinary, and multicultural crews in isolated, confined environments;
- culture and quality of life, including safety and recreation, to be created at space stations, lunar installations, and a Mars/Deimos outpost;
- recruitment, selection, training, support, and evaluation of long-duration spacefarers and their families;
- on-site expert systems and simulations for education and entertainment in orbit, as well as computerized systems for monitoring and diagnosing physical and psychological health;
- art and science of space management on Earth and in orbit;
- effective operation aloft of space commerce, manufacturing, and mining, as well as solar power systems;
- planning, establishing, sustaining, and governing space settlements to ensure safety and security.

Other chapters to come will analyse these matters further. The conclusion now is that such research and exploration can best be accomplished by joint ventures between the public and private sectors, especially among global space agencies. It is the human family that is migrating aloft, not just one or two nationalities. Such international cooperation is essential for current planning to return to the Moon permanently in the next decade!

There has also been a reluctance on the part of legislators to act relative to space exploration and settlement. For example, the *Space Exploration Act of 2002* introduced to the U.S. House of Representatives by Congressman Nick Lampson (D-TX) has yet to be passed into law. H.R. 4742 proposed "to restore a vision for the United States space flight program by instituting a series of incremental goals that will facilitate the scientific exploration of the solar system and aid in the search for life elsewhere in the universe." The bill would have provided modest financial stimulation to private sector exploration initiatives within a new NASA Office of Exploration with its budget. Despite the advocacy of the National Space Society (NSS), that type of helpful national legislation has yet to obtain public support and to be enacted (see Section 3.1 regarding comparable efforts to pass a Space Settlement Act, H.R. 4218, back in 1988).

SETTLEMENT

The concept of space settlement implies not only human relocation beyond Earth, but the establishment there of communities with the necessary infrastructure and supplies until *spacekind* can support themselves. The challenge is far beyond that experienced by terrestrial colonists to new regions on this planet. Yet, there is much to learn from those here who have lived in isolated, confined environments, such as the Arctic [9]. In an interview at the Paris Air Show (*Reuters*, June 13, 2005), Dr. Michael Griffin, NASA administrator, said that his Agency has the funding to put people back on the Moon between 2015–2020, adding: "We should build a lunar outpost similar to all kinds of multinational outposts in Antarctica." Another astute commentator reminds us that there is always a pause between discovery and settlement or development: in the case of the Moon, this "pause" will be almost 50 years!

Essentially, the goal is to free humanity from dependence on one planet, so that we may in time expand outward in the Solar System and possibly beyond. Thus, today's plans for lunar settlement should have that ultimate objective in mind. The initial building blocks of a spacefaring civilization require global support. The National Space Society maintains that the public needs assurance that space activities in general, and settlement aloft in particular, will make a difference in fulfilling long-standing human aspirations, as well as providing direct benefits here to earthkind. With lunar colonization, citizens need to be educated as to the value and payback possible in building a twin economy of Earth–Moon. They need to appreciate the return on investment (ROI) in a necessary space transportation system, one that will produce returns comparable with that of the worldwide satellite industry. The start-up and upfront costs of lunar installations will be staggering, but the eventual payback could be tenfold or more. Initially, with Earth-based replenishments and supplies, a Moon base and facilities can be designed to become self-sufficient, especially with the use of closed-loop systems for life support offworld. Non-terrestrial material (NTM) can be mined and transformed for the benefit of all, whether on or off this planet. Further, microgravity research and science in orbit may produce cost-effective electronics and pharmaceuticals, as well as products and services that we cannot now imagine. A space repair industry may emerge beyond that of satellites, to include spacecraft and space habitats. Expect an astroconstruction industry also to prosper in orbit, then on the lunar and Martian surfaces. Eventually, clean space-based energy may replace polluting fossil fuels on Earth. To offset extraction of scarce or costly natural resources on this planet, offworld mining and fusion energy may become desirable. But, to put in place the necessary space infrastructure for major migration beyond Earth demands unprecedented international collaboration and coordination on a scale never before practiced. Such ventures will require the renewal and transformation of most disciplines and professions, such as engineering, architecture, finance, law, medicine, chemistry, physics, among others. Space technologies, methods, and vehicles will have to be standardized among nations and be compatible if such are to be cost-effective. Space lawyers will have to design new agreements to facilitate economic consistency in many undertakings aloft, from power and communication systems to uniform mechanisms and regulations for unmanned docking, refueling, spacecraft construction and maintenance, materials science, etc.

But, to supplement human exploration capability, the whole field of robotics will need upgrading to meet widespread demand in space enterprises. *Lunar astronomy, science, manufacturing, food production, and environmental protection will challenge our creativity.* For any of this to happen will means redesigning education and human resource development (HRD), so that space enterprises can proceed both on the ground and in orbit. Spacefaring settlers will have to be multi-skilled, competent, knowledge workers if they are to survive and prosper in the microgravity environment. The high frontier will entail reform of multiple human systems, including government practices, for humanity to profit from extraplanetary resources and opportunities. Only then can we anticipate the rewards possible from sustained space commercial and industrial expansion!

Until now those who went into space were short-term visitors: now we are planning long-duration residence in orbit or on the lunar surface. Exhibit 16 summarizes a space psychologist's realistic thinking on the matter.

As more humans go into orbit, there is ever greater diversity emerging among spacefarers, as Exhibit 17 illustrates.

Back in 1951, Wernher von Braun identified this problem when he commented:

"I believe that the time has arrived for medical investigations of the problem of manned rocket flight, for it will not be the engineering problems but the limits of

Exhibit 16. HUMAN CHALLENGES IN SPACE LIVING.

Interview with Dr. Albert A. Harrison, University of California-Davis and author of *Spacefaring: The Human Dimension*. Berkeley, CA: University of California Press, 2001:

- It's extremely difficult to live and work in space ... While the wobbly legs of an astronaut just returned to Earth may be the most obvious side-effect of a year long space mission, simply getting along with other astronauts for months at a time is even harder ... The people going have to be able to get along with one another. In orbit 240 miles above the Earth with little opportunity to get away, living under conditions where they are cramped together, they work out patterns of mutual existence ...
- The challenges of long-term amity can become even more difficult when astronauts come from cultures with different ways of relating to one another. Today's international crews raise the complexity. A lot of effort goes into ensuring that international crews can function comfortably (see Exhibit 17) ...
- We're all happy to see smiling faces of and occasional clowning around of astronauts on the Shuttle or Space Station. We should never lose sight of exactly how dangerous and risky space travel is. Even with the deaths of the *Columbia* crew, there was unanimity of their families, NASA officials, and the greater community in their sentiments that, though the loss of these astronauts was a great tragedy, space exploration must go on ...
- We make tremendous advances in science as a result of the exploration of space. While much of this knowledge is about outer space, it has a deeper, inner significance. After looking down on our planet from orbit, some astronauts have reported a deeper understanding of the interconnectedness of all life on Earth: it's one planet, one people! ...
- Though this more holistic view of Earth may help us survive as a species, space travel might help us ensure that humankind will continue to exist, even with widespread disaster on our home world. As soon as we re able to become a two planet species, not limited to Planet Earth, we can protect ourselves and our future from events that are catastrophic or at the extinction level.

Source: Douglas Vakoch, "Barriers to Space: And Why They Should Be Overcome," *www.space.com*, September 11, 2003 (see also the website on space biology *http://ceres.cals.ncsu.edu*).

the human frame that will make the final decision as to whether manned spaceflight will eventually become a reality."

Well, it did become a reality, and ever since there has been continuing research under way to counter the negative effectives of living in microgravity. For example, bone tissue deterioration or loss can be counteracted by sustained exercise in orbit, so a machine generates the type of resistance that gravity typically provides. NASA

Exhibit 17. Transnational, multicultural crew teams. Artistic rendering courtesy of NASA'S Lyndon B. Johnson Space Center, Houston, TX 77058. The diversity pictured above has also increased significantly with the presence of more women spacefarers.

established a Behavior and Performance Laboratory to deal with psychological and emotional problems aloft, such as depression on long-duration spaceflights or on re-entry. The Agency also funded a Space Radiation Laboratory at Brookhaven National Laboratory. There studies are under way to delimit the risk of cancer offworld due to ionizing radiation, and to set exposure standards for space missions. Currently, NASA has prepared a "Bioastronautics Roadmap", an analysis of critical research and developmental issues relating to human medical needs for exploration and settlement of space (*www.bioastronauticsroadmap.nasa.gov/*). NASA coined the word "bioastronautics" or the study of the biological and medical effects of space flight on living organisms.[1] Your author's concern is that this new term seemingly does not include the psychological aspects of humans living offworld.

---

[1] See *The American Heritage Dictionary of the English Language* (Fourth Edition, 2003) published by Houghton-Mifflin, or *www.nap.edu/catalog/11191.html* or *en.wikipedia.org/wiki/Bioastronautics*. Also the Committee on Review's *Preliminary Considerations regarding NASA's Bioastronautics Critical Path Roadmap*, published by the National Academies Press, 2005 (*www.nap.edu/catalog.php*).

The emphasis now, rightly so, is on the physical dimensions of our living aloft, such as shielding lunar dwellings to deflect the problem of no atmosphere on the Moon. Ways are being developed to repel the impact of cosmic rays and flares on humans moving beyond the Earth's protective atmosphere. On the Moon, for instance, warning systems can be installed to alert lunar dwellers to imminent solar radiation storms, so people there can avoid being on the lunar surface during such occurrences. Similarly, innovative solutions will be forthcoming to cope with other microgravity challenges, such as cataracts, nausea, bone fractures, kidney stones, loss of balance and non-verbal communication skills. Obviously, there is a real need for increased research on health care aloft by medical schools, academia, corporations, space agencies, and government institutes, as well as professional medical associations (see Appendix D). Synergy demands that such researchers share their findings on human spaceflight! Once people begin living offworld for longer and longer periods, such as in space settlements, then behavioral science research and therapies will be essential.

2.2 EMERGING SPACE SETTLEMENT ISSUES

The above section indicates only some of the challenges that lie ahead with reference to permanent space habitation. To elaborate on these *issues* for continuing research, consider

(a) **Relations among crews of diverse composition on long-term flights**. Although there has been much study to date in the field of group dynamics, some of it directed toward life in confined quarters of submarines, there has been little transfer of this information and insight to the study of team behavior among spacefarers, and its implications for future space colonists. Both American and Russian experiences on space stations and shuttles have immediate application to longer missions on the International Space Station and on the Moon. Behavioral science research needs to be expanded on the preparation of international crews for this purpose. A larger challenge is in training multicultural spacefarers to function more effectively as synergistic teams with their ground support staff, whether they go aloft as professionals, tourists, or settlers.

Also transferable to space habitation research are the significant investigations in the field of interpersonal and intercultural communication, particularly in terrestrial isolated and confined environments to space habitation [10]. During the December 1983 *Columbia* shuttle flight of six astronauts, an incident was recorded on television for all the world to observe. Those in flight were overwhelmed by the communications and requests from ground control, particularly from project managers of on-board experiments. Increasingly, there could be similar communication confrontations as larger crews go into space for longer periods, not only among those in flight—but with those mission managers back on Earth. In the next stage of spacefaring, ground control will not only have to give up some of its domination—but more autonomy will have to given to those in orbit. This is especially true of those at a lunar outpost and base. Improved information technology alone is not sufficient; communication specialists need to apply their research to space living, starting with the prototype stations and bases.

At the 37th International Astronautical Federation Congress, Russian scientists presented the "Ethical Problems of Interaction between Ground-based Personnel and Orbital Station Crewmembers" [11]. They viewed long-term space missions as complex socio–human–machine systems whose effectiveness largely depends on the quality of interaction between the subsystems. For the Russians, the ethical aspects concern the human relations which permeate every component of spaceflight and determine mission efficiency. Therefore, in addition to psychological and medical examinations before/during/after manned missions, they utilize socio-psychological and organizational–technical approaches to facilitate interaction, work, and leisure, for improved interpersonal and intergroup relations.

In the U.S.A., a leading researcher on spaceflight communication issues has been Dr. Mary Connors at NASA Ames. She concluded that these issues range from the spacecraft environment to the psychodynamics of communication aloft, recommending that a broad behavioral science perspective be used in investigations. Consider just two items for further ethical research: (1) the monitoring of spacefarers' electronic mail and exchanges; (2) the use of computerized tracing devices to monitor the physical and psychological health of those in orbit.

(b) **Increasing research by Earth-bound scholars on space habitation**, or matters of permanent settlement in the microgravity environment is another issue to address. Human advances in space already alter our terrestrial culture. The more people reside aloft for longer periods, the likelihood is that they will cope and adapt in innovative ways to their low or gravity-free environment. Many anthropologists, for example, seem fixated on the past within terrestrial communities, being reluctant to deal with future space communities. As we will see in Chapter 3, cultural anthropologists could make enormous contributions if they would direct the tools and the skills of their discipline toward the emerging space culture. One facet of culture is how people organize and manage themselves, say at a lunar base. Management scientists also could assist with the enormous task of organizing, transporting, supplying, and administering the material and human resources required for a permanent human community in space. Space management opens new vistas for research and inventiveness, particularly the macromanagement of large-scale enterprises (see Chapter 8). Initially, the operational administration of an International Space Station represents a unique opportunity for management scholars to inaugurate such studies which would later have application to the management of a lunar base.

(c) **Personnel to be transferred to and from orbit** for colonization purposes is another challenging issue to be investigated by social and life scientists. Our experience in manned flights has been so very limited, in terms of both numbers and type of elite astronauts and cosmonauts. In the past 50 years, most of these have been disciplined military males, with fewer scientists and women making it to the "high ground". An occasional politician, a Saudi prince, and one teacher have flown on the Shuttle, while the Russians have had more diverse spacefarers on their orbiting station, allowing guests ranging from French *spacionautes* to a Japanese journalist, and more recently high-paying tourists. In this century, diversity will be the norm as masses of people migrate aloft, starting with contract construction workers, but ending with settlers and tourists. In *Voyagers to the West*, Harvard historian Ballyn reminds us that the colonists to the New World during the 18th century were largely poor, ill-used white artisans and indentured servants, as well as African slaves [12]. The prospects are that space colonists of the 21st century will be more affluent and self-directed, better educated and selected. Expertise is required of specialists in cross-cultural relocation and living in exotic environments, so as to design systems for deployment and support of spacefarers (see Chapter 6). For example, a knowledge culture will prevail in lunar communities, and many aloft will have a transitional experience.

(d) **Global leadership in space settlement and colonization** will become a critical issue throughout this new millennium. At present, no national government or space agency is seriously addressing the matter of governance, so the leadership is likely to come from the private sector. A *case in point* within the U.S.A. was the 100th United States Congress, which passed legislation attached to NASA budget authorization to promote space settlement studies (see Section 3.1). Provisions in this foresighted document included a report every two years to the President and Congress by NASA reviewing its progress toward this end. Mark Hopkins, president of *Spacecause*, described that legislation as having strong symbolic importance, but documented how that bill was not being enforced with reports delayed and de-emphasized. However, in 1995 the NASA Office of Spaceflight Policy and Plans issued a strategic report on *Human Exploration and Development of Space*.

The National Space Society's position paper on *Settling Space: The Main Objective* states that permanent space settlements require a space-based economy to sustain self-sufficient growth aloft. To this end, NSS advocates

- reduction in the cost of access by new space transportation systems;
- exploiting of non-terrestrial material (NTM) and resources;
- standardization of vehicles and systems among nations so that they are compatible and less expensive;
- expansion of microgravity research by robotic and human activities.

Space settlement is an intellectual arena where sociologists, political scientists, architects, and urban planners may come into their own, especially in terms of technology transfer offworld. These and similar habitation issues will be considered in other chapters on human enterprise in space. For the U.S.A., along with its Russian, Canadian, Japanese, and European partners, the initial experimental laboratory for such matters is the now-orbiting International Space Station. The next challenge will be a lunar base, so universities and corporations are examining such habitat realities. For example, the School of Architecture, University of Southern California, offers a seminar in extraterrestrial and extreme environment habitats (ARCH 599). Under the guidance of Professor Madhu Thangavelu, co-author of *The Moon: Resources, Future Development, and Colonization*, students examine such topics as rationale and range of human missions in extreme environments; materials and methods of constructing habitats; life support and environmental control systems; food production and waste management; human factors and communications; ergonomic design of interiors (email: *thangavelu-girardey@home.edu*).

Space habitation is a world issue requiring a systematic mechanism for international exchange of information learned about living and working in an orbital environment. Perhaps the time has arrived for consideration of a world space administration to coordinate exchanges not only among national space entities and private enterprise—but to promote a more comprehensive strategy for extraterrestrial living by our species. Having studied the subject of *governance in space settlements* for some years, the author concludes that space agencies alone will never provide the solution, and that a new structure or authority will have to be created for this purpose in the near future. Two possibilities that could involve both the public and private sectors are a *Global Space Trust* for all international space joint ventures, and a *Lunar Economic Development Authority* just for transportation, industrialization, and settlement on the Moon [13]. Exhibit 18 offers further insights into the varied challenges to be faced in settling aloft in a microgravity environment.

2.3 INTERDISCIPLINARY CONTRIBUTIONS TO SPACE HABITATION

If the next stage in exploration and utilization of the space frontier is to focus on human migration and habitation, then planning should be amplified to include contributions of knowledge from a broader array of human sciences. The late Peter Drucker, management guru, observed that technology can only be improved so much and that financial resources are limited. To expand quality control, productivity, and profitability, he proposed concentration on the unlimited potential of human resources [14]. Until now, aerospace planners and managers of R&D efforts have concentrate primarily on technical considerations, with a narrow view of human factors.

During a NASA summer study on strategic planning for a lunar outpost, your author, serving as a Faculty Fellow, had an opportunity to analyze some of the non-technical aspects involved in putting a human back permanently on the Moon. In Exhibit 19, artist Dennis Davidson illustrates my findings in a modular form, similar to the possible appearance of a future lunar installation. In a sense, it is a conceptual model of some *other* human dimensions beyond the technical to be considered with reference to space habitation, whether in an orbiting station or at a Moon or Mars base. The reader may use this paradigm to examine these ten dimensions (given below). Use imagination to expand on the descriptions in the categories, or even to add new classifications. Then, apply these insights to just one aspect of space

Exhibit 18. OFFWORLD SETTLEMENT CHALLENGES.

Living and working on the Moon or Mars will not be easy, just as the pioneers found out when moving to America's western frontier. Apart from the space suits which must be donned for life support everytime someone walks on the surface, consider these protection issues that planners and settlers will have to resolve:

- The white space suit that astronauts have used for 130 safe space walks has undergone a continuing evolution in terms of life support and communication systems, as well as levels of protection and mobility. Currently, NASA calls it an Extravehicular Mobility Unit, a suit made of 14 layers of nylon and spandex, laced with 300 feet of tubing that constantly circulates cooling water around the astronaut. Some layers prevent oxygen from escaping, others regulate pressurization or provide thermal protection and shield from micrometeoroids. The 300-pound suit, costing $10 million each, is a part of a Portable Life Support Subsystem, including helmet, boots, and gloves ... But, for such garbs to be used on the surface of the Moon or Mars, a radical new design will be necessary. Ed Hodgson's Chameleon Suit, for instance, combines nascent technologies with advanced materials, information systems, and biomimetics. (Source: "A Space Suit for 2040: Smarter, Lighter, Easier to Pack," *USRA Researcher*, Spring 2003, pp. 1–3. The University Space Research Association in Columbus, Maryland is a consortium of universities (*www.usra.edu*).)
- Bigelow Aerospace has been building inflatable habitats for use in orbit. After the success of their Genesis 1 and Genesis 2 inflatable modules, they are working on orbital habitats that will support a crew. Their inflatables are boosted into orbit via Dnepr boosters hired from ISC Kosmotras, a Russian–Ukrainian company. Entrepreneurial leader, Robert Bigelow, announced a new orbital craft called Sundancer that will offer 180 cubic meters of habitable space with a trio of windows for a three-person crew. Eventually, this company hopes to produce habitats that could be inflated for use at a lunar outpost. Currently, it has partnered with Lockheed Martin to use an Atlas V 401 rocket system to boost its capsules. Lockheed Martin, a supplier of launch services, is working with Bigelow to haul expandable space complexes into orbit that carry passengers, as well as cargo. (Source: Leonard David, "Bigelow Orbital Modules: Accelerated Space Plans," *www.SPACE.com*, November 22, 2006.)
- To build things on the Moon or Mars will require a wide variety of spacefarers, including contractors, technologists, software experts, and even craftspeople of all sorts. Whatever facility or system is installed will necessitate humans, along with robots, capable of maintenance, repair, and expansion. Such specialists in the use of tools and instruments will be aloft to provide services relative to building materials, electronics, ventilation, heating, plumbing, among others. We are likely to be using lunar regolith as a primary building material, be it for manufacturing bricks or glass. Given the high cost of space transport, these diverse "technauts" will have to be multi-skilled.

> Thus, we must begin now to prepare such specialists for lunar habitation in a dozen years or more. It is a real opportunity to provide the human resources needed for developing space settlements. (Source: Adapted from G. B. Leatherwood's "Help Wanted: Space Colonists Need to Be More than Astronauts, *Ad Astra* (*www.SPACE.com*, March 30, 2005).)
>
> - Lunar colonists will be living deep underground protected from both meteors and deadly solar flares. The dry lunar soil, low gravity, and lack of atmosphere on the Moon make it ideal for underground habitats. By contrast, permafrost soils may be a problem when settling Mars. Building underground settlements will be difficult if the permafrost is actually made up of liquid water. Soils with high water content are more unstable to seismic shock waves and meteor impacts. (Source: Dr. Thomas Matula in an email to the author, December 11, 2006. For readers who would like to exchange views on these issues with this on-line professor, contact *tommatula@hotmail.com*)

development, such as industrialization, commercial activities, or habitations aloft (for further information, see Chapter 5).

(1) **Physical.** The medical or physiological concerns for survival and development of spacefarers, including *space ecology* (the biological aspects of closed ecosystems); as well as *space ergonomics* (the design of equipment and facilities to ensure both safety and quality of life) [15]. It encompasses all forms of protection and safety practices from radiation and other hazards aloft, to habitats and space suits. This dimension of study and research is concerned not only with weightlessness and ensuring the continuation of human life in a low or zero-gravity environment aloft, but its enhancement through preventative health, fitness, and wellness programs before, during, and after spaceflight. All aspects of *health care* and *life sciences* impact this area of concern, especially telemedicine (see Appendix D).

(2) **Psychological.** The adaptation and behavior of spacefarers, especially emotionally and in small groups, within an alien or hostile orbital environment. Initially, the focus should be on human performance, productivity, crew team morale, and management for long-term space living, including *stress reduction*. Special programs and simulations, some computerized, will be developed to counteract the negative effects of an isolated, confined environment and life style. Eventually, it will extend to the role of other animals who are introduced into space habitats and settlements, some of whom will become human pets. Also, this aspect of study will include *human/machine relations*, such as with robots. In time, all branches of psychology and psychiatry may contribute to improving space living and working, especially *mental health*. Applied and action research would include such dimensions of psychology as experimental, developmental, social, abnormal, clinical, counseling, environmental, industrial/organizational, management, and educational or learning. Special emphasis would be directed to preventing and/or treating depression and

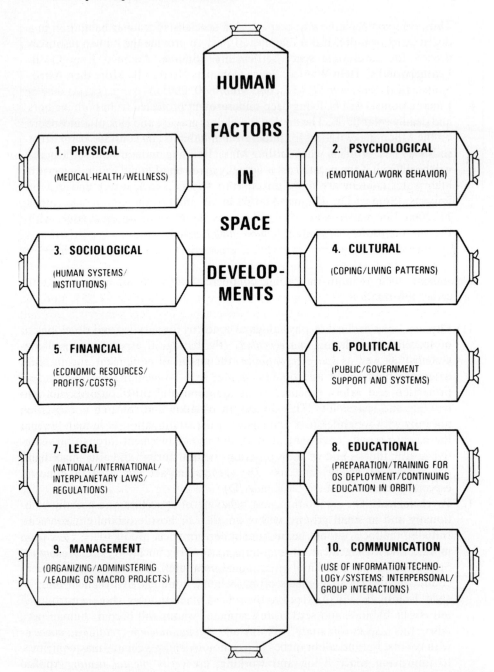

Exhibit 19. Human factors in space development beyond the technical. Source: Illustrated for your author in the form of lunar habitat modules by Dennis M. Davidson, former astronomical artist at the California Space Institute and New York's Hayden Planetarium (*www.novaspace.com*).

alienation, so as to take advantage of the microgravity environment. Particular study should be directed to the phenomenon of culture and re-entry shock, or the *overview effect*. The specialized field of space or astro-psychology is just emerging.

(3) **Sociological**. The study and development of space human systems, whether aloft within spacecraft, stations, settlements, or societies. The evolution of such space groups and institutions may require social innovations back on Earth, such as synergistic efforts by national space agencies or inter-institutional cooperation. Under this category, new sociological methods may be devised ranging from social engineering and computer simulation to adoption of future research approaches for technological forecasting and environmental scanning. Research should also address orbital group behavior, conflict resolutions, and teamwork. A key concern in this area is the differences, relationships, and interactions between *earthkind* and *spacekind*. In Chapter 3, we will discuss developments in the subfield of astro-sociology (*www.astrosociology.com/*).

(4) **Cultural**. The creation of coping skills, customs, communications, and life styles appropriate to the space environment (see Chapter 4) [16]. Initially, this may involve the adaptation or convergence of Earth's many macro and micro-cultures, as is already happening with international participation on both American and Russian spaceflights. Eventually, it will lead to formulation of *unique free-fall cultures*, suitable to different orbiting facilities, institutions, colonizers, and planets. Research is needed here on team, organizational, and family cultures. All of the other nine categories described in this model, plus others, will contribute to this process. Many will receive some new designation like "astro" or "space" (e.g., astrolaw or astro-education, space management or space finances).

(5) **Financial**. The economic dimensions of funding and supporting outer space projects, both in the private and in the public sectors, first on Earth and then in orbit. Forecasts indicate that as much as $270 billion will be invested in space activities by 2010, and as much as $540 billion by 2030! Studies would be devoted to the priorities in such spending. Further provisions have to be made for money-like exchanges and compensations, personal or electronic, to be provided between and among spacefarers, as well as with those back on Earth. Although this exploration began with public monies, there is growing evidence of privatization through investment and profit-making in space enterprises. Besides taxation and allowances, space financing through the public sector may range from bonds and national lotteries, to leasing of real estate and facilities aloft. In the private sector, such funding may occur through issuance of commercial stocks, as well as fees and rentals. Space industrial parks and colonies will require not only new venture capital, but mechanisms for underwriting insurance, joint ventures, or limited partnerships. Obviously, innovative cooperative macro-funding actions are desirable through consortia of government/university/industry, as well as between and among industries, institutions, and nations. Global space trading corporations

and global space trusts may find ways of funding macro-space undertakings (see Chapters 9 and 10). Long-term capitalization of space enterprises means creating mechanism for both inter and intra-governmental funding. Risk factors both on the ground and in orbit call for adequate insurance plans to cover accidents, disabilities, and death. These needs and prospects may require new provisions for national or international charters that help to underwrite the investment costs of space consortia and cooperatives. When space government is eventually established, it may also mean creation of "meta-money and meta-banks". In time, space communities will establish financial regulatory and audit systems. Further, financial analysts will have to design suitable structures for a twin planet and interplanetary economies (see Appendices B and C).

(6) **Political**. The building of national/international consensus and policies to obtain commitment of resources for exploratory, scientific, commercial, and defensive undertakings in space, as well as for the creation eventually of appropriate political systems and procedures of governance in space communities (see Chapters 7 and 9). International or national administrations and commissions may set space goals, but implementation depends on facilitating the political will and spending decisions, particularly with the setting of agency or program budgets. It is in this arena that space agencies must develop new skills, and that space activist organizations become more adept in lobbying and exercizing political power. In such matters, the contributions of statesmen, political scientists, policy analysts, and historians are most needed, in addition to those elected or appointed to political office. Eventually, spacefarers will develop their own governance systems suitable to microgravity realities. In time, human ingenuity will be taxed to create social orders that not only ensure participation for space colonists—but also autonomy from earthly sponsors [17].

(7) **Legal**. The analysis and application of relevant national and international laws, regulations, and agreements relative to space matters, as well as the creation of interplanetary authorities and laws to regulate extraterrestrial conduct of human affairs (see Chapter 9). *Space law*, lawyers, and associations already exist; treaties, conventions, and agreements are in place relative to the exploration and use of outer space, including liability and damage from space objects, establishment of space agencies and contracting with them, as well as commercialization of space activities and services. Currently, the principal applications of space law are to satellites, spacecraft, spaceports, accidents, and debris, as well as to U.N. space treaty obligations. Space commercial law is especially needed to define rights, privileges, and responsibilities from regulation and decision-making to profit sharing for inventors, researchers, investors, and users. *Astrolaw* is the term used to designate the future creation and practice of law in orbit by space settlers themselves. With law comes the need for order, so perhaps in this category should come eventually justice or security systems in orbit, as well as law enforcement and peacekeeping [18].

Sec. 2.3] **Interdisciplinary contributions to space habitation**   85

(8) **Educational**. In general, the preparation of people for the Space Age and space environment; in particular, personnel with their families for successful deployment to and from outer space (see Chapter 6 and Appendix E). The educational programs and technologies to be developed must be both ground-based (e.g., a space academy or university departments and courses), as well as in orbit (e.g., in-flight training or self-learning provisions aloft, or a space university and school systems for the young). Today's career or human resource development methods now should include preparation for space roles and occupations that are recent or have yet to come into existence. Formal systems of education on Earth will have to completely revise their curricula and methodologies to deal with the emerging space realities and opportunities, while space educational systems and universities will have to be inaugurated for the high frontier. On-line and distance education from ground to orbit and back will be established. Satellite communications will then link both ground-based and orbiting or planetary educational programs. Already the harbingers of future space education can be seen currently in aerospace departments within universities, the International Space University, the Challenger Centers for space science for young persons, and even Russia's Star City. Perhaps someday, the U.N., or a consortium of spacefaring nations, will found a Unispace Academy for the training of space travelers and settlers.

(9) **Management**. The organization, administration, and leadership required for outer space systems, projects, and installations, both on the ground and in orbit, manned and unmanned. This ranges from macro to micro-management of space enterprises, including human, material, and financial resources. This requires development of *new* management paradigms, technologies, and information systems, as well as space-relevant organizational models and leadership styles. Global management processes and systems will evolve to administer large-scale projects of great distance and cost in outer space (see Chapter 8). Space enterprises will need a new type of team management, entrepreneurialism, and innovation (see Chapter 9).

(10) **Communication**. The multi-dimensional aspects of human interchange related to space exploration and enterprise, from the marketing of such programs to data banks for the operation of space projects on the ground and in space. The range extends from technology, such as computer and communication satellites, to information systems and artificial intelligence, to human–machine interfaces, to interpersonal/intragroup/intergroup interactions. New kinds of digital, miniaturized communication hardware and software will be necessary. The whole area of cybernation (communication and control in humans and machines) receives new applications in space: *telerobotics* and expert systems being cases in point. Progress in experiments of humans living aloft requires communication systems which provide continuing, non-intrusive monitoring and feedback on experience (see Chapter 3). Innovative adaptations of both communication and information technologies, as well as virtual reality and smart space phones, will enhance living and working aloft [19].

The ten designations above are but convenient categorizations suggested for studying space developments beyond the scientific or technical dimensions. They are by no means exhaustive, indicating only the possibilities for a new taxonomy. For example, should space commerce and industrialization be considered as a separate entity, or does astro-business cut across several other categories, such as sociological or financial or management? Suppose this were an 11th classification, and one wished to include in this paradigm a 12th category: the technical. That would have a more comprehensive model for systematic analysis of space planning, operations, and management.

2.4 REDIRECTING KNOWLEDGE AND WORKFORCES UPWARD

It should be clear by now that opening up the space frontier demands re-thinking as to who and what we humans are, how we learn and disseminate information, and how we prepare future generations to live and work in orbit. Seventeen years ago, *The Economist* did a special survey on space (June 15, 1991), containing this astute conclusion:

> "The harvesting of knowledge from space will be one of the greatest scientific endeavors of the next century ... An age is dawning in which evolution will no longer be a matter of blind adaptation to an environment, but a matter of conscious choice."

Human development in space is not the sole province of aerospace industries, agencies, or career persons in the field. On Earth, we are in the process of creating a *knowledge culture* with *knowledge management* as a new tool [20]. Knowledge is a state of knowing from ideas, information, and understanding gained through experience, observation, study, and experimentation. As a result of our offworld experiences, both past and future, the sum of what humanity discovered, perceived, or learned, already exploding, will double and then triple in this century alone. Spacekind will develop new consciousness, concepts, and competencies that will lead to amplified powers, erudition, and wisdom. Only as we transition beyond this planet can humans fully develop our potential as knowledge workers within an interplanetary, knowledge economy and culture!

This challenge can be best perceived in terms of two emerging realities:

(1) *Refocusing academic disciplines and professions upward.* Advances in high and information technologies have already revealed the culture lag within traditional schools and universities, particularly with reference to studies in science, engineering, and computer programming. Discoveries or developments in space stimulate not only cross-disciplinary endeavors, but require professionals to redefine the very scope of their activities and the learning provided to candidates for their discipline or profession. Space requires a more holistic, multidisciplinary approach to problem-solving and learning. That is already happening in many

Sec. 2.4] **Redirecting knowledge and workforces upward** 87

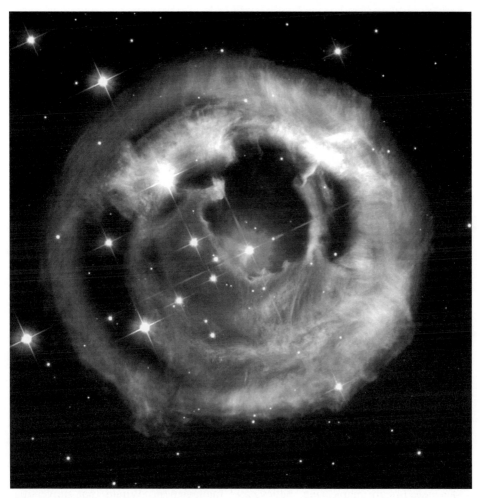

Exhibit 20. Vastness and beauty of space, as seen from photographs of the Hubble space telescope. Pictured are the stars which have fascinated humankind since the dawn of time. And to think that humanity has begun the long journey among these celestial points of light that beautify our nights, inspiring us to dream beyond our home planet. *Ad Astra!* Source: NASA/JPL, Pasadena, CA.

fields, such as in space medicine or psychology. For example, Dr. Stewart Johnson, a professional engineer in Albuquerque, New Mexico, has begun such a re-examination with his wife, Mary Anis Johnson. The Johnsons have jointly prepared a paper, "The Civil Engineer and Space", which they have permitted us to quote below. Their thesis, the effective use of Greater Earth resources (including the Moon and intervening space, particularly solar energy), is a subject of increasing importance to the nation, the world, and the engineer. Among trends cited by this professional couple are the following [21].

- The Civil Engineering Research Foundation will fund research related to space projects in robotic construction techniques, advanced material applications, computer-integrated manufacturing at construction sites. Space exploration and exploitation, especially at a lunar base, is dependent on such developments for habitats and facilities, resource and vehicle planning, and astronomical observatories.
- This convergence of civil engineering expertise and space exploration will benefit both terrestrial and off-terrestrial construction through major technology transfer from the space program to commercial applications. But, it requires cooperative research and education that will alter both engineering and the construction industry.
- The *Mission to Planet Earth* should be broadened to include electric power produced by solar energy transmitted from the Moon (see Appendix B). Similarly, other space/lunar resources have implications for their profession as confirmed in space volumes published by the American Society of Civil Engineers.
- The support of the civil engineering community for utilizing space resources can provide new energy sources, while mitigating environmental damage on Earth, as well as in space. But, it also offers strong motivation for education of more youth in science, mathematics, and engineering: all three are people-serving professions.

(2) *Needs of the emerging space workforce.* To succeed with space development, the workforce must be made up of competent, high-performing, and diverse personnel. The authoritative weekly, *Space News* (February 4–10, **2**(3), 5–10), did a special report on the Space Work Force in the U.S.A. Here are some of its observations that underscore themes made so far in this book.
- The aerospace industry faces two major workforce roadblocks: lack of qualified personnel, and lack of understanding between engineers and management. To correct these deficiencies and achieve an ambitious civil space program, an interdisciplinary curriculum has been designed which combines studies in engineering, policy, and management. Would that universities adopt such integrated programs in engineering, behavioral, and life sciences!
- A NASA study highlighted the anticipated need for scientists, engineers, and technicians increasing 33% by the year 2000, while student interest in these fields is declining in the U.S.A. The Agency's educational affairs division, therefore, is funding more cooperative programs with universities in these fields where the need for qualified graduates has only increased in this 21st century! NASA's civil service personnel, as well as its contract workers has been diminishing. The Agency seeks to obtain and retain a higher quality workforce through changes in Civil Service provisions, and improved compensation/benefits.
- According to the U.S. Bureau of Labor Statistics, 155,000 people were employed in the space industry in 1988. During the 1990s, labor analysts reported that workforce has lessened from corporate cut-backs and mergers, as well as a downturn in market conditions. The Aerospace Industries

Exhibit 21. Orbital repairs. Astronauts Steven Smith and John Grunsfeld use a robot arm to perform servicing tasks on the temporarily captured Hubble Space Telescope. Source: Photograph courtesy of NASA.

Association expects space spending in the U.S.A. to remain at present levels, with a consequent growth in space jobs from maintenance to technical to professional in the 21st century.

o The Planetary Society's executive director maintains that for professional space jobs, an educational background in science, biology, chemistry, or mathematics is basic. Although a bachelor degree is necessary to be hired, a Master, Ph.D., or other specialized training will be necessary for advancement ... Furthermore, the Center for Innovative Technology in Herndale, Virginia, predicts a change in the tools such workers use (e.g., supercomputers, neural networks, robotics, and more powerful orbital telescopes). Exhibit 21 depicts the role of future "technauts" (orbital technicians working with robots both in orbit and on the Moon or Mars).

o For further information consult *www.Vault.com* for the NASA job profile regarding workplace culture, pay, and recruiting; or *www.space-careers.com* on 100 offworld space occupations.

In the transition under way to a new work culture, knowledge and service workers will predominate, especially within space career fields. To prepare a properly skilled global workforce demands synergy not only between and among disciplines, industries, and government entities—but among nations. Further, youth guidance programs worldwide should be providing students with occupational information about such careers (e.g., astrophysics, astromedicine, astrocommerce, space science, space law, etc.)

Outer space is more than a place for travel. Space satellites have already proven the interconnectedness of our planetary systems for our habitability on Earth. The high frontier presents the human organism with a profound need for a new homeostasis! Space habitation demands not only a systems approach, but multinational and multidisciplinary contributions [22]. If there are to be permanent settlements on the inexhaustible frontiers of space, then

(a) biological knowledge must be joined not only to physics and chemistry—but with the social sciences;
(b) engineering must be linked not only to all of these—but information sciences;
(c) robotic science and technology focus on human interactions, plus robots as an extension of human sensory and manipulative capabilities;
(d) new applications of law and political science not only to promote migration and commerce aloft—but to protect human rights and ethics in orbit.

2.5 SPACE IS A PLACE FOR SYNERGY

Obviously, the vastness, complexity, and costs of exploring the space frontier call for *synergy* as an integral norm of human behavior. Space technologies were spawned toward the end of the Industrial Age, a period of rugged individualism and aggressive competition, especially at the national level. Yet, the success of the Apollo missions and lunar landings proved the effectiveness of collaboration between NASA personnel and aerospace contractors, as well as with university consultants. Space achievements contributed to the emergence of a post-industrial, information, and technically oriented society. But, to continue progressing, this new condition requires the practice of synergy, a concept of cooperative or combined action.

Today, we live and work in a global village and economy, an interdependent world brought closer together by advanced forms of communication and transportation. In that environment, *effective* human performance depends on sharing perceptions, insights, and knowledge. We create cultural synergy by capitalizing on people's diversities and talents. Space synergy occurs when disparate groups work together on extraterrestrial macroprojects. The Space Age has demonstrated that more is to be gained by individual, organizational, or national cooperation than simply going it alone. Already, the trend grows dramatically toward greater international cooperation in joint space technological and social efforts.

If spacefaring nations and project sponsors were to become more synergistic, then their enterprisess aloft would be characterized by

- a win/win philosophy of cooperating to mutual advantage;
- promotion of individual growth through team or group development;

- support of programs that advance the common good, while developing human potential;
- values that espouse openness to change, mutual reciprocalness, non-aggressive behavior, and enhancement of human capability (such as through the use of robotics) [23].

Collaboration among countries, institutions, and their representatives will advance the construction of a space infrastructure for the colonization of the Solar System. Here are three illustrations of a *synergistic space ethos* in practice.

First, **synergy among space scientists and other professionals**. This occurs in multilateral, cross-disciplinary, integrated enterprises that prompt scientists and academics, engineers and technicians, government workers and contractors, military personnel and civilians to move beyond traditional boundaries in the *macromanagement* of space projects. The ISY '92 World Congress manifested this when global professional groups gathered together under the joint sponsorship of the International Council of Scientific Unions (ICSU) and its Committee on Space Research (COSPAR), the International Astronautical Federation (IAF), the U.S. National Academy of Sciences and its Space Science Board, plus numerous other associations ... Two other organizations demonstrated this trait by studies and recommendations to promote *international cooperation in space*: namely, the 1993 reports on the subject by the Center for Research and Education on Strategy and Technology; and the American Institute of Aeronautics and Astronautics [24]. Both efforts brought together the best thinking on the subject from a variety of experts representing diversified affiliations. Professional space societies or associations also achieve such aims through local, national, and international conferences, workshops, and publications. The International Academy of Astronautics in Paris, for example, publishes the journal *Acta Astronautica* for original contributions from all fields of engineering, social sciences, and space sciences (*www.elsevier.com/locate/actaastro*).

At the International Space University in Strasbourg, France, there is a conscious policy to promote a multidisciplinary approach to space activities. The faculty and graduate students there already bring the perspective of diversity: engineers, medical doctors, lawyers, scientists, business people, architects, and policy analysts. Further, the classes and projects force students to move beyond their disciplines to learn plasma physics, orbital dynamics, propulsion engineering, and human factors aloft. As far back as 1988, this comprehensive approach was being used to plan an international lunar base. As an international community of scholars, those at ISU have a more global perspective about space exploration and settlement (*www.isu.edu*). This unique university provides a testbed for synergy in space, particularly with education aloft.

Sometimes professional synergy can be stimulated by a common concern, such as *telemedicine*. The University of Texas Medical Branch has the largest and most active telemedicine program in the world. UTMB holds annual conferences on "Pushing the Envelope" for practitioners from many fields of medicine and health care who give presentations on unusual, exotic, and extreme environments, such as space. Such synergistic exchanges have facilitated the use of telemedicine in orbit, as well as on the ground ... Still another manifestation of this collaboration occurs in university space research: for example, the California Space Institute is a research unit for all campuses of the public University of California (*www.calspace.ucsd.edu*), whereas the Center for Research in Science does space studies for the private Azusa Pacific University (*www.apu.edu/cris*).

Synergy is again evident within space agency and interagency activities, such as real collaboration between and among personnel.

- Headquarters and field centers
- Astronauts and cosmonauts
- Natural and behavioral scientists for enhancing manned flight programs
- Space agencies of several nations (e.g., annual ILEWG or International Lunar Exploration Working Group)
- Public and private sectors, as when space agencies and other government departments cooperate among themselves and with universities, corporations, and entrepreneurs.

Lack of cooperation between the entities of a national space program can hinder progress. Before both Japan and India organized their space efforts under a single administration, each had a different experience. Japan's space projects suffered from rivalries and budget competition, while India succeeded with its first launch of Agni, an intermediate range ballistic missile, because scientists in two sister defense research organizations worked together.

Second, **synergy among space advocates**. The pro-space groups worldwide usually operate independently, but increasingly must engage in joint undertakings, such as in support of *macroprojects* like a lunar base or space-based energy. The history of space development advocates is one of missed opportunities by not uniting for greater impact and improved quality of service. Leaders in the National Space Institute and the L-5 Society proved they were synergistic when they merged to form the National Space Society, which has many chapters, some beyond the borders of the U.S.A. (*www.nss.org*). Their affiliated entity, *Spacecause*, also reaches out to enlist the whole space community in their lobbying with the U.S. government. However, to raise the consciousness of any nation's citizenry about the importance of utilizing space resources requires a coming together of many entities in the pro-space movement in more effective educational and media efforts. Perhaps this explains why among the 250 million Americans, only 15 million or so pay any attention to the space program, including the fans of *Star Trek*. Similarly, the British Interplanetary Society has failed to gain the support of its government for vigorous space enterprise. Too many space associations talk only to themselves and serve their own constituencies—they do not reach out and synergistically relate or network with those outside their sphere of influence, especially with the younger generation. A *global space ethos* would motivate space proponents to *unite internationally* so as to make a convincing public case for space investments which strengthen economic, educational, and technological systems, while contributing to environmental wellbeing.

Third, **space synergy between the public and private sector**. Space entrepreneurs and commerce need government cooperation in such matters as funding, licenses, regulations, and laws. One example where this is most evident is in the building and maintaining of spaceports whether within a state, nation, or region. In many parts of the world, as well as within the United States, various entities are planning or actually building such terrestrial spaceports (see Chapter 8). Such undertakings call for the formation of consortia and partnerships whose commitments are confirmed in signed agreements. For example, in 1996, a Canadian company Akjuit Aerospace of Winnipeg, Manitoba, sought an alliance with the Scientific and Technological Center of Moscow to launch Russian Start rockets from northern Canada. A deal was signed to use the Akjuit polar launch facilities with STC Complex launch services and spacecraft to orbit small commercial satellites ... In 2007, New Mexico entered into an agreement to provide a spaceport for Virgin Galactic's aerospace planes and tourists ... Similarly, bi-lateral collaboration between ESA and Japan's NASDA has been under way for 30 years, but now these technical exchanges have taken on long-term cooperative projects, such as those related to the International Space Station, Earth observations, and space transportation.

From its beginnings, NASA has used private contractors to accomplish its missions, and that trend is increasing in this century. In addition, the Agency has undertaken initiatives to involve private enterprise in its ventures, such as the promotion of space commerce and medicine which utilize its technological innovations and research advances. It promoted the formation of the Universities Space Research Association to enable institutions of higher education, their faculty, and students, to participate in cross-disciplinary space research (*www.usra.edu*). NASA periodically issues calls for proposals by the private researchers on space studies. Since it favors acronyms, such cooperative agreements are accomplished under program titles like HEDS (Human Exploration and Development of Space) or HTCI (High Technology and Commercialization Initiative). A recent NASA attempt to improve its public–private partnership is *Dreamtime*: the opening and digitizing of its archives with over 80 years of historic photos, audio and videos, and documents of space and flight (*www.dreamtime.com/home.html*).

Fourth, **space synergy among nations**. This began in 1967 when six nations signed the *Treaty on the Principles Governing the Activities of States in the Exploration and Use of Outer Space, Including the Moon and Other Celestial Bodies*. Eventually, there were 100 signatories to this milestone document which held that space is for the benefit of all peoples. The countries bound themselves to application of international law in space, banning of weapons of mass destruction there, and provision of assistance to all spacefarers, the envoys of humankind!

The scope of human migrations into the universe necessitates transcendence of national and political differences. Photographs and remote sensing from outer space of planet Earth illustrate its unity and fragility, while enabling us to catalog its resources and catastrophes. Those who have been in orbit return with appreciation for the interdependence of our ecosystems, and urge us to think globally beyond local borders and concerns. Arthur C. Clarke who in 1945 first conceptualized communication satellites, astutely observed that we are witnessing the rise of the global family or tribe: humans linked together electronically across the world who are challenged to transcend ancient frontiers in our loyalties and interests. He urged the launching of a satellite, Peacesat, which would be devoted entirely to promoting peaceful, constructive relations among nations and people! A prototype was the successful Comsat which launched the satellite communications industry, and accounts for over 1% of the world's telecommunication infrastructure, a service that brings television and mobile phones into our homes and businesses. Space developments make all this and more possible globally.

Most nations are perceiving that space cooperation is mutually beneficial, but they have not realized that it also means creating new social attitudes, relationships, and systems for mutual achievements. Since the Space Age began, we have had numerous examples of collaboration among nations on the high frontier, such as the Apollo/Soyuz linkup in 1975, Intelsat communications satellite, East/West space science exchanges such as the Intercosmos satellite program, the formation of the European Space Agency, the multinational agreements to build an International Space Station, and the opening up of space facilities to foreign visitors. With the globalization of space undertakings in this decade, a significant synergistic ethos became more evident, as happened within the European Union when ESA became the focal point of multinational collaboration on that continent, as well as with the Americans in both the NASA Shuttle and Station programs, as Exhibit 22 illustrates.

Since 1994, the United States has signed a series of partnership agreements with the Russian Federation. This has resulted in joint missions for exchange of astronauts/cosmonauts on each other's space vehicles; providing financing, equipment, and personnel for completion of the International Space Station; collective research on environmental monitoring and planetary exploration. A new level of synergy was reached in 1995 when NASA's Dr. Norman Thagard flew on a Russian spacecraft to spend 115 days on the Mir orbiting station. Beginning in February 1995, the American Shuttle *Discovery* (STS-63) flew around the Russian Mir, followed by *Atlantis* (STS-71) which actually docked in June 1995 with this orbiting station (see artist depiction of historic rendezvous in Exhibit 23). Later when the Shuttle fleet was downed with the loss of *Columbia* (see dedication page), the Russians carried personnel from other countries, as well as supplies, to the International Space Station. Today, spacefarers from many nations, such as Brazil, routinely fly to ISS, but the Russians are the primary partners with the Americans in this orbiting laboratory.

Instead of the Cold War rivalry epitomized by the early competitive race into space, Russian/American space synergy is evident in both the private and the public sectors. Now, cosmonauts and astronauts train in each country's space facilities, and American entrepreneurs book tourist flights on Russian spacecraft to ISS. We end this section with two examples of this positive trend which need to be expanded.

(1) *The Citizen Ambassador Program* (110 Ferrall St., Spokane, WA 99202; fax.: 509 534-5245) co-sponsors with InfoMOST, Inc. (fax.: 703 448-5669). Delegations of business leaders and investors traveled to Russia to gain better understanding of

Sec. 2.5] Space is a place for synergy 95

Exhibit 22. Space is a place for synergy. Photograph of STS-63 international crew onboard Space Shuttle Discovery's flight deck: (front row, left to right) pilot Eileen M. Collins, commander James D. Wetherbee, and payload commander and mission specialist Bernard A. Harris Jr.; (back row, left to right) mission specialists Vladimir G. Titov (Russian), C. Michael Foale (British), and Janice E. Voss. Source: NASA Johnson Space Center.

the direction and capabilities of its space programs and to meet top-level contacts. Guided by Jeffrey Lenorvitz, former editor of *Aviation Week & Space Technology*, the visitors met with counterpart professionals and companies to examine Russian space facilities, program funding, technology transfer, and how to cooperate for mutual benefit. Apart from visiting cultural sites, the teams conferred with key executives at the Russian Institute for Space Devices, NPO Energia, Russian Space Forces Satellite Control Center, NPO Saturn/Lyukia, and NPO Lavochkin ...

(2) *Cosmos International Center for Advanced Studies* (Volokolamskoe sh.4, Moscow 12571; fax.: 095 229-3237; email: *cosmos@sovamsu.Sovusa.com*). This is a joint-stock company founded by enterprises specializing in aerospace technology. Founders include the Moscow Aviation Institute, Cosmos Concern, Salyut Design Bureau, S. A. Lavochkin Central Research Institute of Machine Building, INTERGRAPH Corp. (U.S.A.), the University of Alabama in Huntsville, and others. Under its director general, Dr. Oleg M. Alifanov, the company promotes: *cooperative international programs in the field of personnel development for aerospace science, technology, and related disciplines; research and*

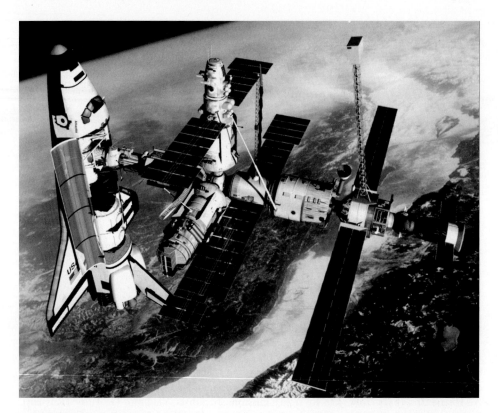

Exhibit 23. Russian/American space cooperation on Station Mir. Artwork of John Frassanito & Associates depicting the Shuttle docking to the Kristall module of the Russian space station in June 1995. The combined nearly 240-tonne spacecraft, Shuttle STS-71 + Mir 18, comprised the largest, human space platform ever assembled in orbit, while the two crews exchanged visits. Source: Courtesy NASA Johnson Space Center.

experiments in these fields to promote new, market-oriented, economic development in Russia; promotion in commercial applications of findings from Russian aerospace enterprises for interested organizations abroad.

Cosmos also fosters human resource development by aerospace education and research through consultation, training, data banks, and improved marketing. Similar agreements should be increased with the Chinese and Indian space agencies by NASA, ESA, RKA, and JAXA, as well as among the academies of science of all six countries. Exhibit 24 confirms the global trends to promote space synergy.

As this chapter is written in the fall of 2007, humanity celebrates the 50th anniversary of the launch of the first satellite, Sputnik, just as we are entering the *international lunar decade*!

Exhibit 24. RUSSIA WANTS TO JOIN NASA MOON PROGRAM.

Rivals for lunar conquest four decades ago, Russia hopes to join the current U.S. moon exploration program, bringing to such a joint venture their considerable space know-how and technology. Negotiations are under way between the space agencies of both countries as to Russia's participation in a mission to send a crew of four to the Moon by 2020, so as to establish a permanent international base camp there. Their spokesman, Igor Panarin, noted that Russia wants to contribute technology, not funds, to the project. RKA Energia has also proposed its own lunar exploration program, but the plan has yet to receive Russian government backing.

The Russian Space Agency envisages an agreement with the Americans comparable with those it has with other nations: such as to help the European Space Agency launch commercial satellites on Russian Soyuz rockets at France's Kourou base in Guiana; to assist the Chinese Space Administration to adapt its Shenzhou spacecraft (based on the Russian Soyuz) to land a robotic probe on the Moon by 2010. Russia wants to cooperate with the international community in space research, especially that related to the Moon.

Source: Adapted from Associated Press report by Vladimir Isachenkov, December 7, 2006 (*www.space.com/news/061207*).

2.6 CONCLUSIONS: SYNERGIZING SPACE EXPLORATION AND SETTLEMENT

Space is a place not only of high risk—but of unparalleled *opportunity if humankind can adapt to cooperative actions there*! It is also a place for joint ventures of high-leverage technology and research in space transportation and construction, energy and information processing, automation and artificial intelligence. There is a need for partnerships regarding space habitats and social innovations that ensure survival and the quality of human life aloft. The high frontier represents an evolutionary departure in human culture that requires the merging of art and science, economics and technology, public and private sectors in the pursuit of free enterprise and human enrichment. The Aerospace Technology Working Group is another example of a forum for space collaboration, between and among NASA, its contractors, and academia. Their recent publication, *Beyond Earth: The Future of Humans in Space*, has this telling proposal by an admired friend, Dr. Buzz Aldrin, which emphasizes the spirit of this chapter [25]:

> "*Space Trips for Peace* would create crews from nations marginally friendly, hostile, or even at war with each other. Space, new to civilization and without terrestrial boundaries and national sovereignties, would be the ideal frontier for demonstrating that people of all cultural beliefs and religious backgrounds, are able to set aside differences and work harmoniously for goals mutually considered

good. The space environment would give humankind an opportunity to establish new precedents and create defining moments in the quest for worldwide peace. The selected space crews and travelers would be fully trained and active during these highly publicized missions."

Perhaps the ideal place to undertake such a synergistic venture would be toward the lunar community to be settled beginning in the next decade!

But, for actions of this nature to occur, individuals, institutions, and even nations must have a vision of what they wish to do and where they want to go offworld. In the above sections, we have reviewed some of the approaches that will contribute to human development aloft. But, to build a spacefaring civilization, there must also be supporting *myths* that harmonize past and future. A myth is one attempt by people to explain unusual experience in their lives, a way to make visions believable. Mythologist Joseph Campbell envisioned the space program as a kind of projection screen on which new mythologies are created and displayed. Commenting on this insight, David Cummings, executive director for the Universities Space Research Association, wrote [26]:

"Human exploration of space, for example, is an extension of the great exploration mythologies of the past, giving cultural guidance about the importance of courage and the spirit of adventure in our lives. The famous view of Earth from lunar orbit gave us another lesson about the importance of living harmoniously with the Earth's environment, as did the exploration of Mars and Venus."

Cummings reminds us that the accomplishments of space agencies show us the power of combined efforts by thousands of gifted, dedicated men and women, scientists and engineers of diverse racial and ethnic origins. Space program achievements also teach lessons about the value of international respect and collaboration. This is the synergistic message we have tried to convey in this chapter. Even apparent failures, as the Hollywood film *Apollo 13* recalls, can be transformed into heroic acts when people on the ground and in orbit work together as problem-solving teams to discover creative solutions! As Lt. Gen. James A. Abrahamson once reminded an audience,

"That incredible spirit of reaching beyond all reasonable bounds is now an accepted characteristic and attitudinal spin-off of the space program."

Yet, the complexities in space ventures and colonization are beyond the capability of a single space agency or nation, or even one discipline or industry; to discover and utilize the resources of outer space demands synergistic relationships! Thus, *space settlement* represents a new direction and vision for humankind [27]. Pioneering offworld implies more than orbital survival; it involves human procreation, distribution, civilization, and destiny. Space resources may not only provide the human family with a means for coping with species' overpopulation, but a way to channel our fecundity outward.

Before extraterrestrial mass migration can occur, technical problems of space transportation and cost of access must be resolved so that *large numbers of people* can be transferred into orbit, a hundred miles or more up through Earth's gravity well and beyond. To that end research proceeds on advanced, reusable launch vehicles, aerospace planes, and even laser propulsion. Creative solutions may call for *in vitro* fertilization to populate other planets, or other innovative strategies such as described in subsequent chapters. However, we transport ourselves to the high ground; the transition of human civilization into space marks an epochal transformation in our species!

MIND-STRETCHING THOUGHTS

Human evolution has reached a critical juncture: escalate earthly rivalries to the frontier of outer space or acknowledge the unique characteristics and potential of the space environment, and consciously choose to build on existing cooperative relationships based on experiences of trust—instead of fear. Political, economic, and other forces have already begun to show that nations find it mutually beneficial to work together on Earth to research and develop the space frontier which is only 100 km above all our heads. There is a growing momentum toward international cooperation in space activities that has literally opened the skies—and our minds—to a new way of thinking.

Recall the beginning of the U.N.'s 1967 Outer Space Treaty, "The exploration and use of outer space ... shall be carried out for the benefit and in the interests of all countries ... and shall be the province of all mankind."

Source: Thomas E. Cremins and David E. Reibel in *Spaceline*, Institute for Security and Cooperation in Outer Space, Spring 1988.

2.7 REFERENCES

[1] Ward, P. E. "Will We Leave Earth for Permanent Space Colonies?" in Velamoor, S. (ed.) *Humanity 3000—Humans and Space: The Next Thousand Years* (section 4.2.1, pp. 47–55). Symposium proceedings published in June 2005 by Foundation for the Future (123, 105th Avenue Southeast, Bellevue, WA 98004; *www.futurefoundation.org*).

[2] Daniels, P. A.; and Hyslop, S. G. *Almanac of World History: Explorers*. Washington, D.C.: National Geographic, 2007, pp. 360–367.

[3] Kasturirangan, K. "India's Space Program," *Humanity 3000—Humans and Space: The Next Thousand Years*. Bellevue, WA: Foundation for the Future, 2005, appendix 3, pp. 275–283 (*www.futurefoundation.org*).

[4] Stuster, J. *Bold Endeavors: Lesson from Polar and Space Exploration*. Annapolis, MD: Naval Institute Press, 1996.

[5] Woodcock, G. R. *Space Exploration: Mission Engineering* (Vol. 1), *Systems Engineering and Design* (Vol. 2). Melbourne, FL: Orbit Books/Krieger Publishing, 1991 ... Surkov, Y.

Exploration of Terrestrial Planets from Spacecraft. Chichester, U.K.: Ellis Horwood, 1981 ... American Astronautical Society, *History of Rockets and Astronautics.* San Diego, CA: UNIVELT, 1977–1994, 15 vols.

[6] Cockrell, C. S. (ed.) *Martian Expedition Planning,* Vol. 107; Clarke, J. D. *Mars Analog Research,* Vol. 111. San Diego, CA: UNIVELT/AAS Science and Technology Series, 2004, 2006.

[7] Harrison, A. A. *Spacefaring the Human Dimension.* Berkeley, CA: University of California Press, 2001 ... Leach, C.; Antipov, V. V.; and Anatoliy, I. G. (eds.) *Humans in Spaceflight.* Waldorf, MD: American Institute of Aeronautics and Astronautics, 1997 ... Tascone, T. F. *Introduction to the Space Environment* and *Space Environmental Hazards.* Melbourne, FL: Orbit Books/Krieger Publishing, 2 vols, 1987, 1991 ... NASA Life Sciences Division. *Exploring the Universe: A Strategy for Space Life Sciences* and *Advanced Missions with Humans in Space.* Washington, D.C.: U.S. Government Printing Office, 1988, 1987 ... The interview with Dr. Anatoly Oganov appeared in "Life Sciences Alive in Russia," *Aerospace America,* March 1997, pp. 42–47.

[8] Bell, S. E.; and Strongin, D. L. "Evolutionary Psychology and Its Implications for Humans in Space;" pp. 77–83; Stewart, J., "Space and Humanity's Evolution," pp. 156–160, in Krone, B.; Mitchell, E.; Morris, L.; and Cox, K. (eds.) *Beyond Earth: The Future of Humans in Space.* Burlington, Ontario: Apogee Books, 2006 ... Cheston, T. S.; Chafer, C. M.; and Chafer, S. B. (eds.) *Social Sciences and Space Exploration.* Washington, D.C.: U.S. Government Printing Office, 1984, NASA SP-192.

[9] Harrison, A. A.; Clearwater, Y. A.; and McKay, C. P. (eds.) *From Antarctica to Outer Space: Life in Isolation and Confinement.* New York: Springer-Verlag, 1991.

[10] Morphew, M. E. (ed.) *Journal of Extreme Performance in Extreme Environments,* **5**, October 1, 2000 and June 2, 2001 (*www.HPPE.org*).

[11] Sulzman, F. M.; and Genin, A. M. (eds.) *Life Support and Habitability Series.* Reston, VA: American Institute of Aeronautics and Astronautics, 1994 (*www.aiaa.org*) ... NASA Life Sciences Division, *A Strategy for Space Life Sciences.* Washington, D.C.: U.S. Government Printing Office, 1988 ... Lorr, D. B.; Garshek,V.; and Cadoux, C. (eds.) *Working in Orbit and Beyond: The Challenge of Space Medicine.* Melbourne, FL: Orbit Books/Krieger Publishing, 1989.

[12] Ballyn, B. *Voyagers of the West: A Passage in the Peopling of America on the Eve of Revolution.* New York: Knopf, 1987 ... Harris, P. R. "Managing Transitions and Relocation in the Global Workplace," in Moran, R. T.; Harris, P. R.; and Moran, S. V. (eds.) *Managing Cultural Differences,* Seventh Edition. Burlington, MA: Elsevier/Butterworth-Heinemann, 2007, pp. 260–305.

[13] Piradov, A. S. "Creating a World Space Organization," *Space Policy,* **4**(2), May, 112–114 (published by Elsevier Science of Oxford, U.K.) ... O'Donnell, D. J.; and Harris, P. R. "Legal Strategies for a Lunar Economic Development Authority," *Annals of Air and Space Law,* **XXI–II**, 1996, 121–130 (published by McGill University, Quebec, Canada; *www.iasl.mcgill.ca*).

[14] Drucker, P. F. *Managing in the Next Society.* Oxford, U.K.: Butterworth-Heinemann, 2002 (*www.elsevier.com*).

[15] Eckart, P. *Spaceflight, Life Support, and Biospherics,* 1996 ... Dewitt, R. N.; Duston, D.; and Hyder, A. K. (eds.) *The Behavior of Systems in the Space Environment,* 1993. Hingham, MA: Kluwer Academic Publishers ... Hunton, L.; Antipov, V. V.; and Grigoriev, A. I. (ed.) *Humans in Spaceflight.* Reston, VA: American Institute of Aeronautics and Astronautics, 1997 (*www.aiaa.org*).

[16] Harris, P. R. "The Influence of Culture on Space Development," *Social Concerns*, **4**, 189–217 ... McKay, M. F.; McKay, D. S.; and Duke, M. (eds.) *Space Resources*. Washington, D.C.: U.S. Government Printing Offices, 6 vols., 1992; NASA SP-509 (*www.univelt.com*).
[17] Robinson, G. S.; and White, H. M. *Envoys of Mankind: First Principles for Governance of Outer Space Societies*. Washington, D.C.: Smithsonian Institution Press, 1986 (see Appendix A of *Space Enterprise*).
[18] Refer to back editions of the journal *Space Governance*, 1993–2006 published by United Societies in Space (499 South Larkspur Rd., Castlerock, CO 80104; email: *jjopc@quest.net*; *www.usis.org*).
[19] Refer to journal *Space Communications*, published by Elsevier Science in Oxford, U.K. (*www.Elsevier.com/journals*).
[20] Harris, P. R. *Managing the Knowledge Culture*. Amherst, MA: Human Resource Development Press, 2005 (*www.hrdpress.com*).
[21] Johnson, S. W.; and Johnson, M. A. "The Civil Engineer in Lunar Industrialization: Space, Energy, Environment, and Education," *Space Governance*, **4**(1), January 1997, 26–29 (available at *www.USIS.org* or from the authors at 820 Rio Arriba S.E., Albuquerque, NM 87123; email: *StWJohnson@aol.com*).
[22] Cox, J. K.; Krone, B.; and Morris, L. (eds.) "Theory and Action for the Future of Humans in Space," *Beyond Earth*. Burlington, Ontario: Apogee Books, 2006, pp. 271–281.
[23] Moran, R. T.; Harris, P. R.; and Moran, S. I. "Leadership in Cultural Synergy," *Managing Cultural Differences*, Seventh Edition. Burlington, MA: Elsevier/Butterworth-Heinemann, 2007, pp. 227–259.
[24] The two 1993 reports cited were *Partners in Space—International Cooperation in Space: Strategies for a New Century* available from CREST (1840 Wilson Ave., Ste. 204, Arlington, VA 22201) ... *International Space Cooperation: Learning from the Past, Preparing for the Future* available from the American Institute of Aeronautics and Astronautics (1801 Alexander Bell Dr., Ste. 500, Reston, VI 20191; *www.aiaa.org*).
[25] Hannon, M. "Cooperative, Worldwide Space Collaboration Epiphany: A Turning Point or Else," in Krone, B.; Cox, K.; and Morris, L. (eds.) *Beyond Earth*. Burlington, Ontario: Apogee Books, 2006, pp. 73–77.
[26] Cummings, D. "Editorial," *USRA Quarterly*, 1995, published by the Universities Space Research Association (Box 39, Boulder, CO. 80306; *www.usra.edu*; email: *researcher@usra.edu*).
[27] Jones, E. M.; Quigg, P. W.; and Gabrynowiez, J. J. (eds.) *The Space Settlement Papers*. Los Alamos, NM: Los Alamos National Laboratory, 1985 (#LA-UR-85-3874) ... Rudoff, A. *Societies in Space*. New York: Peter Lang Publishing, 1996.

SPACE ENTERPRISE UPDATES

- NASA is concerned about lunar dust particles on astronauts returning to the Moon for longer durations. The National Biomedical Research Institute is examining issues of lunar dust deposits in the lungs. Potential solutions involve aero-sprays and drug medications for the lungs, limiting exposure, and other methods for countering the dust's highly toxic substances.
- Altered by absence of gravity, everyday bacteria mutate into a highly lethal bug which may pose a threat to spacefarers on long-duration missions. In pushing the frontiers of biological and space sciences to see their limits, we not only produce menacing microbes, but in the process, we may also discover new therapies for infectious diseases. Microbiologist Cheryl Nickerson of Arizona Southern University at Tempe has experiments on the Shuttle labs for creating lethal bacteria in sealed containers; *Salmonella* grown in weightless conditions, for instance, turns more virulent. She said wherever humans go, microbes go, so novel countermeasures are necessary.
- The magnetic field of Earth protects humanity from radiation in space which can damage or kill cells. Once beyond that shield, people are more vulnerable and need artificial means of radiation protection. Dangerous levels of radiation could bar humans from prolonged activity on the Moon or Mars, unless ways to delimit such risks are devised. Spacesuits and habitats with shielding, for example, protect from dangerous levels of cosmic rays or bursts of radiation. In the U.S.A, the National Research Council has assembled experts from biology and space science to increase the margins of safety. Spacecraft designers and mission planners are currently using high-density plastic to reduce the weights of protective strategies, while science researchers probe new possibilities, such as lava tubes that might be used on the Moon as habitats. Where humans cannot go, there are always robots to do the job for us!

3

Space habitability and the environment

> *Spacefaring is a partnership involving technology and people ... Many believe that space exploration will help us grow psychologically and spiritually ... The human factors in space are the human capabilities and limitations in relations to jobs, machine, and work environment, as well as the personality, interests, attitudes, social relations, and culture of the spacefarers ... Because we have not evolved in Earth orbit, on the Moon and on Mars, we must be inventive just to survive. Habitability depends to a greater degree on the specifications of the spacecraft, and to a lesser degree on the attitudes and expectations of its occupants; it also depends on life support systems that maintain a satisfactory atmosphere, keep temperature and humidity within tolerable limits, guarantee adequate supplies of food and water, and assure acceptable levels of hygiene.*
>
> Albert A. Harrison [1]

By the end of this 21st century, *Homo sapiens* might well be described as an *interstitial species*. We are not just between two time periods, but between two ways of life. Humankind is being transformed from a terrestrial to an extraterrestrial being. In the process of this metamorphosis, our species will be changed both physically and psychologically, especially if genetics are engineered or synthetically induced characteristics are used to ensure survival aloft. Beginning with this new millennium, social practices, institutions, and knowledge organized for living on Earth are being profoundly altered to acclimate to offworld living conditions. The survival of the species in outer space demands significant adaptation to differing environmental realities. Technology and management have already discovered that both with spacecraft and spacefarers, innovative approaches must be created for dealing with the complexities of safely getting and keeping humans in orbit, as well as returning them back unharmed. Limited experience in this Space Age teaches us powerful lessons: (1) the necessity for *synergy* that goes beyond individual organizations, nations, and

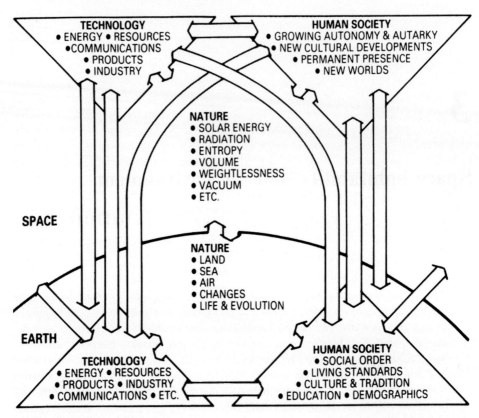

Exhibit 25. Super-ecology: future interactive loops. Source: Jesco von Puttkamer from his "Foreword" to the first edition, *Living and Working in Space*, by Philip R. Harris. Chichester, U.K.: Ellis Horwood, 1992, p. 21.

disciplines, as discussed in Chapter 2; (2) the *interdependence* of all life systems, especially between those of Earth and those considered non-terrestrial.

As a result of putting human beings, our technology and society, into outer space, Jesco von Puttkamer suggests that we will transform not only our images of our species—but of our place in the universe. As he illustrates in Exhibit 25, our perceptions of the latter are further changed by adding space to land, sea, air: a new type of *super-ecology* is being created in our minds. This in turn will introduce many more interactive loops and replace the geocentric view of ourselves and our environment.

Harry Holloway, once a NASA administrator for Life and Microgravity Sciences and Applications, astutely observed in an *Aerospace America* editorial (April 1995):

"We are in the process of becoming a global, spacefaring civilization. And space exploration is helping to blaze the way toward a new paradigm for life on Earth:

life without boundaries. By exploring the planets in space, we can also begin to explore our own internal universe."

This new reality and interdependence also applies to both the non-technical and the technical aspects of manned spaceflight, especially as it involves missions of increasingly longer duration that lead to space settlements. In the search for integrative strategies for living and working in the orbital environment, the concept and processes of the *behavioral sciences* seem appropriate as a way to facilitate interdisciplinary research and development. Dividing knowledge and information into separate fields, academic professions, and departments may have been suitable in past centuries. The latter will become of growing irrelevance in a *meta-industrial work environment and knowledge culture*, particularly as we struggle to cope with extending human life off this planet [2].

The term "behavioral sciences" has been explained by the *Harper Dictionary of Modern Thought* in this manner:

"Those sciences which study the behavior of men and animals (e.g., psychology and social sciences, including anthropology) [3].

It adds that some practitioners believe that for such studies to be scientific, they must be observable and measurable. Behavioral science is relatively new terminology that is much misunderstood. The concept appears to have been originated at the University of Chicago in 1949 when James Grier Miller named a Committee on Behavioral Sciences "an interdisciplinary group of scholars concerned with developing a general theory of behavior." Later, Miller and his colleagues would found there in 1953 an Institute of Behavioral Sciences. When they relocated to the University of Michigan, some of these same scholars in January 1956 created *Behavioral Science*, an interdisciplinary quarterly journal which Miller, the founding editor, edited for over 40 years until new owners renamed the publication *Systems Research and Behavioral Science* [4]. In his inaugural editorial, Dr. Miller wrote of different approaches to the study of behavior: mathematical biology, biochemistry, physiology, genetics, medicine, psychiatry, psychology, sociology, economics, politics, history, philosophy, and others [5]. Among the last-mentioned, I would include law, management/business sciences, communication/information sciences, as described in Chapter 2 (Exhibit 20). Miller sought a unitary behavioral science in which separate skills converge in the scientific study of human behavior. Originally, he hoped this new umbrella designation would cover both social and biological sciences, although many colleagues prefer to restrict the concept to the former. Eventually, his efforts were to climax in living system theory and its application to space habitation, something to be discussed in Section 3.6 ... For historical context, it should also be noted that the NTL Institute of Applied Behavioral Science and its journal of the same name has focused its action research on human relations, group dynamics, organization development and transformation, all research applicable to humans in space.

**Exhibit 26A. UNIQUE SPACE ENVIRONMENT:
USEFUL ATTRIBUTES OF SPACE LIVING.**

- Weightlessness (facilities, special manufacturing activities, construction of very delicate structures, and reliability of operations)
- Easy gravity control
- Absence of atmosphere (unlimited high vacuum)
- Comprehensive overview of Earth's surface and atmosphere, for communication, observation, power transmission, and other applications
- Isolation from Earth's biosphere, for hazardous processes: little or no environmental, ecological, or "localism" issues
- Readily avialable light, heat, power (ten times the rate on Earth)
- Infinite natural reservoir for disposal of waste products and safe storage of radioactive products
- Super-cold temperatures (infinite heat sink near absolute zero)
- Large, three-dimensional volumes (storage, structures)
- Variety of nondiffuse (directed) radiation (ultraviolet, X-rays, gamma rays, etc.)
- Magnetic field
- Availability of many Earth hazards (storms, earthquakes, floods, volcanoes, lightning, unpredictable temperatures and humidity, intruders, accidents, orrosion, pollution, etc.)
- Potentially enjoyable, healthful, stimulating, or otherwise desirable for human wellbeing

Source: Jesco von Puttkamer from his "Foreword" to the first edition, *Living and Working in Space*. Chichester, U.K.: Ellis Horwood, 1992, p. 10.

3.1 SPACE HABITABILITY AND LIFE SCIENCES

After a half century of space exploration, there is consensus among those who have been in orbit or studied microgravity that *outer space is indeed a unique environment to live and work* [6]. Some of the positive and negative aspects are reviewed in the next exhibit (first, refer back to Exhibit 16, and then go to Exhibit 26, above and opposite).

Just remember that since 1961, only about 460 men and women have actually gone beyond Earth into outer space. So far they have represented some 40 nations in the course of approximately 267 journeys offworld provided largely under government auspices. Only two dozen humans have soared beyond low-Earth orbit, and none since the last Moon landing in 1972. Yet, several space agencies are planning a lunar return, and entrepreneurs expect space tourists in other LEOs and GEOs. As hundreds, then thousands, and finally millions move off Earth, out into our Solar System, human habitability aloft will become a critical concern!

Thus, the emphasis throughout this chapter on *space habitability*. In their classic work on *Living Aloft*, Connors, Harrison, and Akins have clarified the meaning of

Exhibit 26B. STRESSORS IN THE SPACE ENVIRONMENT.

- Demanding work
- Unpredictable events and challenges
- Space adaptation syndrome (space motion sickness)
- Unusual photoperiodicity, circadian rhythmn, and work shifts
- Adverse physiological effects (radiation, cardiac, bone, muscle problems, etc.)
- Concerns and occurrences over equipment and system failures
- Concerns and occurrences of accidents
- Fatigue and sleep disturbances
- Concerns and occurrences regarding inadequate performance
- Remoteness, isolation, and break-off phenomenon
- Lack of information or bad news about home (social, national, world events)
- Lack of privacy, personal space, and territory; crowding
- Adverse physical conditions in spacecraft or habitat (air pollution, noise, high temperatures, insufficient power, etc.)
- Interpersonal irritations and misunderstandings among crew, and with ground staff, including negative effects on both groups
- Boredom, long stretches of "unfilled" time

Source: Peter Suedfield "Invulnerability, Coping, Salutogenesis, Integration: Four Phases of Space Psychology," *Aviation, Space, and Environmental Medicine*, **75**(6), June 2005, pp. B61–B66.

this concept [7]:

> "Habitability is a general term which connotes a level of environmental acceptability by potential users. The requirements for conditions to be 'habitable' change dramatically with circumstances. For brief periods, almost any arrangement that does not interfere with the health of individuals or the performance of their jobs would be acceptable. Over the long term, conditions must support not only individuals' physical, but also their psychological health."

Note how Space Psychologist Harrison expands on that definition 20 years later in the quotation at the opening of this chapter. Space habitability, then, implies a quality of life in orbit that ensures both human survival, while developing our species' potential. To achieve the latter, behavioral scientists will have to contribute more than they have so far to the space program and research, especially if successful settlement there is to occur!

Although the Russians have long used behavioral sciences in their space studies, the National Aeronautics and Space Administration has been slow to adopt the term, the concept, or the expertise associated with this integrative strategy. Since the organizational cultures of both NASA and its aerospace contractors have been dominated by science, engineering, and technology, the emphasis has been on human factors from a narrow, industrial engineering and medical perspective. Until recently,

Exhibit 27. NASA PERCEPTIONS OF SPACE LIFE SCIENCE PROGRAMS.

- *Gravitational biology*. Understanding the role of gravity in the development and evolution of life
- *Biomedical research*. Characterizing and removing the primary physiological and psychological obstacles to extended human spaceflight
- *Environmental factors*. Defining the space environment and habitat in which humans must function safely and productively, including air and water quality and the biological effects of radiation fields
- *Operational medicine*. Developing medical and life support systems to enable human expansion beyond the Earth and into the Solar System
- *Biospheric research*. Developing methods to measure and predict changes on Earth on a global scale and the biological consequences of these changes
- *Physiochemical and bioregenerative life support systems*. Assembling the knowledge base needed to design, construct, and operate life support systems and extravehicular suits in space that are independent of major resupply
- *Exobiology*. Exploring the origin, evolution, and distribution of life in the universe
- *Flight programs*. Conducting experiments in space, including the development of facilities and hardware for spaceflight, mission planning integration, and flight plan implementation

Source: Life Sciences Division, *Exploring the Living Universe: A Strategy for Space Life Sciences*. Washington, D.C.: NASA Headquarters, 1988.

even that was generally neglected by the agency, along with bioscience. Instead, NASA has preferred to adopt the designation "life sciences" with a limited concentration on applied medical and biological research relating to spaceflight. This approach has suffered from philosophical differences, as well as inconsistent direction and management [8]. Until the mid-1990s, the NASA management view of life sciences was as illustrated in Exhibit 27. Note that psychology is submerged under "biomedical research", with no mention made of anthropology, sociology, or other behavioral sciences.

Today this space agency has broadened that perspective, with a new emphasis on "Bioastronautics". As mentioned in Chapter 2, that term encompasses the study of biological and medical effects of spaceflight on living organisms (www.bioastronauticsmap.nasa.gov/). There is still no mention of its psychological effects on humans aloft.

3.1.1 Emerging life/behavioral science research

Today, *life science disciplines* are thought to include life support systems, medical science systems, biological science systems, human engineering, and extravehicular systems and techniques. The term "closed loop systems" is used with reference to long-duration human spaceflight requiring controlled ecological support systems.

Both Russian and American space life scientists now share their experiences, and a five-volume series on space biology and medicine has been published by the American Institute of Aeronautics and Astronautics. More recently, the Aerospace Medical Association has studied medical issues related to space passengers [9].

Concern about the long-term requirements of space living relative to biology and behavior goes back to 1960 when a NASA Biosciences Advisory Committee submitted a report urging the agency to undertake fundamental investigations in medical and behavioral sciences on human functions and performance aloft. The issue was not really addressed until 1987 when the National Research Council through its Space Science Board undertook to develop a strategy for space biology and medical science. In the following year's report, NRC was even more candid on the manned/unmanned mission debate:

"The National Research Council has examined prospects for intensive exploration of Mars. The Council concluded that while geologists can learn much from autonomous devices and sample returns, it is insufficient. Ultimately, to resolve important questions and to compare Mars in detail with the Earth will *require exploration capabilities on the surface of Mars possessed only by humans.*" (*Space Sciences in the Twenty-First Century: Imperatives for the Decade 1995–2015.* Washington, D.C.: National Academy Press, 1988).

At the same time, a Congressional Budget Office analysis on NASA's future noted that the agency's Pathfinder technology initiative contained more funds in the 1989 budget for life science research related to manned exploration of the Solar System. Despite rhetoric about the importance of long-duration human missions, at this writing NASA continues to severely delimit its spending on life and behavioral sciences, largely devoting research to the medical and biological studies of spaceflight. Most space agencies worldwide are not doing sufficient research today on human factors related to building a lunar base and settlement. Perhaps this is the area where universities should encourage faculty and students to concentrate their investigations.

In a life science review, "Learning from Living in Space" (*Aerospace America*, March 1995, 24–30), Theresa M. Foley offers these insights:

- "Although research over the last forty years has not found a single, long-term, detrimental health effect when people return to Earth from orbit, there are concerns about extended missions aloft ... Humans returning from spaceflights sometimes fall down or pass out; they walk more slowly as their bodies try to maintain balance in 1 g. They have fewer red blood cells, their bones and muscles are weak, and they are unable to keep their eyes naturally focused on objects as they move. In space, most people experience nausea and discomfort for a few days while suffering from *space adaptation syndrome*. Because they get taller in orbit as their spines stretch, most spacefarers suffer back pains. Since gravity shapes cells and tissues, little is known about the long-term effects of microgravity upon the body chemistry and physiology—such as calcium loss and bone weakening, shutting down of the immune systems and its effects, regulation of body fluids, loss of muscle and protein, etc. ..."

- "NASA finally recognized the importance of such factors to future goals by creating in 1993 an Office of Life and Microgravity Sciences and Applications (OLMSA). Yet, former Apollo astronaut and U.S. Senator, Harrison Schmitt admitted, 'We've wasted at least 25 of the 30 years of manned spaceflight by neglecting to undertake serious, well-planned scientific investigations of the effects of spaceflight on the human body.' He complained that the Agency failed to establish a protocol for life science experiments and to establish a data base on what has been learned. Schmitt also contends that NASA still does not have a good plan in place to take care of sick or injured crew members ... Yet, in 1995 NASA spent $183 million on life sciences as part of its new 'strategic enterprise' called *Human Exploration and Development of Space* ... By 2007, the National Research Council's buzz words were, *Safe Passage: Astronaut Care for Exploration Missions*—actually the title of a book of NRC recommendations (edited by J. R. Ball and C. H. Evans, first published in 2001 by The National Academies Press (*www.books.nap.edu/catalog/10218.html*; or *snha.georgetown. edu/guihs/ vol1no2/astronaut.htm*))."
- "The issue is not just Lower Earth Orbit problems, such as on a Shuttle or Station, but longer missions to the Moon and Mars where high-energy particles can kill human cells." When former NASA Administrator, Dan Goldin admitted that this research must go beyond the experience of four decades which he called 'pale, male, and stale,' and be aimed more at women and minorities, as well as integrating findings with partners in Russia, Europe, Japan, and Canada. One area of study should be on how the reproductive system might work if a woman became pregnant in microgravity.

In March 2007, NASA's chief Health and Medical Officer Dr. Richard S. Williams announced the appointment of a Medical Review Team to conduct a comprehensive review of the Agency's health services, including the behavioral health care available to astronauts. This committee engages in a comprehensive review of the Agency's health care systems, medical policies, standards, and certifications for astronauts. Dr. Richard E. Bachman, an Air Force Colonel and expert in aerospace medicine and extreme environments, leads the team's investigations. Other members represent the fields of clinical psychiatry and psychology, medical legal matters and privacy, as well as astronaut physicians. NASA has an executive from its Office of Safety and Mission Assurance as an *ex officio* representative. Unfortunately, there are no space nurses or other behavioral scientists on the team.

In this 21st century, it is obvious that global joint research in the life sciences should be done by consortia of spacefaring nations, including China and India. Further, such collaborative studies should include behavioral science studies on the psychological and sociological aspects of individual and group behavior in isolated, confined environments aloft. This became evident over ten years ago when then NASA Administrator Goldin publicly confessed Agency inadequacies after

Norman Thagard, astronaut–physician–engineer, experienced adjustment problems during his 115 days on the Russian space station Mir:

> "We put all our focus on the physical well-being of the astronauts and the success of the mission. We neglected their psychological well-beings and Dr. Thagard made it clear to us ... I think this was one of the major findings of the mission. If we expect to send people on missions of two or three years, we had better deal with the psychological aspects in addition to the physiological ones" (*Houston Chronicle*, July 8, 1995).

On Mir and its predecessor station, for example, cosmonauts manifested interpersonal problems by fighting with one another and ignoring ground controllers' instructions when the stress of living and working in an isolated, confined spacecraft for extended periods got to them. Under such circumstances sociologist B. J. Bluth indicates that disturbing characteristics may arise, such as boredom, irritability, fatigue, depression, anxiety, and mood fluctuations [10].

In Europe, the German Aerospace Research Establishment (DLR) is also interested in advanced life support investigations. At the Institute of Aerospace Medicine, Team Leader Marianne Schuber expressed interest in

- joint ground-based research programs within a multinational framework;
- support of ground-based research with already available inhouse infrastructure;
- optimization of simulation methods and image processing;
- support of spaceflight experiments;
- development, design, and preparation of life science experiments;
- development of telescience studies.

For more information contact the DLR Institute of Aerospace Medicine, Linder Hohe 45, D-51147 Köln, Germany; tel.: 49-2203-601-0; fax.: 49-2203-696.212).

Since 1990, the European Space Agency has engaged in isolation research within the confined environment of hyperbaric chambers. This has been part of their Isolation Study for Manned Space Infrastructure (ISEMSI) to prepare people for space station life and work. The word *EMSInaut* was coined for the initial six young European males, future astronauts, who were isolated for four weeks at the Norwegian Underwater Technology Center (NUTEC), and analyzed as to their behavior, group performance, and interaction under these stressful circumstances. This was followed by EXEMI '92 when a crew of four volunteers, one woman and three men, spent 60 days in isolation in the TITAN hyperbaric complex at the DLR aeronautical medicine institute near Cologne, Germany. In 1992, ESA astronauts were prepared for missions on both American and Russian spacecraft, and a European Astronaut Corps and Center was established to prepare Europeans for long-duration crewed missions in the 21st Century. In 1993, ESA's Long Term Program Office under the Directorate of Space Station and Microgravity Studies began *Human Behavior Space Simulation Studies (HBSSS)*. In addition to ESA isolation studies, this endeavor seeks to learn from similar studies in polar stations, underwater habitats, and submarines [11].

Professional associations can also render valuable insights into the challenges of living offworld. This happens through their publications, workshops, conferences, and sponsored research. An example is the Society for Human Performance in Extreme Environments which facilitates collaboration between researchers, practitioners, technologists, and other professionals to improve performance in extremely risky and challenging environments. In addition to sponsoring annual meetings, they seek scholarly contributions from the engineering, behavioral, and life sciences to their excellent journal (*www.hpee.org*) [12]. HPEE aims to lower the boundaries between scientific disciplines for those interested in the performance of people in complex, high-stress, and demanding environments and occupations. This society defines the term *extreme environment* as "settings that possess extraordinary physical, psychological, and interpersonal demands that require human adaptation for survival and performance." Examples of such isolated, confined, or extreme environments include outer space and high-altitude aviation, polar regions and deserts, underground and underwater. Such careers may be found on Earth among aviators, mountaineers, police, fire fighters and forest rangers; emergency service, search, and rescue workers; hazardous materials and anti-terrorist specialists; divers and extreme sport athletes, all such experiences serve as ground analogs for living and working in space. Then, there are those, like in Exhibit 28, who actually get into orbit as professionals or tourists.

Three powerful factors are driving space agencies globally to give more attention to habitability issues:

- *One* is the development of a permanently orbiting International Space Station by 2009. Early on, a British observer, M. H. Harrison of the Royal Air Force Institute of Aviation Medicine, succinctly forecast the issue [13]:
  "When America's Space Station becomes fully operational, it will serve primarily as a laboratory, workshop, and servicing facility. As such, it will provide unique opportunities for scientific studies in the material and life sciences ... The Soviets have also been first to take seriously the psychological aspects of space flight. During their missions of several months duration, problems of boredom, restlessness, and depression have been encountered ... Traditionally, NASA has been dominated by engineers, and this has been reflected in low priority being given not just to space psychology, but to space medicine generally. With a new era dawning of sustained human operations in space, this attitude will have to change."
- *Two* is that planning for the *Vision for Space Exploration* underscores the need for more studies related to long-duration space living and its effects on individuals and groups aloft. Going back to stay on the Moon and developing a lunar base there, as well as undertaking a human mission to the Martian surface requires more information on space habitability than we now possess.
- *Third*, there is increasing public interest worldwide in space colonization, but a

Sec. 3.1]	Space habitability and life sciences 113

Exhibit 28. Shuttle docked at International Space Station. Such orbital spacecraft are laboratories which provide behavioral science information and insights about human life offworld. Computer-generated scenario by John Frassanito and Associates of the Space Shuttle *Atlantis* docking at the newly constructed station to deploy a six-person crew made up of Americans, Europeans, Canadians, Japanese, Russians, and maybe Indians and Chinese. Source: NASA Johnson Space Center.

woeful neglect by both space agencies and scholars to identify what is required for real space settlement. Meanwhile, the private sector moves ahead with rockets and space planes that lower the cost to orbit; and inflatable stations that could become tourist destinations, beginning in LEO and eventually going beyond.

Perhaps the next two exhibits will help to underscore the challenges and risks of offworld migration. As previously noted, exposure to radiation is a hazard in space travel and settlement. Exhibit 29 offers some new thinking on this subject.
The continuing concern to improve space habitability was evident in the following two actions by NASA in 2007.

New systems to help Space Station crews breathe easier
In anticipation of increasing crew sizes to nine at the International Space Station, NASA is testing an innovative oxygen generation system. The hardware is part of the station's environmental control and life support system used to augment the Russian

Exhibit 29. RADIATION CHALLENGES OFFWORLD.

- "Radiation on the Moon and a three-year mission to Mars is dangerous and uncertain. Since the Moon and Mars have no atmosphere and no global magnetic field, astronauts will not have the protection from radiation that we have on Earth and in low-Earth orbit. Travel away from the Earth's surface makes it essential to monitor the types of radiation exposure ... In space we can't predict when radiation events occur, nor their severity, so it's crucial to develop a rugged, light-weight portable system that can make real time measurement of the radiation environments."

  Thus, the United States Naval Academy is developing, in partnership with NSBRI, a radiation and detection system. Called a microdosimeter, the instrument measures radiation doses at the cellular level, and will determine dose levels for scientific and medical purposes. Radiation exposure aloft can lead to fatigue, hair loss, cataracts, vomiting, central nervous system problems, changes in physiology and genetic make-up, and cancer, among other diseases. By integrating such microdosimeter sensors into spacecraft and space suits, spacefarers will be better able to assess risk, receive advance warnings of enriched radiation, and determine safe locations during such periods. Radiation comes from varied sources, such as particles trapped in the Earth's magnetic field, cosmic rays, and energetic solar events. (Source: Dr. Vince Pisacane, professor of aerospace engineering, U.S. Naval Academy, Annapolis, MD and researcher, National Space Biomedical Research Institute.)

- "Flights to the Moon and Mars will be subject to higher levels of space radiation than that experienced on the Shuttle or International Space Station. Thus, radiation detection and protection become critical says Daniel Baker, director of the University of Colorado's Laboratory for Atmospheric and Space Physics. He chaired a 2005 committee report on *Space Radiation Hazards for the Vision of Space Exploration* issued by the National Research Council. Today radiation limits for astronauts vary with age and gender. Currently, by age 55, the total days allowed in space max out at 265 days for male astronauts." (Source: *Space.com/news*, "Report: Space Radiation a Serious Concern for NASA's Exploration Mission," by Tariq Malik, www.space.com/news October 23, 2006.)

- "Life in space is bad for bones. Scientists know that microgravity conditions reduce bone mass. But a new study by Clemson University researchers indicates that bone loss also occurs in mice exposed to radiation similar to that which astronauts will receive during Mars or Moon missions. In mice, as well as in humans, the spongy inner tissue damaged by radiation cannot generally be replaced once lost." (Source: *National Geographic*, February 2007 (www.ngm.com).)

- "Suborbital space tourists would be exposed to minute amounts of additional radiation on their flights; enough to cause only 0.01 fatal cancers per 10,000 passengers ... The space tourism industry does not need to be concerned about a real threat of cancer. The problem though is that people will contract cancer after spaceflight and assume it was caused by radiation exposure during flight, particularly young people. That makes informed consent releases all the more important."

Source: Ron Turner, ANSER Corporation scientist, as reported in "Current Issues in NewSpace," by Jeff Foust, *Space Review*, March 5, 2007 (www.thespacereview.com/article/823/1).

Sec. 3.1] Space habitability and life sciences 115

Elektron oxygen generator. The aim is to generate 12 pounds of oxygen per day, and ultimately 20 pounds daily for a crew up to 11 in number. It is designed to replace oxygen consumed through breathing or used during experimentation and airlock depressurization. This research will help open new pathways for future exploration. The 1,800-pound refrigerator-size component was delivered to ISS during *Discovery*'s STS-116 mission in December 2006, and installed in the station's Destiny laboratory. Since then, hardware and software have been added, and the system became operational after a space walk by Chief Engineer Clayton Anderson during *Atlantis*'s STS-117 mission to install the required hydrogen vent valve. The new system produces oxygen by tapping into the station's water supply and utilizing the process of electrolysis. The oxygen is delivered into the crew cabin, while hydrogen is vented overboard through the vent valve; the hydrogen will also be recyled for water production from carbon dioxide (*www.nasa.gov/station*).

Contract extended for biomedical research

NASA's Johnson Space Center has awarded a five-year, $120-million extension of its cooperative agreement with the National Space Biomedical Research Institute. Houston's NSBRI will continue biomedical research under way with the Agency's Human Research Program in support of long-term human presence in space. These joint studies probe the health risks related to long-duration spaceflight, developing countermeasures to mitigate against them. The projects address space health concerns such as bone and muscle loss, cardiovascular changes, radiation exposure effects, nutrition, physical fitness aloft, rehabilitation, remote medical treatment technologies, as well as neurobehavioral and psychological factors in orbit (*www.nsbri.org*) (for further insight, see Appendix D).

3.1.2 Space settlement

As indicated in Chapter 1, an attempt was made to deal with the orbital habitability issue through a *Space Settlement Act* introduced in the U.S. Congress's House of Representatives in 1988 as legislation H.R. 4218. This measure as enacted incorporated the idea of settlements in space as official government space policy. Originally, it proposed to amend the National Aeronautics and Space Act of 1958, so as to set the establishment of space settlements as a long-term mission objective for that agency. Moreover, the legislation, eventually attached to a NASA budget bill, would have required NASA to be the lead governmental agency in conducting a steady, low-level effort to explore all the scientific, technical, and *sociological* issues relating to the achievement of settling space with humans. When its proponent, Representative George E. Brown, Jr. (D-CA), was unable to get the Act approved *per se* by his colleagues as first written, the wily Congressman did manage to incorporate some of its language and provisions as an attachment to a NASA appropriations bill which the 100th Congress passed in its second session. The wording is reproduced in Exhibit 30 as a model of what governments might do to further exploration and colonization on the high frontier. For the first time on record, this 1990 funding legislation officially recognizes the inevitability of such offworld activity! Unfortunately,

Exhibit 30. SPACE SETTLEMENT ACT: U.S. CONGRESSIONAL LEGISLATION

To authorize appropriations to the National Aeronautics and Space Administration for research and development, space flight, control and data communication, construction of facilities, and research and program management, and for other purposes.

SEC. 217. (a) The Congress declares that the extension of human life beyond Earth's atmosphere, leading ultimately to the establishment of space settlements, will fulfill the purpose of advancing science exploration, and development, and will enhance the general welfare.

(b) In pursuit of the establishment of an International Space Year in 1992 pursuant to Public Law 99-170, the United States shall exercise leadership and mobilize the international community in furtherance of increasing mankind's knowledge and exploration of the Solar System.

(c) Once every two years after date of the enactment of this Act, the National Aeronautics and Space Administration shall submit a report to the President and to the Congress which:

(1) provides a review of all activities undertaken under this section including an analysis of the focused research and development activities on the Space Station, Moon, and other outposts that are necessary to accomplish a manned mission to Mars;
(2) analyzes ways in which current science and technology can be applied in the establishment of space settlements;
(3) identifies scientific and technological capacity for establishing space settlement, including a description of what steps must be taken to develop such capacity;
(4) examines alternative space settlement locations and architecture;
(5) examines the status of technologies necessary for extraterrestrial resource development and use and energy production;
(6) reviews the ways in which the existence of space settlements would enhance science, exploration, and development;
(7) reviews mechanisms and institutional options which could foster a broad-based plan for international cooperation in establishing space settlements;
(8) analyzes the economics of financing space settlements, especially with respect to private sector and international participation;
(9) discusses sociological factors involved in space settlement, such as, psychology, political science, and legal issues; and
(10) addresses such other topics as the National Aeronautics and Space Administration considers appropriate.

Source: From S. 2209 (Section 217), 100th U.S. Congress, 2nd Session, 1988; *http://law2.house.gov/downloads/pls/42C26.txt* (see Hopkins, M. "The Space Settlement Act," *Ad Astra*, January/February 1994, 16... Office of Spaceflight Policy and Plans, *Strategic Plan for Human Exploration and Development of Space*. Washington, D.C.: NASA Headquarters, 1995).

Congress did not provide NASA funding for regular reporting by the Agency on its space settlement efforts, so regular progress reports have largely been ignored.

The failure to get that national legislation passed as originally conceived, along with the inability to get into law the *Space Exploration Act of 2002*, point up the lack of vision and political will among American leaders regarding offworld colonization and industrialization. This second more recent effort had as its purpose:

> "To restore a vision for the United States human space flight program by instituting a series of incremental goals that will facilitate scientific exploration of the solar system and aid in the search for life elsewhere in the universe, and for other purposes" (for further information about both bills, contact Steven Wolfe, Council for a Positive Future: *wolfesm@aol.com*).

U.S. Congressional legislation has not really provided the long-term financial support the space agency needs to carry out its multiple missions. A case in point is the NASA Authorization Act of 2005 which gave the first congressional endorsement of the Administration and Agency space exploration agenda. While it did authorize the spending level requested by the White House to implement the VSE lunar return policy, subsequent budget negotiations actually reduced the Agency funding. Long-term space planning is undermined by political infighting connected with passage of short-term budget allowances.

However, the above space settlement legislation does establish a framework for behavioral science research on space habitability. In the past 40 years, NASA has funded some studies and conferences that resulted in publications, previously cited, that address broader human issues in space exploration.

- *Space Resources and Space Settlements* was a classic under the leadership of Dr. Gerard K. O'Neill which NASA (SP-428) issued in 1979.
- One study produced with Georgetown University resulted in *Social Science and Space Exploration: New Directions for University Instruction* (Cheston, Chafer, and Chafer, 1984).
- Two others in conjunction with the University of California-Davis produced seminal volumes: *Living Aloft: Human Requirements for Extended Spaceflight* [7]; *From Antarctica to Outer Space* (Harrison, Clearwater, and McKay, 1991), which was also underwritten partially by the National Science Foundation.
- Another NASA publication in conjunction with the California Space Institute was released under the title of *Space Resources* [21]; the fourth of this six-volume series, *Social Concerns*, is of special interest to behavioral scientists. The next three chapters will discuss these and other NASA-financed studies on human performance in orbit.

One of the most definitive reports on habitability themes was produced by the NASA Life Science Strategic Planning Committee (1988). Entitled *Exploring the Living Universe*, it provided unusual findings, recommendations, and strategies, along with an extensive bibliography on human spaceflight. Among the former were proposals for increasing research, scientists, facilities, and funding in the space life sciences both on the ground and in orbit. Two of its strategies are of particular significance here.

- *Synergize the presently independent research activities of national and international organizations through the development of cooperative programs in the life sciences at laboratories of both global space agencies and universities.*
- *Complete and consolidate global data base consisting of basic life science information and the results of biomedical studies of spacefarers conducted on a longitudinal basis. This data base should be expanded to incorporate information obtained from all spacefaring nations and be available to all participating partners.*

Imagine the value of such a data bank if it became multinational in scope, containing, for example, the spaceflight experience of the Russian cosmonauts, as well as Europe and China. In fact, the findings from all terrestrial analogs to space living would be a welcome addition to this stored information, whether collected in submarines, on offshore platforms, at Arctic scientific outposts, from foreign deployment experiences, or experiments like Biosphere 2 in Arizona (see Chapter 6) [14]. In 2005, the Human Factors Research Program in the psychology department of the University did review past and present data collection on spaceflight, but unfortunately funding curbs may not encourage this innovative research on human performance aloft to go forward.[1]

The potential contributions that well-funded behavioral science research could make to advancing global space programs are manifold. A sample of habitability issues worthy of well-designed, scientific investigations are

- the sociobiological implications of isolated, long-duration missions and space biospheres;
- the integrative requirements for personnel deployment and habitation for masses of people aloft;
- the training, communication, and perceptions of ground-based support teams of a multi-disciplinary composition;
- the educational strategies, programs, technologies, and institutions that will be required for large numbers of spacefarers both on ground and aloft;
- the reinterpretation of terrestrial knowledge and science for adaptation to extraterrestrial environments;
- the broadened study of space technology and development by behavioral scientists as to their long-term impacts on human culture and society.

[1] For reprints of D. M. Musson and R. L. Helmreich's "Long-term Personality Data Collection in Support of Spaceflight and Analogue Research," *Aviation, Space, and Environmental Medicine*, **76**(6), June 2005, B119–B125, contact David M. Musson, M.D., Ph.D., Human Factors Research Project, University of Texas at Austin, University Station A8000, Austin, TX 78712 (email: *musson@mail.utexas.edu*).

From the perspective of an astrophysicist, Dr. David Criswell presented four space habitat dimensions, including research needs, to stimulate scholars. In Exhibit 31, the Director of the University of Houston's Institute of Space Systems Operations summarizes the type of biospheric issues to be considered with humanity's expansion beyond our terrestrial home.

Criswell has outlined why continued growth of our post-industrial way of life can only occur if humans move off the Earth and permanently settle space. In other research reports, he makes a convincing case that this transition will create new wealth from solar energy and the common resources of our Solar System (see Appendix B). But none of this can happen if spacefarers are not safe and healthy, so life science research is the foundation on which we build settlements aloft, as well as long-duration missions to Mars, and beyond.

Behavioral health planning and research for living and working successfully offworld calls for studies of pre-departure and on-site needs of space voyagers, as well as integration of findings into orbital operations. For instance, the French National Center for Space Studies announced in 2006 that a team of physicians are getting ready for surgery in zero gravity by operating in near-weightless conditions on a man with a fatty tumor: this is being done on an Airbus 300 Zero G making 30 parabolas during flight. The ESA also has a project under way to develop

Exhibit 31. SPACE HABITATION DIMENSIONS.

A. *Uses* —Diverse colonies of people —Solar System travel/interstellar flight —Directed research centers for exploration, theory observation, experiments —Entertainment/Tourism —Macro-machines —Macro-brains (swarm)	B. *Long-term questions* —Maths-based human language —*Homo sapiens* divergence —Si and Ill V alloys in intelligent systems —Sentient/Inert mass ratios for Solar System, Galaxy, universe
C. *Immediate research needs* —NTMs and growing systems (mass multiplication) —Parallelism: how small a first system growth limits criticality —Human extenders: teleoperators/robots —Improved space suits —Food (via chemical systems) —One-sixth (lunar) gravity operations —Independence from Earth —Training technical personnel as physicians	D. *Habitat research needs* —Scaling, life support systems, closure limits, fire hazards, time constraints, control and passive mass uses —Long-term residency: radiation shielding, toxicity infection —Special medical procedures with low-cost Earth to LEO transportation

Source: Dr. David R. Criswell, "Human Roles in Future Space Operations," *Acta Astronautica*, **8**(9/10), 1981, 1161–1171, Pergamon Press.

Earth-guided robots to conduct space surgical treatment. To reduce stress and promote psychosocial adaptation aloft, strategies and programs need to be designed now—not when humanity founds a lunar outpost. That might include a wide-range of activities in space: from monitoring behavioral health to computer-assisted therapy, to sexual intercourse and raising children in microgravity.

Too many space agency administrators and researchers have a myopic view of human needs beyond Earth. For example, NASA has avoided discussions and publications of human sexuality in orbit, a subject that is natural, normal, and inevitable [15]. The topic of sex, pregnancy, and birth in weightlessness has been generally ignored, even by medical researchers. Apart from tourists, astronauts deployed on the Moon for, say, six months or long-duration space settlers would warrant some serious studies, especially relative to family life aloft.

3.2 BEHAVIORAL ANALYSIS OF LIFE ALOFT

A pioneering psychologist in NASA's original experiments with animal behavior in space, Joseph V. Brady, has set forth a research agenda for studying human behavior in space environments [16]. Further, this Johns Hopkins University medical school professor has also theorized on applied behavioral analysis of life aloft. He calls for studies of the motivational factors essential for the maintenance of quality performance in long-duration operational missions. Brady and other researchers are concerned that such investigations on human group maintenance of satisfactory behavioral ecosystems be analytic studies, experimental in nature, utilizing a scientific method that is observable, manipulable, and measurable. Using a behavior management or modification strategy in space studies, Brady advocates B. F. Skinner's contingencies of reinforcement method for empirical analysis of both antecedent and consequent environmental events that influence behavior in confined, isolated situations akin to space living. Brady concludes that human space behavior and environmental action research requires multidisciplinary inputs from wide-ranging fields such as molecular biology, environmental physiology, behavioral biology, architecture, political science, sociology, and others.[2]

Two other factors propelling behavioral science space research are early 21st-century plans for a lunar base and eventually manned mission to Mars. Anthropologist Ben Finney believes that going back to the Moon permanently requires that social research be part of the planning process, and that the space station should be used for prototype studies. He also proposes that Earth analogs and space simulations should be used now to design space communities that will not be inhabited for decades. Relative to the exploration of Mars by spacefarers, psychologists Harrison and Connors [17] have produced insightful papers on the behavioral challenges in such missions. Such an undertaking might involve a mixed

---

[2] For a complete review of Joseph V. Brady's research and proposals, consult his "Behavioral Health: The Propaedeutic Requirement," *Aviation, Space, and Environmental Medicine*, **76**(6), June 2005, B13–B24 (email: *jvb@jhmi.edu*).

crew of men and women, up to 12 people. If it is done with multinational sponsorship, the space travelers are likely to be multicultural in composition. Apart from technical considerations, the duration in orbit would be between two and three years. Harrison and Connors envision this not only as a challenge in human adaptability, but requiring serious preflight analysis of the psychological and social dimensions of those going on a Marsflight. To keep the mission within tolerable human safety and stress limits is not enough. These researchers seek to provide an environment in which spacefarers can be fully productive and happy, where they can prosper and grow, particularly if we expect to establish eventually a base on the surface of Mars or its moons. Therefore, they contend, behavioral scientists need to examine a whole range of issues, from human qualities to behavior, to ensure successful performance aloft [17]. These and other psychological considerations will be identified in the Section 3.3.

Whether the emerging human missions are to the Moon or Mars, greater attention by space planners will have to be directed in the near term to

- architectural and environmental interventions which assure habitation that is both survivable and livable;
- the combination of effects likely to occur in an offworld outpost or base: psycho-physiological, psychological, and social;
- issues of governance, administration and leadership, including command structure and roles.

The impact of living aloft on the evolution of the human species can be perceived somewhat by reviewing the model developed by Gerard, Kluckholm, and Rapoport (see Exhibit 32). Analyzing its multidimensional content takes on added meaning in the context of space culture (see Chapter 4).

It is impossible in this chapter to review the wide-ranging prospects for behavioral science contributions in this regard. To illustrate the possibilities, the author will attempt in the next sections to synthesize studies and views of colleagues from the perspective of anthropology, psychology, and sociology, ending with the living system paradigm which shows much promise for space habitation research. Then, the Chapters 4–6 will expand on these themes, while the contributions of other social sciences will be examined in Chapters 7 and 8.

3.3 ANTHROPOLOGY AND SPACE HABITATION

Anthropology's greatest contribution to space development may be in its unique evolutionary macroperspective on human behavior, especially on our nature as wanderers, explorers, and colonizers. While our technology gives our species the capacity to migrate off this planet, it is our explorer's bent, embedded deep in our biocultural nature, that is leading us to the stars. As leading space anthropologist Ben Finney of the University of Hawaii reminds us, the field is a tool for humanizing space [18].

Exhibit 32. CORRESPONDENCES BETWEEN BIOLOGICAL AND CULTURAL EVOLUTION.

Biological evolution	*Cultural evolution*
Distinct species and varieties	Distinct cultures and subcultures
Morphology, structural organization	Directly observable artifacts and customs distinctive of cultures
Physiology, functional attributes	Functional properties attributable to directly observable cultural characteristics
Genetic complex determining structures and functions	"Implicit culture" (i.e., the inferred cultural structure or "cultural genotype")
Preservation of species but replacement of individuals	Preservation of cultures but replacement of individuals and artifacts
Hereditary transmission of genetic complex, generating particular species	"Hereditary" transmission of idea–custom–artifact complexes, generating particular cultures
Modification of genetic complex by mutations, selection, migration, and "genetic drift"	Culture change through invention and discovery, adaptation, diffusion and other forms of culture contact, "cultural drift"
Natural selection of genetic complexes generally leading to adaptation to environment	Adaptive and "accidental" (i.e., historically determined) selection of ideas, customs, and artifacts
Extinction of maladapted and maladjusted species	Extinction of maladapted and maladjusted cultures
CORRESPONDENCE BETWEEN CULTURES OR SUBCULTURES AND SUBSPECIES OF A SINGLE SPECIES	
Partial isolation of subspecies: "cellulation"	Partial isolation of cultures: "cellulation"
Cross-breeding through migration and limited interbreeding	Cross-breding through diffusion of ideas, customs, artifacts
Hybrid vigor?	Hybrid vigor?

Source: Gerard, R. W.; Kluckholm, C.; and Rapoport, A. "Biological and Cultural Evolution: Some Analogies and Explorations," *Behavioral Science*, 1(1), January 1956, 6–34.

Again we turn to *Harper's Dictionary* for insight into this discipline:

"While anthropology in the past centered its studies around evolution, embracing the biological, prehistoric, linguistic, technological, social and cultural origins about the development of humankind, modern anthropology is more focused on the concepts of biological endowment, environment and culture."

Typically, as an academic subject, the field is divided into sub-specialities, such as physical, social, and cultural anthropology. The latter two areas may yet prove to be the most vital to space studies, as applied anthropologists become less past-oriented and turn their skills toward futures research, especially relative to space communities. Harrison, a psychologist, maintains that modern anthropologists have to become less *Earthnocentric* [19]!

The survival of the wisest in space will depend on the accumulated knowledge and insight of interrelated sciences and multidisciplinary research, as was intimated back in Exhibit 20 and its explanation. To adapt successfully to extraterrestrial living, humankind will require synergistic information, paradigms, and methods from all the sciences, particularly the biological and behavioral. Unfortunately, the comparative studies of our species from an anthropological viewpoint is too often ignored within the space community. The NASA published report on *Social Sciences and Space Exploration*, previously cited, failed to even include anthropology. On the other hand, anthropologists have generally neglected the implications of their field and methodology for space. When a distinguished group of such scholars compiled a volume on *Anthropology Today* (CRM Books, 1971), a picture of an astronaut working in space was used to illustrate the opening chapter on the "science of man", but the contributors failed to discuss the emerging space culture in the 500 pages which followed. One can only speculate how many current textbooks on introduction to anthropology mention the opportunities to apply this discipline offworld.

Interestingly, when a group of anthropologists and other social scientists did publish *Cultures beyond Earth*, their concerns in extraterrestrial anthropology were for human contact with other ET species through interstellar travel. That is still the space focus of many social scientists. Similarly, anthropologists lead in the development of the *Contact* movement which held annual public forums to prepare for "contact" with alien cultures. At these simulations, participants broke up into two groups as a Human Team and an Alien Team to invent possible *contact* scenarios and experiences. *Cultures of the Imagination* proceedings were also published ... As far back as 1966 the American Association of Physical Anthropology organized a symposium and published proceedings in Berkeley, California, on *Man in Extra Terrestrial Environments: The Role of Physical Anthropology*. By the 1990s, Roland A. Foulkes was teaching a course at the University of Florida-Gainesville on *Astroanthropology and Futuristics* in which he used anthropological perspectives and futures research to span pre-human, human, and post-human evolution [20].

Culture is a central concept for space exploration and settlement, as we will discuss in Chapter 4. Therefore, it is puzzling that cultural anthropologists do not focus more research on the concept. Consider its implications for this purpose from this useful insight of the prominent anthropologist Edward Hall:

"In physics today, so far as we know, the galaxies that one studies are all controlled by the same laws. This is not true of the worlds created by mankind. Each cultural world operates according to its own internal dynamic, its own principles, and its own laws—written and unwritten. Even the dimensions of time and space are unique to each culture. There are, however, some common threads that run through all cultures. Any culture is primarily a system for creating, sending, storing, and processing information. Communication underlies everything ..." (*Hidden Differences* by E. T. Hall and M. R. Hall. Garden City, NY: Doubleday/Anchor Press, 1987).

What other roles can anthropology play in space enterprise. In addition to the above prospects, Dr. Ben Finney suggests anthropology can contribute to space planning by [21]

- providing a vision of where we are going extraterrestrially, as well as where we have been terrestrially;
- joining multidisciplinary research teams examining interpersonal and group behavior in the space environment, especially when crews and societies in orbit are heterogeneous and international, requiring skills in cross-cultural relations;
- demonstrating how cultural resources and differences can be best utilized and cultural synergy created among space travelers;
- doing field work on small space communities, particularly examining issues of mating and reproduction, child rearing and education;
- designing cultures and principles for space settlements;
- anticipating possible contact and communication with other ET species in time through interstellar travel or technology, or with descendents of Earth cultures who have developed aloft as independent and quite disparate cultural entities.

Another anthropologist, Namika Raby, reminds us that anthropology can be used for cultural engineering in space colonization [21]. Our terrestrial human experience has used culture for developing rules of behavior, sets of values for judging others, and to provide meaning to group actions. This former World Bank consultant would apply anthropology not only to study humans in orbit—but space agencies on the ground—especially with reference to the impact of bureaucracy on the exercise of power, authority, and decision-making; the emergence and resolution of cultural dissonance and conflict; the bonding goals in human exploration and settlement; the political, scientific, military, and commercial goals of space exploitation. Professor Raby, now teaching at the California State University-Long Beach, would further apply anthropological expertise to better understand subcultures of elite groups, such as astronauts, aerospace engineers, and scientists; as well as differences in the organizational cultures of NASA and its contractors. Dr. Raby believes that her field could contribute much to mission success by welding together and creating effective "space groups". She even thinks that anthropology can help in understanding space failures like the *Challenger* orbiter explosion. During a NASA Summer Study at the California Space Institute, Raby proposed additional anthropological research into

(1) a charter that embodies the central cultural values for promoting cultural synergy among space groups;
(2) enhancing the symbolic valence for bonding space crews at a space station or lunar base, as well as for socially sanctioned outlets for the resolution of conflict aloft;
(3) studying space-living issues of status differentiation, proxemics, structured interactions, rituals and ideology, and other such anthropological concerns.

In an address to the Society of Applied Anthropologists (March 29, 1986), Albert Harrison, psychologist at the University of California-Davis, proposed that anthropologists might engage in such high-frontier studies as [19]

- analysis of the social and technical systems developed to meet the challenges of the nascent space era, as well as creation of new systems to facilitate the transition from visiting space to living there for longer durations;
- minimization of problems and increasing human performance and production in space;
- strategies for improving the quality of life or habitability aloft for spacefarers, beginning with the Space Station and lunar base;
- application of the anthropological method of participant observation.

Dr. Harrison provided pragmatic advice to anthropologists so they could adapt their research proposals to NASA culture and spaceflight settings, urging them to enter their theoretical papers and reports into mainstream scientific literature. He forecasts that anthropologists may gain their greatest achievements by participating on interdisciplinary teams doing research on human habitability in exotic environments as terrestrial analogs for space living. Someday, astro-anthropologists may provide services at a lunar base, as seen in Exhibit 33.

Until we actually establish outposts on the Moon or Mars, we are dependent on analog research in the behavioral sciences. Dr. Jack Stuster of Anacapa Sciences has done just this, so as to extrapolate design and procedural guidelines for future expeditions and voyages of discovery offworld.[3] He maintains that many of the problems encountered by future explorers will be similar to those of their predecessors (e.g., strong-willed subordinates, cultural differences, misunderstandings, communication delays, equipment malfunctions, and weather).

3.4 PSYCHOLOGY AND SPACE HABITATION

Psychologists have been involved in the space program since its inception over 30 years ago, both in the U.S.A. and in the former U.S.S.R. Working in conjunction with medical teams, they have helped to determine if both astronauts and cosmonauts have "the right stuff" and what the impact of spaceflight is on their mental and emotional wellbeing. In collaboration with industrial engineers, psychologists have expanded human factor research about spaceflight. In fact, a whole new specialization of *space psychology* has begun to emerge, among which this author includes himself. Furthermore, university departments of psychology are beginning to offer courses related to the psychological aspects of spacefaring. The Master of Space

---

[3] See Jack Stuster's "Analogue Prototypes for Lunar and Mars Exploration," *Aviation, Space, and Environmental Medicine*, **76**(1), June 2005, B78–B83 (email: *jstuster@anacapasciences.com*).

Exhibit 33. Lunar base activities in the 21st century. In this artist rendering, *closed-loop* life support systems recycle and reuse air, water, food, and waste in an integrated fashion. Source: art by Robert McCall in National Commission on Space, *Pioneering the Space Frontier*. New York: Bantam Books, 1986, p. 2.

Studies at the International Space University has a multidisciplinary program in the Humanities featuring behavioral science [22].

Before proceeding, it might again be useful to go again to *Harper's Dictionary*, previously cited, to obtain a basis for understanding this field:

> "Psychology—the study of mind, behavior, or man interacting with social and physical environment. The ultimate aim is systematic description and explanation of man in the fullness of his powers, as a thinking, striving, talking, enculturated animal."

Dictionary editors Bullock and Stallybrass remind us that psychology has yet to devise central concepts such as the laws of physical science, and can be characterized in terms of either its choice of processes to be studied, or its method of analysis. Most practitioners concerned about input/output metaphors focus their investigations around perception and habitation, and the ways organisms transform energy; for those concerned about response processes, the focus is on expressive movement, motives, drive states, language, and various forms of social behavior. With respect to the methods of analysis, psychology uses a variety of experimental approaches (e.g., clinical, observational) and tools (field studies, mathematical, and computer modeling). As a university subject, psychology has proliferated in its applications: industrial, educational, cognitive, existential, genetic, humanistic, mathematical, informational, organizational, and so forth. Perhaps psychology will only complete itself as a field of knowledge when it converges into the behavioral sciences, possibly within the context of human orbital studies and research in outer space? Harrison and Summit [23] argue that recent reviews of psychology and spaceflight tend to focus on performance decrements, dwindling motivation, emotional stability, social conflict, and other adverse consequences of prolonged missions aloft. They argue for a "third force" of humanistic psychology that describes the psychological benefits of manned spacefaring, such as [23]

— enhanced competence and mastery;
— peak experiences or the overview effect;
— increased ability to cope with stress;
— high social cohesion and teamwork;
— role model and inspiration for others.

These psychologists contend that such an approach would emphasize human talent and resourcefulness aloft which has implications for space mission design.

3.4.1 Emerging contributions of space psychologists

What are some of the unique contributions to be made by psychology in space developments and habitation? Connors, Harrison, and Akins [7] maintain that the answer to that question is in fostering a high level of spacefarer wellbeing and maximum productivity. But, to do that they think that the profession must gain greater understanding of the psychological issues of adaptation to outer-space life. T. S. Cheston [42] postulates that, as a discipline, psychology needs to thoroughly address how crew members can be taught self-generated reinforcement strategies to enhance their sense of personal and professional accomplishment aloft. Apparently in agreement with Joseph V. Brady's position stated above, Cheston reminds us that psychology has already made a remarkable start toward ensuring more effective human performance by manipulating schedules of reinforcement and punishment, but this research should be applied to orbital human factors [24].

Christensen and Talbot [25] did a comprehensive review on the psychological aspects of spaceflight. They have examined both the former Soviet and the American experience aloft with regard to perception, cognition, psychological stability, performance, small-group dynamics, stress, as well as psychological methods and models. Their remarkable Exhibit 34 summarizes the key factors which influence orbital behavior and performance in terms of environment, space systems, and support measures. Furthermore, in this study for NASA's Life Science Division, Christensen and Talbot have offered far-ranging suggestions to behavioral scientists

Exhibit 34. SPACE FACTORS AND HUMAN BEHAVIOR.

A. Psychological, psychosocial, and psycho-physiological	B. Environmental	C. Space system	D. Support measures
Limits of performance (perceptual, motor)	Spacecraft habitability	Mission duration and complexity	Inflight psychosocial support
Cognitive abilities	Confinement	Organization for command and control, division of work, human/machine	Recreation
Decision-making motivation	Physical isolation, social isolation		Exercise selection criteria
Adaptability	Weightlessness	Crew performance requirements, information load	Work–Rest/avoiding excess workloads, job rotation
Leadership productivity	Lack of privacy, artificial life support, noise		
Emotions/Moods, attitudes	Work–Rest cycles	Task load/speed, crew composition	Job enrichment, preflight environmental adaptation training
Fatigue (physical and mental)	Shift changes	Spacecrew autonomy	Training for team effort
Crew composition, crew compatibility	Desynchronization, simultaneous and/or sequential multiple stresses	Physical comfort/quality of life, communications (intracrew and space–ground)	Inflight maintenance of proficiency
Psychological stability	Hazards		Cross-training
Personality variables	Boredom	Competency requirements	Recognition, awards, benefits
Social skills		Time compression	Ground contacts, self-control training
Human reliability (error rate)			
Space adaptation syndrome			
Spatial illusions			
Time compression			

Source: Christensen, J. M.; and Talbot, J. M. "Psychological Aspects of Space Flight," *Aviation and Environmental Medicine*, March 1986, 203–212 (NASA Contract #3924).

for near and long-term research and development, especially in terms of Shuttle/Space Station operations and observed aberrations.

3.4.2 Addressing spacefarers' psychological needs

What are the major psychological issues that spacefarers face, past, present and future? Kanas addressed that question relative to both former Soviet and American space studies, particularly in light of longer manned space missions and more heterogeneous space crews. He identified nine psychological and seven interpersonal issues requiring further research [26].

- *Psychological issues*: sleep problems, time sense disturbances, demographic effects, career motivations, reaction stages to isolation, transcendent experience, postflight personality changes, psychosomatic symptoms, and anxious/depressive/psychotic reactions.
- *Interpersonal issues*: interpersonal tensions, problems resulting from crew heterogeneity, anger displacement on outside personnel, need for dominance, decreasing group cohesiveness over time, task-neutral interactions, and types of leadership.

Analysis of other space studies has provided me with this summary of additional behavioral issues worthy of further research:

- recruitment, selection, training, and orientation of spacefarers, especially those who are not members of the astronaut corps, such as contractor personnel;
- individual and group adjustment to living in a zero or low-gravity, unearthly environment;
- human performance in exotic environments which are isolated, confined, and high-risk settings as an analog to space living;
- human factors in design of space stations and habitats, especially as these become more spacious and comfortable;
- human performance effectiveness and productivity aloft in terms of workload, assignments, and scheduling;
- orbital sleep, rest cycles, and recreation;
- socialization of new arrivals, inflight support, and preparation for re-entry to Earth life and culture;
- coping with problems of extended spaceflight, personality conflicts, and deviant behavior;
- communication aloft among mixed crews of government employees, military, and private contractors with sexual/cultural/national and professional differences, as well as between such diverse crew and ground-based personnel;
- crew morale aloft, and communications with ground-based families and friends;
- inflight biomedical and psychological monitoring of spacefarers' health and wellbeing;
- inflight exercise of authority and decision-making, leadership, and management of space missions and emergencies;

- postflight debriefing and psychological evaluation of crew adaptation and performance. The latter would require scholarly examination of the whole phenomenon of the "transitional experience" or *overview effect*, already well documented, and its implications for those who undergo offterrestrial living [27].

3.4.3 Space psychologists in the future

What are some additional contributions that psychologists may undertake toward the furtherance of space development and habitation? Obviously, orderly investigations of the above issues by psychologists, in conjunction with other behavioral and biological scientists would advance the human presence in space. As Harrison and Connors [17] so rightly conclude, psychologists should provide leadership in helping astronauts and other space travelers maintain a healthy psychological state, examining particularly how to promote an atmosphere of cooperation and mutual support within a group aloft. Furthermore, they urge that such research findings be published in mainstream scientific and space literature. Recall that Oleg Gazenko, when head of Space Medicine in the former Soviet space program, expressed the belief that the limitations of living in space are not medical, but psychological.

The case for interdisciplinary studies can best be appreciated by examining just one of the issues cited above, sleep. Dr. Martin Moore-Ede, professor of physiology at Harvard Medical School, thinks that inadequate investigations have been addressing crew shifts and sleep patterns during long-term space travel (*Omni*, June 1987, 2). Both astronauts and cosmonauts report that they do not sleep deeply in space, and have difficulty in the early stages of missions. Despite the need to develop new sleeping patterns and arrangements more suitable to gravity-free living, Frank Sulzman, when NASA chief of biomedical research, admitted the agency then had no organized programs examining crew schedules, the physiology and psychology of sleep, and human circadian rhythms in space.

With a little imagination, psychologists should be seeking to engage in studies with other social scientists on the design and conduct of (1) new university courses in space psychology, astro-anthropology, and astro-sociology, or better still *space behavioral science*! and (2) multidisciplinary education and training programs for spacefarers. Social and environmental psychologists, in particular, have only begun to apply their expertise to improving the spaceflight experience in the design of space habitats and the organization of space communities. They have largely contracted their services to the public sector, whereas their future may lie with private space enterprise. For instance, Bigelow Aerospace of Nevada is designing and testing an inflatable habitat for orbit named Genesis Pathfinder. Are psychologists involved in these plans for private space stations offworld?

One facet of such research was undertaken for NASA on privacy and space habitat design by Harrison, Sommer, Struthers, and Hoyt [28]. These psychological investigators from the University of California-Davis examined the dimensions of privacy and social contact and their implications for life aloft, especially on a space station. After a review of the psychological literature on the matter, they made 50 specific recommendations to the NASA Ames Habitability Group. The latter, under the direction of environmental psychologist Yvonne Clearwater, were engaged in research for the Space Human Factors Office at Moffett Field, California, to make a space-based workforce both comfortable and productive during long-duration flights (i.e., 30 days or more aloft). Other innovations in behavioral science space research will be discussed in Section 3.6 on living systems.

However, one recent development is in the area of positive psychology relative to spaceflight.[4] This approach contributes to the human experience in space through prevention and countermeasures to personal problems aloft. Dr. Peter Suedfeld, cited above (Exhibit 26B), observes that the emphasis now is on the self-enhancing aspects of living offworld, be they strengths or vulnerabilities. This University of British Columbia psychologist notes that positive psychology studies factors like optimism, success, hope, courage, creativity, altruism, enthusiasm, exaltation, faith, and transcendence. For space travelers, it means focusing research especially on the positive thoughts and emotions experienced, constructive attitudes and mind expansion that may result, peak experiences that come from high-orbital performance, or overcoming stress and challenges. Other dimensions of people who live and work in space to be analyzed are achievement, adaptability, affiliation, coping skills, positive stress, resilience, resourcefulness, self-integration, and actualization.

Astronomer Carl Sagan forecast that *cultural diversity* will be a strength and key to survival in future space societies, each of which would take aloft aspects of their terrestrial worlds such as planetary engineering, social conventions, hereditary dispositions. Psychological selection of spacefarers would, therefore, seek persons more tolerant of differences in lifestyle, race, or religion. Eventually, psychological research should be directed toward the differences in *earthkind* and *spacekind*. In time, the adaptations of the latter will bring about biological, physical, psychological, and social changes that sharply distinguish them from their terrestrial counterparts. For now, such behavioral and biological science research starts on the Shuttle and Space Station, as indicated in Exhibit 35.

3.5 SOCIOLOGY AND SPACE HABITATION

One recommendation of the National Commission on Space's report in 1986 advocated building institutions and systems that make accessible vast space resources and support human settlements beyond Earth orbit, from the highlands of the Moon to the plains of Mars. But, Professor Ben Finney believes that this demands planning for social organizations in space. Gerald Carr, commander of the 84-day Skylab mission, has gone on record that he expects that "the sociological problems will prove to be more difficult to solve than the technological ones" [29]. Currently, ISS is a multinational space station, as will be the case at Moon/Mars bases, challenging

---

[4] See Snyder, C. R.; and Lopez, S. J. (eds.) *Handbook of Positive Psychology*. Oxford, U.K.: Oxford University Press, 2002 ... Suedfeld, P. "Applying Positive Psychology in the Study of Extreme Environments," *Journal of Human Performance in Extreme Environments*, **6**, 2001, 21–25.

Exhibit 35. Research on living and working in microgravity. (Top) STS-120 pilot George Zamka (right) uses a camera, while astronauts Stephanie Wilson (left foreground), ESA's Paolo Nespoli (top), both mission specialists, and Daniel Tani, Expedition 16 flight engineer, share a meal in the galley of the Zvezda Service Module of the International Space Station, while the Space Shuttle Discovery is docked with the station. (Bottom) Astronauts Pam Melroy (left), STS-120 commander, George Zamka (bottom right), pilot, and ESA's Paolo Nespoli, mission specialist, sleep in their sleeping bags, which are secured on the Shuttle's middeck area. Source: NASA Johnson Space Center.

sociology and its practitioners to address offterrestrial social issues and civilization. Few in this discipline have done futures research on *Societies in Space*, the timely title of Alvin Rudoff's book [30].

To comprehend better the meaning of the concept, let us return to the explanation in *Harper's Dictionary*:

"Sociology—the study of societies by observation and description within a coherent conceptual scheme."

From the viewpoint of space settlements, what is lacking in the above description is the *design of future* societies, such as we expect to do in orbit. The dictionary's editors, Bullock and Stallybrass, inform us that sociology is less fully a discipline than its counterparts because of several competing schemes and failure to develop coherent sets of concepts which can be applied and codified as theories. The three principal perspectives for viewing society are

(1) *mechanistic*, analogous to social physics which explains social phenomena and variations by reference to climate, soil, population, or a combination of such physical attributes;
(2) *social evolution*, progress due to evolving consciousness or humankind's material powers;
(3) *systematic empirical enquiry* about the facts of social life. Contemporary sociology is heavily influenced by the teachings of Karl Marx on class structure and ideology; of Herbert Spencer on social morphology of societies in terms of structures and functions; of Emile Durkheim on social solidarity and segmentation; of Max Weber on social action and comparative method, especially in the study of authority, power, bureaucracy.

Sociology has developed analytical tools that may prove useful in space research, such as conflict/evolutionary/ecological/mathematical *models*, case study methods, and scenario development. Sociologists may join forces with biologists in social biology, the application of biology to social problems, such as pollution, ecology, and overpopulation. Or they may participate in social engineering with other behavioral scientists concerned with planning social change. Or they may collaborate with social psychologists in studying human social behavior, by combining biology and social science.

Extending human civilization to outer space is a challenge that may enable sociology to achieve its fullness. B. J. Bluth, a forerunner in the application of sociological methods to space development, stated her case in this way [31]:

"Sociology organizes knowledge to identify and analyze more of the hidden potential in human behavioral systems ... Systems of sociological concepts can be broken into three basic categories: social systems, or systems of ways of doing things; cultural systems, or systems of meaning (e.g., language, values, beliefs, ideas); and personality systems, or systems of need disposition.

Bluth maintains that sociologists can contribute to space living because of their expertise in identifying patterns of relationships between events and systems, and their ability to analyze the consequences of behavior. As one trained in organizing knowledge of human behavioral systems, this sociologist is interested in diverse space activities, such as:

- astronaut survival and safety;
- crew size, compatibility, and constraints;
- quality of space life;

- social processes aloft;
- human experience in orbit, especially with gender-mixed crews;
- the impact of space technologies on Earth's social and cultural systems.

However, Bluth has long-term concerns as a sociologist relative to the design of lunar and space communities, and the intelligent applications of behavior system approaches. When she was a sociology professor at the California State University-Northbridge, Bluth taught courses in astronautical sociology. During a stint at NASA headquarters doing research for the Space Station, Bluth analyzed the Russian analogs that may assist in U.S. planning for long-duration missions, such as on Mir or on the International Space Station (ISS) [32]. Currently, she has been researching and writing on designing for human performance in space.

Another space sociological leader is William E. MacDaniel [33]. While at Niagara University, he taught courses in both future studies and "Living in Extraterrestrial Space". He co-directed there the Space Settlement Studies Project which published a journal and newsletter. As a result of his research at the Johnson Space Center in 1982, this former Air Force jet pilot undertook in 1984 an innovative Delphi scenario study on *Extraterrestrial Space Humanization*. Now a professor emeritus in San Antonio, MacDaniel still believes that sociology can make a unique contribution to the humanization of extraterrestrial space, as well as in social inventions for the space frontier. His studies convince him that ET social organization and culture will differ significantly from terrestrial experience. The unique combination of factors in outer-space living, from zero or low gravity to unlimited solar power, will force human adaptations and alternative "free-fall cultural development". MacDaniel maintains that technologically oriented professionals have influenced space developments, while social scientists have abdicated their responsibilities to participate with research leadership on space settlement and societies.

In light of such observations, an interesting footnote is that a 1983 Smithsonian symposium and publication on human adaptation barely mentions our greatest challenge in that regard—space—and then only in terms of speculation on space colonies [34]. Although it views species' adaptability as a biocultural odyssey, these proceedings are notable for emphasis on the biological bases for social behavior, as well as by the absence of the three social sciences under discussion here. In contrast, Rudoff has offered three frameworks for sociologists to contribute new insight on extraterrestrial human adaptation [35].

(1) Space sociology from a terrestrial perspective
(2) The study of small groups in confined space and its implication for space living
(3) Space sociology from an emerging extraterrestrial perspective.

In "Reflections on the Sociology of Interstellar Travel," NASA's John Mauldin suggests that trips on *starships* with their *microsocieties* in the centuries ahead may involve the lifespan of generations [36]. He envisions both biotechnology and sociobiology impacting the sociology of such a high-technology environment which might involve 30 generations for 1,000-year missions! Mauldin predicts that such spacefaring will push human potential and species' knowledge to the ultimate limit!

Exhibit 36. Lunar habitability. Artist rendering of a lunar worker and robotic helper. Lunar dwellers will always wear a life support system when on the surface, except when they go into a facility that provides such support. Source: NASA.

If the unusual problems of survival in the offworld environment can be partially solved through societal mechanisms and pre-planning or social engineering, then sociologists should be involved. By using their competencies in designing space communities, sociologists will help not only to improve civilization aloft—but may prevent the transfer there of cultural, social, and biological dysfunctions of this planet. As envisioned in Exhibit 36, the knowledge culture on the Moon will find scientists of many disciplines engaged in cooperative work, but always with a life support system, whether above or below the lunar surface.

Perhaps the most notable development in this professional discipline is the emergence of *astrosociology* (*www.astrosociology.com/*). Its leading proponent, Dr. Jim Pass, clarifies the term:

"Astrosociology is defined as the sociological study of two-way relationship between astrosocial phenomena aspects of society and its social organizations. That concept pertains to the study of social conditions, social forces, organized activities, objectives and goals, and social behaviors which are directly or indirectly related to (1) spaceflight or exploration; (2) any of the space sciences (astronomy, cosmology, astrobiology, astrophysics). It includes all outcomes of these phenomena in the form of scientific discoveries and technological applications, new paradigms of thought in the astrosocial and non-astrosocial sectors of society, as well as changes of social norms and values in any of the social structures of a particular society" (*www.allacademic.com/meta/p108789_index. html*).

Since astrosociologists are concerned about both the micro, middle, and macro-aspects of social life beyond Earth, their potential contribution would be in matters of space settlement. As global plans are made for a lunar base by 2020, that would seem to be the place to target their research.! Hopefully, they will also link up with space lawyers to study matters of space governance (see Appendix A).

3.6 LIVING SYSTEMS AND SPACE HABITATION

To delimit the problems of orbiting groups and societies, to promote synergistic relationships and cultures in orbit, there is need first for integrative studies among terrestrial behavioral scientists themselves. Since space experience alters human perception, only the convergence of behavioral science methodology and insight will permit the holistic interpretation as to the meaning of life experienced aloft. One way to achieve this might be if such professionals devoted their research to a summative theme, such as "human sexuality and reproduction in space", a subject generally avoided by space agencies. Another opportunity would be for behavioral and biological scientists to team up in utilizing a common paradigm for studying space habitation issues.

That is what Connors, Harrison, and Akins [7] did in their monumental work on *Living Aloft*, when they adopted the recommendation of the Space Sciences Board to use a systems perspective to analyze spaceflight. Missions were then viewed as comprising highly interdependent components (e.g., technical, biological, social), such that variations in one element have repercussions on the others. These researchers attempted to expand this conceptualization by incorporating features of open-systems theory as devised by James Grier Miller and his colleagues [37]. In "Lunar Bases: Learning to Live in Space," Finney also confirmed Miller's approach, emphasizing that in space studies we are dealing with living systems involving the biological, technological, and social [17]. As an anthropologist, Dr. Finney called for social scientists to work closely with biologists, human factor specialists, architects, and, ultimately, the engineers and managers who conceive, design, and operate the whole space system. Thus, he himself has done such research at the NASA Ames Research Center.

3.6.1 Studying orbital living systems

During the 1984 NASA Summer Study at the California Space Institute, Dr. James Grier Miller, a psychiatrist/psychologist, directed a lifetime of research on living systems toward issues of space habitation. At the same time, another Faculty Fellow in that endeavor, your author, was then examining matters of space management, culture, and people deployment [38]. When NASA officials there requested proposals from both researchers on their mutual endeavors, it resulted in an interdisciplinary, combined project entitled *Living Systems Applications to Human Space Habitation*. Originating through the School of Medicine, University of California-San Diego, Miller served as principal investigator and Harris as senior collaborator, along with a team of 15 behavioral, information, and natural scientists. Unfortunately, the then NASA's Life Sciences Division found the approach too holistic and declined to fund this innovative study. However, the proposal may have contributed to the present thinking at NASA on "systems integration". The project team still envisioned LST applications to isolated, confined environments anywhere, whether in Arctic outposts, submarines, or outer space. To collect and analyze system data in orbit, Dr. Miller advocated the use of identification badges with infrared transponders in the form of microchips, questionnaires, observation, computers for information storage, and artificial intelligence expert systems for analysis [39].

In an article on "The Nature of Living Systems" with his wife Jessie, a psychiatric social worker, James Miller maintained that planning for extraterrestrial living requires a primary focus on the human beings who are to inhabit the projected settlements. They suggested that Living Systems Theory (LST) can facilitate space planning and management because it is an integrated conceptual approach to the study of biological and social living, the technologies associated with them, and the ecological systems of which these are all parts. They described living systems as open systems that maintain a thermodynamically improbable energic state. This occurs through a continual interaction with the environment in which these substances of lower entropy and higher information content input more than they output. The LST school of thought views biosocial evolution moving in an overall direction toward increased complexity.

James and Jessie Miller have developed visual encapsulations of 20 critical subsystems and processes, with some applications to space activities. The paradigm illustrates on its vertical axis eight levels at which living systems have produced cells, organisms, groups, organization, communities, societies, and supranational systems (such as the United Nations). On the horizontal axis, the essential subsystems are depicted as reproducer, boundary, ingestor, distributor, converter, producer, matter/energy storage, extruder, motor, supporter, input transducer, internal transducer, channel/net, timer, decoder, associator, memory, decider, encoder, and output transducer. They conceived of behavior as as an energy exchange within an open system or from one system to another (i.e., the linking of exchanges of information and matter/energy).

The LST strategy, according to the Millers, facilitates empirical cross-level research. Each system can be identified in terms of a set of variables describing its basic processes. At the level of groups or below, these represent aspects of flows of matter or materials, energy, and information or communication. At the level of organization and above, it is useful to measure two additional flows: personnel (individual and group) and money or its financial equivalent (e.g., costs). The Millers have documented a wide variety of LST research applications by behavioral scientists over the past 30 years in a variety of fields from the military to health care.

Before his death in 2002, James Grier Miller was proposing LST application to space habitation, first in terms of the International Space Station and later for a lunar base. Astronomical artist Dennis Davidson created, under Miller's direction, a series of illustrations using LST symbols to explain through color usage how this complex theory would be utilized. To provide the reader with some insight into such applications, we have produced here three summary diagrams, but these are in black and white, and do not reflect the distinguishing and separate color flows. Exhibit 37 explains the living systems symbols. Exhibit 38 visualizes Living Systems Theory applied to the International Space Station, with its five major flows for that open system which may become fully operational with eight spacefarers by the year 2009. Exhibit 39 does the same in the context of a lunar outpost which may be functional by approximately 2020. The complex patterns for a Moon base show a command center, habitation unit, generating station, as well as provisions for storage, solar power, and a nuclear power plant.

With renewed interest in space human factor research by global space agencies, this living systems integrative methodology is useful to collect both subjective and objective data on spacefarers at any orbital locale, employing computers and sensors, as well as a centralized knowledge base for analysis by an artificial intelligence expert system. If the first two prototype applications of LST aloft prove valid and worthwhile, then the studies might someday be extended to a Mars base and other space colonies which emerge in the next century.

Living systems theory has been successfully applied for over four decades in terrestrial settings. Now, it would seem appropriate to continue this strategy of multidisciplinary research to improve extraterrestrial human performance and habitation. By using the LST template and five critical flow measurements in space communities, it would seem that more comprehensive planning, information, and management may result. In terms of this chapter's theme, the living systems conceptualization provides a computer-based framework for analysis of biosocial behavior both on and off this planet. It represents an opportunity to bring together in meaningful investigations the combined talent of behavioral, biological, and informational scientists.

3.7 HABITABILITY AND THE SPACE ENVIRONMENT

Earlier in this chapter it was determined that outer space is an extreme environment for space habitation. But, what happens to that *environment* in terms of human presence there. We have already created the problem of orbital debris, which

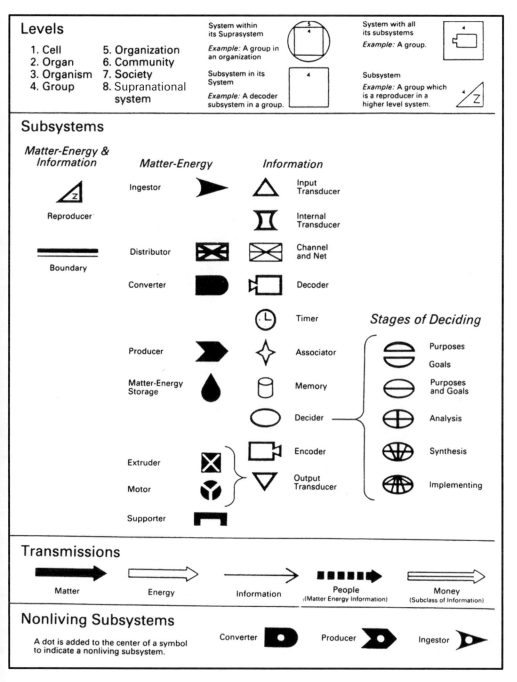

Exhibit 37. Living systems symbols. Source: Miller, J. G. "Applications of Living Systems Theory to Life in Space," *Space Resources* (NASA SP-509, Vol. 4, pp. 231–259). Illustration by Dennis M. Davidson.

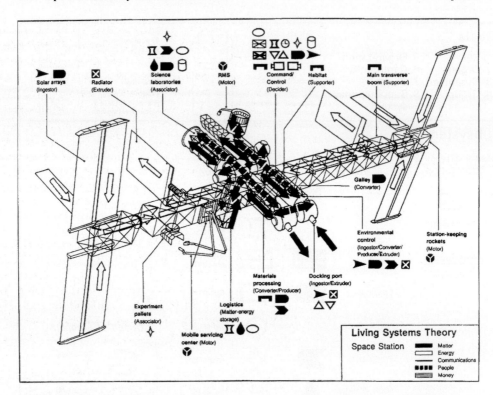

Exhibit 38. Space Station: five living systems flows. Source: Miller, J. G. "Applications of Living Systems Theory to Life in Space," *Social Concerns* (Vol. 4, pp. 231–259) of *Space Resources* (NASA SP-509). Washington, D.C.: Government Printing Office, 1992 (*www.univelt.com*). Illustrations by Dennis M. Davidson, astronomical artist.

increases as we abandon satellites and other spacecraft [40]. China, for example, sent up a guided missile in 2007 to destroy one of its obsolete satellites, thereby causing a serious debris problem as its many parts scattered into the space environment, providing future hazards to space travel. Every spacefaring nation has been negligent about their responsibility for its space trash and rubble. Some 600,000 such objects are now in orbit, according to the ESA's Meteoroid and Space Debris Terrestrial Environment Reference. Our generation is obligated to either remove orbital debris, or at least mitigate its proliferation!

Furthermore, humanity has been remiss in caring for the environment of its home planet. We have befouled and polluted the Earth for centuries, and now face huge problems from global warming, wasteful water practices, and misuse of trees, forests, and oceans. Thus, many worry that humans will now transfer our environmental carelessness and profligate ways to the Solar System. But, as the environmental protection or "green" movement increases globally, there is hope that we may carry this sense of environmental stewardship into orbit. The idea of using space technologies and resources to defend the planet's environment also needs to be

Sec. 3.7] Habitability and the space environment 141

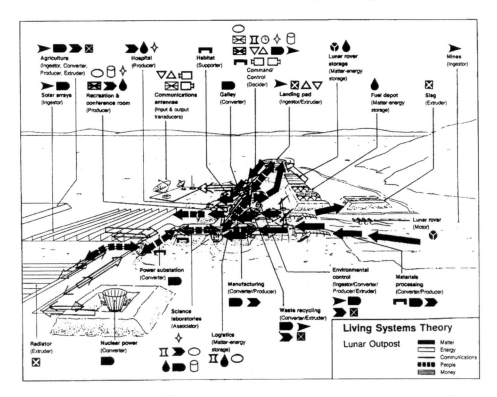

Exhibit 39. Lunar Base: five living systems flows. Source: Miller, J. G. "Applications of Living Systems Theory to Life in Space," *Space Resources* (NASA SP-509), Vol. 4, pp. 231–259). Illustration by Dennis M. Davidson. This and the previous two exhibits lack the original color-coding of symbols.

extended to the preservation of the environment beyond Earth. Organizations like the Natural Resource Defense Council (*www.nrdc.org*) should extend their environmental protection mission offworld. One example is the Sierra Club of San Francisco which published in 1986 the first book to address issues related to the space environment. Entitled *Beyond Spaceship Earth*, the volume represented the concerns of leading thinkers about human impact aloft (*www.UNIVELT.com*).

Another aspect of defending our planetary environment is the problem of objects from outer space striking Earth itself. There are numerous examples of meteorites having struck our planet in the past, altering terrestrial life. Now scientists are trying to prepare against near-Earth objects (NEO) from space striking again, such as asteroids. Exhibit 40 provides some insights on these efforts, especially by detection and deflection.

Two major issues need to be addressed with reference to our planetary environment. One is what is outer space to be used for? Your author agrees with existing space treaties and advocates who maintain that it is part of the common heritage of humankind, and therefore our focus should be on its peaceful use and development.

Exhibit 40. NEAR-EARTH OBJECTS IMPACTING THE ENVIRONMENT.

Since Project Icarus in the late 1960s, the preferred method for dealing with an NEO if it was on an impact course with our planet was to hit it with one or more nuclear weapons. In recent years, however, research into asteroids, including spacecraft missions to or past several asteroids, have challenged that approach. Small celestial bodies are less likely to be integral chunks of rocks than "rubble piles", an agglomeration of smaller objects loosely held together by gravity.

Former astronaut Ed Lu, a director of the B612 Foundation (an organization devoted to the study of asteroid deflection and impact threat mitigation), suggests that if you break a NEO into pieces, you create more space debris on the Solar System scale. He and fellow astronaut Stanley Love, would prefer controlled deflection, and have proposed an alternative, a gravity tractor. This is a spacecraft that would position itself near the asteroid to make a minor change in its orbit, NASA does not favor this "slow push approach", preferring a nuclear detonation above the asteroid threat. At its Jet Propulsion Laboratory, the NEO Program Office has detected about 4,000 NEOs, including 700 large ones with diameters of one kilometer or more. The closest recent encounter came in 2004 when Apophis (99942) was detected and is estimated to have a 2.7% probability of colliding with Earth on April 13, 2029. It is one of 137 risks which JPL estimates have a very small chance of colliding with Earth in the next 100 years! LINEAR in New Mexico and SAFEGUARD in Arizona are two systems currently discovering and monitoring some 100,000 NEOs . In 2005, the U.S. Congress budget authorization bill required NASA to investigate ways to discover 90% of these NEOs by year 2020. An array of ground and space-based shared telescopes would seem to be the best way to discover potentially hazardous objects. However, such enterprises to protect the planet would be very costly and should be done through international partnership.

Source: Adapted from Jeff Foust's *The Three D's of Planetary Defense*, March 19, 2007 (*www.thespacereview.com/article/835/1*).

But, there is another view of space as the "high ground" which must be captured by nations to ensure the protection of its inhabitants. The supporters of this perspective have already wasted millions of dollars on the "Star Wars" program for firing guided missiles, both to and from space, as well as for the proliferation of spy satellites. In 2007, the United States revealed that it is pursuing a multibillion dollar system to develop the next generation of spy satellites. This basic system proposed by the Pentagon planners would invest $2 billion–$4 billion of taxpayer's money for photo-reconnaissance to gather information about adversarial governments and terror groups. Fortunately, many national leaders are questioning the high costs, given the wasteful funding of previous programs of this type. In this misdirection of priorities and resources, citizens wonder if such funds could be spent for better use, such as to address domestic needs, or even to underwrite further utilization of space resources for the benefit of humanity. One can envision this misunderstanding of "defense and security" in space someday escalating into orbital warfare.[5] Planetary defense is more than protecting Earth from near-Earth objects, nuclear catastrophes, and even potential ET aliens—the term needs to be re-defined to emphasize the defense or protection of our environment, whether on Earth or in our Solar System!

The cosmic environment is marked by change and complexity. As knowledge creatures, we must use our talents and resources to further life in the universe, wherever it may be found. Many in the human family are considering how long our tiny and fragile planet will be able to sustain people because of increasing natural disasters, as well as our destructive environmental practices and overpopulation. Some futurists are examining how human culture, economy, and technology can be altered to preserve the quality of *all* life on Earth, and beyond. Still others research the prospects for terraforming other celestial bodies to ensure the survival and progress of our species [41]. Planning and preparations are under way to use the Moon as a new launch pad into the Solar System within 50 years. But, as space experts design a lunar base for the next decade, will they use their technology to enhance the Moon's environment? As is suggested in Exhibit 41, space scientists expect to use lunar regolith and rocks to manufacture everything from radiation shields and habitats, to products hardly imagined at this time.

Now is the time for the global space community to provide safeguards and agreements that protect the offworld environment, beginning with the permanent occupancy of space stations and settlements!

3.8 CONCLUSIONS ON FUTURE RESEARCH DIRECTIONS

Writing on the subject of "Space and Society", Cheston made a case for space social sciences, or *spaceology*, as a new arena of study for integrating the methodologies and insights of mainline disciplines. Subsequently, he elaborated on the concept in this way:

> "The first contours of space social science as a separate entity are only now issuing from the jumble of other disciplines ... Space social science, or spaceology, if we are to coin a word for economy of language, will become a recognized area of study that integrates the methodologies and insights of mainline disciplines ... We might define spaceology as that branch of knowledge that treats of the origin, development and varieties of interactions between human culture and the extraterrestrial environment. Spaceology would draw upon the humanities, social sciences, and natural sciences with equal facility" (Stephen Cheston, "Space

---

[6] For this perspective, see R. C. Anding and T. C. Powell's *An Introduction to Defense: A Study of Warfare Applied to Extraterrestrial Invasion*, 2006 (www.amazon.com), or consult the proceedings of the annual Planetary Defense conferences (www.aero.org/conferences/planetary defense/).

Exhibit 41. Lunar dweller with robotic helper transforming Moon/Mars materials. Source: Artistic rendering by Pat Rawlings. Courtesy of NASA Headquarters Media Services (*www.nasa.gov/multimedia*).

and Society," in Hargrove, E. C. (ed.) *Beyond Spaceship Earth*. San Francisco, CA: Sierra Club Books, 1986, pp. 22–23).

While Cheston envisioned spaceology as a course or curriculum to be taught in higher education, he noted that it might be employed within a consortium of 52 universities, known as the Universities Space Research Association (*www.usra.com*) [42]. It promotes synergistic space projects with government agencies and industry. UNRA's Lunar and Planetary Institute in Houston, adjacent to the NASA Johnson Space Center, also has a division of biomedicine ... Perhaps that is also the place for implementing the following research agenda proposed by conferees at Johns Hopkins University's School of Medicine (see Exhibit 42). In its introduction, Dr. Joseph Brady observed:

> "To some considerable extent at least, the general lack of behavioral science impact on space research initiatives to date can be attributed to the characteristically narrow, oversimplified actuarial approach of the social and psychological disciplines with their traditional emphasis on statistical significance often at

Exhibit 42. BIOLOGICAL AND BEHAVIORAL SPACE RESEARCH CHALLENGES.

(1) To integrate through computerized data banks, worldwide findings and insights of diverse literature on human experience in exotic, isolated, confined environments, whether terrestrial or offworld; then to analyze this environment in preparation for long-duration spaceflight and space settlement.

(2) To further a convergence of worldwide biological and behavioral research within the field of space life sciences on such matters as biospheres and closed systems, habitability and communications, human factors and genetic engineering, telemedicine and orbital health services, radiation and toxicology, immunologic and circadian changes, cardiovascular and regulatory physiology, group dynamics and team building, etc.

(3) To apply such knowledge that may result from the above activities to the development of artificial intelligence and expert systems, diagnostic instruments and simulations, computerized and automated systems which could be used in the assessment and pre-departure orientation of spacefarers, as well as to facilitate their effective performance in orbit and on re-entry to Earth.

(4) To utilize and promote such research as will enhance the *quality of life aloft for spacefarers*, such as by creation of space deployment systems, including comprehensive, in-orbit support services; by designing communications technologies for data collection on human performance in orbit, and recycling of such information in the preparation of future *envoys of humankind* in space.

(5) To foster studies for the creation of a spacefaring civilization, including *governance* of future space societies, human and property rights in space communities, legal/financial systems to facilitate scientific and commercial activities on celestial bodies, relationships between *earthkind* and *spacekind*.

Source: Harris, P. R. *Living and Working in Space*, Second Edition. Chichester, U.K.: Wiley/Praxis, 1996, p. 92 (*www.praxis-publishing.co.uk*).

the expense of biological relevance" (J. V. Brady, *Human Behavior in a Space Environment: A Research Agenda*. Baltimore, MD: Johns Hopkins University School of Medicine, 1982).

Among the insights provided in the above-cited report was the recommendation for establishment of a free-standing Institute for Human Behavior in the Space Environment, in close proximity to spaceflight operations at the Johnson Space Center! There is an idea that the time has more than come for behavioral science and technology data to make significant contributions to extended space occupancy by human systems. Perhaps, then, outer-space studies and research may foster more holistic scholarly efforts in contrast to the present fragmentation so evident throughout the academic and research worlds, as well as through global space agencies!

In another NASA publication cited previously, Cheston, Chafer, and Chafer (1984) insisted that by its very nature, the social science study of space is an interdisciplinary endeavor. They not only called for a coherent course of university space studies in what might preferably be called the *behavioral sciences*—but proposed methods and materials. One recommendation would incorporate futures studies and research methodology in such courses. Space education and investigation might well benefit from the approach used by futurists and strategic planners relative to environmental scanning, identification of trends and issues, and forecasting, along with their tools for surveying, scenario development, and simulation.

Certainly, space is largely a study of the future, of what Thomas Paine, when chairman of the National Commission on Space, described as "the evolution of our species into the cosmos" (*Los Angeles Times*, May 25, 1987, p. 28). Ideally, serious professionals in all fields from anthropology, sociology, psychology, and communications, to economics, political science, management, and law should be addressing how they can apply their expertise to expanding offterrestrial human enterprises in the 21st century. Just think for a moment what the experts in graphic arts, simulations, and gaming might give to space education of the masses. It would be unfortunate for any discipline to miss out in contributing to what Krafft Ehricke so aptly described as the creation of a polyglobal civilization in his "The Anthropology of Astronautics" [43]:

> "The concept of space travel carries with it enormous impact because it challenges Man on practically all fronts of his physical and spiritual existence. The idea of traveling to other celestial bodies reflects to the highest degree the independence and agility of the human mind. It lends ultimate dignity to Man's technical and scientific endeavors. Above all, it touches on the philosophy of his very existence. As a result, the concept of space travel disregards national borders, refuses to recognize differences of historical or ethnological origin, and penetrates the fiber of one sociological or political creed as fast as that of the next."

The 21st century requires integrative and collaborative contributions by both the biological and the behavioral sciences to human habitation of the orbital microgravity environment [44]. The following challenges conclude this chapter with recommendations as to the direction such research should take!
In this millennium, humanity faces the ultimate challenge of whether we can sustain life on our home planet. If not, our species may be forced to seek refuge elsewhere in the Solar System [45]. That is why we should be using space today as a learning laboratory for researching how humans may best live and work in space, as illustrated in Exhibit 43.

THE PROMISE OF ASTROBIOLOGY

Astrobiology is the scientific study of the origin, distribution, evolution, and future of life in the universe. The term was coined by NASA for its multidisciplinary approach to fundamental questions about the living universe ...

Life on Earth has been traced back 3.8 billion years to a period when heavy cometary bombardment brought life-giving water and organic chemicals while it battered the planet with lethal quantities of impact energy ... Researchers are only beginning to probe the adaptability of life to conditions beyond Earth ... Modern science is able to approach this question from many directions. How did life originate on Earth? What are the processes of self-organization that led to the formation of membranes and cells? How did the first living systems acquire the ability to metabolize and reproduce? ... Is there life on Mars or in the ocean of Europa? Where on these planets should scientists search for life and its fossils, and how can they recognize them? Will life be much like terrestrial life in its molecular structure, or will it differ in exciting ways? ...

The first manned outposts now are in orbit, and within the next generation we may move outward to the planets. How will terrestrial life adapt and evolve in extraterrestrial environments? Astrobiology may help scientists chart the future course of life ... (Source: David Morrison, "The Promise of Astrobiology," *Space News*, September 8–14, 2007, 13).)

3.9 REFERENCES

[1] Harrison, A. A. *Spacefaring: The Human Dimension*. Berkeley, CA: University of California Press, 2001, pp. xi–xiii.

[2] Hardy, D. A.; and Moore, P. *Futures: 50 Years in Space, The Challenge of the Stars*. New York: HarperCollins, 2005 ... Zimmerman, R. *Leaving Earth: Space Stations, Rival Superpowers, and the Quest for Interplanetary Travel*. San Diego, CA: UNIVELT/Joseph Henry Press, 2003.

[3] Bullock, A.; and Stallybrass, O. (eds.) *The Harper Dictionary of Modern Thought*. New York: Harper & Row, 1977.

[4] Jackson, M. C.; and Swanson, G. A. "Special Issue: James Grier Miller's Living Systems Theory," *Systems Research and Behavioral Science*, May/June 2006, 445 pp. (official journal of the International Federation of Systems Research published by Wiley Interscience in Chichester, U.K., visit *www.interscience,wiley.com*).

[5] Miller, J. G. "Editorial, Behavioral Science, a New Journal," *Behavioral Science*, **1**(1), January 1956 ... See also Miller, J. G. *Living Systems Theory*. New York: McGraw-Hill, 1978, pp. xv–xvi.

[6] Harrison, A. A.; and Nunneley, S. A. (eds.) *New Directions in Spaceflight Behavioral Health*. Alexandria, VA: Aerospace Medical Association ... *Aviation, Space, and Environmental Medicine*, **76**(6), June 2005, Section II Supplement (*www. asma.org*) ... Lane, H. W.; Sauer, R. L.; and Feeback, D. *Isolation: NASA Experiments in Closed Environmental Living: Advanced Human Life Support System Final Report*. San Diego, CA: UNIVELT/AAS, Vol. 104, 2002.

[7] Connors, M. M.; Harrison, A. A.; and Akins, F. R. *Living Aloft: Human Requirements for Extended Spaceflight*. Washington, D.C.: U.S. Government Printing Office, 1985 (NASA SP-483) ... Kamler, K. *Surviving the Extremes: What Happens to Body and Mind in Extreme Environments*. New York: Penguin Books, 2004, pp. 236–273 ... Aerospace Medical Association, "Medical Guidelines for Space Passengers II," *Aviation, Space, and Environmental Medicine*, **73**, 2002, 11323–11324.

[8] Churchill, S. E. (ed.) *Fundamentals of Space Life Sciences*. Melbourne, FL: Krieger Publishing,/Orbit Series, 1997 ... Pitts, J. A. *The Human Factors: Biomedicine in the*

Manned Space Program. Washington, D.C.: U.S. Government Printing Office, 1985 (NASA SP-4213) ... Nicogossian, A. E. *Space Physiology and Medicine.* Philadelphia, PA: Lea & Feiberger, 1989, Second Edition ... To access NASA human factor studies, contact the Center of Aerospace Information (800 Elkridge Landing Road, Linthicume Heights, MD 21090, or *www.nasa.gov*).

[9] American Institute of Aeronautics and Astronautics, *Space Biology and Medicine.* Reston, VA: AIAA, 1993–1995, 5 vols. (*www.aiaa.org*) ... Aerospace Medical Association Task Force on Space Travel, "Medical Guidelines for Passengers, I/II," *Aviation and Space Environmental Medicine*, **72**, 2001, 948–950; **73**, 2002, 1132–1134; **75**(7), Section II. Alexandria, VA: Aerospace Medical Association (*www.asma.org*).

[10] Bluth, B. J.; and Helppie, M. *Soviet Space Stations as Analog.* Washington, D.C.: NASA Headquarters, 1986 (NAGW- 839) ... Harland, D. A. *The Mir Space Station.* Chichester, U.K.: Wiley/Praxis, 1997.

[11] Harvey, B. *Europe's Space Programme.* Chichester, U.K.: Wiley/Praxis, 2001.

[12] *Journal of Human Performance in Extreme Environments* (HPPE, 2652 Corbyton Court, Orlando, FL 32114; *www.HPEE.org*; email: *society@hpee.org*).

[13] Harrision, M. H. "Space Station: Opportunities for Life Sciences," *Journal of the British Interplanetary Society*, **40**, March 1987, 117–124.

[14] Rummel, J. D.; Kotelnikov, V. A.; and Ivanov, M. V. (eds.) *Space and Its Exploration.* Reston, VA: AIAA Publications, 1993–1995, 5 vols ... Allen, J. *Biosphere 2: The Human Experiment.* New York: Penguin Books, 1995 ... Osland, J. S. *The Adventure of Working Abroad: Hero Tales from the Global Frontier.* San Francisco, CA: Jossey-Bass, 1995.

[15] Woodmansee, L. *Sex in Space.* Burlington, Ontario: Apogee Books, 2006.

[16] Brady, J. V. *Human Behavior in Space Environments.* Baltimore, MD: Johns Hopkins University School of Medicine, 1988 ... "Behavioral Health: The Propaedeutic Requirement," *Aviation, Space and Environmental Medicine*, **76**(6), Section II, June 2005, B13–B24.

[17] Finney, B. R. "Lunar Bases: Learning to Live in Space," in Mendell, W. W. (ed.) *Lunar Bases and Space Activities of the 21st Century.* Houston, TX: Lunar Planetary Institute, 1985, pp. 731–756 ... Harrison, A. A.; and Connors, M.M. "Crew Systems: Integrating Humans as Technical Subsystems in Space," *Behavioral Science*, **39**(3), 1994, 183–212; "Psychological and Interpersonal Adaptations to Mars Missions," in McKay, C. (ed.) *The Case for Mars II.* San Diego, CA: UNIVELT, 1994, Vol. 62, pp. 643–654. The American Astronautical Society through *www.UNIVELT.com* has a Science and Technology Series with numerous volumes on Mars exploration. For example, *Martian Expedition Planning* edited by C. S. Cockell, Vol. 107, 2004 ... *Mars Analog Research* edited by J. D. A. Clarke, Vol. 111, 2006. AAS has also published a series of the *Case for Mars* annual conference proceedings.

[18] Finney, B. R.; and Jones, E. M. (eds.) *Interstellar Migration and Human Experience.* Berkeley, CA: University of California Press, 1985 ... Finney, B. R. "Anthropology and the Humanization of Space," *Acta Astronautica*, **15**, 189–194 (published by Pergamon Press, Oxford, U.K.).

[19] Harrison, A. A. "Beyond Ethnocentrism: Anthropology on the High Frontier," *Space Power*, **7**(3/4), 345–352.

[20] Maruyama, A.; and Harkins, A. *Culture beyond Earth.* New York: Random House, 1975 ... Harrison, A. A. *After Contact: The Human Response to Exterrestrial Life.* New York: Plenum Publishing, 1997 ... *Contact: Cultures of the Imagination*, annual conference proceedings published since 1985 (direct inquiries to Prof. James E. Funaro, Cabrillo College, Aptos, CA 95004) ... Foulkes, R. A. "Why Space? An Anthropologist

Response," *Space Governance*, **1**(2), December 1994, 22–27 (for course curriculum, email *RolandAFoulkes@usa.com*).

[21] Finney, B. R. "Anthropology and the Humanization of Space," in McKay, M. F.; and McKay, D. S.; and Duke, M. B. *Space Resources*. Washington, D.C.: U.S. Government Printing Office, 1992, Vol. 4, *Social Concerns*, pp. 164–188 (NASA SP-509 available at *www.UNIVELT.com*). Also in same volume, Raby, N. "Creating Space Culture: Issues for Consideration."

[22] Santy, P. A. *Choosing the Right Stuff: Psychological Selection of Astronauts and Cosmonautics*. Westport, CT: Praeger, 1994 ... Atkinson, J. B.; and Shafritz, J. M. *The Real Stuff: A History of NASA's Astronaut Recruitment Program*. New York: Praeger, 1985 ... Johnson, M. (ed.) *Cosmonautics: A Colorful History*. Washington, D.C.: Cosmos Books (4200 Wisconsin Ave. NW, Ste. 106-231), 1993, Vol. 1 ... Runge, T. E. "Psychological Factors in Future Space Missions," in Cheston T. S.; and Winters D. L. *Psychological Factors in Outer Space Production*. Boulder, CO: Westview Press, 1983.

[23] Harrison, A. A.; and Summit, J. "How Third Force Psychology Might View Humans in Space," *Space Power*, **10**(2), 1991, 185–203.

[24] Connors, Harrison, and Akins, *op. cit.* [7].

[25] Christensen, J. M.; and Talbot, J. M. "Psychological Aspects of Space Flight," *Aviation, Space, and Environmental Medicine*, March 1986, 203–212 (NASA contract #3924).

[26] Kanas, N. "Interpersonal Issues in Space: Shuttle/Mir and Beyond," *Aviation, Space, and Environmental Medicine*, **76**(6-II), June 2005, B26–B134. "Psychological and Personal Issues in Space," *American Journal of Psychiatry*, **144**(6), June 1987, 703–709 ... Kanas, N.; and Manzey, D. *Space Psychology and Psychiatry*. Dordrecht, the Netherlands: Kluwer, 2003.

[27] White, F. *The Overview Effect: Space Exploration and Human Evolution*. New York: Houghton Mifflin, 1987 ... Harris, P. R. "Managing Transitions and Relocation in the Global Workplace," in Moran, R. T.; Harris, P. R.; and Moran, S. V. *Managing Cultural Differences*, Seventh Edition. Burlington, MA: Elsevier/Butterworth-Heinemann, 2007, pp. 260–303.

[28] Harrison, A. A.; Sommer, R.; Struthers, M. J.; and Hoyt, K. *Privacy and Space Habitat Design Report*. Davis, CA: University of California Psychology Department, 1986 (NASA Grant Report NAG-237).

[29] Carr, G. "Comments of a Space Lab Veteran," *The Futurist*, **15**, 1981, 38.

[30] Rudoff, A. *Societies in Space*. New York: Peter Lang, 1996.

[31] Bluth, B. J. "Sociology in Space Development," *Social Science and Space Exploration*. Washington, D.C.: U.S. Government Printing Office, 1984 (NASA EP-192), pp. 72–79 ... "The Human Use of Outer Space," *Society*, January/February 1984, 31–36.

[32] Bluth, B. J.; and Helppie, M. (eds.), *op. cit.* [10].

[33] MacDaniel, W. E. "Scenario for Extraterrestrial Living," *Space Power*, **7**(3/4), 1988, 365–381 ... "Free-Fall Culture," *Space Manufacturing*. Reston, VA: American Institute of Aeronautics and Astronautics, Vol. 5, 1985 ... *Space Utilization, Sociocultural Problems with Extraterrestrial Populations: A NASA Report*. Niagara, NY: Niagara University, Dept. of Sociology, Space Studies Settlement Project, 1984.

[34] Ortner, D. J. (ed.) *How Humans Adapt*. Washington, D.C.: Smithsonian Institution Press, 1983.

[35] Rudoff, A. "Space Sociology: A Terrestrial Perspective," published paper in the proceedings of the Twenty-third Annual Meeting of the American Astronautical Society, October 1977 (*www.UNIVELT.com*).

[36] Maudlin, J. H. "Reflections on the Sociology of Interstellar Travel," *Ad Astra* (National Space Society magazine), July/August 1995, 48–52.

[37] Miller, J. G. *Living Systems*. Niwot, CO: University Press of Colorado, 1994 (paperback edition available at PO Box 849, zip code 80544) ... See "Special Issue: James Grier Miller's Living System Theory," *Systems Research and Behavioral Science*, **23**(3), May/June, 2006 (*www.interscience.wiley.com*). See there D. Hammond and J. Wilby's "The Life and Work of James Grier Miller," pp. 429–435.

[38] Miller, J. G.; and Harris, P. R. *Living Systems Applications to Space Habitation* (NASA Proposal #886049). La Jolla, CA: University of California-San Diego School of Medicine, 1998/1999.

[39] Miller, J. G. "Applications of Living Systems Theory to Life in Space," in A. A. Harrison, V. A. Clearwater, and L. P. McKay (eds.), *From Antarctica to Outer Space*. New York: Springer-Verlag, 1992 ... Also see "Special Memorial Issue to James Grier Miller's Living Systems Theory," *Systems Research and Behavioral Science*, **23**(3), May/June 2006 (edited by M. C. Jackson and G. A. Swanson) (*www.interscience.wiley.com/journal/srbs*).

[40] Bendisch, G. J. *Space Debris and Space Traffic Management*. San Diego, CA: UNIVELT/AAS, Vol. 112, 2006. This is one of a series of American Astronautical Society volumes offered through *www.UNIVELT.com* on the subject of "space debris" (e.g., conference proceedings: Vols. 100, 1999; 103, 2001; 109, 2004; 110, 2005) ... Campbell, J. D. "Clear Skies: Understanding the Orbital Debris Problem and Its Possible Solutions," *Ad Astra*, **19**(3), Summer 2003, 32–34 (*www.nss.org/adastra*).

[41] Fogg, M. *Terraforming: Engineering Planetary Environments*. New York: Society of Automotive Engineers, 1995.

[42] Cheston, T. S. *The Human Factors in Outer Space Production*. Boulder, CO: Westview Press, 1986.

[43] Ehricke, K. A. "The Anthropology of Astronautics," *Acta Astronautica*, **2**(4), November 1957.

[44] Jakosky, B. *Science, Society, and the Search for Life in the Universe*. Tucson, AZ: University of Arizona Press, 2006.

[45] Calvin, W. "Sustainability on the Home Planet," in Velamoor, S. (ed.) *This Tiny Planet—Workshop Proceedings: The Next Thousand Years*. Bellevue, WA: Foundation for the Future, 2005 (*www.futurefoundation.org*).

Note to readers: On the subject matter of this chapter, see the international journal of Earth–space life support and biosphere sciences entitled *Habitation*, available from Cognizant Communication Corporation (3 Hartsdale Road, Elmsford, New York 10523; tel.: 914/592-7720; fax.: 914/592-8981; email: *cogcomm@aol.com*).

SPACE ENTERPRISE UPDATE

"The militarisation of space: Dangerous driving in the heavens," *The Economist*, January 19th, 2008, pp. 13–14, 25–27:

- The world needs a better code of conduct for spacefarers ... On a good day, spacefaring nations observe certain understandings. On a bad day, it is celestial road rage ... Given the dangers of a clash in space, and the degree to which military and civilian use of space have blurred together, why have the big powers so far failed to negotiate either arms control or simple rules of the road, as they have done on Earth? ... If they cooperate on surveillance of space, the spacefaring countries could also do a much better job of monitoring space debris.
- Space provides the high ground from which to watch, listen, and direct military forces. But the idea that countries would fight it out in space has been confined so far to science fiction. International law treats outer space as a global commons, akin to the high seas. Countries are free to use space for "peaceful purposes" but may not stake territorial claims to celestial bodies or to place nuclear weapons in space. Peaceful has been interpreted as "non-aggressive" rather than "non-military".

Exhibit 43. Work performance in orbit. Source: Original painting by Dennis M. Davidson, and the artwork is owned by Philip R. Harris with rights of reproduction. For other works of this astronomical artist, visit *www.novaspacegalleries.com* or email *info@novaspace.com*).

4

Cultural implications of space enterprise

It is my hypothesis that the fundamental source of human conflict in this new world will not primarily be ideological or economic. The great division among humankind and the dominating source of conflict will be culture. Culture and cultural identities ... are shaping the patterns of cohesion, disintegration and conflict ... Peoples and countries with similar cultures are coming together.

S. Huntington [1]

During the past 50 years, humankind has been extending its presence successfully into outer space either through automation or in person [2]. Landing a "Man on the Moon" through the Apollo 11 mission in 1969 broke our perceptual blinders: in short, we are no longer "earthbound" as our ancestors believed for millennia. Extra Vehicular Activity (EVA) in orbit dramatically reminds us of our new reality, as Exhibits 43 (opposite) and 44 illustrate. Indeed, perhaps the real home and potential of the human species is on the high frontier. Just as application of fire and tools altered our forebears, so space technology and settlement forces modern men and women to change their image of the species. Now we can move beyond the "gravity well", free to explore and utilize the universe to improve the quality of human existence. The technological achievements of global space agencies and private enterprise aloft contribute mightily toward actualization of our potential. Space exploration and exploitation of offworld resources not only alters human culture on Earth—but contributes to the emergence in this 21st century of an entirely new *space culture*, a transformation into a being called *spacekind* or *Homo spatialis* [3].

4.1 CULTURE: A COPING STRATEGY

Culture is a unique human capacity: a coping ability of *Homo sapiens* to the environment which facilitates daily living. Consciously or unconsciously, groups transmit

Exhibit 44. EVA: we are no longer earthbound! Untethered astronaut with a mobile power pack during extravehicular activity in orbit. *Homo spatialis* is emerging! Source: NASA Johnson Space Center.

cultural information and insights to future generations. By understanding the concept of culture, we gain better comprehension of human behavior whether on or off this planet in relation to one's physical environment.

Human beings create culture, or their social environment, in the form of customs, norms, practices, traditions, and taboos for survival and development. The cultural conditioning and lifestyle of particular groups are passed on to descendants. Subsequent generations receive and adapt these "truths" of accepted behavior in society, creating their own situational standards, values, and ethics. Culture is communicable knowledge that is both learned and unlearned, overt and covert in practice, which influences all human systems, including technological. On "Spaceship Earth", culture has been remarkable for its diversity, and those who would be successful here or in orbit need to learn skills to deal with cultural differences and to create cultural synergy [4]. In the last chapter, we listed some of the environmental circumstances that distinguish outer space and influence human adaptations and culture aloft.

Space culture has been in formation for some five decades as humans traveled up more than 100 km into orbit. Depending on one's perspective and academic discipline, we might add to the listing provided previously in Exhibit 20. Certainly, those engaged in life science research aloft have been emphasizing biological factors, while those in the behavioral sciences focus on psychological and sociological influences related to living and working on a space station or lunar base [5]. Although unmanned spacecraft have probed millions of miles around our Solar System [6] only a few hundred humans have actually been in orbit, and these mostly for relatively short durations. For the most part, these pioneering astronauts and cosmonauts came largely from three cultural entities: American, Russian or former Soviet bloc countries, and European. Until the 1980s, these first spacefarers were principally white males from the subcultures of the military and science [7]. The harbingers of future diversity among spacefarers can be seen in the composition of the late *Challenger* Shuttle crew: two women and five men; five Caucasians, one Afro-American, and one Asian-American; six professional astronauts and one civilian, multidisciplinary and multireligious in backgrounds. Also note the diversity in the late crew of *Columbia* orbiter (see Dedication photographs on p. xii). Whether sponsored by space agencies of the U.S.A., Russia, or the European Union, crews are becoming more multicultural and international in make-up (see Exhibit 45). Just as on Earth there are aspects of human experience that cut across cultures, spacefaring is becoming a cultural *universal*, as demonstrated by Japanese, Chinese, and Indians going offworld in greater numbers! It is also evident that the next thousand people in orbit will have to possess greater cross-cultural competencies in dealing with such differences among those in orbit together.

In this century, as we slowly expand our presence upward beyond 100 miles, establish human communities in ever larger numbers, be it on space stations or Moon/Mars bases, the need for more *cultural synergy* is essential! Such synergy optimizes differences in people, fosters cooperation and teamwork, and directs energy toward goal accomplishment and problem solving in collaboration with others. The very complexity of transporting people into space stimulated the development of matrix or team management during the Apollo program. While it takes but hours to place humans in low Earth orbit and only 3 days to reach the Moon, opening up the trade and tourism routes to Mars means venturing many tens of millions of kilometers from the home planet on round-trip journeys of two years or more! That reality is the case for *space as a place for synergy*: only global cooperation will succeed on the space frontier, as our limited experience aloft already indicates. To create successful and productive space habitats and colonies in zero or low-gravity environment will require the practice of both synergistic leadership and multicultural management.

Current research in evolution indicates that harsh environments and new frontiers often result in innovation among species. The pattern of the past reveals

Exhibit 45. Multicultural relations in space. Onboard STS-60, American astronaut Ronald Sega (left) and Russian cosmonaut Sergei Krikalev (right) work on a joint multinational metabolic experiment. Shown on the Shuttle *Discovery*'s mid-deck, this 8-day mission occurred in February 1994. Source: NASA Johnson Space Center.

that creatures are better at inventing and surviving when so challenged to move beyond boundaries, like to outer space. Multinational, multidisciplinary studies are under way to apply terrestrial analogs to offworld living prospects that may accelerate human development. The European Space Agency, for instance, uses the Norwegian Underwater Technology Centre. There scientists from the Russian Institute of Biomedical Problems supervise the ESA astronaut training, such as the German Thomas Reiter, who was aloft for 179 days on Mir. The Japanese were among the first to plan for space tourism, with companies such as Shimizu Corporation designing spaceports and orbiting hotels, while Obayashi Corporation designed a lunar city for 10,000 inhabitants. Unfortunately for the Japanese, they have lost the lead in space tourism to American, Russian, and British entrepreneurs (see Appendix C). Movement from the home planet offworld for whatever reasons, exploration, science, commerce, or settlement, is not only revolutionary—but is transforming our culture and species! Three examples of this are reported in Exhibit 46.

Space scientists, engineers, and supporters are only beginning to appreciate the implications of culture for space developments with reference to:

- human perception and behavior on this planet, particularly in terms of preparation of spacecraft and missions, as well as issues of safety;

Exhibit 46. TRANSFORMATION OF ANCIENT CULTURES.

When humans landed on the Moon in the 1960s, billions of people watched this momentous event on television around the world. Consider how diverse peoples have been affected by the Space Age and its technologies, especially in their perceptions about our home planet.

- *China.* The Chinese culture may go back as far as 4,000 years. A PRC government history project claims the emergence of an early Chinese kingdom during the Stone Age. By the 21st century, China has vigorous space and satellite programs. The nation has orbited its own astronauts, is undertaking major lunar missions, and has signed an agreement with Russia to explore together Mars and one of its moons by 2009. The country is becoming an economic superpower.
- *Native Americans.* In September 2000, 120 locations on Navajo, Hopi, and Havasupai reservations throughout Arizona, New Mexico, and Utah were supplied with new satellite dishes and computer equipment for two-way, high-speed Internet service. In one case that meant that state-of-the-art satellite dishes had to be carried down by packs of mules over eight miles of trails to the bottom of the Grand Canyon! These aborignal peoples are jumping generations of technology in a few hours: they have gone from dial-up telephones that did not work well to mobile phones. The equipment and service provided by Starband is impacting tribal education, law enforcement, health care, entertainment, and many other aspects of Amerindian life. An ancient culture is being enriched for life in this Information Society.
- *United Nations.* UNESCO is using space communications to promote its global missions. Telecommunication enables the agency to transmit a greater volume worldwide of valuable information and images. Thanks to space technology, mass media enhances the rapid spread of education and cultural exchange, particularly in emerging economies. But, developing global culture through space communications requires international cooperation and agreements.

Source: "China Says Its Civilization is 4000-plus Years Old" by Charles Hutzler, Associated Press, November 10, 2000 ... "American Indian Reservations Go Online" by Mindy Sink, *New York Times* News Service, October 6, 2000. Both stories as reported in the *San Diego Union-Tribune* ... *Space Communications and Mass Media*, UNESCO Report #41, 1963 (UNESCO Publications Center, 317 East 34th St., New York, NY 10116).

- alteration of *earthkind* into *spacekind* as a result of living and working in the orbital environment;
- organizational cultures of space agencies and aerospace corporations influence success or failure in the deployment of spacecraft and people into orbit;
- technological systems designed for aerospace flight.

Space technology and achievements also impact cultures on Earth, as the above exhibit demonstrated. Here are more examples of changes brought by space exploration and development:

- the way we perceive our species and planet, especially its environment
- the diversity of spaceflight crews in terms of nationality, ethnicity, and gender
- the expansion of knowledge and expertise, especially with reference to astronomy and life sciences
- the globalization of communication and entertainment, particularly by orbiting satellites
- the multiple applications to improve health care, education, transportation, etc.
- the resources discovered and utilized off the planet helping to increase the quality of life on the Earth.

Another dimension of culture to be impacted by space enterprises is work culture. On Earth today we are in transition from a post-industrial work culture to a knowledge culture [8]. Knowledge economies, workers, and industries are dominant with special emphasis globally on high-tech and information technologies. Such trends will only accelerate as we develop a space culture with its astroscience and astrobusiness. The space frontier will create new knowledge, resources, opportunities, and wealth. Space workers will be forced to innovate and learn new multiple skills. Operating beyond Earth will facilitate self-reliance and equality, as well as democratic practices and institutions. Hopefully, the pioneering settlers will leave behind negative cultural baggage, and create human cultures suitable to the orbital environment and experience.

Chapter 3 suggested that space planners and administrators would benefit immensely by utilizing more effectively the insights and resources of behavioral scientists. Thus, strategic planning, especially for settlements and long-duration missions, would be enriched through a more comprehensive use of *systems analysis*. A special issue on human factors in *Earth–Space Review (ESR)* confirmed these observations [9]. In writing there on "Safety Cultures and the Importance of Human Factors," Dr. John K. Lauber highlighted *ergonomics* as that aspect of human factors concerned about our adaptations of machines to the abilities and capabilities of people. He described current efforts at *crew resource management* with its focus on team-oriented approaches in aerospace systems design ... In the next article in *ESR* on "Cultural Issues and Safety," Capt. Daniel Maurino of the International Civil Aviation Organization admitted that only recently has ICAO examined cultural issues related to accidents and safety management. He wisely observed that aviation technology is designed, culturally speaking, within a narrow segment of international industry, while it is used worldwide without little consideration for cultural differences. Maurino called for research that can help the ICAO develop strategic solutions to cross-cultural safety issues related to high-technology aerospace endeavors ... In the next piece of that *ESR* issue, Paul Sherman and Earl Weiner made a case for studying the interplay between culture and person–machine interaction because aerospace ventures are increasingly multinational and robotic in scope ... Still another article in the same journal on "Implementing Human Factors Training for Space Crews" by a team of psychologists from the German Aerospace Research Establishment (DLR), decried that human factor problems of manned spaceflight until recently focused on biomedical issues, especially related to weightlessness. Manzey, Horman, Fassbender, and Schiewe maintained in another *ESR* feature that the demands of today and tomorrow's space missions necessitate expanding human factor consideration to include behavioral and psychological issues, particularly to address greater crew heterogeneity and the rising risk of psychosocial problems arising from differences in crew roles, tasks, ethnicity, as well as personality attitudes, traits, and values.

Exhibit 47 offers a further example of cultural differences and synergy when space agencies seek to work together.

4.2 EMERGENCE OF A NEW SPACE CULTURE

The above input underscores why your author proposed almost 25 years ago in a NASA Summer Study at the California Space Institute, that a new paradigm is necessary for comprehending human factors in space developments [10]. Exhibit 20 previously described this broader approach that goes beyond technical systems, so important to the engineering subculture. That model in the form of a lunar habitat proposes ten other dimensions to consider and defines each: physical, psychological, sociological, cultural, financial, political, legal, education, management, and communication. Each of these components, explained in Chapter 2, would be important to the development of a spacefaring civilization. For example, under *legal* (#7), a body of space law has been developing with the leadership of the International Institute of Space Law; eventually spacefarers will formulate aloft their own *astrolaw*. Now, United Societies in Space, Inc., as explained in Chapters 7 and 10, proposes that a space *Metanation* be founded under the U.N. Trusteeship Council for New Territories which could act on behalf of the "common heritage of humankind" principle, while administering a series of space authorities. The first might be a Lunar Economic Development Authority with the power to issue bonds that might finance lunar industrialization and settlement (category #5, financial), so as to facilitate private investment [11].

The dozens of inhabitants on orbital platforms, such as Skylab, Spacelab/Spacehab, Salyut, Mir, and now ISS, are precursors of a future human type called *spacekind*. Whether as astronauts or cosmonauts, these *envoys of humankind*, a U.N. designation, learn how to cope with *free-fall culture*. These short-term missions, usually up to a year or less, have been mostly enjoyable and productive, though high-risk experiences. As far as we know only one space mission to date has been aborted by illness when a Russian developed the first psychological case of *space culture shock* (Soyuz T-14, 1985). When Vladimir Vasyutin was reported listless, fatigued, uninterested in his work and spent long hours gazing out the window, ground control ordered the mission terminated and evacuated the station; once back home, the cosmonaut recovered quickly. There will be many more who will suffer

Exhibit 47. SEEKING SPACE ORGANIZATIONAL SYNERGIES.

In the last decade, the American and Russian space agencies began to undertake more joint efforts. As a result, each organization's culture was changed by the other, even among the astronauts and cosmonauts themselves. When female astronaut Shannon Lucid returned on the American Shuttle after six months aloft on Mir with two male cosmonauts, she observed how all three crew members marveled at their shift from antagonism to cooperation, and how they learned about each other's culture and competencies. And the process is expanded as people from some dozen nations collaboratively construct and operate the International Space Station.

One American contractor, Bruce Brandt, emphasized a similar multicultural experience on the ground when his company, Rockwell International, assigned him to Russia, to jointly design a system for NASA's Orbiter to dock safely with the RKA's Mir space station (*Aerospace America*, November 1995, pp. 4–7). Brandt made 30 visits to Kaliningrad on this mission, learning from the many political, technical, and cultural challenges he met in working with Russian aerospace engineers. His and his engineering team experience was that they were:

- initially naive and filled with anxiety of the unknown;
- rushing to conclude business and go back home;
- facing unexpected language barriers requiring special expertise by their translators (e.g., technical terms having different meanings in English and Russian);
- having long cross-cultural discussions with apparent agreement, only later to be surprised by their counterparts' different understanding;
- trying over two years to define mutual roles and responsibilities before RSC-Energia and Rockwell representatives were able to sign a contract allowing the former to produce space hardware for the latter.

Yet, that contract eventually became a prototype for further aerospace agreements between businesses in both countries. The Americans gradually learned they could not impose U.S. government specifications and standards on their Russian contractors. So, Rockwell incorporated the essence of applicable specifications in their Russian procurements, and the two entities became partners in building high-fidelity, integrated engineering units and systems to replicate form, fit, and functions acceptable in the United States. But, the Americans also benefited from cultural learnings about Russian values for both family and workplace, especially the attitude of equality on the job. The "Yankees" were impressed by Russian engineering skills, design elegance, and the quality of docking mechanisms. Professional contacts grew into personal friendships and mutual respect. Thus, both organizations were able to blend their technical expertise and assets for the benefit of humanity!

The cross-cultural enterprise reached a climax when a picture was taken on June 29, 1995 of Astronaut Commander Robert Gibson aligning the *Atlantis* Orbiter with Mir, and shaking hands in the mating tunnel with Cosmonaut Commander Vladimir Dezhurov. The media sent that photograph around the world, symbolic of what international cooperation can accomplish over national conflict. Today, the original partners have been joined by Europeans, Canadian, and Japanese partners in synergistic efforts to complete construction of the ISS, and hopefully soon build a lunar base. In his memoirs, Bruce Brandt summarized the lesson best: "The alternatives to non-cooperation are competition, duplication, and isolation. These lead to mistrust, greed, and wars, both literal and figurative. A far better path is to build an international asset that may be viewed with pride, one that should foster oneness of purpose for years to come!"

Source: Philip R. Harris "Russian/American Cultural Synergy in Space," in Elashmawi, F.; and Harris, P. R. (eds.) *Multicultural Management: Essential Cultural Insights for Global Business Success*. Burlington, MA: Elsevier/Gulf Publishing, Second Edition, 1998, Appendix B, pp. 287–290.

such "culture shock" aloft, a disorientation within an alien space environment. While 20th-century spacefarers were counted in the hundreds, this new millennium will see that number expand to thousands and thousands, as bases are built on planets and asteroids or on orbiting colonies [12].

Actually, present day spacefarers are laying the foundation for tomorrow's space culture. Astronauts and cosmonauts have been forming a culture of high achievement and an ethos of "failure is not an option." But space organizations have to be careful not to promote the super-hero myths that create too much pressure and stress on such human beings with needs and frailties. Those who are now going offworld take pride in their self-discipline and, as astronaut Jerry Linenger remarked, setting "a goal and it's just going, going, and you let nothing stand in your way." Then, when the goal is achieved, doubts set in about what to do next? Do we want to create an orbital culture of overachievers who might suffer personality breakdowns while aloft? Those who sponsor people beyond Earth need to focus on their human relations with one another, so that those in this extreme, isolated environment become supportive and helpful with one another. A spirit of "can do" needs to be balanced with effective teamwork and group concern for each other's wellness. The Russians use the word *aesthenization* to describe mental health problems that may arise on long-duration spaceflights of six weeks or more in orbit. Preventative health care might avoid such breakdowns (see Appendix E).

With permanent extension of human presence beyond Earth, a true space culture will emerge, quite distinct from its terrestrial counterparts. Humanity can learn from the mistakes in its past by utilizing the high frontier as a living laboratory to promote peaceful, synergistic societies. Imagine space communities which have cultural norms that support cooperation instead of competition, group development over excessive individualism, mutual help in place of aggressive behavior. Such a space culture has a better chance for survival and development, in contrast to the 1587 "lost colony" of our English forebears at Fort Raleigh, Roanoke Island, Virginia! The hostile orbital environment requires a collaborative, regulated society to ensure the safety of the commonwealth: individuals will be regulated for safety's sake and that of their community ... Trends all point to the first real space settlement which is likely to take place on the Moon within the next two decades [13]. The lunar environment will cause multiple adaptations of our Earth-bound culture. Remember, it lacks atmosphere and "weather" familiar to *earthkind*; various kinds of cosmic and solar radiation require protective cover: the Moon exposes future *Selenians* or lunar dwellers to 25 rem–50 rem a year and other risks. Despite the Clementine unmanned probe which mapped the Moon in great detail, there is still much about our sister "planet" that we do not yet know: such as its south pole and crater formation, the extent of available water and other natural resources. The rationale for returning to stay on the Moon ranges from scientific to commercial, as well as using it as a launch pad for further exploration of celestial bodies (see Chapter 10 and Appendix B).

Exhibit 48 illustrates ten characteristics of an emerging lunar culture—yes, that is the Moon in the background. Let us briefly review these classifications which can be used to analyze *macrocultures*, such as a nation, or *microcultures*, such as a national space agency.

(1) **Sense of self and space** to be experienced in this extreme environment. Culture contributes to one's sense of self-identity and acceptance, providing a sense of life space or comfort within one's group. Lunar culture offers an environmental reality of no atmosphere and one-sixth gravity, with a large sense of space and vast view of the home planet and the universe. This has been described as the "overview effect" with a perspective of seeing Earth as a single ecosystem [14]. Furthermore, the microgravity environment alters one's sense of space (e.g., what is up and down?).

We can only speculate as to whether a closed or open society will emerge on the Moon, but personally I recommend the latter: an open, friendly, informal, synergistic, and supportive community devoted to common welfare. Lunar pioneers are likely to be self-reliant, competent, high-performing knowledge workers in the beginning. However, we can also expect re-entry problems for long-term settlers on this frontier who wish to return to the home planet.

(2) **Communication and language** to be utilized aloft. Culture groups develop verbal and non-verbal communication systems. Lunar settlers will utilize computers, satellites, and other information technologies extensively in their communications with ground "control" and families, as well as on the lunar surface. In the beginning, multiple languages may be used, principally English and Russian, but in time, terminology will change and expand because of the environmental circumstances (e.g., facility construction, lunar "laws", new terms, and concepts). And when e-spacefarers go to Mars, they will innovate with communication at or beyond line-of-sight distances, modulation frequencies and time message delays, transmitter power, data compression, and network development.

The aerospace industry and agencies have already spawned their own language. The U.S. Air Force published a pamphlet, *Can You Talk the Language of the Aerospace Age?* The 33-page publication has a glossary of terms ranging from *ablation* and *abort* to *zero gravity* and *zero length*.

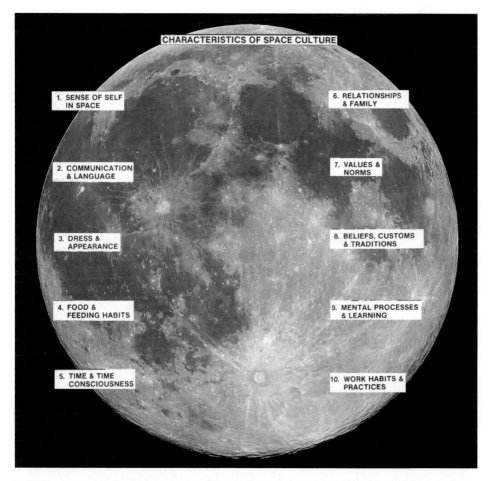

Exhibit 48. Characteristics of space culture. Characteristics of space culture such as will be created on the Moon. Illustrated for the author by Dennis M. Davidson, with a photograph of the Moon as the background. Source: Harris, P. R "Global Leaders and Culture," in Moran, R. T.; Harris, P. R.; and Moran, S. V. (eds.) *Managing Cultural Differences: Global Leadership in the 21st Century*. Burlington, MA: Elsevier/Butterworth-Heinemann, seventh edition, 2007, pp. 6–13 (*www.books@elsevier.com/business*).

Imagine what new words, acronyms, signs, symbols, and gestures that spacekind will develop to communicate more effectively in orbit!

(3) **Dress and appearance** to survive and function. Culture has always been expressed in outward garments, adornments, and decorations. Astronauts and cosmonauts use space suits with life support systems for EVAs and wear mission patches, but clothing design today emphasizes using materials that offer more flexibility and lower costs. For example, the Hamilton Sundstrand space suit is an extravehicular mobility unit which tracks and

164 Cultural implications of space enterprise [Ch. 4

displays vital signs. It has 18,000 parts to help orbiting astronauts to survive with oxygen, water, temperature control, and CO removal. Its eleven layers also enable spacefarers to cope with temperatures ranging from $-250°F$ to $+250°F$, as well as to protect them with a puncture-resistant outer layer that shields against micrometeoroids traveling at 17,000 miles per hour. Its undergarment has 300 feet of thin tubing that circulates cooling water, so temperature can be adjusted by the wearer. A special mesh-lined bag provides drinking water every eight hours in zero gravity. Its helmets have three visors for adjustment to changing light, and its backpack has a complete life support system. No wonder the suit costs millions of dollars!

Lunar dwellers will wear similar clothing and headgear when they go outside the life support and protection of surface vehicles and underground habitats. Such a lifestyle on a sustained basis will alter human culture, as Exhibit 49 illustrates.

(4) **Food and feeding habits** will gradually change in orbit. Food preparation, diet, and eating procedures of groups of people set them apart from one another. Space agencies have pioneered in food technologies and compositions, altering earthkind's eating intake as a result. Transportation costs offworld necessitate advances in hydroponic farming, organic gardens, and seed development which will permit Moon dwellers to sustain themselves on lunar-grown foodstuffs within closed biological systems. Since microgravity affects taste buds, Moon

Exhibit 49. Dress and appearance of humans and robots aloft. Offworld planetary exploration and exploitation requires that spacefarers wear such protective suiting with life support systems, both on their backs or in the robotic vehicles which assist them. Source: Artistic rendering, NASA Johnson Space Center.

dwellers can anticipate new foods, packaging, and preservation, as well as changes in diet, food preparation, and the manner of dining. Today's astronauts on the Shuttle or Space Station are testing out the type of food sustenance to be refined on celestial bodies and for interstellar spacecraft travel.

(5) **Time and time consciousness** to be adopted. Culture influences both one's sense of time and method of tracking it. People in some cultures are very precise about time, while others are more casual. Being "on time" for meetings may also differ, especially in cultures where the young are expected to arrive first and elders last. Earthkind have set their time sense and clocks quite differently: some by the seasons, by daylight and standard time, by an 8-hour or 24-hour day, etc.

Chronologists who specialize in study of the body's internal clocks consider factors like sleeping patterns, body temperature, chemical composition of blood serum and urine, peak periods of energy, and even type of work schedules. Psychosomatic medicine examines human performance in orbit, including how to trick the body's internal clock to suit the needs of long-duration space travelers. Further, after three months in space, research shows that astronauts lose sleep and sleep less soundly because of lack of gravity. Space sleeplessness is caused by changes in the brain's endogenous circadian pacemaker and rhythm. The body clock aloft may be altered by light hitting the eye's retina, thus determining the time of day. Microgravity causes spacefarers to expend less energy than on Earth, and not tire as easily.

Since lunar explorers will be in touch with "ground control", in the beginning it is probable that a 24-hour time system will be adopted, but will it be set to Greenwich mean or Moscow time? Because the Moon has 14 days of light followed by 14 days of darkness, that may influence future concepts of "day, night, and seasons", The lunar environment will affect circadian rhythms, as well as work schedules and sleep patterns. Time schedules affect performance and productivity, so *Selenians* may learn to slow their biological clocks and adapt their biorhythms, creating in the process a new time sense, one likely to be less exact and more relative. Leisure time in lunar conditions offers unprecedented opportunities to develop human potential.

(6) **Relations and families** to be maintained and established beyond Earth, and with people on the ground. Cultures fix human and organizational relationships by age, sex, status, degree of kindred, as well as by wealth, power, and wisdom. The first lunar dwellers are likely to be knowledge workers or *technauts* with expertise in a variety of technologies and disciplines, so professional relationships will be important. Pre-departure training is likely to emphasize team relations and group dynamics, so as to encourage teamwork and camaraderie aloft.

Since the microgravity environment favors the aged and disabled, our attitude aloft toward such persons may change. Many astronauts are in their mid to older years. In 1962, John Glenn made history as the first American to orbit the Earth; in 1998, Senator Glenn again broke records as being the oldest individual in space at age 77. His mission on STS-95 studied the effects of spaceflight on seniors. With 40 years of physiological data on Glenn, NASA considered him to be an ideal subject to analyse his orbital aging and sleeping processes. The University of Texas Medical Branch found this senior astronaut did well on his experiments.

Initially, husband–wife couples may be encouraged to go aloft, but long-term lunar dwellers are likely to alter their understanding of marriage, partnerships, and families. Considerable research is needed on sexual activities and reproduction in weightlessness, as well as the problems of pregnancy in space and differences in children born there ... At the start, support services from Earth will likely facilitate linkages and communication exchanges with families and friends on the home planet. But, as sexual relations and lunar families develop in time, new generations of spacekind will vary significantly both physically and psychologically from earthkind; hopefully, interdependent relationships will prosper among both types of human groups.

(7) **Values and norms** to be cultivated. Based on their needs and philosophy of life, cultural groups have always set unique priorities and standards. On the Moon, the need system will be focused on survival and development until a lunar infrastructure is in place. In such a high-risk and high-performing environment where life support is paramount, competence may become the new norm, regardless of gender, race, color, creed, sexual preferences, or national origins. Thus, spacefarers there may value their fellows and their expertise more than sponsors located remotely on the home planet.

Lunar living may dictate a commitment to shared values. These may include minimizing violence; promoting physical/psychological wellbeing and safety of settlers; guaranteeing a level of social and political justice; maintaining and improving ecological quality and the environment. New consensus will emerge on synergy, harmony, and nurturing within an isolated community, plus we may expect redefinitions of ethics, astrolaw, and governance principles. Terrestrial experience indicates that an effective space culture should value such qualities as openness, flexibility, adaptation, collaboration, equality, and talent. Will such be adopted in orbit?

(8) **Beliefs, customs, traditions** to be fostered offworld. Culture manifests people's attitudes and outlook on life, motivating behavior influenced by spiritual themes, philosophies of life, and moral convictions. With a lunar population that is likely to be international in composition, a multicultural society will develop with diverse belief systems and a new sense of "cosmic consciousness". The view from the Moon will have a transforming effect that questions earthly customs and traditions, substituting mindsets and practices more appropriate to the new order and reality. Living among the stars may encourage spirituality, changing attitudes toward life, death, and the hereafter.

(9) **Mental processes and learning** to be cultivated. The way people think and learn varies by culture because of different emphasis on brain development and education. Space culture may cause humanity to focus on whole—not split—brain development [15]. With a well-educated population at the start, lunar colonists will utilize a wide range of communication technologies to further information sharing and knowledge development. A lunar university and

educational system may draw on terrestrial data transmission and encourage self and small-group learning through media, such as computer-assisted learning, teleconferencing and electronic networking. Cross-training in multiple disciplines may be common, with heavy emphasis on the use of simulations, virtual reality, and artificial intelligence (see Exhibit 50).

One scenario would have the International Space University in France establish an on-line educational capability on the space station and on the Moon as a start (*www.isunet.edu*). Eventually, great world universities might develop offworld campuses and educational programs in orbit ... Another possibility is that terrestrial agencies and organizations might create a single *Unispace Academy*. There both space travelers and their families could receive pre-departure training for both short and long-term assignments aloft. The curriculum should include leadership selection and development, as well as role clarifications and responsibilities. In addition to that ground-based preparation, this global Academy might provide further training and education in orbit. Finally, for those returning to Earth, this facility could offer re-entry orientation and continuing educational services while back on Earth.

(10) **Work habits and processes** to be established in orbit. One way of analyzing a culture is to examine how a society produces its goods and services, and conducts its economic affairs. On the space frontier, the work culture will be *meta-industrial*, featuring extensive use of high/information technology, as well as robotics and automation. Lunar vocational activities are likely to include habitat and facilities construction; human/material transportation; regolith mining, materials processing; helium-3 energy production and power beaming for solar energy; health and emergency services; in addition to scientific research and astronomical observations, whole new industries are likely to be founded (see Appendices). New career and work processes will emerge (e.g., operating space bubble machines and hemispherical concentrators; manufacturing from local resources thin metal mirrors for dish antennas and solar collectors). The principal ally to these multiskilled "ET" workers will be "tin collar laborers" or robots who will one day assist in the construction and maintenance of space stations and orbital complexes (see Exhibit 51). Obviously, offworld technologies will influence emerging space culture.

Of course, we can expect the tourism industry and its allied requirements for services to promote innovative occupations (Appendix C). There will also be a need for "peakeepers", officers concerned about security, safety, conflict resolution, as well as maintaining law and order. In this emerging knowledge culture within an extreme environment, workers will likely have an ethic of competence and high performance in many skills and disciplines [16].

Because culture is multifaceted and pervasive, other categories could be added to these ten classifications. Culture, whether on the ground or aloft, is a complex system of interrelated parts that must be understood holistically. Another model for creating or examining space culture is to use systems analysis. For example, plans for the first lunar settlement might employ these eight classifications of its culture:

Exhibit 50. Education on the Moon. Whimsical artistic rendering of the first educational facility on the Moon attended by *Selenians* (a designation from Greek mythology of *Selene*, the Moon goddess). Realistically, educational buildings there will probably be underground, covered by lunar regolith. The backpacks hold life support and computer systems, not books! Source: NASA Johnson Space Center.

Sec. 4.2] **Emergence of a new space culture** 169

Exhibit 51. Orbital work habits and practices. Astronauts Rick Mastracchio (right), STS-118 mission specialist; and Clayton Anderson, Expedition 15 flight engineer, participate in the mission's third planned EVA (extravehicular activity) as construction and maintenance continue on the International Space Station. The blackness of space and Earth's horizon provide the backdrop for the scene. Source: NASA Johnson Space Center.

(a) *kinship* systems or familial relationships and networks;
(b) *economic* systems or the way that society produces and distributes its goods and services;
(c) *educational* systems or the way people acquire learning and expertise, either formally or informally;
(d) *association* systems or network of social groupings, whether personal or electronic;

(e) *health* systems or the way people encourage wellness, or care for those who are ill and disabled, as with victims of disease, accidents, disasters;
(f) *political and governance* systems or the means for governing, exercising power and authority;
(g) *recreational* systems or the way people socialize and use leisure time, such as through athletics, performing arts, reading, or whatever promotes relaxation and fitness;
(h) *spiritual and ethical* systems, or the way people satisfy their non-physical needs and develop a quality of life, whether through religion, meditation, or other supernatural endeavors.

While the above may be the main considerations in cultural analysis, one could add others for consideration, such as *communication, technical,* and *legal* systems.

It is not acceptable for one form of earthkind's culture to be imposed on spacekind, who will create their own adaptations. Krafft Ehricke, the late visionary rocket scientist, left us a blueprint for a lunar city which he called *Selenopolis* after the Greek moon goddess of mythology. He viewed lunar settlement and industrialization as a process of socio-psychological development and anthropological divergence which would cause humanity to transcend global confines to create a unique biosphere, featuring new life styles and social structures. This, Ehricke concluded, would lead to a *polyglobal* civilization with no limits to growth [17]!

4.3 CULTURAL INFLUENCES ON AEROSPACE ORGANIZATIONS

Culture has already impacted our future in space. One application is the organizational cultures within the principal proponents of space technology and travel [18]. These systems or institutions influence what goals to set, what missions to pursue, what launch systems and spacecraft to use, as well as what people to send, and how they should be trained and supported. Too often, space administrators, scientists, and engineers have ignored cultural factors while designing space vehicles, systems, and programs. The schemata outlined in Exhibit 52 may be applied to the analysis of organizational cultures like those of national space agencies, such as NASA, CSA, ESA, RKA, CCTV, and JAXA. The same dimensions illustrated above also apply to aerospace contractors and private space enterprises, such as Lockheed Martin in the U.S.A., Innovation Enterprise Lunokhod in Russia, CNES/Arianespace in France, Alenis Spazio of Rome, or Spar Aerospace of Toronto. Such institutional cultures contribute to setting objectives, missions, roles, expectations, obligations, and boundaries which affect personnel morale and performance. These pervasive *microcultures* influence decisions, planning, operations, research and development, as well as all activities going into space transportation systems and missions. The space agencies, like their corporate partners, have their unique sets of values, myths, heroes, rites, rituals, and communication systems. NASA, as a case in point, has its own space legends, beliefs, symbols, customs and practices which are embedded deeply in the system. Similarly, in China, the military run the space program with limited civilian

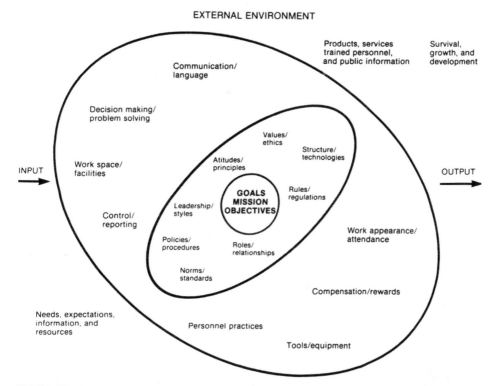

Exhibit 52. Aerospace organizational culture. Whether in space or on the ground, aerospace institutions and teams develop a unique culture. Institutions, like individuals, operate within both a physical and psychological space, indicated above by the circles. Some of its dimensions have been illustrated above for the author by Dennis M. Davidson. Source: Harris, P. R. *Living and Working in Space*. Chichester, U.K.: Ellis Horwood, 1992, p. 143.

or corporate input: all military organizations have unique microcultures. In a number of contemporary management books, the research reported supports the conclusion that excellent organizations have strong functional cultures, constantly subject to renewal. Thus, it would be wise to heed the counsel of management consultants Peters and Waterman when they said [19]:

"In the very institutions in which culture is so dominant, the highest levels of true autonomy occur. The culture regulates vigorously the few variables that do count, and provide meaning. But within those qualitative values (and in almost all other dimensions), people are encouraged to stick out, to innovate."

In the early decades of the Space Age, many public agencies and private corporations in the aerospace field were quite innovative, but in time became overly bureaucratic. In the public sector, the space administrators have found some of their budgets cut and functions stripped away, sometimes given to other government agencies; in the private sector, the aerospace industry has experienced downsizing and restructuring, with more of the innovation coming from start-up entrepreneurial companies. Today most space enterprises, both public and private, are involved in the process of *globalization*, which means more international cooperation in joint ventures and agreements. So, organizational cultures and systems are becoming more *transnational*, each influenced by their counterparts in other countries.

As a demonstration case, consider the increase in multinational agreements among global space administrations, particularly with reference to the building of the International Space Station. Such an undertaking means long-term missions, something the Russians have much experience in because of the Mir station which was in orbit for more than a decade. Thus, the Russian Federal Space Agency has agreements with NASA and ESA, as well as China and India, to exchange personnel and equipment, and to conduct joint projects. But again, the organizational cultures of the partners in each undertaking influence the outcome of such endeavors. Such multicultural interactions among the sponsors of any such space *macroproject* are bound to influence and change the organizational culture of each participant. For example, NASA's culture aimed to have only professional astronauts on ISS, whereas the RKA has been more entrepreneurial, enabling five wealthy civilians to go to the station on their Soyuz TMA spacecraft. With cosmonaut training provided at their Star City near Moscow, the 13 days in orbit cost $25 million and such working guests assist with ISS experiments!

Take another application, the use of behavioral scientists in planning and participating in space missions. As indicated in the previous chapter, NASA's organizational culture previously downplayed the role of psychiatrists, psychologists, sociologists, and anthropologists, and focused on physical issues with the help of medical personnel and biologists. Now that the "hard science/engineering" subculture's influence is diminishing as the U.S. agency places more emphasis on life science and biospheric research. NASA's culture is further impacted by the outlook and interaction with its new Russian partners. Research psychologist Vadim I. Gushin from Moscow's Institute for Biomedical Problems observed:

"The Russian system of spaceflight psychology was created to give cosmonauts psychological support based on observed data of crew members' psychological state and work capability ... Information about the cosmonauts' interests, tastes, family, and friends forms the basis for psychological support measures in space ... There are also psychological experiments using questionnaires and special devices to test on board" [20].

But Gushin admitted that cosmonauts, like their astronaut counterparts in the West, try to avoid inflight psychological investigations and that the data collected are largely anecdotal. This Russian scientist advocates making crew members allies of behavioral researchers by training them through preflight simulations in astute observations of symptoms among their colleagues which might undermine safety, health, performance, or mission success (see Exhibit 53).

Exhibit 53. Learning from Russian spaceflight experience. With their Mir space station in orbit for well over a decade, we have much to learn from Russian long-duration missions. Here we see (left to right) cosmonaut Pavel V. Vinogradov, Mir-24 flight engineer, cosmonaut Salizan S. Sharipov, Shuttle payload specialist representing the Russian Space Agency, cosmonaut Anatoliy Y. Solovyev, Mir-24 commander (wearing the space helmet), and astronaut Andrew S. W. Thomas. Source: NASA Johnson Space Center.

My point is that NASA and RKA organizational cultures are mutually influenced by one another when the two agencies engage in joint ventures, or with other international partners. The same might be said for other space agencies in the world who form partnership agreements. As space organizations go global, they become less provincial and myopic, more cosmopolitan and ready to learn from their "foreign" allies, as Exhibits 47 and 53 confirm. And the future for such cultural synergy is promising. The planned retirement of the American Space Shuttle fleet by 2010 means that Russia will become the principal carrier of crews and cargo to the International Space Station until NASA's new Orion capsule is ready by 2015. By then the reliable Soyuz spacecraft will need replacement after 50 years of service. RSC Energia proposes to build a reusable orbiting Kliper, a six-seat, more comfortable craft that could cut the costs of space transportation by two-thirds. But, like NASA, will their government fund five such vehicles at $1.5 billion?

To ensure successful spaceflight, NASA culture was once renowned for high standards of safe and efficient operations. What was questionable in that culture was the emphasis on excessive redundant systems. But organizational culture impacts relations and activities between a space agency and its contractors. During the Apollo mission days, contractors' use of matrix management also affected NASA administrative practices. More recently, investigations into the loss of two orbiters and their crews revealed deficiencies in Agency culture, prompting changes in work policies and the work environment. The first catastrophic event was the Shuttle *Challenger* which blew up after launch in January 1986, with the loss of seven astronaut lives. The complex investigation by outside experts centered on the primary cause: the failure of O-rings that were supposed to seal various parts of the solid rocket booster, but did not because of cold weather and ice conditions at the Kennedy spaceport which contributed to their disintegration. But their comprehensive report also uncovered failures in the management and engineering cultures of both the Agency and its prime contractor, Morton Thiokol. Launch decision-makers failed to receive or heed the warnings of their concerned engineers at lower levels of the organization. NASA's bureaucracy with so many rules and regulations, as well as worker specializations, hampered information from getting to top management, hindering their timely response. So, administrators and managers promised to correct the technical causes for the disastrous mishap, and reform their organizational cultures. But the organization could not control political leaders in Congress who did not provide sufficient funding to properly maintain the aging Shuttle fleet, so as to meet the schedule for completing the ISS.

Similarly, the independent board which examined the Shuttle *Columbia* tragedy in 2003 again found problems with NASA management culture, especially related to information flow and communications about safety and risk problems from engineers at lower levels. With regard to the horrendous loss of the STS-114 orbiter and crew of seven, the inquiry report made five technical recommendations for improved Shuttle operations, as well as correction of specific management failures. During the period when the whole Shuttle fleet was grounded, NASA then made major cultural changes and organizational progress in risk reduction. Behavioral Science Technology, a private company contracted to fix the culture at various installations, confirmed this in a 2005 worker survey, but their findings revealed that many technicians were still afraid to speak their minds on such issues. Some still feared reprisal if they challenged center management, and complained of stress caused by competition for jobs and resources among the Agency centers. Overall, though, the Agency is transforming its culture and is taking on new missions, but this is still a work in progress. People were finding it easier to bring up bad news and to get the problem fixed, while minority opinions were being sought at meetings ... At the end of that same year, an Agency Management Conference of risk managers, workers, and former astronauts agreed that headway was being made in improving organizational communications across the centers, but information still did not filter down to personnel below the management level. With a mandate to return to the Moon permanently by 2020, one chief engineer in charge of safety-minded Technical Authority, Christopher Scolese, worried about individual workers "buying in" to safety measures at the grass-root level. John Tinsley, mission support director for NASA's Office of Safety and Mission Assurance, then added that their focus should also be as much on the quality of performance and hardware in their complex systems, as it now is on safety [21]. The conclusions of a book on *Reinventing NASA* have been partially repeated in Exhibit 54: however, the book ignored the whole issue of Agency culture!

Exhibit 54. RENEWING NASA CULTURE.

Agency culture. The largest point being exercised here is that reforms imposed over the past decade or so are drastically altering how NASA conducts its business. That surface reality though has obscured the underlying continuity with which the Agency aggressively pursues the larger agenda. There is no deeply hidden master plan passed on by secret ritual; rather NASA since its inception has been, comparatively speaking, up front regarding its active pursuit of human spaceflight and beyond. Its adaptability in pursuing that goal has been held. Repeated congressional and presidential frustration of NASA's efforts have not deterred that thrust from continuing as the Agency's major priority ...

NASA's difficulties have been compounded by its certainty (others say arrogance) that the organization alone truly knows the best way to achieve that illusive goal and remains the agent best equipped to do so. (Source: Roger Handberg, *Reinventing NASA: Human Spaceflight, Bureaucracy, and Politics*. Westport, CT: Praeger, 2003.)

Astronaut Corps culture. In a 2007 report on astronaut health, a special panel appointed by NASA revealed that, on two occasions, astronauts were allowed to fly after flight surgeons and other astronauts warned that the individuals were so drunk, they posed a safety risk. Despite the 12-hour "bottle to throttle" prohibition, the investigators found "heavy use of alcohol" prior to launch. The Agency is still questioning the accuracy of these anecdotal reports, and its own internal investigation found no support for these accusations. Indeed, such deviant behavior was out of character from most previous spacefarer flights. However, if flight surgeons' concerns over astronaut-impaired performance were ignored by high-ranking NASA officials, then that points to weakness within the Agency's culture. The medical specialists reported that all NASA leadership wants is a "go" signal for flight. They do not want to hear of crew unfitness for duty or behavioral problems.

This is the same organizational syndrome evident in *Columbia*'s final flight: the bosses only wanted to hear positive news about fuel-tank insulation foam, and so seven astronauts were killed. One of the investigators of that disaster, Nobel prize–winning physicist Douglas Osheroff explained it: "NASA has had a history of ignoring indications that something is wrong, and even though the odds were with NASA, they have lost ... It always seems to come down to schedule and pressure to launch which led in large part to *Columbia*'s demise."

In the "brotherhood of space" many fail to speak up for fear of ostracism by colleagues. As a result of reforms, in the flight-readiness reviews conducted before shuttle launches, engineers and surgeons have been encouraged to speak up, even with dissent, on matters of flight safety. It's hard to bring about change in the Agency's 50-year-old culture and traditions.

Source: Marcia Dunn, "Claims of Drunken Astronauts Rattle NASA Brass," Associated Press, July 27, 2007; "NASA Problems Often Arise from Disregarding Underlings," Associated Press, July 29, 2007, as reported in the *San Diego Union-Tribune*.

For NASA to achieve its present goals of planetary exploration and human settlement on the Moon and beyond, then the Agency will need more political autonomy, a five-year budget allocation, and increasing involvement with private enterprise beyond the traditional aerospace industry ... But other national space agencies are also seeking to transform their organizational cultures. Russia was conditioned by almost 90 years of centralized planning. When the Soviet Union collapsed in 1991, its space organizations suffered severe cuts in financial support and public adulation. It pushed RSC Energia toward the market economy and entrepreneurialism, as well as more international collaboration. Currently, the Japanese are struggling to formulate a new JAXA organizational culture out of the remains of two previously independent space institutions. Similarly, ESA continues to formulate a unique multicultural organization culture, made up of the national cultures of the European Union members, as well as coping with organizational culture of various national space agencies.

Once again, the culture of any institution impacts behavior and performance within any system, so leaders in space enterprises have a serious responsibility to ensure a satisfactory and productive work environment! This is a critical factor because space exploration and development is a high-risk, complex, and costly endeavor as Exhibit 55 attempts to illustrate.

4.4 COSMIC CULTURES

Presently, there are two schools of thought concerning extraterrestrials: (1) eventually we will encounter other species and cultures within the vast universes; or (2) humans will evolve into these diverse ETs with their unique cultures, beginning on the Moon and Mars. The scientists and supporters of the Search for Extraterrestrial Intelligence (SETI) of Mountain View, California, take the former position (*www.teamseti.org*) [22]. Annually, they have brought together space scholars and enthusiasts with anthropologists to discuss future encounters with peoples from offworld exotic cultures (*www.contact[-conference.org*). The SETI members seek to uncover the universals of culture, so as to construct messages that might be intelligible to independently evolved civilizations. Thus, SETI researchers have erected radio transmitters in Puerto Rico that send interstellar signals which they hope may be received by a society in other cultural worlds out there. Should they exist and receive our signals, how can we communicate further? Some maintain through mathematics, while cognitive scientists argue that mathematics is an artifact of human culture. Our senses or capacities for vision, space, time, hearing, smell, etc., may be quite different from such aliens. However, the SETI community is convinced that their experimental searches of the Galaxy are worthwhile, and their speculations will help us deal better with spacekind—even if it is us!

As we develop hypotheses about extraterrestrial biological and social entities, psychologist Albert Harrison believes that it should be done within the framework of science, viewing the evolution of life and civilizations [23]. For this, he recommends using the insights of living systems theory of the late James Grier Miller as a point of

Exhibit 55. Work culture on the Moon. Working on or under the lunar surface will not only be risky, but technically oriented. This artistic rendering not only depicts the constant view of Earth in the background, but the need for a strong interdependent team culture among the "technauts" there whose lives and welfare depend on each other's morale and performance, as well as robotic helpers. Source: NASA Headquarters, Vision for Space Exploration Office.

departure across species and cultures. LST stresses continuities among the physical, biological, and social sciences within the context of analyzing biosocial systems of different size and complexity. This approach is useful for future comparative studies of lifeforms and cultures, whether on the Earth or off this planet. In humanity's journey among the stars, Dr. Harrison speculates that, in time, extraterrestrials may eventually help us solve problems of earthkind. Since the 1960s, astronomers like Frank Drake and Carl Sagan were convinced that life is a frequent rather than a rare occurrence in our universe, so it is plausible that in other places in the Galaxy and Solar System, such life might contact and communicate someday with us!

4.5 CONCLUSIONS ON EMERGING SPACE CULTURE

Culture is a powerful concept and influence on the future of space programs and development of the high frontier. Those engaged in human enterprises aloft would benefit from social science knowledge and research on the subject. Specifically, behavioral scientists, especially cultural anthropologists and organization development specialists, can make contributions to the global space program by

(1) *Increasing research on intercultural relations on the International Space Station.*[1] This facility can become a human relations laboratory engaged in constructing prototypes for culture and behavior that will be likely a decade from now at a lunar base.
(2) *Planning for the preliminary cultures of space settlements*, beginning with communities to be formed around 2020 at a lunar base and industrial park.
(3) *Consultation with aerospace agencies and corporations* on renewing their organizational cultures, so as to be more effective in space industrialization and colonization through inter-institutional cooperation.
(4) *Assisting aerospace designers and engineers* to understand *how* (a) cultural differences between designers and users, particularly when the latter are "foreigners", may undermine safe operations of spacecraft and facilities; (b) complex automated systems reflect cultural assumptions that may lead to conflict, misunderstandings, and mismatch between the systems' and users' cultures which may increase risk and even lead to tragic failures [24].

To this writer, it seems unlikely that global space agencies will provide the initiative and innovation required to fully promote space settlement and commercialization. But, it is more likely that private enterprise and non-profit organizations will be the force behind this, particularly in the formation of space culture. Furthermore, catalysts for lunar activity will likely come from world commerce, which appreciates the return on investment that can be realized by utilization of space resources, such as space-based energy. Your author agrees with Professor Haym Benaroya of Rutgers University, previously cited, who concludes that space and planetary science, concerned about the accumulation of knowledge and understanding, must be decoupled from space engineering and industrialization which have more specific, immediate goals and realistic schedules; they are two different—yet parallel—cultures.

The payback for investing in space exploration. Those within the space community

---

[1] For further information on this see J. B. Ritter's "Cultural Factors and the International Space Station," *Aviation Space and Environmental Medicine*, **76**(6), June 2005, B135–B144 ... Sindal, G. M. "Culture and Tension during Intercultural Space Station Simulations," *Aviation, Space, and Environmental Medicine*, **75**, 2004, C44–C45 ... Kraft, N. O. "Intercultural Issues in Long-duration Spaceflight," *Aviation, Space, and Environmental Medicine*, **74**, 2003, 575–578 ... Kring, J. P. "Multicultural Factors for International Spaceflight," *Journal of Human Performance in Extreme Environments*, **5**, 2001, 11–32 .... Leon, G. R. "Women and Couples in Isolated Extreme Environments: Applications for Long-duration Missions," *Acta Astronautica*, **53**, 2003, 259–267 ... Santy, P. A. *et al.* "Multicultural Factors in in the Space Environment," *Aviation, Space, and Environmental Medicine*, **64**, 1993, 196–200 ... Kelly, A. D. "Crewmembers Communication in Space: A Survey of Astronauts and Cosmonauts," *Aviation, Space, and Environmental Medicine*, **63**, May 1992, 721–724.

worldwide must recognize that we still have global cultures which are terrestrially oriented. Even though there is some enthusiasm for space films and television programs, opinion polls confirm public opinion is not generally supportive of investing in outer space. The average person does not perceive space as an investment whose resources will have a vast payback. The media and political leaders can do much to educate humanity as to the necessity of going offworld, and the values of developing a twin-planet economy of Earth–Moon. Modern marketing methods can do much to help people appreciate the benefits that space activities can bring to them. The public, for instance, is intrigued with the beginnings of space tourism. Space activities need to engage two institutions in supporting space developments. One is education where professors and teachers are preparing the next generation of spacefarers. The other is professional societies and associations regardless of the discipline—such profession or career will be eventually transformed through its "astro" applications and occupations. The field of medicine is one example where physicians and researchers have embraced *aerospace medicine* through their specialized professional organizations, conferences, and publications. Civil engineering, like aerospace engineering, is beginning to explore its future role in the construction and maintenance of space infrastructure. The travel and energy industries are also probing space prospects. With the expansion of knowledge on this new frontier, every type of vocational pursuit should be exploring their opportunities beyond Earth! Then we will begin to create an interplanetary culture [25].

Finally, perhaps the sheer challenge and complexity of space migration may turn the human race and cultures away from inward, self-destructive tendencies toward outward self-actualization. *Human emergence* may truly occur when we leave our cradle Earth, to settle our Solar System [26]! "Crawling" began a few hundred kilometers upward into low-Earth orbit (LEO), and "walking" begins when our species regularly, safely, and economically extends its personal presence 36,000 kilometers above into geosynchronous orbit (GEO). By practicing cultural synergy, we earthlings may mature through the formation of space civilizations that permit us to step into the universe and a new state of being!

The overall goal for the study of human behavior in space is the development of empirically based scientific principles that can identify the environmental, individual, group, and organizational requirements for long-term occupancy of space by humans.

Committee on Space Biology and Medicine. *A Strategy from Space Biology and Medical Science.* Washington, D.C.: National Academy Press, 1987, p. 169.

4.6 REFERENCES

[1] Huntington, S. *The Clash of Civilizations and the Remaking of World Order*. New York: Simon & Shuster, 1996.

[2] The American Astronautical Society has a 27-volume series on the history of spaceflight. Contact *www.UNIVELT.com* for a listing of available titles ... Neal, V.; Lewis, C. S.; and Winter, F. H. *Spaceflight: A Smithsonian Guide*. Washington, D.C.: Smithsonian Institution Press, 1995 ... Freeman, M. *How We Got to the Moon: The Story of the German Space Pioneers*. Washington, D.C.: 21st Century Science Associates, 1993 ... Surkov, Y. A. *Exploration of the Terrestrial Planets from Spacecraft*. Chichester, U.K.: Wiley/Praxis, 1997.

[3] Finney, B. R.; and Jones, E. M. (eds.) *Interstellar Migration and the Human Experience*. Berkeley, CA: University of California Press, 1985 ... Robinson, G. S.; and White, H. M. *Envoys of Mankind: A Declaration of Principles for Space Societies*. Washington, D.C.: Smithsonian Institution Press, 1986 ... Harrison, A. A. *Spacefaring: The Human Dimension*. Berkeley, CA: University of California Press, 2001, ch. 14.

[4] Moran, R. T.; Harris, P. R.; and Moran, S. V. *Managing Cultural Differences: Global Leadership Strategies for the 21st Century*. Burlington, MA: Elsevier/Butterworth-Heinemann, 2007, Seventh Edition (*www.books.elsevier.com/business*). Check also the other 12 titles in the Managing Cultural Differences Series.

[5] Sulzman, F. M.; and Genin, A. M. (eds.) *Life Support and Habitability: Space Biology and Medicine*. Reston, VA: American Institute of Aeronautics and Astronautics, 1994, Vol. II ... Tascone, T. F. *Introduction to the Space Environment*. Melbourne, FL: Orbit Books/Krieger Publishing, 1988 ... Freeman, M. *Challenges of Human Space Exploration*. Chichester, U.K.: Wiley/Praxis, 2000.

[6] Ride, S.; and O'Shaughnessy, T. *Voyager: An Adventure to the Edge of the Solar System*. New York: Science Publishing, 2006 (especially suitable for children).

[7] Atkinson, J. D.; and Shafritz, J. M. *The Real Stuff: A History of NASA's Astronaut Recruitment Program*. New York: Praeger Publishing, 1985 ... Harvey, B. *Race into Space: The Soviet Space Programme*. Chichester, U.K.: Ellis Horwood Publishing, 1988 ... Oberg, J. *Star-crossed Orbits: Inside the U.S. Russian Space Alliance*. New York: McGraw-Hill, 2002.

[8] Harris, P. R. *The New Work Culture ... Managing the Knowledge Culture*. Amherst, MA: Human Resource Development Press, 1998, 2005.

[9] "Special Issue on Human Factors," *Earth–Space Review*, **1**, January/March, 1995 (published by Gordon & Breach Science Publishers, PO Box 90, Reading, Berkshire RG1 8JL, U.K.) ... Lane, H. W.; and Schoeller, D. A. *Nutrition in Spaceflight and Weightlessness*. Washington, D.C.: CRC Press, 2000.

[10] Harris, P. R., "Impact of Culture on Human and Space Development," *Acta Astronautica*, **36**(7), October 1995, 399–408 (journal of the International Academy of Astronautics).

[11] O'Donnell, D. J. "Founding a Space Nation Utilizing Living Systems," *Behavioral Science*, **39**(2), April 1994, 93–116 (journal of the International Society for Systems Science) ... O'Donnell, D. J.; and Harris, P. R. "Space-Based Energy Needs a Consortium and a Revision of the Moon Treaty," *Space Power*, **13**(1/2), 1994, 121–131 (journal of the SUNSAT Energy Council) ... O'Donnell, D. J.; and Harris, P. R. "Strategies for Lunar Development and Port Authority," *Space Governance*, **2**(1), June 1996 (journal of United

Societies in Space, Inc., USIS, 499 Larkspur Rd., Castle Rock, CO 80104; email: *djopc@quest.net*).

[12] O'Neill, G. K. *The High Frontier: Human Colonies in Space*. Princeton, NJ: Space Studies Institute (PO Box 82, zip code 08542) ... Savage, M. T. *The Millennial Project: Colonizing the Galaxy in Eight Easy Steps*. New York: Little, Brown & Co., 1994 ... Krone, R.; Morris, L.; and Cox, K. (eds.) *Beyond Earth: The Future of Humans in Space*. Burlington, Ontario: Apogee/CG Publishing, 2006.

[13] Schrunk, D.; Sharpe, B.; Cooper, B.; and Thangavelu., M. *The Moon: Resources, Future Development, and Settlemennt*. Chichester, U.K.: Springer/Praxis, 2008, Second Edition (*www.praxis-publishing.co.uk*) ... Mendell, W. W. (ed.) *Lunar Bases and Space Activities for the 21st Century*. Houston, TX: Lunar Planetary Institute, 1985.

[14] White, F. *The Overview Effect: Space Exploration and Human Evolution*. New York: Houghton-Mifflin, 1987 ... Hsu, F. "Managing Risks on the Space Frontier: The Paradox of Safety, Reliability and Risk," in Krone, R.; Morris, L.; and Cox, K. (eds.) *Beyond Earth*. Burlington, Ontario: Apogee/CG Publishing, 2006.

[15] Springer, S. P.; and Deutch, G. *Left Brain, Right Brain*. New York: W. H. Freeman & Co., 1993, Fourth Edition.

[16] Harris. P. R. *The Work Culture Handbook*. Mumbai, India: Jaico Publishing, 2003 ... Shusta, R. M.; Levine, D. R.; Wong, R. Z.; Olsen, A. T.; and Harris, P. R. *Multicultural Law Enforcement: Strategies for Peacekeeping in a Diverse Society*. Upper Saddle River, NJ: Prentice Hall, 2007, Fourth Edition (*www.prenhall.com/criminaljustice*).

[17] Ehricke, K. A. "Lunar Industrialization and Settlement: Birth of a Polyglobal Civilization," in Mendell, W. W. (ed.) *Lunar Bases and Space Activities of the 21st Century*. Houston, TX: Lunar Planetary Institute, 1985, pp. 827–855.

[18] Schein, E. *Organizational Culture and Leadership*. San Francisco, CA: Jossey-Bass Publishing, 1985.

[19] Peters, T. J.; and Waterman, R. H. *In Search of Excellence*. New York: Harper & Row, 1982.

[20] Gushkin, V. I. "Problems of Psychological Control in Prolonged Spaceflight," *Earth–Space Review*, **4**(1), January/March 1992, 28–31 ... Kanas, N. "Psychological Support for Cosmonauts," *Aviation, Space, and Environmental Medicine*, **62**(3), May 1991, 53–5.

[21] Dunn, M. "Skepticism Remains as NASA Makes Progress on International Culture," Associated Press, 2/21/05 ... Malig, T. "Work Remains for NASA Culture Change," *www.spacecom.com/news*, 12/7/05 ... Handberg, R. *Reinventing NASA: Human Spaceflight, Bureaucracy, and Politics*. Westport, CT: Praeger, 2003.

[22] Vakoch, D. "Culture in the Cosmos," *www/space.com/searchforlife/seti_culture_070503.html*, May 2, 2007 ... Maruyama, M. "Settlements in Space," in Fowles, J. (ed.) *Handbook of Future Research*. Westport, CT: Greenwood Press, 1978.

[23] Harrison, A. A. *After Contact: The Human Response to Extraterrestrial Life*. New York: Plenum Publishing, 1987.

[24] Sherman, P. J.; and Weiner, E. L. "At the Intersection of Automation and Culture," *Earth–Space Review*, **4**(1), January/March 1995, 11–13.

[25] Blagonravov, A. A. (ed.) *Collected Works of K. E. Tsiolkovski*, Vols. 1, 2, 3, NASA TT-F-236/237/238. NASA Center of Aerospace Information, 1965 (Document Request, 800 Elkridge Landing Rd., Linthicum Heights, MD 21090; fax.: 301/621-0234; order #N65-21736/32975 and 32765).

[26] Harris, P. R. *Toward Human Emergence*. Amherst, MA: Human Resource Development Press, 2008 (*www. hrdpress.com*).

SPACE ENTERPRISE UPDATE

Change in NASA's culture is evident in two environmental assessments for 2008. One was on the potential impact on people, communities, and Agency field centers resulting from the retirement of the Space Shuttle Program in 2010. Another purpose was to ascertain the effects of disposal of Shuttle assets, such as real property (buildings, structures, and land), and personal property (flight hardware, parts, and materials). The second assessment was relative to the Shuttle's replacement and transition to the Constellation Program (the NASA program to create a new generation of spacecraft for human spaceflight, consisting primarily of the Ares I and Ares V launch vehicles, Orion crew capsule, and Altair lunar lander), including the environmental impact of its preparation and implementation. Public comments were invited on the drafts for both reports (*www.nasa.gov/mission*).

5

High-performing spacefarers

When we look into the sky, it seems to us to be endless. We breathe without thinking about it, as is natural. We think without consideration about the endless ocean of air. Then you sit aboard a spacecraft, you tear away from Earth, and within ten minutes you have been carried straight through the layer of air, and beyond there is nothing! Beyond the air, there is only emptiness, coldness, darkness. The boundless blue sky, the ocean which gives breath and protects us from endless blackness and death, is but an infinitesimally thin film. How dangerous it is to threaten even the smallest part of the gossamer covering this conserver of life!

Vladimir Shatalov, Russian Cosmonaut

Human performance in space, it is worth repeating, continues to be accomplished in two ways. The first is *unmanned* by extending ourselves out in the universe through automated spacecraft [1]. The premier achievement to date is undoubtedly the grand planetary tour of the two Voyager spacecraft. Leaving our Solar System at over 50,000 kilometers per hour, these space vehicles have traveled more than 8 billion kilometers from our earthly home where no member of our species has yet gone. The odyssey began over 30 years ago, in 1977, when NASA's Jet Propulsion Laboratory scientists launched the most sophisticated robot spacecraft ever built, each weighing nearly one ton at launch and made up of 65,000 parts. Of the two spacecraft in the $865 million joint mission, Voyager 1 flew past Jupiter and Saturn, making a close pass of Saturn's largest moon, Titan, and was allowed to end its planetary sojourn by zipping out to the stars. Voyager 2 continued outward using the gravitational field of Saturn to fling it on to an encounter with Uranus in 1986. As a result of thousands of images sent back by Voyagers 1/2 from throughout the Solar System, 20 new moons were discovered. The project scientists, under the leadership of chief scientist Edward Stone, vice president of Caltech, have been virtuosos of these complex computerised space probes, each with circuitry equivalent to 2,000 color television sets. Through this unique human–machine interface, their explorations expanded eventually to Neptune and Triton, in 1989, from which radio transmissions traveling at 300,000 kilometers per second reached JPL in Pasadena, California, four hours and six minutes later! Planetary scientists and astronomers extend our horizons beyond our Earth into the universe. For example, as this is being written, the media are reporting another planet discovered trillions of kilometers away which is *earthlike* in its environment: average temperatures are estimated to be between 0°C and 40°C, and any water there would be liquid. But, with our present spacecraft it would take tens of thousands of years to reach that celestial body!

But, marvellous human-mechanical performance feats are being continually improved on. As Exhibit 56 illustrates, the journey begins in Earth orbit.

Another milestone was ESA's Ulysses launched in 1990 from the Shuttle *Discovery* and monitored by NASA ground-tracking facilities in Madrid, Spain, and Goldstone, California. Following an encounter with Jupiter, this unmanned spacecraft was catapulted into an orbit from which it could survey the Sun's high-latitude regions and poles. It completed its northern polar pass in the summer of 1995, having made its first, southern polar pass a year earlier. It gradually descended in solar latitude, traveling out to Jupiter's orbit by 1998. Then it headed back toward the Sun on a high-latitude trajectory, returning to the south polar regions in late 2000 and early 2007, with further flights over the north pole in 2001 and early 2008. What a tribute to the human mind which conceived and executed this flawless mission! ... In November 1995, the European Space Agency launched an armada of unmanned scientific spacecraft to study further the Sun, our local space environment, and the far reaches of the universe. As a result, both European and American scientists use two observatories, called ISO and SOHO, to examine the inner workings of stars and galaxies and of our own Sun ... Next a study team at the Massachusetts Institute of Technology proposed a new strategy that could represent a profound change in the space program's culture, according to Stephen Bailey, then manager of the Johnson Space Center's team studying Mars robotic missions. At a 1991 meeting in La Jolla, California, America's leading space scientists concurred on the approach suggested by Rodney Brooks and colleagues from the MIT mobile robotics laboratory: that is, 20 small Martian probes launched from Earth in batches of 5, beginning in 1999. Some would contain small-scale surface rovers, mobile robots able to operate independently of direct control, not bound by NASA traditional criteria of 99%-plus reliability. The multiple simpler systems operate at significantly lower cost—yet, with robots of increasingly sophisticated behavior. By adding one capacity at a time, the scientists hoped to evolve the activities of these roving robots as they learn from coping with the rigors of spaceflight and hostile environments (later, Exhibits 62/68 illustrate how prototype unmanned robots might operate on the lunar surface).

But, our concern in this chapter is for that *other* form of human performance in orbit: "manned" missions, past, present, and future. As James E. Davidson of Microsat Launch Systems reminds us, a frontier is an imaginary boundary surrounding human activity. Davidson in a commentary on "More than Missions" suggests that space exploration offers an opportunity to push beyond that boundary 600 km

Exhibit 56. Deploying Eureca from the Space Shuttle. The European Retrievable Carrier is an automated, unmanned platform designed to carry instruments and experiments for life sciences, crystal growth, and astronomy from 6 to 18 months in orbit. Retrieved after its first mission in June 1993, it is reloaded at its control center in Darmstadt, Germany, and then redeployed from the Shuttle Orbiter. Data collected are shared by scientists all over Europe. Source: European Space Agency, Paris.

up on the space frontier by

- providing people on the ground and aloft with education and incentives to invest and create wealth through innovation and labor in building space infrastructure and homesteading;
- establishing outposts and permanent settlements on the Moon and Mars, so that their pioneering inhabitants are capable of innovative research and industrialization.

186   **High-performing spacefarers**   [Ch. 5

5.1   EXTENDING HUMAN PRESENCE ALOFT

The first human being to safely leave our planet provided a legacy of high performance in space on April 12, 1961, commemorated today as a national holiday in Russia. Yuri Gagarin then was a 27-year-old Russian, a major in the Soviet Army, who flew a 5-ton space capsule, Vostok 1, for 1 hour and 48 minutes aloft. After liftoff, these were the memorable words of this top-performing cosmonaut:

"Everything is working perfectly ... I feel fine. I am in a cheerful mood."

This technological feat of launch and return from orbit was not only significant for humanity—but it galvanized the Americans to meet the challenge with a series of manned missions called Mercury, Gemini, and Apollo which culminated in the Moon landings of a dozen humans (see Exhibit 57)!

Exhibit 57. Peak performance: humans on the Moon. Apollo 17 extravehicular activity (EVA) on the lunar surface by the first scientist there, astronaut Harrison "Jack" Schmitt. Dr Schmitt, a civilian geologist with a Ph.D. from Harvard University, is pictured collecting rock samples by a huge, split boulder during the third Apollo 17 EVA at the Taurus-Littrow landing site. The photograph was taken by mission commander, Eugene Cernan. Source: NASA Headquarters. The last of 12 humans to visit there left a plaque with this inscription "Here man completed his first exploration of the Moon. 1972 AD. May the spirit of peace in which we came be reflected in the lives of all mankind." Will humanity return permanently by 2020, as planned in the Vision of Space Exploration program?

Because of humankind's accomplishments in space for the past 40 years, our species is poised now for dramatic offworld growth in the 21st century. A gigantic leap in the development of human potential is likely to occur among spacefarers in this millennium! For those who live and work in space, the zero or low-gravity environment offers both problems and opportunities for peak performance comparable with the first lunar landings. The challenges inherent in long-term space travel and the construction of habitats on the high ground will necessitate removal of Earth-based physical and psychological blinders. Despite numerous hazards of operating on that frontier, the work there extends individual capabilities and ingenuity. For planners of space transportation and living systems, the demand will be to move beyond technical and economic consideration, so as to ensure more than survivability. To take full advantage of our humanness in such circumstances means that organizational sponsors, be they space agencies or private enterprise, must create a situation that ensures crew safety, productivity, and quality of life through exceptional team performance, and the human/machine interface. True space settlement means confronting also a variety of psychosocial issues that range from preparation and training, to scheduling and "re-creation", to space ergonomics and ecology. The next hundred or thousand people in space will continue to be highly selected and probably top performers in many ways. The prospects for extraterrestrial achievement and accomplishment are enormous.

The deaths on returning from orbit of the three crewmembers of Soyuz 11, as well as the *Challenger* and *Columbia*'s 14 astronauts, confirms that space exploration is a high-risk activity. It demands high-quality functioning by all involved in building the space infrastructure and making it suitable for human occupants. Since the return to flight of the Shuttle *Discovery* in 1988, the subsequent operation of the Shuttle fleet has placed renewed emphasis on safety. As Exhibit 58 indicates, the international record for all types of spacecraft resulted in 3,569 successful launches into orbit from 1957 to 1993, a tribute to extraordinary human and technical performance. From year 2002 through 2005, there were a total of 208 successful space launches worldwide. The largest number of spacecraft safely lifted into orbit were by Russia with 80 and the U.S.A, with 78. Interestingly, the remaining launches were by France/ESA with 50, China with 14, Japan with 6, India with 5, Sweden with 2, and Israel with 1. Today many smaller Asian nations are also launching communication satellites and missiles.

As this was written another high-performing space mission was under way. On August 7, 2007, the Shuttle *Endeavour* was launched for its 20th orbital flight.[1] The diverse crew of seven included a Canadian physician, a chemist with sign language competence, a competitive sprinter and long jumper, an educator, and a commander whose twin brother is also a Shuttle pilot. Among the five men and two women was Barbara R. Morgan, the back-up for the "first teacher in space", Christa McAuliffe who died in the *Challenger* accident in 1986. Morgan, now a

---

[1] Dunn, M. "Teacher Ready for Historic Trip to Space," Associated Press Release in the *San Diego Union-Tribune*, August 7, 2007, pp. A1–A10 ... Leary, W. E. "Teacher-Astronaut to Fly Decades after Challenger," *New York Times*, August 7, 2007, p. A.11.

Exhibit 58. SUCCESSFUL SPACE LAUNCHES 1957–1993.

Year	Russia/ CIS	U.S.A.	China	Japan	ESA	India	Israel	France*	Australia*	U.K.*	Total
1957	2										2
1958	1	5									6
1959	3	10									13
1960	3	16									19
1961	6	29									35
1962	20	52									72
1963	17	38									55
1964	30	57									87
1965	48	63						1			112
1966	44	73						1			118
1967	66	58						2	1		127
1968	74	45									119
1969	70	40									110
1970	81	29	1	1				2			114
1971	83	32	1	2				1		1	120
1972	74	31		1							106
1973	86	23									109
1974	81	24		1							106
1975	89	28	3	2				3			125
1976	99	26	2	1							128
1977	98	24		2							124
1978	88	32	1	3							124
1979	87	16		2	1						106
1980	89	13		2		1					105
1981	98	18	1	3	2	1					123
1982	101	18	1	1							121
1983	98	22	1	3	2	1					127
1984	97	22	3	3	4						129
1985	97	17	1	2	3						120
1986	91	6	2	2	2						103
1987	95	8	2	3	2						110
1988	90	12	4	2	7		1				116
1989	74	18		2	7						101

Year	Russia/CIS	U.S.A.	China	Japan	ESA	India	Israel	France*	Australia*	U.K.*	Total
1990	75	27	5	3			1				116
1991	59	18	1	2	8						88
1992	54	28	3	1	7	1					94
1993	47	23	1	1	7						79
Total	2,415	1,001	33	45	57	4	2	10	1	1	3,569

*Note that France, Australia, and the U.K. did not launch their own satellites during this period. France and the U.K. are members of the European Space Agency, which built the Ariane launch vehicle that first lifted off in 1979. Italy conducted eight launches from 1967 to 1975, and one in 1988, using U.S. Scout rockets from the San Marco platform off the coast of Kenya. Since these launch vehicles were American, NASA includes them within the U.S. launch statistics and they are treated the same way here.

Despite some failures, high performance is manifested in the 3,569 successful launches during the period 1957 to 1993, covered by the data in this exhibit. The majority of these launches were to low Earth orbit (LEO). "CIS" refers to the Commonwealth of Independent States established by Russia from former republics of the U.S.S.R. Since 1993, until 2008, there has been a comparable record of mostly successful launches. In 2004, 2005, and 2006 there were 26, 27, and 29 launches respectively by Russia/Ukraine; 18, 13, and 18 launches respectively by the U.S.A.; 8, 5, and 6 launches respectively by China; and 4, 8, and 11 launches respectively by the rest of the world including Europe. Source: U.S. Congressional Research Service.

mother of two, stayed in the Astronaut Corps so that on this mission she could teach the world's youngsters what it is that motivates her:

> "I believe in my heart that space exploration is the key for all of us, especially our young people to keep their futures open-ended."

At 55 years old and after 22 years of NASA waiting/training, Barbara operated a robot arm and oversaw the transfer of cargo from the Space Shuttle *Endeavour* to the International Space Station during the STS-118 mission.

In addition to the six hours she spent teaching aloft, on her return Morgan also conducted an interesting interactive lesson for young viewers based on her videos. She also distribute some 10 million basil seeds she had on her spaceflight, so teachers and students on the ground could raise them in growth chambers of their own design. This is the type of equipment settlers will take to the Moon or Mars to grow fresh food ... During their time in orbit which lasted 14 days, this competent crew installed by means of the robot arm and space walks a 5,000-pound girder on the station, an 11-foot extension of its solar power network, replaced a failed steering gyro, helped repair Russian computers, and transferred 10,000 pounds of cargo. Now that is top orbital performance!

5.1.1 Human orbital expansion

During the next 50 years, those persons who actually get to work in low or geosynchronous orbit will be engaged principally in research and development, or in construction of facilities that will benefit those who will follow. Scientists, engineers, miners, and the military are likely to dominate the space scene in the near term. Astronauts and cosmonauts who pilot spacecraft will be outnumbered by mission specialists, contractors, and tourists who represent a wide variety of professional fields and disciplines, often under the sponsorship of nations, consortia, corporations, or even universities. The initial 21st-century visitors from Earth to the high frontier will expand beyond elite professionals and cosmonauts to include contract workers or "technauts" sent to build or service space stations, as well as eventually Moon and Martian bases. Beyond the research and development phase, we can then expect occasional "space citizens" to encompass a wider range of the global community, whether politicians, teachers, journalists, artists, or even tourists to LEO. Many will simply seek the microgravity experience, as is happening with people who are paying $3,950 to the Zero Gravity Corporation for a series of plunges from a high-altitude, refitted jet airplane that offers 25 seconds of weightlessness with each plunge . The first disabled person to do this was the noted astrophysicist and author, Dr. Stephen Hawking on April 26, 2007. Hawking, 65, wheelchair-bound for most of his life by Lou Gehrig's disease, unable to use hands, legs, or voice, experienced what he called "the bliss of weightlessness!" This great thinker predicts that humanity has to go into space as a precaution against any calamity on the home planet.

Expanded human presence in space takes place in two general ways. First, as indicated in the introduction to this chapter, by those who remain physically on Earth, but extend themselves by unmanned spacecraft deep into the Solar System; or those others on the ground who provide the technology and support services for the envoys of humankind aloft. Second, by those who are actually launched beyond our gravity forces into orbit for increasingly longer duration flights. The complexity and scope of space activity requires in either circumstance people of high competency, both on Earth and in orbit. (Ironically at the very time that NASA is having difficulty recruiting the best engineering graduates from American universities, the Russians have serious difficulties in financing and employing their large workforce of space scientific and engineering elite.) Furthermore, for progress to be made in recruiting skilled space professionals in the coming decades, considerable alterations must occur in space planning and management of human resources for the knowledge work culture [2].

More extended and permanent human occupation of space will bring role change from space operations of the past: the Space Station or Moonbase will function with less ground control and more autonomy for crewmembers. The extensive use of automation and robotics will enhance human capabilities (e.g., by teleoperations, especially telemedicine). Advances in communication technology for and from space will produce new capabilities resulting in expanded information and insights, while innovative expert systems or applications of artificial intelligence will enhance performance. Space industrialization and defense systems will engender novel technologies and occupational fields. Gradually, the numbers of people living, working, and traveling in space will escalate from the hundreds, to thousands, to millions until literally a mass migration occurs by year 3000. In the process, humans themselves will be changed, especially the way they behave and perform.

5.1.2 Offworld multiculturalism

The growing multinationalism and multiculturalism of spacefarers is evident in the pictures of European Space Agency astronauts reproduced in Exhibit 59. The ESA Astronaut Corps came into being in 1978, when three European astronauts were selected to train for the first Spacelab mission on the U.S. Space Shuttle in 1983. Ulf Merbold (Germany) subsequently flew on that mission.

ESA astronauts flying on both American and Russian spacecraft engage in a variety of activities, from scientific experiments and manufacturing high-tech materials to deploying the Eureca platform and satellites. The longest manned

Exhibit 59. ESA multinational astronauts. ESA's Astronauts Corps fly on both American and Russian spacecraft, as demonstrated by the German Ulf Merbold (pictured top right), the most experienced European astronaut, who flew two Space Shuttle missions (STS-9 in 1983 and STS-42 in 1992) and visited the Russian Space Station Mir in 1994. Dutchman Wubbo Ockels (pictured left) first flew on Spacelab D-1 in 1985, while the Swiss Claude Nicollier (pictured bottom right) helped to deploy the Eureca platform from the Shuttle's payload bay in 1992, and retrieve and deploy the Hubble Space Telescope during the crucial first Servicing Mission in 1993. Source: NASA Johnson Space Center.

mission in ESA history blasted off from the Baikonour Cosmodrome to the Mir space station. Called EuroMir '95, it began when ESA astronaut Thomas Reiter lifted off with cosmonauts Sergei Avdeev and Yuri Gidzenko; during the 179 days aloft, Reiter undertook two spacewalks (including one of over five hours), and discussed *living and working in space* via live video link-up with the fourth Information Forum for Young Europeans (EURISY) meeting in Noordwijk, Holland. Further internationalization was evident when the U.S. Shuttle *Atlantis* also docked (for the second time) at the Russian station carrying four Americans and one Canadian. Reiter's return to Earth, by Soyuz spacecraft, occurred in Kazakhstan on February 27, 1996.

5.1.3 High-performance offworld norm

Why the necessity for exceptional performance in space? Because of the risk and cost of orbital activities, humans have to function at their best, especially during the early stages of space exploration and development. Perhaps it would be useful now to define the term *high performance* for its application to space work. Obviously, top performance in any job, whether on this planet or in orbit, is marked by accomplishment which is consistently above average. That is, the worker's behavior is deemed to be in the upper levels of productive effort, characterized by [4]

- high output in terms of quantity and quality;
- individual or team performance manifesting high levels of commitment, competence, and concern for completing tasks and achieving objectives;
- people compete against themselves—not others—to achieve their potential.

Such behavior was very much in evidence among the 40,000 workers involved in the Apollo mission series that took us to the Moon, as well as among their counterparts who kept the Mir space station safely functioning in orbit for 13 years until it was decommissioned [5].

In a survey of management literature on high-performing work environments, your author identified what organizations and their leaders do to achieve such a state among their members. Applying these findings, for example, to any space organization and its space missions, high performance would result *when*

- there is joint goal setting by both managers and workers, so that their targets and objectives are set beyond current levels to stretch human efforts without sacrificing safety requirements;
- standards of excellence are established which are based on norms of competence, collaboration, and consensus—rather than undue concern for minimum wages, hours, and other benefits so prominent in the disappearing bureaucracies of the industrial work culture;
- an open communication system functions which espouses authentic interactions, feedback, and positive re-enforcement;
- management redirects work habits and activities from ineffective to effective means, helping employees to work smarter and learn from mistakes;

- the new work culture capitalizes on human assets and potential by encouraging flexibility, responsibility, innovation, autonomy, and risk-taking among personnel, while maintaining accountability and being results-oriented.

This researcher found that many of the above qualities and practices were present in NASA and its aerospace contractors during the Apollo era of the 1960s, but diminished in the decades that followed (see Chapter 8). The Agency's current administrator, Michael Griffin, has yet another management reorganization under way to improve performance. Similarly, the new Russian Federal Space Agency is attempting to replace the Soviet centralized planning and managed operations by becoming more market and people-oriented. Furthermore, the success to date of European, Chinese, and Indian space agencies demonstrate high performance by their employees.

Other strategies used by high-performing organizations can also be adapted by the space industry and agencies both on the ground and aloft.

- Recruiting, promoting, and rewarding top performers as policy, and then using them as role models for ordinary employees.
- Altering organizational structure and practice so that it is more de-centralized in terms of decision-making and responsibility, more mission-oriented and responsive, while reducing the levels of management.
- Making work more meaningful and fun by cultivating informality, fellowship, and team effort within a context of productive achievement and joyful accomplishment.
- Emphasizing human resource development, while trying a mix of benefits, rewards, and incentives for talented performance.

Strategic planning, bold vision, future orientation, and high purpose capture the imagination of personnel, pushing them on to metaperformance. Exhibit 60 illustrates such work in action.

5.1.4 Exceptional ground support services

Many of the above-proven approaches are directly applicable to ground-based operations that support space missions, and eventually can be adapted to space-based conditions. For example, in a NASA study on *The Human Role in Space*, S. B. Hall concluded that the human being represents a remarkable and adaptable system [6]. He recommended that systems designers develop improved workstations in space which enhance human productivity and value to the mission under way. Currently, human factor engineers are using biodynamic animated computer models to help design crew workstations for the future orbiting laboratory. One of my concerns as a psychologist is that industrial engineers in general, and NASA in particular, have been too narrow in their human factor perspective relative to providing a stimulating work environment for increased performance. Under the leadership of Dr. Albert A. Harrison, broader alternatives are being offered with reference to crew systems and

Exhibit 60. Astronauts Eileen Collins, commander, and James Kelly, pilot, of 'return to flight' mission STS-114, after docking their Space Shuttle Discovery, work as a team in the Destiny laboratory of the International Space Station (August 5, 2005). Source: NASA Johnson Space Center.

support. This space psychologist believes that high performance aloft reduces risk, while raising crew morale. The immediate impetus for ensuring optimum orbital performance comes from prospects of long-duration missions, as well as the needs for the International Space Station, plus planned bases on the Moon and eventually Mars. Professor Harrison, of the University of California-Davis, envisions several dimensions for enhancing human performance and settlement aloft [7]:

(1) *Protection* from acceleration, vibration, temperature extremes, near-vacuums, some poisonous atmospheres, radiation, and impact of meteorites and space debris.
(2) *Improving quality of life* in terms of support systems, equipment, supplies, habitats, amenities, communication, and social relationships.
(3) *Accommodating diversity* among spacefarers in terms of differences in sponsor, role, gender, physiques, culture, national customs, and language.
(4) *Inventing cultural adaptations* to the space environment to ensure survival and growth, as well as to meet emerging needs and tastes (such as creating new sociopolitical institutions and customs).
(5) *Promoting self-sufficiency in orbit* because of the high transportation costs from the home planet; apart from better communication systems and technologies, this

implies learning to use space resources and engaging in *terraforming* of asteroids and planets.

The matter of human performance has long been a centerpiece of behavioral science research; unfortunately, little of this has been centered on functioning in space until recently.[2] Usually, psychologists have examined three interactive forces: the *person*, the *task*, and the *environment*. Positive changes in any of these factors normally result in greater productivity, while negative influences tend to decrease individual effort. The co-authors of *Living Aloft* have proposed that this paradigm be employed with reference to performance on the Space Station: namely, that the model should focus on the difficulty of tasks in orbit, the harshness of the environment, and the spacefarer's fitness, all of which impact performance. These three dimensions are dynamic, multifaceted, and multidimensional factors to be considered in all space work. Harrison rightly advocates that relative to space planning and management, consideration should be directed to these realities for improving human productivity in orbit [8]:

- Space workers' characteristics which include *competence* (combination of ability, experience, and training); *motivation* (amount of energy and diligence that the person brings to a task); *interest* (degree of ego involvement and sense of participation experienced in the work, so that the task receives preference over other activities).
- The task's characteristics, whether on Earth on in space, include *complexity* (level of demands that the task places on the performer's perceptual, cognitive, and motor skills); *duration* (time required to complete the task); and *repetition* (relative frequency with which the task is performed over a period of time). Such variables combine to determine both satisfactory performance and motivation. Harrison observes that complex, challenging, and significant tasks that are consistent with personal values and interest are associated with high commitment and motivation.
- The environmental characteristics may be analyzed at several levels: primitive in terms of lines, forms, shapes, colors, textures, contours and patterns; and higher or more synthetic levels which considers such factors as *legibility* (extent to which the environment provides an easily understood and coherent frame of reference); *arousal potential* (extent to which the environment increases or decreases drive or arousal of people's feelings and energies); *preferability* (extent to which environmental design elements are consistent or inconsistent with personal tastes and values).

---

[2] In addition to those studies cited in the last chapter, some recent examples of space behavioral research are evident in such reports as B. S. Caldwell's "Multiteam Dynamics and Distributed Expertise in Mission Operations" ... M. Shepanek's "Human Behavioral Research in Space" ... W. E. Sipes; and S. T. Vander Arks' "Operational Behavioral Research and Performance Resources for the International Space Station Crews": all in *Aviation, Space, and Environmental Medicine*, **76**(6), June 2005, B145–B153, B25–B30, B38–B41.

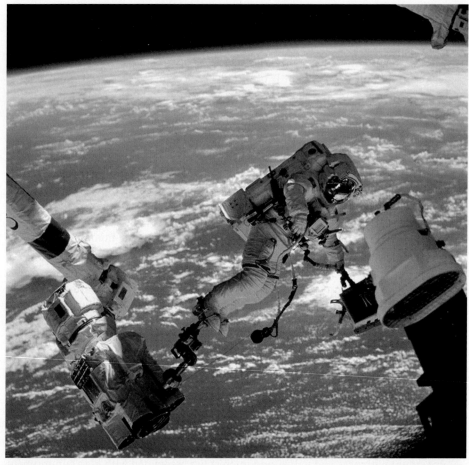

Exhibit 61. High performance offworld. Astronaut David A. Wolf, STS-112 mission specialist, anchored to a foot restraint on the Space Station Remote Manipulator System (SSRMS) or Canadarm2, carries the Starboard One (S1) outboard nadir external camera. The camera was installed on the end of the S1 Truss on the International Space Station (ISS) during the mission's first scheduled session of extravehicular activity (EVA). Source: NASA Johnson Space Center.

Harrison suggests that the perceptual processing of the primitive levels of reaction leads to properties which, in turn, later impact performance. Thus, this person/task/environment construct is useful for studying the performance of *Homo cosmonauticus* or the spacefaring human, as illustrated in Exhibit 61.

5.2 LEARNING FROM HUMAN OFFWORLD PERFORMANCE

During the past 40 years of manned spaceflight, some insight has been gained about human performance in space. Unfortunately, it has been largely about astronauts and cosmonauts on short-duration missions, and is of limited value for those who will live for much longer periods aloft. But, we have learned that, deprived of Earth's gravity for extended spaceflights, the body gradually acclimatizes to microgravity, but cardiovascular deconditioning sets in, human bones lose calcium, a demineralization process begins, and the "space adaptation syndrome" of nausea, disorientation, or discomfort can be overcome by most spacefarers [9].

The mission records for days aloft were first established by Skylab (three astronauts in 1974 for 84 days) and Salyut 7 (three cosmonauts in 1984 for 237 days). Between the Skylab days and the era of the Shuttle–Mir link-ups, American experience in space was limited to orbital Shuttle flights. Thr longest such mission was the STS-80 flight of *Columbia* in November 1996 when the crew of five spent 17 days 15 hours aloft, carrying out many tasks including deploying two satellites and successfully recovering them after they had performed their tasks. The first American to carry out a long-duration mission aboard Mir was Dr. Norman Thagard, who spent 115 days aloft in 1995. Currently, Russian cosmonauts hold the long-duration honors for stays in Earth orbit. Space medic Dr. Valeri Polyakov spent 438 days in orbit (over 14 months) between January 1994 and March 1995. His combined space experience is 679 days (over 22 months). By contrast, the record for the longest single space mission by an American is 215 days, helf by Michael López-Alegría. Such timespans offworld will constantly be extended to years and eventually decades. At present, assignments are for six months aboard the International Space Station, as well as at a lunar outpost when it is completed by 2020.

Our extraterrestrial human experience to date proves that steps can be taken to improve crew efficiency, ranging from automation and environment control systems, to exercising and scheduling alterations. In 1987, some of this accumulated information benefited two Soviet cosmonauts on their Soyuz TM-2/TM-3 missions. Aleksander Aleksandrov stayed in orbit for 160 days, and his companion, Yuri Romanenko, spent 326 days aloft. During this time, Romanenko grew one centimeter taller and lost 1.6 kg of body weight, and experienced fatigue, listlessness, and homesickness during his long flight. In December 1988, cosmonauts Vladimir Titov and Musa Manarov spent over a year in space on the Mir station, returning in the Soyuz TM-6 capsule with their three-week visitor, Frenchman Jean-Loup Chrétien. Radio Moscow reported that immediate medical check-ups on landing showed them feeling well, and subsequent physical evaluation of the two spacefarers at Star City revealed no serious difficulties after their prolonged weightlessness for 366 days. This was partially attributed to exercise twice a day by Manarov and Titov on both a treadmill and an exercise bicycle. They also had special space suits which forced blood to concentrate in the lower parts of their bodies, simulating the effect of gravity, so that they would not become accustomed to weightlessness. Despite these precautions against calcium loss in long spaceflights, they did shrink in size as their legs lost calcium, but managed to return to Earth in better shape by use of vitamin supplements. During their 12 months plus on the station, the then record-holders conducted extensive astrophysical studies, technological experiments, and medical tests, as well as studying sources of radiation and taking more than 12,000 photographs of the Earth's surface (*Los Angeles Times*, December 22, 1988, p. 8). In January 1994, cosmonaut physician Valery Polyakov, 51 years of age, blasted off for 438 days in space, circling the Earth over 7,000 times in Mir. Such accounts confirm *high performance in space* by both astronauts and cosmonauts, feats which will be continuously surpassed by spacefarers to follow! These duration records for orbital assignments made by space pioneers are the precursors for long-term colonists in space.

NASA director of life sciences Arnauld Nicogossian confirmed the paucity of knowledge concerning human physiology and performance in the weightless environment. Unless research is expanded on the biomedical, psychological, and biospheric factors of spaceflight, long-duration manned missions may be jeopardized. Nicogossian, the author of *Space Physiology and Medicine*, made these pertinent observations [10]:

- ground training in biogenic or autogenic feedback may precondition some people to spaceflight so they may master motion sickness aloft;
- after four days in space, the body adapts to weightlessness, yet we do not fully understand the influence of visual perception;
- drugs have different effects aloft than on the ground, so use of that therapy for motion sickness or other ailments has to consider issues of metabolism in space;
- although the Russians have had missions as long as 237 days or more, NASA is presently confident operationally with scheduling for 90 days aloft in terms of providing adequate medical support;
- a health maintenance facility is proposed for inclusion on the International Space Station to deal with incapacitating illness or injury, as well as dental problems—all of which impact performance;
- ignorance exists on closed environments for long-term flights, especially relative to out-gassed material and their contribution to headaches, irritability, drowsiness, and depression; meal times have been demonstrated as of psychological importance for on-orbit socialization;
- insufficient research has been done on matters of decor (color scheme, number of windows, etc.) and privacy influences on performance;
- inadequate data on human factors exist for a manned Mars mission.

5.2.1 Improving human factor research

This sobering assessment highlights what this writer has already reported in Chapters 2 and 3. NASA has begun to address some of these issues by expanding research in human factors, especially functional esthetics, as well as in life sciences, especially with reference to life support systems for the Space Station [11]. The prospect of building a Moonbase within a dozen years has prompted some physicians, such as Ron Schaefer, to examine future medical care for lunar dwellers. With proper selection and effective paramedic services, Dr. Schaefer feels that explorers' illness and accidents can be coped with adequately for up to six months' stay. Beyond that timespan, he is concerned about the treatment of acute illness, severe injuries, and chronic conditions. Schaefer anticipates problems on the lunar surface related to bone demineralization, cardiovascular deconditioning, trauma, decompression sickness, and radiation poisoning. Obviously, the performance of lunar workers would deteriorate with any such disabilities. Thus, ongoing orbital research on the physiological effects of microgravity, as illustrated in Exhibit 62, provides significant insights for future long-duration missions.

In preparation for living and working on the future space station and bases, it is wise to listen to those who have actually been in orbit. In the past, flight surgeon W. K. Douglas [12], for example, interviewed ten astronauts on their spaceflight performance. Their comments confirm somewhat the validity of the above observations. Furthermore, they provided this McDonnell-Douglas researcher with these recommendations for operational changes [12]:

(1) spaceflights have been constructed by NASA so that everything is dictated by checklists, which take away one's ability to think;
(2) a workday of 14 or 15 hours in space leads to inefficiency and mistakes;
(3) space station design should take advantage of a real and unique environment, such as delta temperature and pressure;
(4) the suit used for extravehicular activity (EVA) should be better designed and have more provisions (e.g., from honey water to a toolkit);
(5) crew shifts should allow for group sleep, exercise, and play, as well as permit work occasionally on off-duty time;
(6) zero-g environment can be optimized by pre-departure training and conditioning, so that space sickness and other problems can be eliminated or limited.

5.2.2 Learning from spacefarers

It is useful to seek feedback on experinces from those who have worked in orbit when designing space deployment systems, as will be discussed in Chapter 6. It also demonstrates the importance of the above model proposed by A. A. Harrison, which utilizes the dynamic interplaying forces of person, task, and environment as a means for maximizing performance whether on space platform or at a Moon/Mars base.

Hall [6] is also concerned about the amount of stress which human operators in space can handle, and urges more research on "cognitive ergonomics" or mental workload assessment and its effects on human–machine systems , especially when there is overload (see Section 5.7). Spacefarers will be extremely vulnerable to disruptive internal and external forces; stress may be reduced when crews can control and adapt technology to meet contingencies. Within the living systems of an orbiting station, for instance, the behavior of the individual can affect group performance, contributing to either mission success or mission failure. Since open or living systems methodology is more holistic, its advocates hypothesize that it can contribute to the control of stress in space habitats and enhance human performance by diagnosing system pathologies which lead to inefficiencies and errors (see Chapter 3). In the decades ahead as we get into long-duration spaceflight, it is also vital that surveys be made of such spacefarers, so that we may improve pre-departure orientation and training for extended habitality aloft.

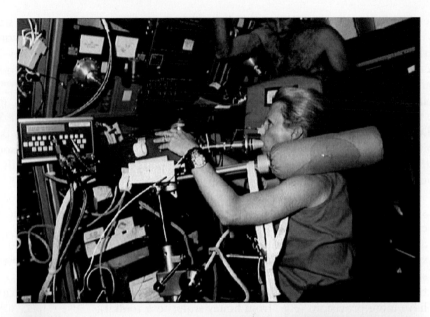

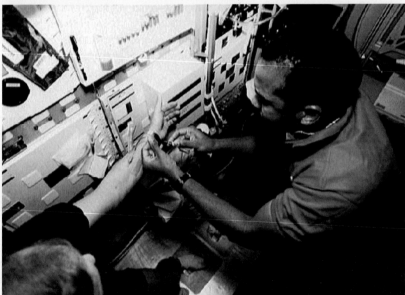

Exhibit 62. On-orbit performance research. The onboard scenes shown in these photographs are of physiology experiments investigating microgravity conditions aboard the Shuttle Orbiter. For example, in one, astronaut-physician Dr. Rhea Seddon, on the bicycle or ergometer, breathes into the cardiovascular unit during an STS-40 spaceflight ... In 1993, during the 10-day STS-55 10-day Spacelab D-2 mission on *Columbia*, Dr. Bernard Harris, an African American physician, draws blood from Hans Schlegel, a payload specialist from the German Aerospace Research Establishment (DLR). Source: NASA Johnson Space Center.

As has been mentioned, the actions of space agency administrators and managers may also impact the performance of both ground controllers and orbiting spacefarers. Such was the case in 1992 when Russia was in transition from the old Soviet system to the new market-oriented economy. At the time, both mission control workers and cosmonauts at the orbiting Mir station demanded a salary increase, and threatened to strike. A banner was erected *Our Work Is Cosmic, Our Pay Should Be Cosmic*, and it was televised. Even state-run media joined in with such commentary as: *One of the most prestigious jobs on Earth has become one of the lowest paid*. The bosses at RSC Energia who then ran the national space program got the message, and compensation did improve, so performance levels remained high.

Besides prior manned mission experience, there are also terrestrial analogs to draw on relative to performance in extreme environments, as will be discussed in Chapter 6. Other helpful information about human performance may be gathered from analyzing how workers fare at remote, alien, or foreign locations on this planet. Social scientists have sometimes described these as "exotic environments marked by severe climate, danger, limited facilities, isolation from family and friends, and enforced interaction with others." In a review of such research literature, Helmreich and associates [13] discovered that a high level of performance under such circumstances may be dependent on the cohesiveness of the isolated and confined work group. The investigations of Furnham and Bochner [13] on psychological reactions to unfamiliar environments also provide performance data for those on extended sojourns and may help to prevent or lessen *space culture shock*. Such developments explain, in part, why the National Science Foundation joined with NASA in sponsoring a 1987 conference on "The Human Experience in Antarctica: Applications to Life in Space" [13]. Exhibit 63 seeks to convey the interplay on varied systems on the Moon that ensure survivability and high performance.

5.2.3 Advantages of diversity and teamwork

The best hope for the future of extraterrestrial performance would seem to be in multilateral research on the subject by various space organizations. For example, in the mid-1990s, agreements have been worked out among the American, Canadian, Russian, European, and Japanese space agencies to share not only their facilities and equipment—but also their findings on humans aloft. Twenty years after the historic Apollo 18 and Soyuz 19 linkup, in June 1995 a linkup of the 37.2-meter-long *Atlantis* Shuttle with the Mir station formed the largest spacecraft ever to orbit the Earth at that time. The synergy lasted for five days as six American astronauts, including two women (physician Dr. Ellen S. Baker and biomedical engineer Dr. Bonnie Dunbar) collaborated with four Russian cosmonauts while conducting medical experiments together. As the *New York Times* (July 11, 1995) wrote:

> "Experts say the new East–West teamwork is an ideal remedy for deficiencies of each side, creating a synergy that will produce the first real understanding of man's destiny as a space traveler. It is also seen as a prerequisite for sending astronauts aloft for many months at a time aboard an International Space

Exhibit 63. Interdependent systems on the Moon. This artistic rendering shows the interplay on the lunar surface of systems that ensure survival and affect performance: the human system, life-supporting space suits, and robotic helpers, including the vehicle that can sustain crews with its life-protecting systems. Source: NASA/John Frassanito & Associates.

Station, which NASA and its Russian counterpart want to build from 1997 to 2002 and operate at least until 2012 ..."

"Dr Laurence R. Young, a professor of astronautics at the Massachusetts Institute of Technology, who has also trained as a NASA payload specialist, said 'With the combined program we hope to get the best of both worlds—Russian length with American strength ... We stand the chance of combining the successes of two national traditions. It's a turning point for the life sciences'."

Dr. Norman Smith, a Van Nuys, Californian management and systems consultant, commented in a memorandum to your author (5/11/86) that the human must become the central focus of the space program. Spacefaring extends human capabilities, even when accomplished by our artifacts, such as remote-sensing satellites. As the duration of space missions increases, surviving and living there become more complex. Because of the unknowns involved, Smith believed that manned offterrestrial systems have to be devised in such a way as to deal with changing conditions, emergencies, and opportunities as they are confronted aloft. He observed that humans are learning systems which operate simultaneously on the physical or biological, psychological, and social levels. For a human to perform effectively in space over long periods, the total systems design must provide for crewmembers at all three levels. In examining with James Grier Miller and others how living systems theory may facilitate such space planning and management, Smith also cited the possible build-up of stress at any of these interacting levels during extended space missions. Performance impairment, he noted, may occur with stress caused by the human–machine interface, group interactions, and disparities between the social and technological systems.

Although human performance in preparation for and in space has generally been outstanding during the last four decades, we need to *learn from our failures and mistakes*. Some of this has been caused on the ground by inadequate or incompetent space workers, as well as fraudulent contractors. Some of the insufficiencies happened in orbit, as in the traumatic flights of Apollo 13, and the Shuttles *Challenger* and *Columbia*, though on the first-mentioned its high-performing crew managed to return home safely. A recent media report about astronaut Bonnie Dunbar, for instance, indicated that she nearly died in space because of a sloppy medical experiment. That problem on Shuttle *Discovery* was caused by deficient laboratory procedures and protocols that permitted Dunbar to inject herself with a drug that triggered a life-threatening allergic reaction. Misguided investigators got more interested in their experiment than in their human subject (*CBS News*, July 7, 1995)... Similarly, one of the major findings from Dr. Norman E. Thagard's 115-day orbital mission in 1995 was NASA's neglect of the psychological aspects of his preparation and support for isolation and confinement aloft. At other times, problems aloft may be compounded by human relations and failures, as was described in the previously cited book, *Dragonfly*, about the problems and heroics of astronauts and cosmonauts coping with a fire on the Mir station [5] ... Sometimes it is a technological defect which is overcome by human performance. As ESA's Ulf Merbold commented, "It's amazing what humans can do when technology fails." He was recalling his sojourn on Soyuz TM-20, when the spacecraft's automatic docking mechanism would not work and there was only three minutes left to dock manually, which is exactly what cosmonaut Alexander Viktorenko did, driving in at high speed to complete the docking successfully. Another example of technical glitches occurred in June 1997 when the unmanned Progress M-34 cargo ferry of $1\frac{1}{2}$ tons rammed Mir because of a 50% electrical power drop on the Russian station. Despite the stress, the Russian–American crew coped well with this latest in a series of accidents on that aging station.

One learning from spaceflights of the past half-century is that women can do as well as men in orbit, and sometimes even better (see Exhibit 65).

Multidisciplinary studies of human performance in space are just beginning and open up vast research possibilities [13]. Missions aloft will be more international and varied, both in scope and in composition of personnel. The next wave to create a space civilization will have even more diverse backgrounds and expertise, so trainers, commanders, and ordinary spacefarers will require cross-cultural preparation. Twenty-first century space settlement and industrialization will be followed by interplanetary exploration in this millennium. Author Alcestis Oberg [16] envisions that these pioneering performers on the high frontier will become ancestors of more remarkable descendants to come, as different from us today as our generation is from the people of the Middle Ages—*spacekind*!

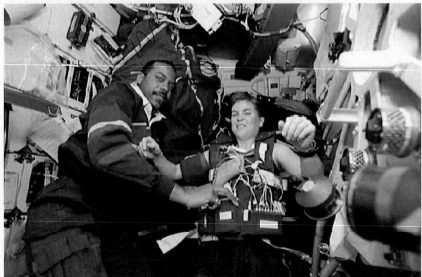

Exhibit 64. Crew diversity and ingenuity. Two more spaceflight scenes which emphasize the globalization of space missions. (Top) ESA astronaut Ulf Merbold is pictured between crew partners Talgat Musabayev and Elena Kondakova onboard Mir in November 1994 ... (Bottom) In this *Discovery* Orbiter photograph, payload commander and physician Bernard Harris, with his test subject, astronaut Janice Voss, who wears a biomedical harness for the experiment to check her muscle responses to microgravity. These middeck Spacelab experiments in the Shuttle Orbiter's cargo bay also included James Weatherbee, commander; Eileen Collins, pilot; and Russian cosmonaut, Vladimir Titov. Source: *Ad Astra*, June 1995 and NASA Johnson Space Center.

Exhibit 65. HIGH-PERFORMING FEMALE SPACEFARERS.

Cosmonaut **Valentina Tereshkova** became the first woman to fly in space on Vostok 8, June 18, 1963. Numerous of her gender have demonstrated exceptional abilities in orbit, such as Dr. **Sally Ride**, the first American women to go offworld. While none of ladies caused problems in orbit, four women astronauts have died aloft in the service of humanity: **Christa McAuliffe** and **Judy Resnik** (*Challenger* Flight 51-L, January 28, 1986); **Kalpana Chawla** and **Laurel Clark** (*Columbia* Flight STS-107). In our previous volume on *Living and Working in Space*, the author cited two females as "Interplanetary Women of the Year for 1996." Since they are symbolic of all women who have gone aloft, these two represent the courage, competence, and caring of female space professionals:

- Dr. **Linda M. Godwin**, NASA mission specialist, engaged in a space walk within the cargo bay of the Shuttle Orbiter *Atlantis*, on March 27, 1996, while it was docked to the Mir station. Godwin, a physicist. was teamed up for this EVA with Lt. Col. Michael Clifford to install four panels used in environmental experiments to catch cosmic space debris and to test coating materials for the future ISS. This was the first time in 22 years since the demise of Skylab that Americans worked outside a space station.
- Dr. **Shannon Lucid** was dropped off on that same *Atlantis* mission STS-76 at the Russian orbiting apartment and laboratory. She remained aboard Mir over six months with cosmonauts Yuri Onufrienko and Yuri Usachev. Lucid, a chemist, got along well with her colleagues while conducting experiments and establishing an endurance record as the first American and woman to be in orbit so long. After this 188-day sojourn, the astronaut said she felt fine physically, with no ill effects from her long time aloft. At 53, she proved that gender and age are no barrier to living and working in space. Shannon was born in Shanghai, China, where her parents were missionaries; six months after her birth the family was interned in a Japanese prison camp for several months. While aloft, she kept in touch via electronic mail with her husband, a petrochemist. and three children! (Source: Philip R. Harris, *Living and Working in Space*. Chichester, U.K.: Wiley/Praxis, 1996, Second Edition.)

And in this 21st century, women continue to astound us with their performances in orbit:

- Dr. **Peggy Whitson**, a biochemist, was a project scientist for the Shuttle–Mir Program, as well as a deputy chief of NASA's Medical Science Division. In 2002, she spent six months on the International Space Station as its first science officer, conducting 21 research investigations. On the station, she orbited the Earth at 17,500 mph while living and working in a weightless state. Peggy was innovative from the way she "loose-bungeed" herself to the walls for a satisfactory sleep, to adjustments in her laboratory where she worked five to six hours most days. Whitson, 46, was perfect for this job, one that requires self-direction. She trained at Star City outside Moscow for her next six-month assignment in September 2007 as commander of ISS. During this mission, she

> broke the record for the most spacewalk time by a woman. The girl from an Iowa soy farm is modest about her achievements and many awards, observing that *everything done in space is worth the risk, because it is one more step of exploration.* (Source: *San Diego Union-Tribune*, September 21, 2006, p. E.3.)
> - Commander **Eileen Collins** retired as an astronaut in April 2006 after logging 872 hours in space. She was the first woman to pilot and command a Space Shuttle. She flew the first flight into space after the *Columbia* disaster in 2003, performing a series of unprecedented twist-and-flip maneuvers to photograph the Shuttle's belly for damage. She was the first female pilot on a Shuttle *Discovery* flight in 1995 which successfully docked with the Mir space station. For that feat she was honored by the National Women's Hall of Fame (see Exhibit 59)... The Elmira, N.Y. native studied mathematics and economics at Syracuse University, and received Masters degrees from Stanford and Webster universities. After serving as a math and pilot instructor at the U.S. Air Force Academy and Vance Air Force Base, Eileen was attending the Air Force Test Pilot School at California's Edwards Air Force Base when she was selected for the Astronaut Corps in 1990. During her NASA career, she had also worked in engineering support and as a mission control communicator. Now, at 49. Collins intends to pursue a new career in the aerospace industry. (Source: *America On-line*, May 1, 2006.)
> - Astronaut **Sunita Williams** reached a milestone in June 2007 with the longest single spaceflight by a woman—195 days. Williams, 41, lived on the International Space Station since December 2006, and surpassed Shannon Lucid's record of 188 days at the Mir station in 1986. When Mission Control congratulated her for this feat, "Suni" replied: "It's just that I'm in the right place at the right time. Even when the station has little problems, it's just a beautiful, wonderful place to live!" She then returned on the *Atlantis* Shuttle flight through the intense heat of re-entry to Earth.

Source: *San Diego Union-Tribune*, June 17, 2007, p. A6. Also read *Women Astronauts*, by Laura S. Woodmansee (2002) and *Women in Space: Cool Careers on the Final Frontier*, by Laura S.

5.3 CREW SYSTEM PRODUCTIVITY

With the globalization of space involvement, crew diversity will become routine. Simultaneously, their sponsors shift studies toward human performance aloft and enhancement of quality of life there. Dr. Mary Connors [8] has provided another helpful conceptualization in the term *crew system*, which she defines as the combination of human and technical subsystems. This NASA Ames behavioral scientist described crew systems as wholes, entireties, or totalities whose overall performance cannot be understood from looking at one subsystem alone. Together Connors and Harrison examined the phenomenological, structural, and functional dimensions of crew systems, so as to obtain a better fit among human, automation, and communication in orbit [14].

The next few hundred persons in space are likely to be crewmembers of a spacecraft or base under the immediate supervision of an American, Russian, European, or Chinese space agency. Although an expanded and mixed group of mission specialists, more varied in sex, race nationalities, and competencies than previously, their performance may set the mould for the thousands who will follow them into the Solar System by establishing the norms for long-duration space adaptability. Preliminary research indicates that a generalist is preferred, one who has multiple skills or is cross-trained in the required fields of expertise.

Experience from almost five decades of manned spaceflight has offered some clues as to what fosters productive behavior. In *The Human Role in Space* study (THURIS) by Stephen B. Hall [6], the following insights were offered:

- hardware and systems improvements do enhance human performance;
- procedure and operational changes will allow for more effective use of the human element in human–machine systems;
- traffic flow through manned modules will be a consideration for improving human productivity;
- better designed and implemented storage and stowage abets productivity, permitting quicker retrieval and replacement;
- other architectural factors have been identified which can advance performance by changes in habitat module dimensions and surfaces, windows, and partitions, configurations and spatial requirements, noise control and restraint systems, as well as improved anthropometrics (e.g., workstations, common areas, waste management, etc.).

Hall's McDonnell-Douglas Corporation research under contract to NASA undertook analysis of past flights, discovering a number of ways to maximize crew productivity in future manned systems. His findings ranged from improved personal hygiene equipment and provisions to better lighting, health maintenance, food and water systems, communications systems, and even housekeeping. The THURIS investigators made recommendations, based on such experiences as Skylab and Soyuz, for both preflight and inflight training, inflight maintenance, planning and scheduling activities, organization and management. The ground support group, for instance, should be composed of mission control and payload operations personnel, as well as principal investigators of experiments on board the spacecraft, but they should function in such a manner as to monitor flight activities while allowing for greater crew autonomy. Ground communication and authority impact human productivity both on the ground and in space. *The Human Role in Space* analysis pointed up many other systems improvements that enhance space performance: computer modeling, IVA (internal vehicular activity), IVA/EVA interfaces, remote systems management, among others.

Having examined both the U.S. and the Russian space programs to the mid-1980s, the THURIS conclusions recommended more provisions for crews to deal with inflight contingencies, in-orbit assembly, and *in situ* emergency management. This analysis confirms that the more crews are permitted to participate in onboard research, working together with principal investigators on the ground, the more they are energized and the success of mission objectives is enhanced. To make the most of the remarkable and adaptable human system in space, the *THURIS Study* (Vol. II) offered criteria of performance, cost, and technological readiness, plus descriptions for the 37 generic activities identified so far for those who work in space.

As indicated above [12], in 1984 another McDonnell-Douglas researcher, William K. Douglas, interviewed about 10% of those astronauts who have been in space, concerning performance issues which have implications for a manned space station. This expert sample offered valuable commentary, especially for working on ISS, such as:

- *Space station management and performance.* The station commander should have final authority, while ground control provides strategic planning and resources, such as data analysis. Orbital veterans also counseled that communication from the ground should be informative—not protective. They advised that while the commander has the ultimate responsibility, that person will have to be a strong but flexible leader, able to deal with diversity, to listen for varied input, and to obtain consensus. A matrix-type organization was implied when they proposed that the commander delegate some responsibilities to a chief scientist.
- *Improved productivity aloft.* Suggestions from the astronauts were summarized in the last section, but ranged from equipment redesign to provisions for a multipurpose toolkit, from work schedule routines to crew cohesion (assigning and rotating them as a unit). They warned that long-duration flights are a new experience and people will be unwilling to make the sacrifices and put up with the inconveniences present in past missions of short duration. These spacefarers observed that personality and human relations issues among the crew will have greater impact on group productivity; optimal stays in orbit have yet to be determined before fatigue, inefficiencies, and mistakes undermine performance.

Interestingly, at a 1991 NASA/NSF conference on isolation and confinement, Dr. William Douglas raised additional issues of what could undermine performance aloft. He carefully examined reports and volumes like *Mutiny on the Bounty* (e.g., C. Nordoff and J. N. Hall, 1932), comparing them with interviews and writings of astronauts, especially in the books of H. S. F. Cooper, entitled *A House in Space* (Bantam Books, 1978) and *The Flight that Failed* (Dial Press, 1973). In examining aberrant behavior caused by psychological and sociological aspects of spaceflight, he focused on the 84-day mission of Skylab which led to a so-called near-revolt of its crew. Crewmembers are described by their own as lethargic, negative, irritable, complaining, bitching, grumbling, and given to bursts of anger. Some of this was caused by the burdensome schedule provided by ground control, climaxing in a request from mission control on the fourth Sunday that they had some additional tasks to perform. Having already worked the three previous Sundays, when the crew was meant to relax, Mission Commander Carr said that he had had enough. The first "mutiny" in space had occurred when Col. Carr responded, "We better not work today. We better do our own thing and get some rest." Douglas concludes that NASA schedulers and experimenters are not always aware of the physical and psychological stressors that occur aloft, and the need to provide enough time for astronauts to gaze out the window, to enjoy the magnificence of the Earth and the heavens.

The conclusion is that performance might very well be enhanced on long-duration missions if there is *less control* from the ground. For crews of a space station or Moon/Mars base, it might be better to mutually set broad mission objectives and milestones before launch, but let the commander on site have the leadership responsibility for the accomplishment. The spacefarers would set their own schedule and task assignments. An example of this occurred in September 2006 while six astronauts from Shuttle *Atlantis* had to work non-stop for a week installing a $17\frac{1}{2}$-ton addition to the ISS. When the winglike solar panels were safely functioning, NASA controllers let them sleep an extra hour before awakening them the next morning with the music *Twelve Volt Man*. Then, they had a full day of unloading supplies for the station, including 90 pounds of oxygen. Since it is difficult to estimate how much time such work requires in space, why not let the crew set their own pace, and allow extra time for orbital relaxation?

5.3.1 Improving orbital performance

In several volumes produced by the Presidential Commission on the Space Shuttle *Challenger*, there are also recommendations to improve personnel performance in the space program, both on the ground and in orbit. Two areas of the report underscore the above astronaut observations, namely management and safety. The Commission's proposed reforms adopted after endorsement by three generations of astronauts include [15]

— more access for the crew to flight decision-makers regarding spacefarer feedback, especially on safety matters;
— role elevation of the flight crew operations director;
— appointment of a NASA safety director who reports directly to the agency administrator;
— encouragement of senior astronauts to enter management posts.

Alcestis Oberg also interviewed space experts and astronauts, as well as reviewing the space literature. The following were among the most interesting performance insights obtained [16]:

- **Engineering concerns** (power, propulsion, protection, etc.) and mission objectives, a healthy and happy environment for people, may initially drive the design of space structures and habitats.
- **Performance criteria** should begin with basic competencies: ability to contribute to group survival, to service mission objectives, and crew needs, to operate coolly

under life-threatening circumstances (thus, a track record of being able to function with "grace under pressure" is an important selection factor).
- **Long-duration spaceflight** does present physical challenges to human performance that may cause disorientation, vestibular dysfunction, and bodily degradation. Protection is required from hazards like space sickness, toxic and noise contaminations, inadequacies of life-support systems, solar radiation, zero temperatures, and meteoroids. Furthermore a continued supply of adequate food, water, and shelter must be provided in orbit. However, the real limitations, as the Russians discovered, may be psychological (sensory deprivation, spatial restrictions, isolation, lack of privacy, and dysfunction caused by prolonged confinement which, in turn, may lead to insomnia, depression, hostility, and carelessness). For long-duration flights, new interpersonal skills will be required; for example, cosmonauts who have flown the longest valued "professionalism and kindness".
- **Group composition and interactions** impact space performance (such as the presence of women and their numbers, training and emphasis on cooperation and synergy, group cohesiveness and camaraderie, style of leadership, ground-based communications.
- **Zero-g work habits** involve developing new skills and instincts (putting on a space suit and operating effectively in it, piloting spacecraft, handling tools, or practicing a scientific speciality in a new way).

The Aerospace Medical Association has consistently published studies on improving spacefarers' performance, the most memorable was published in June 2005, "New Directions in Spaceflight Behavioral Health," a special supplement of *Aviation, Space, and Environmental Medicine* edited by Albert A. Harrison, Ph.D. and Sarah A. Nunneley, M.D. (*www.asma.org*).

5.3.2 Countering the downside aloft

All of the negative aspects of space living also have their positive sides or compensations; by careful planning, the downside may be countered. Here are some examples of what has been learned from spaceflight which could be used to increase productivity:

- *Space nausea*, for instance, may be relieved by medication and by curbing excessive exercise before departure.
- *Vestibular system balance* in the inner ear is influenced by information from the eye; in zero-*g* the vestibular organ fails to work at first because lack of gravity sends mixed signals as to what is up or down, but after two days in orbit the brain learns to ignore the input from the vestibular system and accept the information from the eye as correct; on return to Earth, astronauts are visually programmed, and it again takes two days to readapt to maintain an upright body position (blindfolds can accelerate the process).
- *Bone loss in microgravity* causes osteoporosis. Data reveal that in sustained space travel upward of 3% of total skeletal bone mass is lost within the first two months

aloft. Mechanical forces, such as hydrodynamic flow and elongational strain, play a role in the normal physiology and pathology of blood vessels, bone, and muscles. Dr. John Frangos, professor of bioengineering at the University of California-San Diego, hypothesized that the hydrodynamic forces in bone induced by compression might stimulate bone formation, so his research team developed both *in vitro* models and *in vivo* studies, which support that hypothesis. To increase blood volume before entering the Earth's atmosphere, antidotes are proposed, such as salt tablets and more fluid intake, as well as vigorous exercise before re-entry.

- *Decompression sickness* prior to an EVA which requires a spacesuit can be countered by pre-breathing pure oxygen from one to four hours to purge nitrogen from the body or by gradually lowering the pressure in a space suit or lunar habitat.
- *Computer monitoring* of medical patients in space by an onboard Health Maintenance Facility helps. This includes measuring and administering intravenous fluids over extended periods and using exercise devices to counter bodily changes. In addition, on the International Space Station, NASA provides another Human Research Facility for data collection in the orbital laboratory which includes scientific instruments for use by crewmembers with paramedic training. Both Lockheed-designed facilities and information systems are linked not only to the large Space Station Data Management System, but to a Life Science Medical Operations Computer used by consulting flight surgeons on the ground. These systems collect unique performance records to improve future mission productivity and quality of life aloft.

Experience has demonstrated that humans generally find space work exhilarating, and over periods of many weeks can maintain high level of work competence. Many astronauts and cosmonauts, like the first American woman aloft, Sally Ride, have written positively in books about the spacefaring experience [17]. The uniqueness, importance, and challenge of exploration are among the many factors driving spacefarers to peak performances. Yet, in the immediate decades ahead, more studies need to be directed toward broader issues of crew performance. Analysis by traditional human factor specialists will have to be combined and integrated with broader research by behavioral scientists. Combined life science experiments of global space agencies are necessary for future joint missions regarding the safety, wellbeing, and productivity of space explorers. The fact that only 22 astronauts and cosmonauts to date have lost their lives in the performance of space duties is remarkable. Continuing research and precautions by space organizations, however, should not only ensure survival, but also consider how people can become more productive and comfortable aloft.

As far back as February 1966, the U.S. Space Science Board issued a report on advanced manned programs. Because of the unknown stresses and unexpected responses of the space environment on crew health and performance as the duration of flights increased, the Board urged ground-based simulated research. Even then their findings underscored that psychological and behavioral investigations were largely being neglected in the manned space program [18]. There is sufficient evidence that today this situation is still deficient in the American, but not in the Russian and European space programs. Thus, the urgency for *more comprehensive international* human studies takes on new meaning in light of current planning for a space station, lunar outpost, and manned Mars mission. Perhaps the time has come for national space agencies to support a single Institute for Human Behavior Research in the Space Environment, possibly in conjunction with the International Space University at Strasbourg. A positive step in this direction was taken in preparations for ISS construction. Six space agencies and seven U.S. research institutes cooperated in March 1998 for a Shuttle mission dedicated to neuroscience research. The scientists sought to learn more about the effects of:

- prolonged weightlessness on astronauts' motor skills;
- reduction in work rate in microgravity;
- rapid reactions to emergency situations and safety problems;
- subtle loss of eye–hand coordination;
- switching from using balance organs in the inner ear to strictly visual clues;
- adaptation of vestibular system and syndrome.

The multinational scientific team supervised 26 experiments aloft conducted by 7 astronauts in orbit. It is this type of synergistic, international research that should be ongoing through an institute such as the one the author proposes above, but which also includes cross-cultural behavioral science investigations.

Wubbo J. Ockels, a former ESA astronaut (see Exhibit 58), has made an impressive case as to why increased life science research should be conducted in space, as well as on the ground. Dr. Ockels, ESTEC chair of aerospace engineering at the Technical University in Delft, observed that for two billion years life has evolved subject to Earth's gravity. Such a relationship, he observed, is a major scientific and philosophical question. Thus, he proposes more orbital studies into this relationship because [19]

- microgravity adds a new dimension for humans that is five to six magnitudes greater in its effects;
- limited spaceflight findings justify further studies of the effects of weightlessness on body process and immune system behavior, especially with regard to our perceptions of up and down, and the difference between horizontal and vertical (anisotropy);
- effects of improvements are indicative of our flight with gravity, so how does life interact with gravity and is it a necessity for life?

Ockels pursues such matters related to European manned spaceflight through the ESA's Space Research and Technology Centre (ESTEC) associated with his university in the Netherlands.

In the above context, it is no wonder that the American Medical Association passed a resolution in June 2007 supporting human space travel, citing *potential future benefits to medicine and advances in patient care.* The AMA's House of Delegates affirmed that medical research was under way on the Space Shuttle and International Space Station. NASA's space exploration has furthered progress in medical science, from diagnostics and telemedicine to a Space Shuttle–derived heart pump. Currently, research is under way on astronauts in orbit relative to sleep, nutrition, immune system, isolation/confinement, and countermeasures for loss of bone mineral density and muscle strength, as well as other health challenges encountered in microgravity (see Appendix D).

5.3.3 Human sexuality and family offworld

There are many performance areas related to the orbital environment yet to be researched, such as human sexuality and family life. Georgetown University researchers Dr. and Mrs. Angel Colon have suggested five such areas for investigations, as illustrated in Exhibit 66 [20].

To appreciate the significance of the above five areas for further study, let us focus on just one dimension: circadian cycles [21]. Before sending settlers and families to a lunar colony, more research is needed on how this would affect the body's internal clocks. Today's spacefarer on the Shuttle takes 90 minutes to make one orbit around the Earth, so that crewmembers pass through 16 sunsets and sunrises

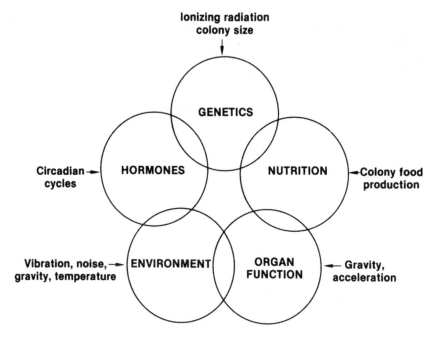

Exhibit 66. Human factor research areas for space family living. Source: Dr. and Mrs. Angel Colon, The Children's Medical Center, Georgetown University Medical School, 3800 Reservoir Rd. NW, Washington, D.C. 20007 [20] (email: *tir.na. nog@comcast.net*).

each day in outer space. This disorients the space traveler's SCN (suprachiasmatic cell nuclei); this hormone ebb and flow cycle influences sleep and wakefulness. Thus, astronauts aboard a Shuttle sleep almost two hours less than when back on the ground, and tend to increase the intake of sleeping pills. Dr. Charles Czeisler who runs the Circadian, Neuroendocrine, and Sleep Disorders Medicine Section at the Brigham and Women's Hospital in Boston, is investigating the effects of melatonin on Shuttle crews. This hormone, made in the brain's pineal gland, affects biological rhythms, so is used in the treatment of people suffering from jet lag and shiftwork. Obviously, such experiments have implications for both long-duration missions and space settlement. It is particularly relevant to living and working on the Moon which for the most part has two weeks of light and two weeks of darkness, except for the poles where there is almost perpetual light. Malapert Mountain at the South Pole has been suggested as a lunar base site because it not only has 90% sunlight, it also faces Earth, thus facilitating continuous communications. Psychologically, the ability to see the Earth may prove important to lunar dwellers [22].

To its credit, NASA recently held a workshop on human reproduction in space; and Ray Noonan wrote a doctoral dissertation on sex in space, a subject generally avoided within global space agencies.[3]

5.4 ORBITAL TEAM PERFORMANCE

The dangers and complexities of living and working on the space frontier make group cooperation and cohesion necessary for survival; this also applies to high performance. As woman astronaut Rhea Seddon so aptly put it: "We are no longer the solitary warriors of the early space program; we are more team players." In his interview with veteran astronauts to improve manned space station planning, flight surgeon Douglas [12] also learned that they recommended that

- crew selection include some peer review and input;
- combined groups should work together for a long period of time before launching;
- it is desirable for crews to have some group dynamics training;
- a lack of psychiatric or psychological support for both astronauts and their families in the past was a NASA shortcoming.

In his conclusions, Dr. Douglas suggested that future station crews would benefit by survival training as a means of fostering social interaction. In a survival situation, even simulated, this physician rightly maintained that people have to learn to cooperate, as well as to make allowances for the emotional needs and personality idiosyncrasies of their colleagues. During such training, he proposed it is easier to identify insensitive, self-centered, and selfish persons who should be excluded from

---

[3] See R. J. Noonan, *Philosophical Inquiry into the Role of Sexology in Space Life Sciences Research and Human Factors for Extended Spaceflight.* New York: New York University, 1998.

the crew. Since human performance is dependent on more than mechanical or environmental improvements, this specialist in aerospace medicine suggested that some type of psychological orientation would be valuable for future station crews, so that the spacefarers could deal with individual and group needs at the human level. It is in this area that behavioral scientists could contribute with their training methods for crew team building, decision-making, and stress reduction.[4]

The International Space Station is an ideal site for human relations research that centers on team performance in an isolated confined environment. It is already happening informally (as Exhibit 67 confirms), but studies should be planned and organized investigations.

5.4.1 Orbital team building

Management scientist Dr John Nicholas of Loyola University, Chicago, has published several interesting papers on space station crew performance and productivity [23]. He is particularly concerned about interpersonal and group behavior aloft, and how pre-departure training can facilitate team performance. Nicholas contends that earthbound analogs of group behavior studies in isolated environments are consistent with reports from Soviet Salyut experiences; for example, 30 days into the station mission, interpersonal hostilities grow between crewmembers, as well as between crew and ground control. However, that has not been borne out in the multinational crews aloft for six months on ISS. However, his arguments for more research in these areas are valid. He calls for further research into

(a) crew group interactions and their impact on output, such as performance and morale;
(b) group communication with controllers;
(c) group norms in orbit, such as treatment of outsiders or visitors;
(d) group leadership in terms of designated commander, and real leader(s) who emerge on long missions.

After a careful literature review of early station experiences in the 1980s, Nicholas maintained that the current manned space program focus is on individual astronaut and environmental inputs, which is insufficient. The professor wisely observes that attention to human factors, task design, and mission planning cannot alone ensure effective group interaction, and may result in lowering performance levels. Dr. Nicholas rightly advocated that space station crews be trained in interpersonal, emotion support, and group interaction skills.

Dr. Nick Kanas believes that physicians assigned to the Space Station need special skill training for dealing with psychological and social problems aloft [24]. He is convinced that when space settlements are established, space psychology and psychiatry will find fertile fields for practice there.

---

[4] For example, Orasanu, J. "Crew Collaboration in Space: A Naturalistic Decision-Making Perspective," *Aviation, Space, and Environmental Medicine*, **76**(6), June 2005, B154–B163.

Exhibit 67. ISS TEAMWORK.

In March 1998, the first crew assigned to the partially constructed International Space Station flew on a Russian Soyuz spacecraft. The team consisted of astronaut William Shepherd, Commander, and cosmonauts Yuri Gidzenko and Sergei Krikalev. They had trained together at the Gagarin Cosmonaut Training Center in Star City which included language requirements, roles, and responsibility. The term "expedition" was to be the flight designation, and this was Expedition 1. Although all three spacefarers were fluent in English, Russian was the operative language, since the hardware and controllers were from Russia. This and subsequent station crews worked well together, as the following anecdotes indicate:

- *Expedition 13.* March 2006 launch: Pavel Vinogradov, Russian commander; Jeff Williams, American flight engineer; Thomas Reiter, ESA astronaut; Iranian-born Anousheh Ansari, first woman space tourist, joined the Expedition 13 crew on Soyuz TMA-9. Key milestones according to Col. Williams, who was also ISS science officer were a host of experiments, including one dubbed SPHERES which tested the ability of two free-flying satellites to determine their position in space, setting an orbital photography record for the station at the 240,000 mark. During the six months' posting, he spent 19 hours outside the station, and racked up an orbital living record of 193 days.
- *Expedition 14.* September 2006 launch: NASA astronauts Michael Lopez-Alegria, Commander, and Sunita Williams, engineer, completed a record spacewalk of six hours and 40 minutes on February 8, 2007 which included a photo survey of the Station Shuttle docking port. Also aboard for a six-month stay were cosmonaut Mikhail Tyurin and ESA astronaut Thomas Reiter. In the course of this expedition, there were 14 crewmates aboard this orbital laboratory engaged in outfitting ISS, adding research facilities while accomplishing many scientific experiments relative to nutrition, oxidative stress, radiation exposure, and sleep patterns. During his stay on ISS, Reiter conducted 19 experiments, focusing on areas such as human physiology and psychology, microbiology, plasma physics, and radiation dosimetry as well as technology demonstrations.
- *Expedition 15.* December 2006/April 2007 launches: Fyodor Yurchikhin, Commander from RKA; flight engineers Oleg Kotov and Clayton Anderson; Sunita Williams remained on the station. Kotov, the Soyuz commander and a physician, took the lead in conducting Russian scientific experiments. Before the departure of Tyurin and Lopez-Alegria, the joint crews celebrated Russia's Cosmonautics Day, honoring Yuri Gagarin's historic first launch on April 12, 1961 that signaled the beginning of the era of human spaceflight. Another American space tourist, billionaire Charles Simonyi, accompanied the replacement team and returned with the two being relieved. The twin launches and dockings by both Shuttle *Discovery* and Soyuz TMA-10 were exceptional parts of this expedition.

Perhaps Ukraine native Col. Oleg Kotov, 41, best summarized the high performance of these three expeditions with this comment: "Mankind always has to have an idea or a goal for the general imagination to move forward!"

Source: News reports from November 21, 2006 to April 8, 2007, on *www.space.com/mission launches.*

The current research of Dr. Jennifer Boyd Ritsher confirms the points raised in Chapter 4 (email: *ritsher@itsa.ucsf.edu*) [25]. She concludes that cultural–psychological factors are especially important for mission success on the International Space Station. Dr. Ritsher is concerned about cultural bias in assessment methods, culture-related risks in relations, and mental health with international partners, as well as better methods for improved prediction, prevention, and performance of crews on long-duration missions.

Similarly, Dr. Barrett S. Caldwell has focused on multi-team dynamics and distributed expertise in ISS operations (email: *bscaldwell@purdue.edu*) [26]. As an industrial engineer, he views space exploration as an evolution in awareness of social and socio-technical issues related to performance and task coordination. He confirms your author's position that spaceflight operations, both on the ground and aloft, create a unique environment in which to apply the classic expertise of group dynamics. Such knowledge is particularly relevant to communications, group process, skill development and sharing, and time–task critical performance. Multi-teams of mission control and flight crews offer an opportunity for study that will improve knowledge sharing and task sychronization, as well as morale. Dr. Caldwell is examining issues of distributed coordination by multiple teams, first perceived when matrix management was created during the Apollo mission era. The complexity of space team performance is increased by the complexity of the technology being used within an extreme environment. That is why intra-team training (i.e., between the ground-control team and the orbiting team of spacefarers) is called for, especially on long-duration missions.

5.4.2 Crew-training strategies

Apart from issues of mental health and therapy, there are forms of *group learning* sessions that can improve personal and vocational performance. As someone associated with the approaches of sensitivity, organization development, and leadership training by the National Training Laboratory (NTL Institute of Applied Behavioral Science), it is obvious to your author that educational programs have been successfully designed to benefit group interaction and performance among average executives, managers, and professionals. In modern industry, there is another group technique called *quality circles* that has immediate implications for space agency and aerospace workers here on Earth, and possibly downrange for those in orbit. Quality circles were invented in the U.S.A., and then transported successfully to Japan before being re-exported to American organizations. This team strategy could do much to counteract the type of inadequate workmanship that went into solid rocket booster O-ring seals (Presidential Commission on Space Shuttle *Challenger*, 1986 [15]). It could ensure high quality in the construction of future orbiting stations and planetary bases. The International Association of Quality Circles (801-B West Eighth St., Cincinnati, Ohio 45203) has defined the term as follows:

> "A quality circle is a small group of people who voluntarily meet on a regular basis, to learn and apply techniques for identifying, analyzing and solving work-related problems. This process is also known as employee involvement."

There is a whole body of social science research literature and technology on group dynamics and leadership development that is directly transferable to preparing people for long-duration spaceflights. Among those on work performance, there are studies and techniques that center on such issues as *team development* or team building, self-directed teams, collaboration in organizations, and improving work groups. My own research and experience as a behavioral science consultant confirm that *team* development and management are central to the new space work culture [27]. If major corporations can utilize *team building* with their personnel, surely it is safe for space organizations to consider such training for both its ground-based and its space-based groups. In the practice of matrix management, a process pioneered by NASA and its aerospace partners in the 1960s for the Apollo missions, team development is standard practice with project teams in such contractors as Hughes Aircraft and TRW Systems. Exhibit 68 demonstrates the convincing case for team building among spacefarers.

Long-duration mission research confirms the importance of teamwork, co-ordination, and conflict resolution for the survival and performance of space crew operation in their high-technology orbital environment. Teamwork leverages human performance through the sharing of talent and by cooperative action. When functioning effectively as a team, people focus more brainpower and energy on task achievement. Today, high-output management and the high-tech work environment both require the development of group interaction skills, especially when dealing with cross-skill training. This is exactly the situation one is likely to meet in space, but with the added dimension that cohesive teamwork is necessary for survival.

High team performance (e.g., on a space station) means that the work unit is a cohesive group with a common purpose; that members develop mutually helpful relationships in the accomplishment of agreed goals, objectives, and tasks. We have seen such prototype behavior among astronauts on Shuttle flights, as well as among cosmonauts at the Russian station. For example, a week into his 1994/95 mission on Mir, ESA astronaut Ulf Merbold reported a potentially disastrous loss of electrical power. The problem was caused by having six spacefarers on board when the life-support system was only designed for five [28].

> "We lost all systems. We could not use the radio. My colleagues calmly analysed the situation and managed to reactivate one of the little Soyuz TM capsules that are attached to Mir. Soyuz has its own batteries and we used the capsule's power to communicate with the ground."

Then the cosmonauts maneuvered the station using the reaction control thrusters on the Soyuz TM spacecraft to orient it so its solar arrays would point toward the Sun, but the power loss meant less time for experiments, some of which had to be canceled. The Russian space agency said the problem originated from excessive power consumption by video and photographic activities.

The challenge is to sustain such creative problem-solving, fellowship, and co-ordinated efforts on long-duration missions. Let us assume that a station or lunar base has a crew of eight divided into two shifts or teams of four. Before launching,

Exhibit 68. Orbital team satellite repairs. The historic first salvage mission of a satellite into the payload of the Shuttle *Discovery*; the first of two such recoveries on mission 51-A. During the 8-day mission beginning on November 8, 1984, two astronauts demonstrated superb teamwork by retrieving two communications satellites for repair. Again in 2007, the continuing teamwork between astronauts was evident in difficult and delicate repairs made to ISS solar panels. Source: NASA Johnson Space Center.

group dynamics or team training could ensure that these people were functioning and achieving together as a unit.

Two British management consultants, Dave Francis and Don Young, described the characteristics of a high-performing team which are pertinent to space crews [29]:

Output   Combined results produced beyond any individual contributions.
Objectives Shared understanding of purpose and mission
Energy   Motivated members who take strength from one another, promoting group synergy.

Structure   Creative mechanisms evident for dealing with organizational issues (procedures, roles, control, leadership, etc.);
Atmosphere   Manifest spirit and culture of mutual interpersonal risk-taking and confidence-sharing.

5.4.3 Strengthening team culture

Before going aloft, such characteristics can be cultivated in groups by skilled consultants to create a *team culture*. Crew leaders on the ground can learn to facilitate, foster collaboration, seek consensus, and even promote participative decision-making when appropriate. On the ground, crewmembers can be trained in human relations and shared leadership behaviors. The latter operates at two levels: task and maintenance. *Task leadership* refers to getting things done effectively on schedule, and includes such activities within the group as initiating and proposing tasks, defining and solving problems, seeking and obtaining relevant information, clarifying and elaborating, analyzing and interpreting, action planning, and implementation. *Maintenance leadership* in teams contributes toward group cohesion and morale, while promoting supportive behaviors such as showing acceptance and friendliness, learning to share oneself and others, giving recognition and empathy, harmonizing or reconciling differences, lessening tensions and resolving conflicts, exploring differences and feelings, negotiating and obtaining compromises, communicating and receiving feedback sensitively. All crewmembers can learn to exercise their unique talents for one or both forms of leadership, and at times to exercise both sets of behaviors. There is little evidence that any space agency is systematically providing such behavioral science training to spacefarers as a regular part of their training.

In a sense, space missions encompass a sophisticated form of project team management. Top performance can result when the groups involved receive both technical and interpersonal training. Technical skill development centers around exercising task leadership in defining and analyzing the work to be performed in space; in planning for and wisely utilizing resources; and in setting performance objectives, priorities, and standards. In time, such space team management could include budget development, funding and allocation; recruiting and developing team members; establishing controls and meaningful supervision; facilitating communication, reporting, and evaluation systems [30]. On the other hand, *interpersonal skill development* for group maintenance leadership involves team building, which helps members to be more authentic with one another, more experimental and flexible, more spontaneous and sensitive, more collaborative and mutually supportive. For extended stays in space, ground-based research should combine both sets of skills to strengthen self-confidence, improve group morale, and enhance problem-solving capabilities among crewmembers. This is but another example of the contribution that behavioral scientists could make to improving the quality of space life, as suggested in Chapter 3.

There is ample evidence from behavioral science research that high team achievement and synergy occur when members

- take interest in both individual/group accomplishment;
- tolerate ambiguity and seeming lack of structure, especially during times of uncertainty;
- give and take feedback freely and non-defensively;
- contribute toward an informal, comfortable, and non-judgmental atmosphere in the group;
- are capable of establishing short intense work relationships, and then disengaging on mission completion;
- encourage group participation, consensus, or decision-making;
- can manage change and promote innovation;
- value listening and authentic communication;
- can periodically clarify roles and relationships.

To avoid psychological and social disasters on future long-term space missions, especially at a lunar base, it would behove organizations sending spacefarers into orbit to build on such behavioral science research with groups and to enlarge its "human factors" training. If so, sponsoring organizations should give consideration to such matters as described above. Then, they could incorporate these strategies into

(a) the recruitment and selection of future crewmembers;
(b) the preparation and training of spacefarers;
(c) the evaluation of space station management experience with a view to developing future space settlements.

Hooper described in detail *The Soviet Cosmonaut Team* of the past, including the program of the Gagarin Cosmonaut Training Center; Bluth and Helppie identified the many practices on Russian stations that contribute to more effective crew performance [31]. These range from policy and environment to technology and organization. Under the latter, these researchers discovered that the general two-year training program for cosmonauts includes psychological training, on which much emphasis is placed. When preparing for specific missions, intensive group dynamic preparations are given to station crews before departure. The latter include development of joint action skills, particularly in critical and stressful situations, group interactions, and games to enable cosmonauts to work together as a team. Crews are selected for compatibility, and then trained to work together effectively. Not only is a psychological support team provided on the ground, but behavioral scientists may visit the station. ESA and NASA astronauts who go aboard for long missions should benefit from similar preparation, and such an approach could be adapted for those assigned to the International Space Station.

In summary, the aim is to develop a space team worker who is more concerned with crewmembers as a whole than his or her own ego needs. Such a spacefarer would be capable of building on individual strengths, while delimiting weaknesses; of facilitating team achievements, while minimizing frictions and antipathies. It is management that installs the mechanisms and means for fostering space team skills and culture which optimize individual and group performance. Within the global space field or industry, the trend today is to move beyond teams to building *collaborative organizations*, whether on the ground or in orbit [32]!

5.5 HUMAN/MACHINE INTERFACE ALOFT

Another way to improve human performance in space is through superbly crafted machines, tools, and habitats. In this category is the extensive utilization of automation and robotics (A&R), and the preparation of human operators or monitors, so that information systems and "tin-collar" workers can be used effectively. In this high-technology environment, whether on the ground or in space, the new technologies transform work, roles, and relationships, while helping people to be more productive. When due regard is given to work ergonomics, so new computers, robots, cybernated or expert systems can be designed to enhance the human performer. Robots in space, for instance, can take over undesirable, dangerous, tedious, or automatic jobs, they can go where humans cannot or dare not go. Exhibit 69 illustrates these possibilities on the Moon, where artist Pat Rawlings envisions an exploration program employing robotic rovers to conduct a lunar site survey. Automated workers help in selecting a suitable place for a piloted landing. Launched on expendable rockets, the proposed Artemis lunar lander is the size of a compact car.

Automated spacecraft made possible the growing communication satellite industry, and remote sensing may be the next lucrative space market. Unmanned spacecraft have already proven how they can probe the far corners of the universe at immense distances without jeopardy to human sponsor and supervisor. For example, in 1997 the Cassini spacecraft was launched by NASA on a 3.5-billion-kilometer journey to the planet Saturn. Standing two stories tall, this huge spacecraft used the decay of radioactive plutonium-23 to generate the electrical power that would be required for its 7-year journey to Saturn and subsequent exploration of the planet, its rings and satellites; the mission was both complex and costly at $3.4 billion. Cassini carried the European Huygens probe, which was released to land on the surface of Saturn's big cloudy moon Titan in January 2005. Pictures of this remarkable world were sent back to Earth, and broadcast to millions via the Internet and high-definition TV, For the vast majority, who cannot travel into space, robots are the next best thing. As *Cassini* set off on its journey, the *Galileo* spacecraft was orbiting Jupiter, returning astonishing images of this gas giant planet and its diverse satellites. Also in 1997, JPL landed the *Pathfinder* craft with its robotic rover on the Martian surface, and *Global Surveyor* successfully entered orbit around the Red Planet. In 2008, JPL scientists enjoyed further success when the *Phoenix* lander touched down in the Martian Arctic to collect and analyze samples of dust and ice. Such feats demonstrate that exciting, scientifically meaningful, and less expensive space exploration can take place without the presence of humans, except by extension of our brain power. Already plans are underway for more sophisticated roving vehicles that will be intelligent and highly mobile; they will seek out samples, on Mars for example, for subsequent return to Earth. There are also plans for balloon probes and even robotic airplanes.

5.5.1 Orbital team mates

But when humans are in orbit, people worldwide really get excited, especially when they team up with robots. The automatic arm developed by the Canadians for use

Sec. 5.5] Human/Machine interface aloft 223

Exhibit 69. Human/Robotic lunar prospects. With Earth in the background, artist Pat Rawlings envisions a near-future exploration program when robotic rovers conduct a lunar site survey for piloted landings. Through its telescopic eyes, millions of earthly eyes would be able to witness humankind's return to the Moon. Source: NASA Johnson Space Center.

on the Shuttle has demonstrated its potential in launching satellites or for space construction. In 2006, mission specialists on STS-115, Daniel Burbank and Steven MacLean, were both spacewalkers and robotics officers. From Shuttle *Atlantis*, they delivered port-side trusses and solar panel wings to the International Space Station. McLean from the Canadian Space Agency used the robotic arm with its 50-foot sensor boom to move 17.5 tons of girders, batteries, and other hardware to the orbiting station, then empty of humans ... Space has become a laboratory for pragmatic tests of teleoperations (manipulator control by a human operator by telepresence); and telepresence (when a remote systems operator feels as if he or she is actually present in the workplace through sensory inputs and feedback).

The Universities Space Research Association under the leadership of Dr. David Criswell conducted a NASA-funded workshop and report on *Automation and Robotics for the National Space Program* [33]. It was congressionally stimulated and mandated with a view to construction of the Space Station. That significant report's foreword by futurist John Platt viewed a space station as a new tool for augmenting human capabilities. Dr. Platt described how the unique environment in space can be used for creating new structures, such as automated factories powered by solar energy and operated continuously and unattended while economically processing asteroidal or lunar materials. The noted professor and humanist wrote that the design and maintenance of balanced ecological systems for space habitats provide learning regarding the ecological systems of our own planet. If A&R in space is human-centered, Platt believes these systems can greatly amplify the powers of human beings and their technology. In the findings section of this California Space Institute report, expert panelists concluded that by building A&R into space station design, there would not only be many spin-off applications on Earth, but the knowledge gained will be used to accomplish, better and at lower cost, an even wider range of space missions. For our purposes here, their most astute observation is:

> "Most important, humans can progress to increasingly higher-level tasks as automated systems take over routine jobs. Such achievements require fundamental knowledge of how to organize exceedingly complex, even somewhat 'self-aware' systems. NASA Space Station programs can demonstrate complex service functions in 'smart' computer systems or extend the electronic reach of humans."

Finally, the Criswell report also envisioned that by 2010, robot teams will be competent aides to humans, carrying out work assignments with high-level crew supervision. These automated workers may even be human-like in appearance, capable of human-like dexterity and grace, and of moving their mechanical arms quickly, smoothly, and precisely while performing required tasks deftly and unerringly. Robots are expected to perform payload deployment, inspection, maintenance, and repairs. Human communication with them will be through efficient, user-friendly interface equipment, sometimes at a detailed, high level of exchange. NASA's Technical and Management Information System (TMIS) uses advanced computing technology for space station design and management, believing A&R technology is for "people amplification". The next step after intelligent robots is likely to be androids: in 2007, a Japanese inventor announced his *Geminoid*, a robot which replicates his own voice and appearance!

Rather than replace people, NASA seeks a human/machine system that integrates standards and expands human capabilities so that A&R operators can move to supervisory and managerial positions. In this way, increasingly sophisticated A&R applications have assisted in the building of the ISS, furthering the technological base, while enhancing human involvement. For example, NASA's Office of Aeronautics and Space Technology has been working on telerobotic control by means of which the human operator wearing a special helmet will someday be able to operate a remote manipulator system or orbital maneuvering vehicle. Work under way at NASA field centers and among station contractors indicates a variety of uses for automation and robotics, ranging from electrical power systems, docking and payload inspection to station monitoring, navigation, laboratory management, and virtual reality systems. John Hodge, when director of the NASA Space Station Task Force, stated:

> "Our experience in space tells us that some tasks are best performed by automated equipment and that other tasks require the unique responsibilities of Man. Tasks that are routine and which can be preprogrammed, tend to be best performed by machines. Tasks that require initiative and judgment are more suitable for Man. The Space Station will have plenty of both kinds" (*Aviation/Space Technology*, Spring 1986, 23).

Steven Skaar of the University of Notre Dame and Carl Ruoff of Caltech's Jet Propulsion Laboratory maintain that space robots differ from their terrestrial counterparts because of weightlessness, vacuum, thermal environment, and the need to minimize mass. Thus, there is an economic tradeoff to consider when it comes to using manned vs. telerobotic space operations, dynamics, and control [34].

Harrison and Connors suggest that crew systems depend on the success of human and technical combinations. In their previously cited study [7] for NASA Ames Research Center, they explained that this means combining human and artificial intelligence; matching technical systems to behavioral propensities of their human users; and coordinating efforts by means of telecommunications. For these behavioral scientists, technical systems must meet the user's emotional, social, and cognitive requirements. In this fusion of "humans and machine", they view a crew system as a *social cyborg*, a social system with expanded capabilities, an entity which can survive and perform where anything else would falter.

The Economist (February 1, 1986, p. 11) pragmatically editorialized that although many jobs can be done by robots, taxpayers will not support a space program that denies them the thrilling spectacle of human beings cavorting in weightlessness or drifting on slender tethers far above a distant Earth. While some space work may be symbolic, other jobs can only be performed by humans. Yet, continuing innovations with automation and robotics in space will not only have direct benefits for improved quality of life on Earth, but contribute to the actualization of human potential and performance. Perhaps the best example of a human/machine interface was when astronauts launched and repaired the automated Hubble Space Telescope, as shown in Exhibit 70.

The public is not fully aware of the extent to which robots are already being employed in space exploration. Robotic systems fall into three categories: extra-vehicular robotic servers, science payload servers, and planetary surface rovers. Such systems significantly improve mission operations, safety, and cost-effectiveness. These automated extensions of human intelligence learn from their experience, becoming self-maintaining and self-replicating in time. The technology drivers for these systems include enhanced collision detection and avoidance, advanced local proximity sensing, task-level control workstations, improved command and control architecture, and there is much more to come. In the near future, expect *Robonauts* to assemble large structures at Gateway L1, a point of gravitational balance between Earth and the Moon. On the lunar surface, anticipate that mini-robots containing

Exhibit 70. SERVICING HUBBLE IN ORBIT.

In April 1990, NASA launched the Hubble Space Telescope from the Shuttle's payload bay into an orbit about 600 kilometers above the Earth. Weighing as much as two adult elephants, the spacecraft has already completed about 93,000 trips around our planet, producing an unprecedented three-quarters of a million photographs of 24,000 celestial objects and phenomena. John Grunsfeld of the Space Telescope Science Institute, command center for the orbiting observatory, said that:

> "Hubble has transcended being an astronomer's tool for science, becoming an icon for science around the world."

After a power failure aboard HST, NASA administrators decided to decommission the aging spacecraft, so as to save costs. Grunsfeld led the global protest for another Shuttle mission to repair the telescope which is in a different orbit from that of ISS. The current administrator, Dr. Michael Griffin, announced the 2008 risky mission by astronauts to fine-tune Hubble so that it can continue to function another five years at the minimum. The spacefarers will do five spacewalks to install instruments that include a new camera and spectrograph which will use ultraviolet vision to study the galaxies and changes in the universe's structure. Already Hubble has helped astronomers calculate the age of the universe (13.7 billion years old); detected proto-galaxies that emitted their light when the universe was less than a billion years old; proved the existence of super-massive black holes in the centers of many galaxies; showed that the process of forming stars with planetary systems is common throughout the galaxy; and has even deteted the atmosphere of an exoplanet.

New tools had to be invented, like a mini power screwdriver, so the astronauts can make complex repairs to the power supply. The "service team" have been practicing for many hours in a simulated environment. In addition to Grunsfeld, the STS-125 *Atlantis* repair crew will consist of mission commander Scott Altman, plus Gregory Johnson, Andrew Feustel, Michael Good, Michael Massimino, and K. Megan McArthur. Peak performance by these seven may keep HST in orbit until 2021 when it is hoped to have an observatory functioning on the Moon. This paradigm-changing, mind-blowing orbital telescope is history which humankind in the future will marvel at for its advancement of science, especially astronomy.

Source: Adapted from Jenna Bryner's "Hubble Hugger Eager to Fix Mind-Blowing Telescope": *www.space.com/news/science/astronomy*, January 1, 2007.

expert systems will construct many structures there, as well as do the principal mining and manufacturing.[5] In the 21st century, a blend of humans and robots will transform the Moon into a laboratory of learning, science hub, and jump-off point for further exploration of deep space [35]. Mindstretch in Exhibit 71 as *The*

---

[5] See "Lunar Robotic and Communications Systems," in D. G. Schrunk, B. Sharpe, B. Cooper, and M. Thangavelu, *The Moon: Resources, Future Development, and Settlement*. Chichester, U.K.: Springer/Praxis, 2008, ch. 4/Ap. A.

Exhibit 71. CAN ROBOTS BE TRUSTED?

- Dr. Vinton Cerf, one of the Internet founders, has been working with NASA to develop an interplanetary internet (IPN). The delay-and-destruction protocol requires provisions for time delays in space communications over long distances. Building such networks in outer space has been tested on the Mars Reconnaissance Orbiter, a probe that went around the "Red Planet". Such concepts are being adopted as part of a 40-year plan for deployment of space platforms. IPNs will eventually consist of interconnected planetary internets.
- After decades of speculation, the idea of space elevators is being researched, and NASA is providing an annual $400,000 prize fund to encourage the technology of building a lift from the Earth's surface into space. The concept calls for an orbiting satellite to be linked to the planet's surface by a cable, so that vehicles could climb up and down. To and from GEO would require a phenomenally strong, yet light-weight cable. By whisking cargo and maybe people in this manner, the cost of access into space would be dramatically cut, even if the elevator costs some $10 billion to build. Two companies have entered the competition and hope to make the space elevator commercially viable. In February 2006, LiftPort conducted its first test with hot-air balloons secured from a cable on which robots climbed up and down. The other firm is X Tech Projects, which is studying how to get a cable to reach GEO some 22,250 miles away!
- Fuel cells capable of powering portable electronic devices are finally heading toward the marketplace, and may someday be used by spacefarers. Each fuel cell is about the size of a cigarette pack, weighs 150 grams, and generates electricity by combining oxygen from the air with an internal fuel. It can be plugged into hand-held devices for 30 hours of talk time, or twice as much for playback time. This form of battery augmentation may someday be used in orbit for astronauts' laptop computers.

Source: Adapted from *The Economist Technology Quarterly*, June 10, 2006, Insert-pp. 3, 4, 33 (*www.economist.com/technologyquarterly*).

Economist, a respected business publication, reports about robotic possibilities on the space horizon.

5.6 HUMAN RESOURCE DEVELOPMENT OF SPACEFARERS

If history is an indicator, the experience in orbit itself advances one's personal and professional growth. As the numbers going to the high ground increase exponentially, more comprehensive and systematic human resource development (HRD) will be needed for spacefarers on the ground before departure, while in orbit, and on re-entry. HRD refers to organized activities that produce behavioral changes within a specific timeframe, such as happens through education and training, be it in groups or self-study. The American Society for Training and Development (ASTD) described human resource areas as concerned with improved and increased quality of work life, relative to productivity, job satisfaction, career development, and readiness for change. According to ASTD, the HRD arena encompasses a wide scope of organizational activities: selection and staffing; human resource planning, organizational/job design; organization development; training and personnel development; union/labor relations; employee assistance, compensation and benefits; personnel research and information systems. These activities on the human side of enterprise impact performance. Relative to present and future space programs, whether under government or industry sponsorship, HRD may be viewed in terms of space workers on Earth or on stations and settlements in outer space. Essentially, such human asset management and development is concerned with facilitating high performance among space workers, while sustaining knowledge and talent management [36].

With regard to the latter, your author has proposed that within the emerging space culture, consideration be given to establishing *space personnel deployment systems*, which will be discussed in Chapter 6. Essentially, this is an HRD strategy comparable with an Earth-based analog of foreign deployment systems used in the relocation of people to overseas locations and foreign cultures. Space personnel deployment systems (SPDS) are planned and ordered means for the exchange of people to/from Earth and the high frontier. As Chapter 6 will explain, these systems are meant to both ensure the safety, satisfaction, and development of spacefarers, while furthering their space missions and settlements. This approach centers around four components: assessment and orientation before departure into orbit, in-space services, and re-entry counseling. Chapter 6 will include further information on human performance in space within the context of the SPDS paradigm.

Obviously, national space agencies already have elements of such systems in place for their astronauts or cosmonauts. However, some of the approaches are either too narrow in scope, such as the de-briefing process, or unsuitable for larger groups of civilians who will eventually go into orbit. The contract workers who maintain the International Space Station, and the varied visitors to it, make that station a perfect laboratory opportunity to study ways for improving living conditions not only for station inhabitants, but for many who will migrate in the next century to other space bases and colonies. Under the Living Systems research strategy described in Chapter 3, the author was among a group of behavioral scientists who also propose development of a data or knowledge bank with a pattern recognition and retrieval system relative to the human role, performance, potential, and challenges when living and working in space. In addition to storing findings from Earth-based analogs of significance for orbital living, this information system could be a comprehensive global storehouse on the human experience and habitation in isolated, confined environments, such as space. By using the Living Systems template of Dr. James Grier Miller for cross-disciplinary analysis, such a framework and databank would enable planners and researchers to have easy access to the knowledge necessary for operating and managing human systems in space.

Space agencies already have prototype HRD programs which include data gathering about spacefarers. But these public organizations have been circumspect in their analysis and release of information about the human experience in space during the past 40 years. Admittedly, NASA has been rather forthright about medical and biological knowledge gained from previous spaceflights, and does publish an internal history. Yet, this same agency has limited the study or release of information on the psychosocial experience of its personnel in space. NASA generally has also limited access to the astronauts by social science researchers, even its own psychologists and psychiatrists; the agency has failed to exploit the data it has collected which could improve spaceflight and living for others to follow. There are indications that transcripts of crew communications going back to the Mercury flights have yet to be fully analyzed from a behavioral science context, so as to obtain clues to improve future missions and avoid tragedies. In the course of my own research, I have proposed that NASA sponsor an anonymous mail survey of past and present astronauts who have flown or are flying in space. Using questions based on both space and ground-based deployment analogs, valuable input could be provided from this expert group for the next generation of spacefarers. If NASA cooperated with the Russian Federal Space Agency in a comparable study of cosmonauts, then the combined findings would have global implications for the HRD of future space travelers! Although professionally a flight surgeon, W.K. Douglas [12] did attempt to gain such insights from ten astronauts, albeit with a very limited sample and some inadequacies in terms of methodology. But, he fully understood that pilots and astronauts have often been described as individuals who distrust or even dislike psychologists and fear physicians. Nevertheless, Dr. Douglas was able to sample astronaut opinions, beliefs, and experiences with a view to improving human performance on manned Space Station missions. The point is that persons who have been in orbit are a resource, both for information and as trainers, mentors, and consultants for preparing others to go aloft. We should capture their insights through instrumented surveys (i.e., surveys using instruments for data gathering), audio and video cassettes, as well as computer systems for the purpose of further space HRD. From these data, simulations, virtual reality, films, and other training aids could be developed for use with future spacefarers.

5.6.1 Educating for space

Michael Wiskerchen, director of the California Space Grant Consortium, believes that the success of human migration into space depends on an evolving educational system that emphasizes the influence of space developments on Earth and its inhabitants. He points to one strategy for human capital development: namely, the National Space Grant and Fellowship Program established by the U.S. Congress and managed by NASA. Today this is a network of 52 space grant networks in each state, serving 500 colleges/universities. Its aim is to advance the nation's science and technology competence, particularly with regard to aerospace research. The space grant model of human development is concerned with education and training of students from K-12, higher education, and life-long learning. It not only provides incentives for these people to become aerospace scholars and researchers, but fosters a student–mentor program for this purpose. For over 15 years this consortium has furthered aerospace human capital development. But the strategy needs to be expanded beyond technical education for space to involve ever larger numbers of teachers and students in enlightening and preparing the next spacefaring generation [37]!

Many innovations and a variety of undertakings are occurring to further space HRD, as indicated by the sample below:

- *The National Aerospace Training and Research Center* opened January 2007 in Southampton, Pennsyvania. NASCAR is testing exciting applications like the Space Training Simulator, a configuration that models spaceflight profiles of launches and re-entry. STS is intended to prepare crews and passengers for commercial space travel. Private and government space travel providers will be able to obtain training and research assistance at the new Center. Its G-FET II Human Centrifuge can achieve nine Gs or nine times the normal force of Earth's gravity (*www.etcusa.com*).
- *The National Space Biomedical Research Institute*, a consortium of 70 institutions studying health risks in long-duration spaceflight, is sponsoring Graduate Education Programs that broaden students' academic and career skills in space life sciences. NSBRI in Houston encourages leveraging existing academic institutional offerings to develop cost-effective approaches that prepare graduates for entry-level positions at NASA and other space organizations. The focus is on Ph.D. or equivalent students in the fields of biomedicine, engineering, or other fields applicable to space life sciences (*www.nsbri.com*).
- *NASA Interns*, a variety of internship opportunities at its various space centers and laboratories are available for a summer or more. Placements range from the NASA Robotics Academy and GSFC/Johns Hopkins Applied Physics Laboratory to Summer Institute of Engineering and Computer Application, and the Undergraduate Student Research Program. For further information, contact
  — *http://university.gsfc.nasa.gov/application*
  — *www.oai.org/education/student/lercip.html*
  — *http://psu.edu/spacegrant/highered/research.html*
  — *htpp://learn.jpl.nasa.gov/higher_ed/graduate.html*
- *Space Resources Roundtable Inc.* is a non-profit organization whose vision is to promote the development and utilization of space resources for the benefit of humanity. Its members constitute a broad spectrum of technically oriented, professional space experts, planetary geologists, mining and aerospace engineers, energy specialists, economists, and space legal experts. The Colorado School of Mines sponsors the SRR annual conference for the continuing education and enjoyment of participants. The Roundtable is also contributing to the National Research Council Solar System Exploration Decadal Study, while supporting the World Space Congress held every ten years. For information, contact Dr. Alex Ignatiev, Director Space Vacuum Epitaxy Center (Physics Dept., University of Houston, Houston, TX 77204; tel.: 713/743-8135).
- *Space Corps for Volunteers* is a proposal for NASA Educational Enterprise put forth by Kathleen M. Connell. The strategy would be based on the Peace Corps model, and seek volunteers among children, youth, and adults from global spacefaring nations. This space cadre would aim to foster (1) migration of life

from the home planet; (2) the search for life in the universe; and (3) utilization of space knowledge and resources for the benefit of humanity. Participants would use group meetings, conferences, and publications, especially via the Internet, to promote these goals. This educational enrichment program would be constituted as an independent, non-profit foundation working in schools, colleges, and community and senior centers to engage the public in age-appropriate space learning and expeditions. The Global Space Corps might develop an alliance with the existing *Challenger Centers*, utilizing as mentors space experts, advocates, and spacefarers. The strategy would be funded by public monies, foundation gifts, corporate sponsors, and professional aerospace associations. If readers like the idea, then lobby for the plan with national legislators, space agencies, and societies, as well as with large foundations, plus the United Nations Committee on the Peaceful Use of Outer Space (UNCOPUOS) and the International Space University. To pioneer public initiatives and private ventures, such as this proposal, for the benefit of humanity, contact the Connell Whittaker Group, LLC (*www.thecwgroup.com* or email: *kconnell@cwgroup.com*).

5.6.2 Troubles in paradise

As far as the public knows, during the past 40 years there have been few public scandals in the American and Russian space agencies concerning astronauts or cosmonauts. That began to change in this decade, starting with problems on the station Mir, no longer in orbit. In 2007, NASA was faced with a bizarre encounter in which an experienced astronaut, Capt. Lisa Nowak, may have experienced a mid-life crisis or emotional breakdown when she seemingly threatened and possibly attacked a female Air Force captain whom she perceived as a rival for the affections of a fellow male astronaut (see Section 6.2.4, "The case of the emotionally unstable astronaut"). This incident prompted two independent studies on astronaut health and behavioral assessment. The first internal review at the NASA Johnson Space Center was on behavioral medicine procedures, especially with astronauts after re-entry from spaceflight. The second was broader and conducted by outside experts who formed the Astronaut Health Care System Review. This latter July 2007 report revealed allegations of improper alcohol abuse by astronauts prior to spaceflight. Subsequently, the Agency questioned two astronauts about drinking while flying aloft in contravention of established rules against such imbibing. NASA administrator Michael Griffin said the charges were based on anecdotal reports that have never been substantiated. Officials also revealed that some equipment destined for the space station had been sabotaged by one of the contractors' employees.

Obviously, NASA still has organizational culture problems. Top management avoids "bad news" that might delay flights and warnings about the condition of these two miscreants had been ignored. Further, administrators seemingly are more concerned about technological issues than human needs and failures. As a result of these findings, NASA undertook these remedial actions: (1) annual flight physicals for all astronauts will now include a behavioral health assessment; (2) adoption for the Astronaut Corps of an official Code of Conduct; (3) continuing internal reviews and surveys of astronauts and flight surgeons to provide feedback on performance problems. Copies of the two reports will eventually be released to the public (*www.nasa.gov/audience/formedia/features/astronautreport.html*).

Spacefarers, at the end of the day, are human beings who bring with them both talent and limitations. Performance stress brought on by both organizational and orbital environments can have negative impacts. Longer duration flights on the International Space Station, and a future lunar base, require a behavioral strategy by both ground and on-site health teams which should include psychologists, nurses, and possibly social workers, as well as physicians (see Appendix D).

5.7 SPACE ERGONOMICS AND ECOLOGY

In this chapter's introduction, attention was directed to the impact of environmental factors and the human/machine interface on performance. Dr. Peter Hancock, when at the USC Institute of Safety and Systems Management, expressed concern that some designs proposed for the Space Station are unacceptable from an ergonomics and living environment viewpoint; that there is the potential there for design failure that would make the *Challenger* disaster look like a minor accident [38]. The development of the whole space infrastructure opens up a new field of applications for ergonomics and ecology.

The Harper Dictionary of Modern Thought describes *ergonomics* as the study of the physical relationships between human and machine relative to the reduction of human strain, discomfort, and fatigue, including the layout of machine tools, seat design, and even the positioning of dials on the dashboard. Workplace ergonomics defuses employee and union opposition to using new technologies by designing equipment and workstations that are more compatible with human needs and safety. In orbit, ergonomics extends from spacecraft and space habitation to space suits and use of automation/robotics. Thus, space ergonomics can be defined in the context of human performance as it relates to the design, positioning, and functioning of equipment, furnishings, machinery, technology, vehicles, or habitats which affect human survival, productivity, and comfort in space. In September 1995, for example, astronauts from the Shuttle *Endeavour* took spacewalks to test gear designed to protect against the intense cold aloft. In a situation comparable with the photograph in Exhibit 67, Dr. Michael Gernhardt and James Voss hovered at the end of the robot arm that was raised 10 meters above the cargo bay. Working in temperatures ranging from $-60°C$ to $-90°C$, the astronauts successfully tested new heated gloves, thermal socks, boot inserts, and redesigned long underwear for future use in constructing the Space Station. On an EVA six months prior, the fingers of orbiting workers froze, so NASA produced these battery-heated gloves and a $10.4 million space suit in the knowledge that hundreds of hours would need to be spent in spacewalks to build that facility. Undoubtedly, such R&D investment will have terrestrial spin-offs in equipment useful to workers in harsh environments on Earth, as well as in the future on the Moon as the Exhibit 72 illustrates.

Exhibit 72. Lunar ergonomics. Not only will a lunar infrastructure have to be built if people are to live and work on the Moon, but special protective suits and gear will have to be ergonomically designed. In this artist's concept, two astronauts watch carefully as their drill bores down into the top layers of the lunar surface. As the two crewmembers wait for their sample, they use their unpressurized rover as a communication link to relay frequent status reports back to their base. Source: NASA.

In the long history of human performance, the Space Shuttle is a milestone in building and operating the most technically complex machine ever built. Up to now, the ultimate test has been constructing and managing the International Space Station, which is presently in orbit (see case study in Section 8.8). Since 1997, 16 nations have been cooperating and investing in this space macroproject. Since the U.S.A. is providing the leadership in the venture, its Congress, concerned for astronaut safety, has directed that as much as possible of the construction work on this facility be accomplished though automation and robotics. NASA announced that the facility would provide for the physical and psychological wellbeing of the visiting spacefarers with highly reliable and space-maintainable life support systems. As Wendell Mendell of the Johnson Space Center noted, the orbiting station contain amenities for comfortable—yet productive—working and living conditions. The 17 nations of the European Space Agency have already spent over $10 billion on the Space Station program. Furthermore, ISS additions are being made to improve the working environment aloft. In 2001, for instance, the Americans installed their Destiny lab module ... In 2007, the Shuttle *Atlantis* delivered a second scientific laboratory of ESA's called Columbus. In 2008, the Japanese provided a huge laboratory, Kibo (meaning "hope" in their language), which completed the additions.

At MIT's Laboratory of Orbital Productivity, Dr. David L. Akin and his group are presently studying design approaches that best fit the relationship between "human and machine". Their premise is that humans with their superb self-adaptive system are a valuable resource in space. The group are examining those functions which can be done by robots alone, by machines under astronaut supervision, and finally only by humans. Numerous operations on the future station are being analyzed to find the optimal human–machine mix [39]. Both ergonomics and ecology are parts of NASA planning for the next major space enterprise: a lunar outpost or base, to be followed by a larger settlement on the Moon.

NASA seemingly employs the term "decor" rather than "ergonomics" with reference to the Space Station. A. A. Harrison, in a proposal for the agency on this subject, defines decor as the environmental design choices that remain viable given the constraints imposed by the station's anticipated weight, volume, and function. He inserts under this decor heading critical performance variables that affect the efficiency, speed, and quality with which work gets done; the accumulation of fatigue and the restorative effects of rest and leisure; and the continuing commitment to mission activities. To Harrison, as a psychologist, decor refers to elective and optional variation in form, shape, color, patterns, and the like which affect the overall environment and performance. It impacts both work and other living activities (such as rest, recreation, and self-maintenance). Therefore, if the environment and tasks in spaceflight are to be matched, Dr. Harrison believes that there are advantages in making habitat interiors definable and redefinable; that is, a flexible environment which provides occupants with an array of alternatives to meet a variety of needs for solitude, interaction with a limited number of persons, or open interactions with all crewmembers. Space habitability specialists must be sensitive to everything from the color-coding of displays and the rounding of equipment edges to providing attractive and functional meeting areas.

Dr. Yvonne Clearwater of the NASA-Ames Research Center joined with Dr. Albert Harrison in examining crew support for a 1 to 3-year possible expedition to Mars. Given the prolonged isolation, confinement, hardships, and dangers of such a roundtrip for spacefarers, they reviewed the prospects of this in the context of habitability and food; physical and psychological health; interface, automation and robotics; as well as crew autonomy. In considering selection and training of astronauts for these journeys, they recommended extensive use of simulations, whether by technical means or in analogous environments. Clearwater and Harrison urge high-quality habitats and environmental design features that assuage stress and increase comfort aloft [40].

In a lecture at the California Space Institute in La Jolla, Dr. Mary Connors revealed that increased crew size, heterogeneity, and mission duration are among the forces moving NASA to make use of behavioral science resources. Research on exotic environments (like outer space) indicates that they provide both stressors and stimulations, and downside risks and rewards. Among the latter is total immersion in one's work, so persons can be judged primarily on their performance. In addition to further study of all aspects in space communications and selection/training, this human factor scientist urges investigations relating to environmental manipulation.

5.7.1 Space ecology

While space ergonomics impacts productivity aloft, human performance in orbit will not be optimized without consideration of a twin concept: *space ecology*. Eugene Konecci, when director of Biotechnology and Human Research for NASA, described the latter as *bionomics*, the branch of biology that deals with the relations between organisms and their environments, which in this context is space or the zero-gravity environment. Konecci, a professor at the University of Texas in Austin, explained space ecological systems in the context of physical, biological, chemical, and situational environments required for orbital, lunar, and interplanetary systems (see Exhibit 73). These systems are dependent on the length of the manned mission or time in orbit, as well as the constraints of the spacecraft or living module [41].

Ecology comes from a Greek work meaning "home or living place", so here we would be using it with reference to a space vehicle, station, or habitat, which are closed ecological systems. Ecologists are concerned with why a living organism can exist under one set of environmental conditions and not another.

Even the littering of our Solar System with space junk is within the purview of their ecological regard. To prevent space from becoming an orbiting junkyard, builders of the International Space Station are trying to develop technology to clean up these hazardous projectiles, such as a "sweeper" that would collect debris caused by explosions, collisions, and mechanical or human errors. At the Johnson Space Center, the Advanced Program Office is also attempting to "harden" the material that goes into spacecraft and stations, so that they can better withstand external impacts, as well as to design shields, and to create warning systems to avoid debris collision.

In the Shuttle Orbiter, for example, miniaturization is a continuing factor; all materials and supplies have to be carried onboard. With a Moon or Mars base, some of the latter may be transported in unmanned cargo carriers, and on-site materials may be utilized (such as lunar dirt or regolith for radiation protection of habitats). With the erection of a station platform or other large space structures, a whole array of ergonomic and ecological issues emerges. Waste management is a critical factor here (one estimate is that $4\frac{1}{2}$ tonnes of waste a week will be generated on the station). Rather then dump waste into space, ecologists are examining such solutions as returning it to Earth by Shuttle, unmanned rockets, or tethers; recycling it for other uses; or storing it (the Shuttle's external tank if taken into LEO could be used for such purposes). The ecological logistics involved in a mission to Mars are staggering to conceive: a three-year roundtrip with a crew of three, the Russians estimate, would require $4\frac{1}{2}$ tonnes of food, 10 tonnes of oxygen, and 17 tonnes of water.

Waste management and toilets have been problematic since we began to send people into orbit. The *New York Times* (September 18, 1995, p. A6) once had a headline, *For Space Shuttle Astronauts, A Troubled Mission's Messy End*. This Associated Press story went across the world detailing how a clogged toilet caused Lt. Col. James Voss and Dr. Michael Gernhardt to become "plumbers" on this *Endeavour* mission. When they were unable to bypass the clogged filter, the explorers had to drain urine and other waste water from a storage tank into smaller emergency containers. Obviously, as the number and diversity of spacefarers increase, more complex and sophisticated life support and waste management systems will have to be designed and developed.

Research on ground-based groups in exotic environments has evolved an ecological model that seems to add another ecological dimension that fits within the situational category of Exhibit 72 under "personalities". Irwin Altman, an expert in bioastronautics, posits a fundamental interdependence under such circumstances among group members and between group members and their environment [42]. Applied to space living, the ecological model, in this sense, would be concerned with the processes and products of social interaction, especially as to how they affect performance aloft. Interpersonal "processes" here include verbal and non-verbal communication, and manipulation of environmental props; "products" refers to levels of intimacy and degree of interpersonal accommodation, attraction, conflict, and the like. This theory of interpersonal dynamics based on investigations with isolated and confined groups warrants further study for its applications to performance in space. It is an illustration of how behavioral science perspectives can be aligned with biological insights to create a synergistic "envelope of life" system that might enhance performance in the remote and alien environment on the space frontier.

5.7.2 Habitation ergonomics and ecology

Similarly, in the design of habitats for the Moon, Mars, or other planets, planners must follow sound guidelines for space ergonomics and ecology. When architect Constance Adams set out to design such an installation by using inflatables, she collected data about personal habits and socialization patterns of astronauts and cosmonauts actually living aloft. She called this *socio-ergonomic analysis*, a way of understanding how in closed habitats, members of a group behave as if they were a single entity [43]. The input was then used by her colleagues at Houston's Johnson Space Center to design, construct, and inflate Transhab. This was a prototype of an inflatable, habitable structure with a closed-loop, advanced life support system which recycles air, water, and waste. The bioplex's eight air-tight chambers simulate conditions of life in space that enable crewmembers to co-exist safely and happily offworld. It includes sleeping, working, and living quarters for six spacefarers, featuring an interior trusswork system that locks into position, so that rooms and equipment can be re-arranged to help the inhabitants feel more at home! This research built on previous work of NASA architect, Kris Kennedy, about using inflatables for lunar exploration habitats [43]. Bigelow Aerospace is currently using such insights to construct orbital stations and someday hotels, or even lunar habitats!

The space culture involves an integrated environment for life and work. Those concerned about human performance there cannot afford to ignore matters of space ergonomics and ecology, both fields of knowledge likely to grow in importance. Environmentalists and conservationists have only begun to direct their attention aloft, as can be seen by the Sierra Club seminal conference and proceedings on

Exhibit 73. Four factors within the environment of outer space which impact human performance aloft.

	Space environment
Physical	Temperature, pressure, humidity, radiation, gravity (acceleration), dust, air ionization, vibration, acoustic energy, illuminations.
Biological	Microorganisms, metabolism, waste disposal.
Chemical	Atmosphere, water, food, drugs, odors, toxicology, fuels, fire.
Situational	Missions, logistics, cabin space, tasks (physical and mental), personalities (on ground control and aloft), emergencies.

Source: Eugene B. Konecci, "Space Ecological Systems," in Schaefer, K. E. (ed.) *Bioastronautics*. New York: Macmillan, 1964, p. 281.

Beyond Spaceship Earth. Within that seminal volume on environmental ethics in the Solar System, there are chapters on orbital pollution, space exploration and environmental issues, extraterrestrial ecosystems, military/nuclear use of space, and the social/physical environment of space stations and colonies [44]!

Another dimension of space ecology may be the use of space technology to benefit the home planet, such as through Earth-sensing and observation. This is the whole purpose of a NASA program called "Mission to the Planet Earth (MTPE)". Many are concerned about global environmental change, such as warming and pollution of the planet, and point to the growing role of space developments in understanding our own world [46]. The U.S. Global Change Research Program seeks to describe and understand the interactive physical, biological, and chemical processes that regulate the total Earth system, including the Moon, as influenced by human action. The MTPE centerpiece is the Earth Observing System, a series of satellites, designed to gather data for observing global warming, ozone depletion, deforestation, and other planetary problems observed from outer space. In other words, space technology and research also benefit the home planet [45]!

5.8 SPACE PERFORMANCE RESEARCH ISSUES

In Exhibit 44 we noted some biological and behavioral space issues requiring further research. This review on offterrestrial human performance led me to certain conclusions on areas for further interdisciplinary research on that subject:

• The need to broaden both the perspective and data gathering beyond the traditional, narrow, human factor and life science approach of industrial engineers.

Space-living improvements might be forthcoming by better utilization of biological/behavioral sciences. More attention should be directed to the space biosphere or envelope of life in terms of both physical and psychosocial considerations.
- The complexity of long-term space living implies that more study should be directed toward social and group interaction among spacefarers, as well as the possibility of their experiencing "space shock" (a form of culture shock manifested through depression, withdrawal, and other alienated behaviors.
- The value of creating knowledge bases or data banks on the global experiences of humans who have livid and worked in space, or in similar ground-based exotic environments, so that cross-disciplinary insights will evolve and expert systems develop.
- The necessity for designing a more complete space personnel deployment system to accommodate a larger and more diverse population aloft, including a feedback mechanism so that those who actually live and work on the high ground may contribute to the preparation of future spacefarers (e.g., broadening of the debriefing process and using astronauts as trainers/educators). This will be discussed further in Chapter 6.
- The expansion of cross-functional investigations into the human role and performance in space. Such research should address the issues raised in this chapter, such as crew productivity, team performance, human/machine interface, human resource development of spacefarers, space ergonomics and ecology, as well as topics discussed elsewhere in this volume (e.g., space culture and macromanagement). Particular emphasis should be directed to human and robotic interactions, especially when artificial intelligence is extended in such automated systems.
- The use of the International Space Station, as well as future bases on the Moon or Mars, should become laboratories for expanded human factor research. These sites then would devote more comprehensive studies aimed at improving the quality of space life, work, and performance. This would enhance the space explorer's environment and experience not only in these prototype communities—but also in space settlements and colonies to come.

In critiquing this chapter, sociologist William MacDaniel raised two other matters for future research [48]:

- The concept of productivity in space needs to be defined. From his perspective, productivity is a product of the total social and cultural environment in which people live, and is not to be limited to on-duty activity. Since productivity standards may or may not be internalized by individual spacefarers, overemphasis on team productivity could contribute to stress. MacDaniel points out that while on Earth we can get away from stressful work situations, but that is more difficult aloft; therefore, the space culture should provide its pioneers with relief from job-related strain when not on duty. A space station or lunar base culture and social structure, for instance, might contain values, attitudes, and norms that support productivity and competent performance, while also offering mechanisms for tension relief and relaxation from intensive duty activity. Such an environment should lessen the prospects for the individual or the group of becoming dysfunctional. Perhaps the Earthly dichotomy between work and play needs to be eliminated in space, so that productivity aloft encompasses leisure activity as well as duty hours?

- The role of conflict and competition among spacefarers requires more study, especially since they impact performance. MacDaniel, a professor emeritus from Niagara University and former military jet pilot, speculates about space workers who are trained in cooperation for survival aloft, who then must return to a home planet culture which emphasizes individualism and competition. He also inquires if a modicum of competition and conflict are necessary for innovation, invention, and original thinking in space communities? Would dependency on a norm of survival cooperation lead to a stagnant and uninspiring environment? However, to me, the latter would not be the same as the synergistic society aloft put forth in Chapter 2. I agree with MacDaniel that a cooperative community in space implies development of a wholly different set of cultural values, norms, and attitudes than is typically experienced now on this planet. Perhaps it is one of the lessons to be learned by *earthkind* from emerging *spacekind*. Furthermore, conflict to behavioral scientists is not necessarily a negative experience, but represents energy to be utilized and channeled.

Exhibit 74 offers some additional recommendations.

Readers may wish to compare the above research issues with those reported in Exhibit 32 ... The 21st century challenge of offworld living is to innovate and to achieve levels of human potential currently not attainable on Earth.

Exhibit 74. STRATEGIES FOR LIFE SCIENCE RESEARCH.

- Synergize the presently independent research activities of national and international organizations through development of cooperative programs in life sciences at space agencies and university laboratories.
- Complete and consolidate the unique national database consisting of basic life science information and results of biomedical studies of astronauts and cosmonauts on a longitudinal basis. This database should be expanded to incorporate information obtained by all spacefaring nations and be available to all participating partners.
- Increase the fequency of life science data acquisition on the Space Shuttles, Space Stations, and international missions ... The knowledge obtained by space life sciences will play a pivotal role as humankind reaches out to explore the Solar System.

5.9 CONCLUSIONS ON SPACEFARERS' PERFORMANCE

Superb human performance aloft has been manifest among both cosmonauts and astronauts from the first man in space, Yuri Gagarin in 1961, to the first loss in flight of both a spacecraft and its seven crewmembers with the *Challenger* and *Columbia* tragedies in 1986 and 2003, to the latest "manned" mission to be launched. One obvious conclusion from this chapter's review is that space planners should call on those astronauts and cosmonauts who have been in orbit for their feedback. Further, these experienced spacefarers are valuable resources to be utilized after their retirement. They have practical guidance to offer as both consultants and trainers of those being prepared for living and working offworld.

With the emergence of space stations, the capacity to stay longer in orbit or to have permanent habitation on the high frontier is now feasible. Each pioneering effort contributes to the migration of our species beyond this planet, as does each successful orbiting performance by forerunners of tomorrow's space colonists. Today space scientists are working on the next big challenge: lunar human performance and settlement!

Charles Darwin once observed that the human mind evolves in response to our environment. The further exploration of space with consequent expansion of human presence there will advance our psychosocial evolution and may eventually even alter the biological evolution of the species [48]. As Robert Powers reminded us, "human dreams and ideas can span centuries before being translated into realities; the future Martians may be living among us now as our children and grandchildren [49]!" Will this become a new age for our cosmos in the epic of evolution [50]? Finally, it is worth recalling the words of the ancient sage, Lucretius:

> "The mind wants to discover by reasoning what exists in the infinity of space that lies out there, beyond the ramparts of this world."

5.10 REFERENCES

[1] Westdick, P. J. *Into the Black: JPL and the American Space Program, 1976–2004*. New Haven, CT: Yale University Press, 2006 ... Surkov, Y. *Exploration of Terrestrial Planets from Spacecraft: Instrumentation, Investigation, and Interpretation*. Chichester, U.K.: Ellis Horwood, 1990.

[2] Harris, P. R. *The New Work Culture ... Managing the Knowledge Culture*. Amherst, MA: Human Resource Development Press, 1998, 2005 (www.hrdpress.com).

[3] Harvey, B. *Europe's Space Programme: To Ariane and Beyond*. Chichester, U.K.: Springer/Praxis, 2003.

[4] Harris, P. R. *High Performance Leadership*. Amherst, MA: Human Resource Development Press, 1994.

[5] Harland, D. M. *The Mir Space Station*. Chichester, U.K.: Wiley/Praxis, 1997 ... Burrough, B. *Dragonfly: NASA and the Crisis aboard Mir*. New York: HarperCollins, 1998.

[6] Hall, S. B. (ed.) *The Human Role in Space: THURIS Study*. Park Ridge, NJ: Noyes Publications, 1989 ... See also Committee on Space Biology and Medicine/Space Studies Board, *A Strategy for Research in Space Biology and Medicine into the Next Century*. Washington, D.C.: National Academies Press, 1998, pp. 194–230.

[7] Harrison, A. A.; and Connors. M. M. "Human Factors in Spacecraft Design," *Journal of Spacecraft and Rockets, AIAA*, **27**(2), 1994, 479–482 ... Harrison, A. A. "Humanizing Outer Space," *Space Governance*, **1**(2), December 1994, 11–13, 20 (journal of United Societies in Space, *www.usis.org*).

[8] Connors, M. M.; Harrison, A. A.; and Akins, A. R. *Living Aloft: Human Requirements for Extended Spaceflight*. Washington, D.C.: U.S. Government Printing Office, 1985, ch. 24 (NASA SP-483) ... Harrison, A. A. *Spacefaring: The Human Dimension*. Berkeley, CA: University of California Press, 2001.

[9] Buckey, J. C. *Space Physiology*. Oxford, U.K.: Oxford University Press, 2006 (*www.us.oup.com*) ... Harvey, B. *Russia in Space* ... Hall, R.; and Shayler, D. J. *The Rocket Men*. Chichester, U.K.: Springer/Praxis, 1997, 2001.

[10] Nicogossian, A. E. *Space Physiology and Medicine*. Philadelphia, PA: Lea & Feiberger, 1989, Second Edition.

[11] Foley, T. M. "Learning from Living in Space," *Aerospace America*, April 1995, 24–33 ... Schaefer, R. "Moon Docs among Moon Rocks," *Ad Astra*, **2**(3), February 1991, 8–11.

[12] Douglas, W. K. *Human Performance Issues Arising from Space Station Missions*. Huntington Beach, CA: McDonnell Astronautics Company (NASA 2-11723 contract) ... For more recent studies, refer to Nicogossian, A. *et al. Space Biology and Medicine*. Reston, VA: American Institute of Aeronautics and Astronautics, Vols. 1–5, 1993–1995.

[13] Helmreich, R. L.; Wilheim, J. A.; and Runge, T. W. *Human Factors in Outer Space Productions*. Boulder, CO: Westview Press, 1983 ... Furnham, A.; and Bochner, S. *Culture Shock: Psychological Reactions to Unfamiliar Environments*. New York: Springer-Verlag, 1987 ... Harrison, A. A.; Clearwater, Y. A.; and McKay, C. F. (eds.) *From Antarctica to Outer Space*. New York: Springer-Verlag, 1991.

[14] Harrison, A. A.; and Connors, M. M. "Crew Systems: Theoretical and Practical Issues of Humans and Technology." Paper presented at the American Group Psychotherapy Association Conference, San Antonio, Texas, February 1991, 10 pp.

[15] Presidential Commission on Space Shuttle *Challenger*. *Report of the Presidential Commission on Space Shuttle Challenger*. Washington, D.C.: U.S. Government Printing Office, Vols. 2/3, 1986.

[16] Oberg, A. *Spacefarers of the '80s and 90s: The Next Thousand People in Space*. New York: Columbia University Press, 1985.

[17] Hurt, H. *For All Mankind*. New York: Atlantic Monthly Press, 1988 ... White, F. *The Overview Effect: Space Exploration and Human Evolution*, 1987 ... Ride, S.; and Okie, S. *To Space and Back*. New York: Lee & Shepard Books, 1986.

[18] Pitts, J. A. *The Human Factor: Biomedicine in the Manned Space Program to 1980*. Washington, D.C.: U.S. Government Printing Office, 1985 (NASA SP-4213).

[19] Ockels, W. J. "Liftoff: Why Conduct Life Science Research in Space?" *Ad Astra*, **1**(5), May 1989, 3 and 22,

[20] Colon, A. R.; and Colon, P. A. "The Psychosocial Adaptation of Children in Space: A Speculation," *Journal of Practical Applications in Space*, **3**(3), Spring 1992, 5–22.

[21] "Science and Technology: Slave to the Rhythm," *The Economist*, February 15, 1997, pp. 77–79 ... See also Mallis, M. M.; and DeRohia, C. W. "Circadian Rhythms, Sleep, and Performance in Space," *Aviation, Space, and Environmental Medicine*, **76**(6), June 2005, B94–B107.

[22] Schrunk, D.; Sharpe, B.; Cooper, B.; and Thangavelu, M. *The Moon: Resources, Development, and Settlement.* Chichester, U.K.: Springer-Praxis, 2008, Second Edition.

[23] Nicholas, J. M. "Interpersonal and Group Behavior Skills for Crews on Space Stations," *Aviation and Environmental Medicine*, January 1989, 22 ... "Small Groups in Orbit: Group Performance on Space Station," *Aviation and Environmental Medicine*, October 1987, 1009–1013.

[24] Kanas, N. "Interpersonal Issues in Space: Shuttle/Mir and Beyond," *Aviation Space and Environmental Medicine*, **76**(6), Section II Supplement, June 2005, B126–B134 ... "Psychological and Personal Issues in Space," *American Journal of Psychiatry*, **144**(6), June 1987, 703–706.

[25] Ritsher, J. B. "Cultural Factors and the International Space Station," *Aviation and Space Medicine*, **76**(6), Section II Supplement, June 2005, B135–B144.

[26] Caldwell, B. S. "Multi-team Dynamics and Distributed Expertise in Mission Operations," *Aviation and Environmental Medicine*, **76**(6), Section II Supplement, June 2005, B145–B153.

[27] Center for Collaborative Organizations, University of North Texas (PO Box 13587, Denton, TX 76203; email: *workteams@unt.edu*) ... Harris, P. R. *The New Work Culture.* Amherst, MA: HRD Press, 3 vols., 1989–1998; also from same publisher, *Managing the Knowledge Culture*, 2005, ch. 4.

[28] Bellevag, I. "A Western View of Life on Mir," *Ad Astra*, May/June 1995, 21; *op. cit.* [5].

[29] Francis, B.; and Young D. *Improving Work Groups.* Indianapolis, IN: Pfeiffer/Wiley, 1980 ... See also Orasanu, J.; Fisher, U.; Krafft et al. *Assessing Team Interactions for Long Duration Space Missions.* Alexandria, VA: International Aerospace Medicine Conference/ASMA, May 2004.

[30] Sears, W. H. *The Front Line Guide Series: To Communicating with Employees; to Management Style; to Thinking Clearly; to Building High Performance Teams; to Mastering the Manager's Job.* Amherst, MA: HRD Press, 2008, five compact volumes.

[31] Hooper, G. R. *The Soviet Cosmonaut Team.* Woodbridge, U.K.: GRH Publications (36 Bury Hill, Melton) ... Bluth, B. J.; and Helppie, M. *Soviet Space Stations as Analogs.* Washington, D.C.: NASA Headquarters (Contract NAGW-659), 1986.

[32] Beyerlein, M. S.; and Harris, C. *Guiding the Journey to the Collaborative Organization.* Denton, TX: Center for Collaborative Organizations, University of North Texas, 2004 ... Beyerlein, M. S.; Friedman, C.; McGee, G.; and Moran, L. *Beyond Teams: Building Collaborative Organizations.* Indianapolis, IN: Pfeiffer/Wiley, 2004.

[33] Criswell, D. R. (ed.) *Automation and Robotics for the National Space Program.* La Jolla, CA: California Space Institute/University of California-San Diego, 1985 (NASA Grant NAGW 629) ... Redmond, C. (ed.) *Robotics Handbook: Version 1.* Washington, D.C.: NASA Headquarters/Office of Advanced Concepts and Technology, 1994.

[34] Skaar, S. B.; and Ruoff, C. F. (eds.) *Teleoperations and Robotics in Space.* Alexandria, VA: American Institute of Aeronautics and Astronautics, 1994 ... Also see Bode, M. E. (ed.) *Robotic Observatories.* Chichester, U.K.: Wiley/Praxis, 1995.

[35] Rather, J. D. "Revolution Technologies for Affordable Lunar Development by 2016," in Durst, S. M.; Bohannan, C. T.; Thomason, C. G.; Cerney, M. R.; and Yuen, L. (eds.) *Proceedings of the International Lunar Conference.* San Diego, CA: UNIVELT/AAS, Vol. 108, 2004, pp. 177–186 ... "Waiting for the Space Elevator," *Technology Quarterly/The Economist*, June 10, 2006, pp. 3–4.

[36] Rothwell, W. J.; and Kazanas, H. C. *The Strategic Development of Talent*, 2006 ... *Planning and Managing Human Resources*, 2006, Second Edition. These and other HR resources are available from HRD Press, 22 Amherst Rd., Amherst, MA. 01002

(*www.hrdpress.com*) ... For information and publications, contact the American Society of Training and Development, 1640 King St., Alexandria, VA 22313 (*www.ASTD.org*) ... For cross-cultural HRD, contact the Society for Education, Training and Research International, 808 17th St. NW, Washington, D.C. 20006 (*www.SIETAR.org*).

[37] Wiskerchen, M. J. "Space Exploration and a New Paradigm for Education and Human Capital Development," in Krone, R. (ed.) *Beyond Earth: The Future of Humans in Space.* Burlington, Ontario: Apogee/CGPublishing (Box 62034, Burlington, Ontario, *www.apogeebooks.com*), 2006, pp. 105–113 ... Also see Harrison, A. A., *op. cit.* [8], chs. 6–11 ... Freeman, M. *Challenges of Human Space Exploration.* Chichester, U.K.: Springer/Praxis, 2000.

[38] Hancock, P. A. "The Principles of Setting Stress Standards" in Egberts, R. E.; and Egberts, C. G. (eds.) *Trends in Ergonomics: Human Factors II.* North Holland, the Netherlands: Elsevier Science Publications, 1985 ... See also the annual proceedings of the Human Factors and Ergonomics Society (*www.hfes.org*).

[39] Gregory, W. H, "Researchers at M.I.T. Say It's a Matter of Ease," *Commercial Space,* **2**(2), Summer 1986, pp. 58–66. For further information on planning for ISS and a lunar outpost, contact the Office of Advanced Concepts and Planning, NASA Headquarters (*www.nasa.gov*).

[40] Clearwater, Y. A. "Functional Esthetics to Enhance Well-being in Isolated Confined Environments" in Harrison, A. A.; Clearwater, Y. A.; and McKay, C. P. (eds.) *From Antarctica to Outer Space.* New York: Springer-Verlag, 1991, ch. 30, pp. 331–348 ... Clearwater, Y. A.; and Harrison, A. A. "Crew Support for Initial Mars Expedition," *Journal of the British Interplanetary Society,* **43**, 1990, 513–518.

[41] Konecci, E. B. "Space Ecological Systems," in Schaefer, K. E. (ed.) *Bioastronautics.* New York: Macmillan, 1964, p. 261.

[42] Altman, I. "An Ecological Approach to Functioning in Isolated and Confined Groups," in Rasmussen, J. E. (ed.), *Man in Isolation and Confinement.* Chicago, IL: Aldine, pp. 241–270.

[43] Kennedy, K. "Space Inflator," *Metropolis,* July 1, 1999, pp. 78–83.

[44] Hargrove, E. C. (ed.) *Beyond Spaceship Earth: Environmental Ethics in the Solar System.* San Francisco, CA: Sierra Club, 1986.

[45] Cockell, S. E. *Space on Earth: Saving Our World by Seeking Others.* New York: Macmillan, 2007.

[46] Johnson, R. G. (ed.) *Global Environmental Change: The Role of Space in Understanding Earth.* San Diego, CA: UNIVELT/AAS, Vol. 76, 1985.

[47] Benson, J. "Conversations with Charles Kennel, NASA Associate Administrator for Mission to Planet Earth," *Aerospace America, AIAA,* August 1995, 10–12.

[48] MacDaniel, W. E. "Scenario for an Extraterrestrial Civilization," *Space Power,* **7**(3/4), 1988, 365–381.

[49] Harris, P. R. *Future Possibilities: Toward Human Emergence.* Amherst, MA: HRD Press, 2008.

[50] Powers, R. M. *Mars: Our Future on the Red Planet.* New York: Houghton-Mifflin, 1985 ... Also see Clarke, J. D. (ed.) *Martian Analog Research.* San Diego, CA: UNIVELT/AAS, Vol. 111, 2006.

[51] Chaisson, E. J. *The Epic of Evolution: Seven Ages of the Cosmos.* New York: Columbia University Press, 2006 ... Larsen, K. M. *Cosmology 101.* Westport, CT: Greenwood Publishing Group, 2007.

SPACE ENTERPRISE UPDATES

- First-generation American astronauts are not common. Like Leroy Chiao whose parents fled from China in 1949. From building model airplanes as a boy, their son Leroy went on to a five-hour spacewalk outside the International Space Station on January 26, 2005.
- An example of high performers on ground support teams is Nurse Laura Steinmann who was a NASA coordinator for family services. In the Astronaut Family Support Center at Johnson Space Center, she worked with exceptional people involved in spacefaring, and hopes her grandchildren will someday be among them.
- Google co-founder Sergey Brin paid $5 million to reserve a seat wth Space Adventures for a future private spaceflight to the International Space Station. That company has already sent five tourists to ISS with the help of Russian space training and equipment.
- The National Aerospace Training and Research Center in Pennsylvania, U.S.A., has been training Sir Richard Branson, founder of Virgin Galactic Spaceflight, and members of his crew. The preparation is for their orbital flight next year on Burt Rutan's SpaceShipTwo. NASTAR uses the STS-400 centrifuge to simulate comparable g-forces as will be experienced aloft.
- Japanese scientists and origami masters hope to launch a paper spacecraft from orbit and learn from its trip back to Earth. The Japanese space agency has feasibility studies under way to create an origami plane that would have implications for the design of re-entry vehicles or space probes from the upper atmosphere. A prototype has already survived severe testing at the University of Tokyo's Department of Aeronautics and Astronautics.

6

Orbital deployment systems and tourism

> *It is our human destiny to live, work, and play among the stars in permanent space communities ... It is our responsibility now to begin creating a spacefaring civilization!*
>
> National Space Society

Since ancient times, humans have left their homelands to explore and migrate to far places. Embarking on a *hero's journey* is the way that renowned mythology expert Joseph Campbell describes the relocation challenge of living or working in a foreign region [1]:

> "Furthermore, we have not even to risk the adventure alone, for the heroes of all time have gone before us. The labyrinth is thoroughly known. We have only to follow the thread of the hero path, and where we had thought to find an abomination, we shall find a god. And where we had thought to slay another, we shall slay ourselves. Where we had thought to travel outward, we will come to the center of our own existence. And where we had thought to be alone, we will be one with the world."

For the expatriate does indeed dare to go outside the known boundaries, and may in the process experience a "shock" to his or her psychological construct, our perceptions of our "world" as we know it. But consider the demands on the terrestrial who goes *offworld* into outer space beyond gravity! Surely that is the ultimate hero's collective voyage and adventure when we fly out of our galaxy into the *island universe* [2]. Through telescopes, astronomers have already preceded actual human presence, as when exoplanetary scientists from the University of Geneva discovered in 2006, a seemingly "earth-like" planet dubbed *Gliese 581 c*, only 20 light years away in the constellation Libra. But within the human family today, only a select and small group of people have flown into orbit. Whether these are called astronauts or cosmonauts, our "manned" extraterrestrial experience is limited to 47 years with only some 400 humans, mostly Americans, Russians, and Europeans, who have stayed aloft for relatively short periods. Each year the record for space pioneers is extended. Presently, aboard Russia's Mir station, Dr. Valeri Polyakov holds the lead with 438 days aloft, surpassing 366 days in an orbital environment for cosmonauts Vladimir Titov and Musa Manarov. Before long, we will be measuring space travel and expatriation aloft in terms of years and decades. But if human presence beyond Earth is to be significantly extended and expanded, then behavioral science research on personnel deployment and acculturation must be increased, in tandem with life science research on support systems and the orbital environment. This chapter will consider some of the relocation issues and strategies that will be necessary before a mass human migration into space can take place, possibly toward the end of this 21st century. Policies, practices, and programs that have been suitable in the past for small elite assemblage in orbit will not be sufficient as missions multiply in duration and complexity, crew size and heterogeneous composition, and diversity of both required personnel and their competencies.

If we examine the near-term plans of spacefaring nations, the mission forecasts are now for crew rotations after 6 months on the International Space Station, then at lunar outposts by 2020, as illustrated in Exhibit 75. To build the required infrastructure for this, the growing population aloft will begin with a variety of contractors and specialists, as well as with scientific and technical representatives from international partners in these macroprojects, eventually expanded by tourists, ending with committed space settlers and colonists. The near-term endeavors center on utilizing these orbiting laboratories for studying personnel deployment, life sciences, and other human factors. Not only do we need to design improved human habitation systems for long stays beyond our planet's atmosphere, but we still require the most economical, reusable spacecraft for round-trips to LEO and GEO [3].

Humanity's next giant leap

To plan for humankind's next "giant leap" into the Solar System, NASA first embarked on "Project Pathfinder", development of the enabling technologies for such missions, including engineering systems for effective performance and good health of human inhabitants. Now it is implementing "Vision for Space Exploration" so as to put our species back on the Moon permanently by the end of the next decade. Beyond hardware, software, and engineering concerns, research should also be expanded on "peopleware", not only in terms of pre-departure selection and training of space explorers, but their support on the high frontier, physically, psychologically, socially, and financially. Furthermore, what guidance and sustenance systems should be provided to the family dependants left on the ground while the spacefarer is away for longer periods? Eventually, some families will go aloft, and require special preparation and support services. Finally, what assistance should be rendered to space travelers and their families to help them cope with re-entry and consequent problems of terrestrial adjustment? Some insights to answer these critical questions are provided in this chapter.

Introduction 247

Exhibit 75. Technauts' lunar deployment. Contractors and technologists at work on the Moon. Artistic rendering of 21st-century lunar workers in protective space suits and habitats to provide life support, while assisted by robotic heavy equipment. Source: NASA Johnson Space Center.

Wally Schirra, former astronaut, raised the fundamental question most simply:

"How long can humans endure in space?"

Speaking before the National Conference of the Aviation/Space Writers' Association in San Diego (May 13, 1987), Schirra urged more research on human survival issues, such as what would be involved in manned Mars spaceflight of several years.[1] Dr Bruce Cordell when at General Dynamics Advanced Space Systems also addressed this matter of 2-year to 4-year missions to Mars and its moons. He noted that NASA planners are concerned about such spacefarers in terms of these five dimensions [4]:

- *adaptation and readaptation* to space and planetary environments, particularly with reference to physiological effects (e.g., from atrophy of bones and muscles, need for gravity stimulation);
- *toxicological safeguards* against possible crew poisoning or exposure to toxic substances (e.g., from life support systems, which need to be more bioregenerative);

---

[1] Wally Schirra died on May 3, 2006. The Navy captain wrote books for Apogee on his experiences on Sigma 7, Gemini 6, and Apollo 7, as well as co-authoring *The Real Space Cowboys* (www.cgpublishing.com).

- *radiation risk assessment and management* because of solar particle events or galactic cosmic radiation (the radiation environment of Mars, for instance, is more than 100 times greater than Apollo astronauts encountered on the Moon);
- *health maintenance capability* for coping with medical emergencies and maintaining wellness (provisions must be made to cover everything from safety and diagnosis of illness, to integrating visitors and coping with death);
- *psychological issues and human factors* related to extended spaceflight by mixed crews as regards gender, nationality, and disciplines (the unprecedented experience of isolation and confinement requires programs in place to counteract impairment of mental/emotional performance).

The concerns of this book include all of the above and more if space habitation on a permanent basis is to be realized. Before we get into interstellar travel [5], we can learn much from many precedent experiences on the ground, so as to better prepare to send population masses aloft.

6.1 TERRESTRIAL ANALOGS

There is a growing body of literature and experience on the terrestrial relocation and adaptation of people to strange and alien places in the interest of adventure, science, defense, or trade. Part of this information relates to exotic environments which are defined as remote, harsh, and potentially dangerous [6]. Some of the studies report various personnel problems of living in Antarctica, a natural prototype for living in space [7]. But much of the insight to be gained comes from foreign deployment of executives, technicians, scholars, and volunteers on our home planet [8].

There is more to be learned from the human experience, whether undersea on a nuclear submarine, or in a remote national park, or overseas on an international assignment, offshore rig, at a polar science or military base; all of which can help us in the management of people in orbit [9].

To illustrate this point, consider the research on expatriates in the world of business conducted by the following:

- Dr. **Joyce Sautters Osland** of the School of Business Organization, University of Portland. As part of her doctoral dissertation, she developed an interview protocol which she utilized with 35 returned U.S. business executives from overseas assignments of at least 18 months. The process included an *Awareness of Paradox* instrument for data gathering. Osland found
   o a lack of adequate information on the expatriate experience;
   o the majority of U.S. companies do not properly prepare personnel for assignment abroad;
   o human resource specialists lack expertise on how to prepare, support, and debrief expatriates.

As a result of this investigation, her book, *The Adventure of Working Abroad*, provides [10]

- understanding of the *transformational* nature of this experience;
- words from other "heroes" about their international assignment and repatriation;
- a framework for helping current and return expatriates make sense of their cross-cultural adventure;
- ideas for improving the orientation of those going on such foreign assignments.

• Drs. **Stewart Black** and **Hal Gregersen** of Dartmouth's Amos Tuck School of Business Administration, along with Dr. **Mark Mendenhall** from the University of Tennessee, have researched and published for years on global assignments, premature returns, and turnover within multinational corporations. In their book together on the subject, they offer [11]
  - a guide to managing all phases of the global assignment process from employee selection to repatriation;
  - qualifications necessary for foreign placement, and methods for training candidates to work effectively abroad;
  - solutions for typical problems and conflicts faced by expatriates and their families as a result of such deployment;
  - ways to ensure that knowledge gained in cross-cultural experience is not lost to the sponsoring organization upon the return of the expatriate.

These scholars focus on *new mental roadmaps and behaviors* required for assignments to alien environments, and emphasize the need to reward expatriates while abroad, as well as to facilitate their adjustment and retention upon return to the home culture and organization.

• Dr. **Rosalie L. Tung**, now a professor at Simon Fraser University in Canada, studied 80 U.S. multinational corporations, and discovered that more than half had failure rates on expatriate assignments of 10%–20% (7% reported a recall rate as high as 30%). Those Americans who were unable to perform effectively abroad had adjustment problems centering on these human behavior areas [12]:
  - inability of the manager and/or spouse to adjust to a different physical or cultural environment;
  - family-related difficulties;
  - personality or emotional immaturity of the overseas manager;
  - inability to cope with managerial responsibilities abroad and lack of motivation to work there;
  - lack of technical competence by the multinational manager.

Tung then compared her American results with comparable surveys which she conducted of 29 West European and 35 Japanese transnational firms. The former reported an average expatriate failure rate ranging from under 5% to 10%, while the latter companies reported largely under 5% (only 14% of the Japanese had failure rates as high as 19%). She discovered the following reasons for lower failure rates among European and Japanese expatriate managers:
  - long-term orientation regarding overall planning and performance assessment;

- more rigorous training programs to prepare candidates for overseas assignments;
- provision of comprehensive expatriate support systems;
- overall better qualification of candidates for overseas work including foreign language capability;
- moral support both from the international orientation of the sponsoring organization and from the family (especially among Japanese nationals).

These findings are consistent with similar relocation studies which attribute such maladjustment and high premature return rate on foreign postings to (a) poor selection and training for the overseas assignment; (b) overemphasis on technical competence at the expense of other qualities that ensure effective cross-cultural performance; (c) short duration of such assignments and over-concern with repatriation [13]. Torbiorn emphasizes that the matter of premature return for terrestrial expatriates is a critical managerial issue that has serious implications both for assignments to polar regions and into the space frontier. Based on the place and circumstances of foreign assignments, researchers estimate that it costs sponsoring organizations from $50,000 to $300,000 per employee or family to bring such persons back to a home base ahead of schedule [13]. The Business Council for International Understanding, on the other hand, maintained that from their studies international personnel who go abroad without proper cross-cultural preparation have a failure rate on such assignments ranging from 33% to 66%, in contrast to less than 2% of those who had such training. Such findings, plus the right costs of relocation and premature return, have caused a whole new industry of relocation specialists to emerge. The Employee Relocation Council is the clearinghouse for such professionals and their consulting firms, and their organ of communication is the magazine *Mobility*. Furthermore, researchers have developed instruments, from paper and pencil questionnaires to computerized expatriate profiles, to use in assessment of those being sent away from their home culture [14].

Unfortunately, there has been limited "technology transfer" of the above research insights to those agencies and corporations who currently send personnel to the Antarctic or into space. For example, whether in polar outposts or in orbit, there do not appear to be any studies comparing U.S. and former U.S.S.R. expatriates in both arenas relative to their pre-departure screening and training, on-site support, premature return rates and costs, and re-entry provisions? The U.S. Navy, for instance, seemingly has done the most comprehensive investigations of its Antarctic winter-over personnel, but the focus has largely been on follow-up studies related to mental and physical health on return. Palinkas *et al.* [15] did review data collected from Antarctic research stations of different nationalities relative to sociocultural influences on psychosocial adjustments.

NASA has generally centered its investigations on limited human factors and medical reports, rarely comparing the results of its experience with the Russian experience aloft as did Bluth and Helppie [16]. From the perspective of deploying people to polar or space sites, the sponsors have much to gain by examining the issues and findings in the foreign relocation or expatriate literature. The movement of a large number of people around this planet has spurred increasing interest and investment by international corporations and associations in the phenomenon of culture/re-entry shock. When people are rapidly transported from their home culture to strange and alien environments abroad, they may experience severe disorientation, confusion, and anxiety as discussed in Chapter 5 [17]. There is increasing evidence on the psychological reactions of spacefarers in unfamiliar microgravity environments and the shock that may impact their psyches, called the *overview effect* [18]. Admittedly, the length of stays in Antarctica or space until now have been too short for studies to emerge of "polar or space" cultural shock, though both astronauts and cosmonauts aloft have given indications of the phenomenon. There is also a neglected area of terrestrial investigation that has implications for space living: returning home from a cross-cultural experience [19]. Ample reports exist from the Peace Corps, the military, and transnational organizations that some personnel experience "re-entry shock" when re-assigned from a host to a home culture. It is reasonable, then, to project that re-adaptation to Earth after long-duration spaceflights will necessitate innovative programs to cope with the physical and psychological effects of living aloft.

Basically, a major deployment issue then becomes how to reduce stress on a person when leaving or returning to a home base, whether on this planet or another [20]. The traveler's sense of identity is threatened when they are removed from the comfortable and familiar and thrust into the unknown and uncertain. Such expatriates, especially when away for many months or years, usually go through a transitional experience that may include phases of growing awareness or differences, rage, introspection, and eventually integration or disintegration. As a result, many multinational businesses and foundations have developed relocation services to facilitate acculturation of their personnel to new cultural changes and challenges.

Although it is unlikely for the new millennium that spacefarers will have to deal with extraterrestrial "foreigners" or aliens, they will have to increasingly cope and adapt to the unique cultural environment of space (see Chapter 4 and Exhibit 48, which illustrate some of the dimensions in the context of humans from a lunar base). What has been learned so far about both terrestrial and space relocation is transferable to long-term space habitation and interstellar migration [21]. From this knowledge base, strategies, policies, and programs can thus be devised that facilitate the extraterrestrial deployment and delimit the psychological shock of isolation, loneliness, and strangeness of prolonged living on the high frontier.

When people enter dangerous and exotic environments, the primary concern is for sheer survival and then safety, as well as the factors which contribute to this for mission achievement. Priorities are productivity and satisfaction, conditions that further adjustment and quality of life. Whether in remote terrestrial or offworld environments, planners should be studying ways to reduce social and health problems (both mental and physical), while optimizing performance in isolated and confined circumstances. In 2002, NASA issued a significant report on its experiments in closed-environment living and life support closed systems [22].

The next four examples illustrate the possibilities of applying findings from ground-based human factor research to the preparation of spacefarers.

6.1.1 Polar regions as laboratories for space

On Earth, the polar regions have long been recognized by behavioral and biological scientists as natural laboratories for studying how humans cope and perform in remote, extreme environments. We have yet to appreciate fully the learnings to be gained in this regard by investigations of native peoples in the Arctic (known as Eskimos, Inuits, or Indians). We are more cognizant of the adaptive skills of Europeans and North Americans who have gone into these far northern regions as explorers, scientists, workers, or settlers. Even with modern comforts and new technologies, the situation is still harsh and hazardous for such hardy persons [23].

An article on "Frozen Assets" in *Canada Today* (1986) contained two vivid anecdotes that highlight the challenge of isolated and dangerous living that eventually may have similar replications in space.

- *Scientists in the Polar Continental Shelf Project.* In 1986, this Canadian project supported 229 field groups, including 10 from U.S. universities, at permanent bases in Resolute Bay and Tuktoyaktuk. The scientists live there from the beginning of March until mid-October in 12 pre-fab buildings close to an aircraft runway. The most ambitious site is on an ice island that is four miles long and 148 feet thick, weighing a billion tons. This natural base enables them to conduct long-term surveys of oil reserves and to study the Arctic Ocean and structure of the continental plate by monitoring ice conditions, water currents, and the shape of the ocean floor.
- *Miners on Little Cornwallis Island.* Since 1981 these Canadian miners worked at the Polaris lead and zinc mines, some 875 miles south of the North Pole; they were among the northernmost and richest mines in the world. The $150 million investment was expensive because the 200 workers had to be protected from loneliness and isolation, as well as the elements. Despite temperatures of 50 below, winds at 70 mph, mining work at 500 feet below surface, and severe outside travel, there were waiting lists of hundreds to apply for positions from cooks and miners to metallurgists and geologists. This was due, in part, to very high pay, two-week vacation flights to civilization every ten weeks on assignment, and luxurious living conditions and food in the residence building which included a basketball court, pool, saunas, lounges, and jogging tracks. The annual turnover of personnel was 12%. The mine was closed in August 2002 due to depletion of the ore body, and a two-year decommissioning and reclamation program was completed in September 2004.

In a few decades we may be reading comparable stories about scientists at a lunar outpost or workers at an asteroid mine. The survival and adjustment of such space pioneers could be enhanced if contemporary scholars would devote more investigative time on how people get deployed to polar regions, how they survive and perform when there, how they re-adjust after they return home from such assignments. Presently, Antarctica may be the best deployment laboratory for behavioral research analogous to the challenges inherent in space habitation. Scientists, technicians, military personnel, and visitors who go to these South Polar research stations represent many different nationalities and sponsors. In the case of the U.S.A., for instance, the sponsors range from government agencies, such as the National Science Foundation, to defense services, such as the U.S. Navy, to private contractors, such as ITT. People who "winter-over" there experience unusual living conditions in small groups; psychological stress is exacerbated by prolonged isolation in a harsh environment, resulting for some in impaired health and performance [23]. Since the turn of this century, humans have been coping there with the windiest, coldest, and harshest environment on Earth [24].

In a lecture at the University of California-San Diego Medical School where he serves on the faculty, Dr. Lawrence A. Palinkas stated that the Antarctic people experience may serve as a model for the Space Station, facilitating future commercialization and colonization of the space frontier. As a sociologist, Palinkas (see [15]) has analyzed data collected on thousands of naval Antarctic personnel, particularly the long-term effects of isolation and exposure to extreme environments, the relationship there of stress and illness, and the health behaviors of enlisted naval servicemen. Palinkas notes that similar types of stressor exist in Antarctica as in space, both in terms of *external* factors (inability to personally maintain contact with family and friends, pressures from real or imagined unpleasant events at home, feelings of rejection through delays in relief parties or shortages in supplies or interference in station routine by outside authorities) and in terms of *internal* factors (lack of privacy and cramped quarters, lack of stimulation and boredom, emotional and sexual deprivation, uncertainties over status and role). The "winter-over syndrome", a form of culture shock, may prove to be comparable with that to be experienced with long-term space living: depression, hostility, sleep disturbance, impaired cognition, and physical and psychological distress. Coping with six months of light and darkness in Antarctica has its counterpart with the light/darkness periods on the lunar surface. Despite the risks and the fact that some took six months to readjust after their return, naval personnel generally were able to perform well, stay healthy, and experience no major adverse effects after the Antarctic assignment.

In 1987, the National Science Foundation joined with NASA to sponsor a conference on "The Human Experience in Antarctica: Applications to Life in Space." The conference organizers, Dr. A. A. Harrison of the University of California-Davis, and NASA-Ames scientists Drs. Y. A. Clearwater and C. P. McKay, edited a significant volume of proceedings [5] which expands our understanding of Antarctic research beyond survival to how humans may thrive under comparable conditions with those found there and elsewhere. Such studies may prove to be prototypes which contribute to maximizing human performance in outer space. There is much to be learned from the contemporary Antarctic selection programs for personnel, their on-site needs and support, and their re-entry problems and experiences. Furthermore, a body of research literature on this subject has been building up for three decades on the emotional and mental health of Antarctic personnel. Both past and ongoing behavioral research of such human experience, combined with analysis of previous investigations of both foreign and space relocations, should provide immeasurable insight for the improvement of large-scale space deployment.

However, the scope of such human behavior studies in the South Polar regions should not only be expanded—but become more systematized and comparative, especially among international sponsors. For example, within the various institutions sending Americans to Antarctica, there does not seem to be synergistic planning and analysis of deployment issues among the sponsoring agencies. Sharing and cooperation on recruitment and selection, for instance, seems to be very limited between ITT, an NSF contractor, and the Navy, or even with other contractors? ITT Antarctica Services advertises in the press for those with engineering or mechanical backgrounds who can manage plant facilities in remote, self-sustaining research stations; who can exercise organizational and leadership skills consistent with the requirements of a small, closed community; who have the education and academic interest to interface successfully with diverse scientific investigators; who have the flexibility to react and manage effectively the fast-paced, changing, and unusual site operations on a 7-day per week basis. This advertisement for a year's Antarctic assignment is not unlike those that may appear someday in the media for technical personnel to go to a Space Station or base on the Moon or Mars.

One wonders what does a corporation like ITT do in its assessment process to ensure that persons with such qualifications are indeed hired? Furthermore, what does it do to assess the effectiveness of that new hire once on site, and how does it help that employee on return to normal civilian life? Is there a system to this process? Is it applicable not only to space, but to other organizations who send personnel to the Antarctic? NASA has begun some joint human factor and life science research with NSF's Polar Biology and Medical Program, but is the information and insight shared with other public and private agencies sending personnel to the North or South Pole? How much ongoing behavioral data-gathering is occurring at polar research installations of other international partners? Are sponsoring entities even aware of the cross-cultural literature on relocation? To illustrate the applicability of this to living in closed communities of Antarctica and space, consider that the Canadian International Development Agency has a pre-departure program for its volunteers which instills these seven skills [25]:

- to communicate respect;
- to be non-judgmental;
- to personalize knowledge and perceptions;
- to convey empathy;
- to practice role flexibility;
- to demonstrate reciprocal concern;
- to tolerate ambiguity.

6.1.2 European isolation studies

There are numerous research endeavors under way in Europe relating to isolated, confined environments (ICEs) which provide insights for spacefaring [26]. One was a simulation experiment which took place in 1989 under the leadership of COMEX, a French diving company in Marseilles. With the participation of the European Space Agency (ESA), the study involved divers in a hyperbaric chamber. Another took place in the U.K. in conjunction with the JUNO mission to fly a British cosmonaut aboard the Mir space station. In this 1990–1991 undertaking, the focus was on selection and training from 13,000 hopeful British applicants to 150 chosen to take a succession of psychological and medical tests. The latter group took an executive physical, followed by tests of symbolic, mechanical, and spatial reasoning abilities. From these findings, 35 then underwent a day of further psychological tests covering personality, working style, and public presentation skills. By this means, hopefuls were reduced to 22, who received five more days of intensive medical tests, further reducing the number to 16. These were permitted to take the stress tests, centered around a centrifuge, at the Institute of Aviation Medicine in Farnborough, U.K. As successive individuals were eliminated, the assessment process climaxed for 10 with final examination by a team of Soviet doctors who came to London. From these, two were finally picked: a woman and a back-up man. Scheduled to actually fly on the eight-day, Anglo-Soviet mission, beginning May 12, 1991, was Helen Sharman, a 27-year-old research chemist, who trained in Star City with Russian crewmembers Anatoli Artsebarski and Sergei Krikalev. JUNO project manager Christopher Hayes admitted that the private enterprise did not fully raise the $12 million required by the former U.S.S.R., but the Moscow Narodny Bank had committed $9.75 million toward underwriting the project (*Space News*, March 3, 1991, p. 2). Sharman was the prototype of civilian space tourists who would come over a decade later, paying up to $25 million for a comparable Russian spaceflight to the International Space Station.

The most significant ESA-sponsored project is the Isolation Study for European Manned Space Infrastructure, pictured in Exhibit 76. Called ISEMSI, it was partially funded by the contractor, the Norwegian Space Center. The first in a series of research studies began from September 17 to October 15, 1990 with six young European males confined in the hyperbaric chambers at the Norwegian Underwater Technology Centre (NUTEC). Within a chamber comparable with a space station environment, this experiment studied the psychological and physiological impact of isolation and confinement and their implications for long-term space missions. Over four weeks inside the two cylinders, participants conducted 37 experiments, including an operation simulation, for some 23 different groups, coordinated by ESA's directorate of Space Station and Microgravity. Under the leadership of their Long Term Program Office (LTPO), those in the facility had contact with the external world only through communication links (audio/video) while following a schedule such as that envisioned for the station aloft. The group aged 25–30 included five engineers/physicists, a physiologist, medical doctor, and professional pilot; the national composition was two Norwegians, two Italians, one Swede, one Dutchman, one Frenchman, and one German. They described themselves as *EMSInauts*, coined from the acronym ISEMSI (the European Program of Isolation Study for Manned Space Infrastructure). Throughout, they were on duty for a 13-hour day and their words, movements, and behavior were monitored; the crew measured medical, physiological, and environmental parameters. They spontaneously organized themselves into very structured patterns and roles, allowing time in the morning for a group meeting on the day's work assignments and in the evenings for recreation. When the group emerged from their 28-day confinement in Bergen, Norway, they appeared healthy,

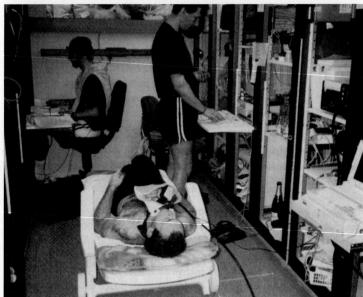

Exhibit 76. ESA ground-based isolation studies. In 1990, the first of 37 experiments involving 23 groups was undertaken by the European Space Agency in an Underwater Technology Center at Bergen, Norway. Called *ISEMSI-90* (Isolation Experiment for the European Manned Space Infrastructure), the *EMSInauts* conducted isolation studies for four weeks to prepare for future long-duration missions in space. In a later experiment, EXEMI '92, a crew of one woman and three men (pictured above) spent 60 days in isolation in the TITAN hyperbaric complex at the DLR aeronautical medicine institute near Cologne, Germany. Source: ESA.

and reported that

- they stepped into the chamber as individuals, but came out bonded as a team;
- cultural differences, such as diversity in diet, did not interfere with crew cohesion, as they gradually became an entity working together and trusting each other;
- minor conflict within the group centered around tidiness, and a mutiny almost arose when the crew objected to the demands of external principal investigators;
- what they missed most in isolation prompted the response "The Dutch landscape" by one of the team.

In this experiment to simulate the astronaut environment in orbit, a total of six astronauts was considered ideal from the viewpoint of group dynamics on a Space Station. Women were excluded in the first experiment in order to reduce the number of parameters, but the *EMSInauts* were unanimous in recommending, "Next time we would like a mixed crew." Jacques Collet, head of ESA's Long Term Program Office, said that Europe had to achieve autonomy in manned spaceflight because the next century will be centered on human presence in space. He announced then that ISEMSI will be followed by other experiments to study the logistics and operational requirement of ground/crew relations. Raymond Fife, head of Space Station and International Affairs at the Norwegian Space Center, noted that NUTEC had acquired expertise and technology in offshore oil operations which could be applied to manned space missions. The medical computer system developed at the Université Paul Sabatier in Toulouse, France, and other telemedicine innovations are already being applied to the offshore oil industry, and could be utilized aloft.

Subsequently, the European Space Agency sponsored a project *EXEMI '92*, a 60-day multidisciplinary scientific study in isolation and confinement (again see Exhibit 75). A crew of four, one Swedish woman and three men from Austria, the Netherlands, and France, lived in the TITAN hyperbaric complex at the DLR aeronautical medicine institute near Cologne, Germany. ESA-LTPO then sponsored a symposium on the *Study of Human Behaviour in Extended Spaceflights* (UNESCO, Paris, December 1–2, 1993) to report their results, inviting others who have researched similar objectives in underwater habitat, submarines, polar stations, etc. The assembled scholars exchanged on such aspects as group assessment and dynamics, individual perceptions and performance, psychological/physiological changes in isolation, and international experiences. ESA/NUTEC continues such studies, expanding on Raymond Fife's observation:

> "Working and studying the extreme environments of inner and outer space create synergy effects which benefit both environments while producing spinoffs for everyday life on Earth, such as improved equipment, medical procedures and instrumentation."

But the European Space Agency discovered the limits to ground-based simulations, so they have expanded their in-orbit studies of ESA astronauts onboard both

American and Russian spacecraft. Projects to prepare for long-duration manned missions, such as Euromir, will be discussed further in Chapter 8.

6.1.3 Biosphere 2 (private enterprise)

One of the most interesting, comprehensive, and longest experiments in isolated, confined environments was conducted under *private* auspices in Oracle, Arizona (see Exhibit 77). Located in the American southwest near the foothills of Catalina Mountains on SunSpace Ranch, just 50 km north of Tucson, the enterprise was funded by Texas billionaire Ed Bass, a co-designer of Disney's Epcot Center. Under the name Space Biosphere Ventures, scientists, architects, ecotechnicians, and entrepreneurs put together a $150 million project. The founders considered the Earth the first biosphere, and called their undertaking *Biosphere 2* [27]. On a 12,000 m^2 area of land, they built a glass-enclosed ecological system, with laboratories or "biomes" for rainforest, savannah, marine, marsh, desert, agriculture, and human habitat. In Spring 1991, this miniature world was tested for its ability to recycle and maintain environments (air, water, and nutrients) supporting 4,000 inhabitant species of plants

Exhibit 77. Biosphere 2. Exterior view of the 3.14-acre (1.27 hectare) glass-enclosed living laboratory in Oracle, Arizona, USA. Privately financed isolation studies have been carried out at this enclosed facility, in research which may someday be applicable to life in space colonies. The scientific focus at Biosphere 2 has shifted away from human subjects between 1991 and 1994 to closed ecological systems with plants and animals, as well as a tourist attraction and conference center. The University of Arizona assumed management of Biosphere 2 in June 2007. Source: Biosphere 2. For more information, see *http://www.b2science.org/*

and animals, including a crew of eight people. Built supposedly to last 1,000 years, the initial test period was for two years to determine whether the inhabitants could live in harmony within their biosphere. Open to public tours, the facility encompasses the Biospheric Research and Development Center, as well as a conference center and gift shop. To cover escalating costs, the venture makes additional income from tourism, media, and spin-off technology (*www.biosphere.com* ... *www.thepepper.com/tucson_biosphere.html* ... *www.b2science.org/*).

In her recent book on *Biosphere 2: Lessons Learned*, Jayne Poynter discusses the implications of the original experiment regarding negative psychological and interpersonal adaptations.

The project's public relations promoted this enterprise as a world of discovery, an excursion into the future, the genesis of a space colony. It contains a miniature mountain, a desert, forests, and a 9 million–liter saltwater sea. The original aim was to test the ability of lifeforms to survive and thrive in a sealed environment for the extended time period. The whole endeavor was envisioned as a prototype for human habitation in space, a learning laboratory for other biospheres on space stations, the Moon or Mars. Dr. Walter Adey, director of the Smithsonian Marine Systems Laboratory and a consultant on the program, commented that scientists try to make simple rules for complex processes, so Biosphere 2 was a serious attempt to step beyond this and work with organisms and ecosystems as the complex systems that they really are. Mark Nelson, chairman of the Institute of Ecotechnics, the London-based ecological organization that conceived and originally managed the research and development, maintained:

> "The technology for producing oxygen and other necessities of life are readily available for sustained human exploration of the heavens. The science of biospheres, understanding how biospheres operate on a planetary scale, as well as our microscale at Biosphere 2, is one key to opening this frontier today in preparation for future frontiers on Earth and in space."

Prior to entry, the first crew of eight, dubbed *biospherians*, trained together as a team for some months, including survival trips to remote areas. The four men and four women were airlocked into Biosphere 2 on September 26, 1991, beginning continuous testing of the regenerative life support systems in five biomes. Only one, Roy Walford, a pathologist from the University of California-Los Angeles, was a trained scientist. Two thousand electronic sensors linked to one of the most advanced artificial intelligence computing systems assisted biospherians in monitoring the ecosystem within the 90% closed environment. Inside the air-conditioned enclosure, the plan was for the crew to grow, harvest, and eat their own food (about 2,500 Calories a day), select their own menus from the 26 crops to be raised, and to drink recycled water. They enjoyed television and videos, as well as other creative hobbies, and were supposed to exchange only information and energy with the outside world during their 24-month stay within this interior space. Project secrecy was such that little was known as to how the initial crew were selected and trained, as well as how they were to be monitored and supported. But, by October 1991 one crew member had left the airlock for medical treatment and reportedly carried in extra supplies on her return. By December 1991, project operators conceded that they needed to pump in extra air, and were coping with problems within the enclosure, such as the spread of plant rust disease. The first mission ended in September 1993, and the crewmembers emerged haggard, having lost an average of around 11 kg during their two-year isolation in an artificial environment. Among the serious problems discovered in Biosphere 2 were (1) self-sufficiency from framing did not live up to expectations, partially because the external El Niño climate effect reduced sunlight from the outside and the closed ecosystem could not supply 2,500 Calories per day for the biospherians; (2) the oxygen levels within the biosphere itself dropped below 14.5% and was deemed hazardous. The pioneer mission was also marked by controversy over management's concealment of setbacks, as well as differences with the scientific community over the experimental nature of the project. The special independent committee of ten scientists who were to oversee the simulated "space habitat" resigned because of lack of cooperation by the operators, who were not listening to the advisors. To address the Biosphere 2 scientific shortcomings, a panel of scientists under Tom Lovejoy, a Smithsonian Institution ecologist, studied the whole project, issuing a report and recommendations which were acted upon (*New Scientist*, February 8, 1993).

A second mission was undertaken in February 1994 with a crew of seven, but the time enclosed was curtailed. Five men and two women stayed in isolation for only six months, but emerged from the sealed laboratory healthy and well-fed. By April 1994, six officers of the original operating company were relieved of duties. The new administration created a research consortium with scientists from Columbia University's Lamont-Doherty Earth Observatory which decided that the emphasis at Biosphere 2 was to be on research in environmental sciences, especially global climate change, biodiversity, and sustainable agriculture. Jointly, a non-profit institute was created to open Biosphere 2 for research by universities, national laboratories, and private teams. Dr. Bruno Marino, a biochemist from Harvard University, was appointed in August 1994 as its new scientific director. From both a tourism and an ecological perspective, Biosphere 2 was somewhat successful. The hope was that its experiments would eventually contribute to NASA's Controlled Ecological Life Support System, which is developing biogenerative life support systems to produce, process, and recycle biomass. The Biosphere 2 hardware may someday be tested in an orbiting space biosphere. Some other positive by-products of this venture include

- A consortium of Japanese corporations and NASDA planning to build Biosphere J ("j" is for junior since it would be a quarter the size of its Arizona counterpart). It has yet to be built in northern Japan.
- Biosphere 2 patron Edward Bass donating $25 million to Yale University to form an *Institute for Biospheric Studies (YIBS)* to conduct further ecological research.
- Spin-off technologies that will have terrestrial as well as in-space applications.

In the 21st century, Biosphere 2 has become a popular tourist destination that runs educational programs related to global warming. Since July 2007, Biosphere 2 has been managed by the University of Arizona. Grants by the Philecology Foundation and other organizations are supporting research programs and the costs of operating the facility. Biosphere 2's research arm, dubbed B2 Earthscience, is led by Travis Huxman, UA associate professor of ecology and evolutionary biology.

Plans are under way for another terrestrial analog in 2010, to simulate lunar living, but this one will be aimed at tourists. Entrepreneur Michael Henderson is creating "Moon Casino Narrows" on an artifical island that will house hotel attractions, cruise ship terminals, and sporting and real estate facilities. The future site has been narrowed down to Singapore, Thailand, or the Bahamas. The gravitational attraction has as one of its consultants, Col. Rick Searfoss (USAF Retd.), a former NASA astronaut and Shuttle Commander. The project hopes to build on public interest in near-term U.S. and China lunar missions.

6.1.4 Design for extreme environment assembly

Prof. Larry Bell of the University of Houston's College of Architecture and other colleagues in the U.S.A. and Canada, Russia and the Ukraine, the U.K. and Japan first came up with a brilliant concept for professional cooperation relative to studies on exotic environments. They conceived an international organization that would sponsor conferences every two years for representatives from academia, industry, commerce, government, and consultancies. Called the *International Design for Extreme Environments Assembly (IDEEA)*, it was open to those interested in research, planning, and operations. The focus was on sustainable development in extreme, harsh, and hazardous conditions, as well as the impact of extreme environments on environmental change and policy. An environment is defined as *extreme* when the circumstances of the locality threaten, disrupt, or cannot normally sustain physical, social, institutional, and economic life. Such conditions may be caused by

- extremes in climate, geographies, topographies;
- environmental change, hazards, or contamination;
- chronic poverty, natural disasters, political/civil conflict.

Extreme environments, such as Antarctica or outer space, present severe operating conditions and drive science and technology development, causing the invention of new designs, management methods, and models which can be utilized elsewhere.

The *IDEEA* participants were also concerned about (1) application of innovative, advanced, and intermediate technologies in such settings; (2) transfer of knowledge, methodologies, approaches, materials, technologies, and experiences from one setting or location to another; (3) identification of positive initiatives for global security and environmental policy with a network of organizations, governmental bodies, and individual leaders. This strategy was uniquely multidisciplinary, bringing together a cross-section of both researchers and practitioners. Thus, there were exchanges on a central theme that drew together a wide variety of specialists, such as architects and construction experts, behavioral scientists and medical professionals, lawyers and political scientists, as well as members from various national space agencies [28].

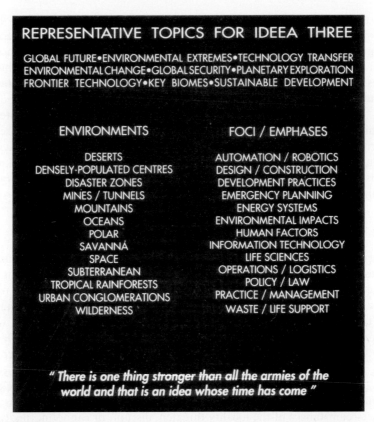

Exhibit 78. IDEEA conference themes. The scope of themes and emphasis at such multi-disciplinary meetings is illustrated above for the Third *International Design for Extreme Environments Assembly*. Source: *IDEEA UK* Secretariat, Oxford School of Architecture, Oxford OX3 OBP, U.K.

IDEEA ONE was convened at the University of Houston in November 1991, attracting almost 500 attendees from a dozen countries. The exciting presentations resulted in a 958-page volume of proceedings ... *IDEEA TWO* was held in October 1993 at McGill University in Montreal, Canada, and saw an expansion of networking and information sharing for problem solving in extreme environments, and another publication ... *IDEEA THREE* was canceled in March 1996 by the Oxford School of Architecture, Oxford, U.K. [28]. The range of topics that were to be discussed is depicted in Exhibit 78. Unfortunately, like so many space organizations that come and go, this was an idea ahead of its time and is now defunct.

However, it contributed to pushing the envelope of knowledge, and has been succeeded by a comparable entity, *The Society for Human Performance in Extreme Environments* founded in 1997 by M. Ephimia Morphew-Lu (*www.hpee.org*). It publishes a journal and sponsors annual meetings to bring together life scientists, behavioral scientists, engineers, and other like-minded professionals. HPEE fosters interdisciplinary collaboration by providing an international forum for scientists, practitioners, operational personnel, students, and other professionals interested in human performance in complex, high-stress, demanding environments and occupations.

6.1.5 Mars virtual explorations

The advocates for near-term human missions to the planet Mars have set up an "underground group" that has undertaken numerous conference, books, and other activities to promote this cause. Led by the energetic Dr. Robert Zubrin,[2] an aeronautical engineer and founder of Pioneer Aeronautics, the group organized itself into the Mars Society, dedicated to furthering the exploration and settlement of the "Red Planet" (*www.marssociety.org*).

One of its projects has been simulated missions to isolated spots on Earth where they replicated and studied the real-life challenges for the first pioneers on Mars. From 1999 to 2002, their unique ground missions involved building habitats, donning space suits, and testing next-generation exploration equipment [29]. In their Earth-bound "space program", they had polar bear encounters, experienced crew tensions, coped with near-disaster and system failures, while inspiring people worldwide through television documentaries and publications. The "would-be space explorers" trekked to parched deserts, frozen wastes, and volcanic moonscapes, learning and collecting data for use in future manned missions offworld [29].

In 2006, the American Astronautical Society published a volume on *Mars Analog Research* with a supplemental CD [30]. It is the most comprehensive review of diverse investigations and studies to prepare for human, scientific exploration of Mars in the decades ahead. This research covers six broad areas: expeditions, field science studies, engineering studies, human factors, analog field facilities, and cultural issues. Led by Dr. Jonathan D. Clarke, from Macquarie University, Sydney, Australia, there were some 45 contributors to this seminal book (edited by Dr. Clarke), including reports by Euro-MARS and MARS-OZ in Australia. The publication is part of the AAS Science and Technology Series (Vol. 111 [30]), an organization that has already published ten other volumes making the case for Mars exploration, all based on technical meetings and in conjunction with its partner UNIVELT, Inc. and Series Editor Robert H. Jacobs (*www.univelt.com*).

In concluding this opening section on terrestrial analogs, we have reviewed five professional activities that have great implications for the high frontier. But ground simulations, such as Biosphere 2 and ESA isolation studies, cannot test for all the effects of closed ecosystems within a microgravity environment, such as the impact of radiation, ozone sunlight levels, severe cold, and soil nutrient conditions. However, there are important contributions to living and working in space which come from the accumulation of knowledge and experience about isolated, confined, or extreme environments. Learning to construct and manage habitat biospheres under such

---

[2] A DVD entitled *The Mars Underground* by Robert Zubrin is available from Apogee Books (*www.cgpublishing.com*).

circumstances allows for the transfer of technology and insight aloft to the high ground. Ground-based research adds to the ever increasing knowledge of humankind's collective experience in orbit to date.

6.2 SPACE PERSONNEL DEPLOYMENT STRATEGIES

In sending humans offworld, especially for long durations, the author has hypothesized that a "systems" approach should be developed for such voyagers based on the previous terrestrial analogs and our limited experience aloft [31]. Incorporated in this system would be careful evaluation and selection of every would-be spacefarer, as well as preparation in interpersonal and intercultural relations and team building. Such an approach should contribute to mission achievement while preventing or reducing depression, withdrawal, hostility, paranoia, and other mental health problems that may afflict orbital travelers.

A space personnel deployment system (SPDS) may be defined as a planned and orderly means for the exchange of people to/from Earth and the space frontier. The system described goes beyond what is currently in place for elite astronauts and cosmonauts on relatively short-duration missions. SPDS is intended for all "those others" who are going into orbit as contractors or settlers. The essential components for this would consist of four stages as are illustrated in Exhibit 79: namely, assessment and orientation before launch, in-space support services, and re-entry policies and programs. The time devoted to each stage would differ with space tourists and short-duration and long-duration spacefarers. We shall now examine each element in the model, attempting to answer four key questions within each phase (who, what, how, and why), as well as considering some research issues involved with each stage of this deployment system:

When the Soviets selected their first cosmonauts, they recruited 20 who were to be brave, reliable, physically fit, not prone to panic, capable of mental endurance, and familiar with "things up there". Later selections, like their U.S. counterparts, put more emphasis on engineering backgrounds, persistence, and more stable personalities. A high rate of attrition was built into recruitment, so about three times the number required were selected. Gradually selection went beyond the military and pilots to include civilians, especially scientists, with a wide mix of age, experience, and background. When man's first home in space, the Salyut station was launched in April 1971, crewing policy changed to an all-Soviet resident crew, and visiting Intercosmos crews with internationals and persons with more diverse specializations. (Source: *Race into Space: the Soviet Space Program* by Brian Harvey. Chichester, U.K.: Ellis Horwood, 1988.)

Sec. 6.3] SPDS stages 265

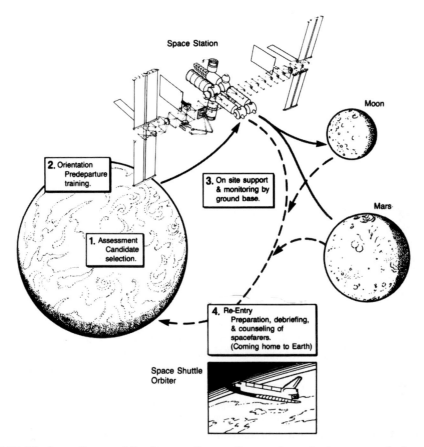

Exhibit 79. Space Personnel Deployment System. Source: the author's concept of a four-stage system for sending people to/from space is illustrated here by Dennis M. Davidson.

6.3 SPDS STAGES

6.3.1 Stage One: Assessment

Assessment aims to recruit and select the most suitable people to go aloft, while screening out potential misfits who would jeopardize their own or others' wellbeing, as well as to ascertain the level of candidates' competencies and coping skills. This is not a new concept, as Dr. James Grier Miller wrote in his *Assessment of Men: Selection of Personnel for the Office of Strategic Services* (Reinhart & Co., 1948)—just that we are now doing such evaluations more scientifically and systematically.

Space developments in the 21st century include a variety of organizations, in addition to national space agencies, sending people into orbit. In time, we can expect sponsors to range from aerospace contractors and commercial enterprises, to various public agencies and the military, media networks and tourist firms who are already experimenting in sending civilians aloft. Eventually, non-profit associations with special interests may wish to underwrite their own space community, just as happened among the early American colonists. Whatever institution sends a human onto the high frontier, it should exercise a measure of responsibility in the spacefarer's selection, preparation, and wellbeing, particularly with reference to that person's impact on the space group of which he or she is a part. Pre-departure evaluation of candidates delimits costly care and rehabilitation later offworld. In the future, the selection and training issues may move beyond the control of individual sponsoring organizations. Some world entity may have to be established, such as a *Global Space Trust* or *Metanation*, an *Orbital* or *Lunar Development Authority*, which regulates and monitors worldwide both space assessment and access [32]. Just as trading companies were established in the 16th century to sponsor and select settlers for the New World, expect global trusts and world corporations to be formed to send their representatives and colonists offworld. The International Space University, for instance, already has in its credo that it will be a part of humanity's movement into the cosmos, so undoubtedly this entity will have admission and training standards set for ISU spacefarers. Many such norms may come from among their own global alumni.

Various space agencies have assessment programs for those accepted into their Astronaut or Cosmonaut Corps A possible model may exist in the European Astronaut Centre, which is responsible for defining the selection, qualification criteria, and coordinating the training of future European astronauts from many nations on that continent [33]. The Centre in Cologne, Germany, also selects candidates for flight or ground activities, such as equipment development, or for special training to liaise with astronauts in orbit, whether they fly on American or Russian spacecraft. Since the Centre is responsible for human safety aloft, it has the authority to define rules and ensure their observance. In its supervision of Europe's manned space program, EAC activities include crew assignments, instructor selection, management of training facilities, and external relations. Under the ESA, this regional center represents member countries in their participation on NASA's Shuttle flights (Spacelab) and the International Space Station, especially with reference to the European-crewed orbiting laboratory, Columbus, and the planned spaceplane/mini-shuttle, Hermes. The same responsibility was extended to Europeans who once flew on Russia's Mir space station. Exhibit 80 provides an overview of the methodology (dating from 1990) in terms of planning, management, and resource utilization at ESA's European Astronaut Centre in Cologne, Germany, as the Agency seeks new talent to bolster its Astronaut Corps for future manned missions to the ISS, the Moon, and beyond. According to André Ripoli, former head of the European Astronaut Centre, the candidates for astronaut training must be citizens of ESA member states, healthy, efficient human beings who are both good professionals and operators with a technical or scientific background. Pre-selection begins in the member states, who put forward three to five candidates; the ESA selection board then choose the persons to be trained for specific assignments, such as laboratory or mission specialists. Other formal requirements for ESA astronauts are

- preferred age (27–37); height (153 cm–190 cm); ability to speak and read English and/or Russian; physical fitness and psychological aptness, including a sound health history and normal weight;

Sec. 6.3]   SPDS stages   267

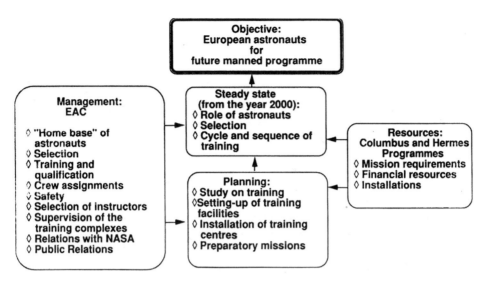

Exhibit 80. ESA astronaut deployment strategy. It is now time for the European Space agency (ESA) to seek out new talent to bolster its Astronaut Corps for future manned missions to the ISS, the Moon and beyond. Consequently, ESA will be searching in each of its 17 Member States for the best candidates to make this vision a reality. The methodology outlined here (dating from 1990) sets forth some objectives in terms of planning, management, and resource utilization at ESA's European Astronaut Centre in Cologne, Germany. Source: European Space Agency.

- university degree or equivalent; professional experience in a relevant field (e.g., natural science, engineering, medicine, as a pilot).

The typical ESA astronaut is expected to be able to stay up to six months in orbit, be a career person eligible to fly over a 20-year time period. It is no wonder, then, that the German Aerospace Research Establishment (DLR) recommends that astronauts get both basic human factor training and special psychological preparation for specific crew missions [34].

Obviously, the screening and training requirements for space tourists and short-term assignments (e.g., 5-day to 15-day spacecraft flight), would be less than that required for long-duration missions. In attempting to evaluate personnel for six months to a year or more aloft, the following guidelines, developed by the author from terrestrial experience with foreign deployment systems, may prove useful, possibly for testing at the International Space Station or a lunar base in the next decade [35]:

- **Who?** Screening requirements of space travelers and settlers would apply to all who utilize public or privately supported space transportation (whether space agency personnel, contract workers, visiting professionals, public servants, media representatives, military members, foreign dignitaries, and/or even tourists). When the time arrives that family members accompany such personnel into space, then they too should be included in the assessment process. For the

21st century, at least, the principle proposed here initially is *selectivity*, the time for experimentation with this process is in the immediate century ahead.

- **What?** The aim of the evaluation should be to determine the suitability and adaptability of space candidates to deal effectively with the new environment and situation aloft; both physiological and psychological evaluations should determine the individual's capability to deal with differences and difficulties in space travel, to live in outer space under constrained conditions, and to cope with its stressors. The first concern is whether the candidate is physically fit for the ardour of space travel and living. The stressors include confinement and dependence on life support systems; lack of privacy and sleep disturbances; high work load interspersed with monotony; reduction in leisure time activities and social contacts. Essentially, the process seeks to identify proneness to "space culture shock", as well as areas for special training so as to improve coping, stress management, and human relations skills when in orbit.
- **How?** Assessment might be conducted through a specialized center that would utilize a variety of disciplines and means for evaluation purposes, such as psychological interviews and group meetings, instrumentation and testing, and simulations (live gaming, computer cases, and virtual reality exercises). It is conceivable that expert systems could be designed for this purpose which are based on accumulated research about the qualities that make for success within the isolated, confined environment. In the beginning, the objective might be to choose those with competencies that ensure survival and mission accomplishment, but eventually the aim may be to group spacefarers by personalities and interests that share values, goals, and concerns for a specific type of settlement.
- **Why?** Recruitment and selection should not only seek appropriate space "pioneers"—but determine special counseling and training needs of those chosen, particularly with reference to the family that is left behind on Earth. Obviously, the first concern is to exclude from space communities those who would be unsatisfactory for such an assignment, subject to premature return symptoms, or likely to become a disruptive influence. Ultimately, the evaluation should choose people who can contribute the most to the emerging community in orbit. The United Nations long ago indicated that spacefarers were the *envoys of humankind*; therefore, we should choose people to go aloft who will be representative and diplomatic!

Initially, the sheer cost of transporting people to and from space, plus orbital risks, should justify a careful assessment policy and program; certainly, the expenses associated with premature offworld return would warrant investment in a deployment system. Outside of space agency personnel, mission specialists are the prototype of other civilian space workers to come. Preferences may, for example, be given to those with special qualifications, such as:

— veterans of some type of service (astronaut or cosmonaut, military or peace officer training, Peace Corps or humanitarian service abroad, extreme environment worker or researcher, Foreign Service or expatriates, space tourists, etc.);

Sec. 6.3]   SPDS stages   269

Exhibit 81. Future crew prototypes. In June 1995, there was a docking of the Shuttle *Atlantis* to the Russian space station. The STS-71 crew contained two cosmonauts when the link-up was made with Mir 19, which already had one NASA astronaut onboard. The photograph of the combined Russian/American crew of ten, including two women, is symbolic of the diversity and synergy sought today in mission personnel selection. Source: NASA/*Countdown*, May/June 1995, p. 54.

— those committed to long-term space living and willing to assume inherent risks;
— healthy, well-balanced married couples with dual competencies;
— professional specialists whose expertise is required in orbit.

Exhibit 81 provides a view of heterogeneous crews of the future, a mixture of gender, nationalities, and professional backgrounds.

The assessment process also involves *screening out* certain types of candidates, such as those with personality problems. Regarding long-duration spaceflights, NASA has expressed concern about excluding those who (a) lack social competencies or human relations skills; (b) lack communication skills for reliable interaction among the crew, and between the crew and ground control; (c) tend to be perfectionists and autonomous. Pre-departure evaluation has difficulty in predicting "wild cards" that may arise aloft in crew relationships, such as may be caused by jealousy or sex/love affairs.

NASA WOMEN ASTRONAUTS

Interestingly, back in 2001, NASA attempted to fly an all-women Space Shuttle crew for the purpose of scientific study. The Agency was concerned about whether prior research on males was applicable to females (such as bone loss dangerous to women prone to osteoporosis, and whether increasing their calcium intake might lead to kidney stones). There were other gender-specific research concerns, such as menopause on long-duration flights, pregnancy, and eventually even child birth aloft.

Women were not admitted to the Agency's Astronaut Corps until 1978, and female nurses have yet to fly into orbit. In 1983, the first women were assigned to the Space Shuttle. No lady piloted a spacecraft until 1995 when Eileen Collins commanded a Shuttle flight. Susan Helms became the first woman to stay aboard the International Space Station, remaining aloft for 163 days in 2001 as part of the Expedition 2 crew. Shannon Lucid was the only American woman to live on the Russian Mir station, remaining aboard for 179 days in 1996. In 1999, of the 119 astronauts in the Corps, only 29 were women. But by 2007, four females had given their lives in space service as a result of the two Shuttle accidents on *Challenger* and *Columbus*. Gradually, male dominance in U.S. spaceflights lessened, and today females hold a variety of important positions both on the ground and in orbit.

Given that context, there was a backlash by women astronauts to the plans for all-female flights, even as a one-time experiment. Their attitude was they had worked so hard and long to get where they were in NASA, the women were against segregation into all-female crews. They were insulted by any policy or strategy in the Corps that smacked of gimmickry or implications that their skills somehow did not measure up. Former astronaut Kathryn Thornton, the first American woman to have an EVA walk, protested that the Agency should simply collect female scientific data on regular mixed gender missions, while assigning more women astronauts to shuttle flights. Thus, the ladies quiet opposition killed off the mission for studies of all female-crews. [Perhaps the change in NASA culture is most evident when in 2007, two women simultaneously were together in leadership positions, astronaut Pamela Melroy commanded the Shuttle *Discovery* Mission STS-120, while her counterpart Peggy Whitson was in charge of the ISS expedition in orbit.—P.R.H.] (Source: Marcia Dunn, Associated Press in the *San Diego Union-Tribune*, April 11. 1999, p. 2.)

Whitson's flight up to the ISS was reported thus: A Russian spacecraft soared from Kazakh steppe toward the International Space Station carrying a Malaysian, a Russian, and Peggy Whitson, the American who will become the first woman to command the orbital outpost ... Whitson of Beaconsfield, Iowa, is making her second trip to the station as a NASA astronaut. (Source: *The San Diego Union*, October 11, 2007, p. A12. For gender-related patterns aloft, see also G. R. Leon's "Men and Women in Space," *Aviation, Space, and Environmental Medicine*, **76**(6), June 2005, B84–B88.)

> **PRESENT SELECTION PROCEDURE**
>
> Currently, NASA selects for two types of astronauts:
>
> (1) *Pilots* who serve as Shuttle commanders (onboard responsibility for crew), as well as pilots who assist in controlling and operating the vehicle and in deploying/retrieving satellites. Such candidates must have a minimum of an undergraduate engineering or science degree and preferably advanced degrees; 1,000 hours of pilot-in-command of jet aircraft with flight test experience considered desirable; ability to pass a Class 1 space physical; height 64–76 inches (limit is 6 foot 4).
> (2) *Mission Specialist* who works with commander and pilot relative to crew planning and operational activities, consumables usage, conducting experiments and payload handling; persons with unique expertise who do not have to be agency personnel, but may be sent by a contractor or foreign government. Required to have undergraduate and preferably an advanced degree, especially in engineering or biological/physical science, or mathematics. Minimum of 3 years professional experience in speciality unless an advanced degree. Must be able to pass Class II space physical, and have height between 58.5 and 76 inches. Military personnel may apply through their branch of the Armed Forces, but civilian teacher program is presently suspended. There are no gender or age requirements.
>
> Those selected for either program are Astronaut Candidates who must successfully complete one year of training, including interviews and orientation. Civilian candidates who are not finally selected as astronauts may continue in other positions. (Source: *Astronauts' Selection and Training*, AHX Astronaut Selection Office, NASA Johnson Space Center, Houston, TX 77059.)

Obviously, such selection categories are changing. Working and living on ISS is one example. With 16 national partners on that macroproject, there is more diversity in choosing station inhabitants. The Russians already have their own selection and evaluation programs, and are responsible for bringing space tourists aboard. The first female tourist, Anousheh Ansari, made it into orbit because Russian selection standards deemed a Japanese entrepreneur, Daisuke Enomoto, physically unfit for both training and the 10-day journey. In other words, the nature of the missions will cause alterations in traditional, short-duration flights to orbit. Plans under way for a lunar base in 2010 will bring further modifications in the choice of people to gain access to projects and facilities on the Moon. The need there for more experts, contractors, and specialized workers will alter and expand the whole personnel deployment system.

Assessment research needs

The most immediate need is for information sharing among worldwide space agencies on current and future practices in crew selection and training. Because of the proclivity in the past to give preference to pilots with flight experience, many speculate as to whether this should remain a factor in future spacefaring candidates, such as a "civilian-in-space" program. The rationale was that such "flyers" come from a stressed background and are used to dealing with life-threatening circumstances, so therefore are better able to cope with the microgravity environment. Apparently, aviation provides a three-dimensional experience, teaches crew coordination, accustoming one to be concerned about displays and consumables. This is a matter for further study, along with the neuropsychological tests which the U.S. Air Force use to select their personnel for spaceflight. For the manned spaceflight engineer (MSE), the military sought qualified personnel with proven performance and initiative, self-confidence and perseverance, as well as technical competence. Are these still valid criteria for other space travelers? The Russians have already entertained on their Mir station a variety of professionals from Japanese journalists and a British female chemist, to diverse international astronauts from ESA/NASA, as well as from other countries in Eastern Europe, Vietnam, India, etc. All received additional training at the cosmonaut center in Star City outside of Moscow. Russia has promoted a similar approach to its paying guests on the International Space Station [36]. Gradually, NASA too has expanded participation by various internationals on its Shuttle and ISS, but it is still hesitant about including "civilians or tourists". It is just a matter of time before national space agencies will no longer be able to control access to space, so research is necessary on some basic global criteria for selecting the future *envoys of humankind*.

As mentioned previously, *The Human Role in Space* project was an early attempt to examine human qualifications for space activities, including the *sensory/perceptual*, *psychomotor/motor*, and *intellectual competencies* required. The study was deficient in ignoring psychosocial factors. Harrison and Connors (cited in the last chapter) reviewed the experience of humans living successfully in exotic environments and identified those socially desirable traits that have implications for the selection of spacefarers: *personal competence*, *emotional stability*, and *social versatility*. The top performers in such situations manifest strong task or mission orientation; coping skills with depression, stress, and anxiety; appropriate group interaction skills, being neither too introverted or too extroverted.

Today's payload specialists, usually scientists, are more like tomorrow's average spacefarer: typically, they are chosen by their companies or a foreign space agency; most women and internationals who have flown the Shuttle came aboard under this category. Their profile includes high levels of graduate education and scientific accomplishment: a generalist with multiple skills or expertise in a required specialized field; ability to pass NASA's medical and psychological screening; and capable of learning quickly in their intensive training programs. One example of an industry astronaut was Charles Walker of McDonnell-Douglas, who returned from orbit to become president of the National Space Society. With the construction of a lunar base, anticipate that contract workers or "technauts" will follow the pioneering astronauts and cosmonauts, followed by settlers, including eventually families.

Once the criteria for spacefarer assessment have been agreed on, a related issue is which astronauts are selected for specific missions and who does the selection? In Chapter 5, we reported on Dr. W. K. Douglas' interviews with astronauts. On the subject of crew selection, there were mixed suggestions as to who should make the evaluation and choice, ranging from an agency committee or management on the ground to the crew commander for a particular mission. For long-duration flights, these spacefarers seemed to prefer peer review, so as to ensure *compatibility*. They recommended that astronauts should have had experience in training and working together before going into orbit. They thought a commander should have a veto over group membership, including mission specialists. They envisioned that space station crews would be selected for their skills, motivation, and physical/mental health. One commented that *sensitive, intelligent people* are needed for longer flights aloft—not "macho-man" types.

But the real opportunity for further research is evaluation relative to groups who have passed ground screening and are actually working together aloft. This would be a performance assessment that goes beyond their work productivity to their teamwork and morale contributions. That is, how do they apply their interpersonal, intercultural, and team skill training as they actually interact within self-directed teams? In analyzing crew selection for a future Mars expedition of two years in orbit, Filbert and Keller suggested four categories of critical skill requirements [37]:

(1) competencies in navigation and piloting;
(2) competencies in equipment operation and maintenance;
(3) competencies in scientific study and analysis;
(4) competencies in medical/dental diagnosis and treatment.

Note the complete absence of psychosocial and team skills, so essential for a successful mission of this length and risk! Furthermore, researchers indicate that social competency should be a dimension to assess in crew composition.

Although American/Russian/European/Canadian selection criteria and experience for spacefarers are valuable, more broad-based policies and procedures are required for larger, more diverse space populations. The potential size and number of stations/bases/settlements in this century demand a more sophisticated system for choosing and evaluating candidates, so expanded assessment research is in order. This is particularly necessary on the matter of multicultural crews, such as was the case with the STS-71/Mir crews shown in Exhibit 81 [38]. Research is particularly necessary on families going into space as settlers: not only how to evaluate them as potential colonists, but also how to prepare them for this vital role! With the emergence of space tourism, studies will focus on short-term, civilian space adjustment.

> **SPACEKIND**
>
> We may talk about exploring the Solar System, but the Earthmen aren't going to do it. It may be done by lunar colonists or other space settlers. They are going to become another breed of human beings. They are going to live a totally different life. They are going to be a different kind of people psychologically and they are going to be the cutting edge of humanity. They will be the ones who get industry out into space. They are the ones who will build the solar powered ships. They are the ones who will create a whole new kind of technology, different from anything we've got here. (Source: From a symposium address of the futurist writer, Isaac Asimov, quoted in *Lunar Development Council Newsletter*, January 1987, No. 1, p. 1.)

Exhibit 82. Preparing spacefarers. The training of spacefarers includes instruction on life support systems and the donning of spacesuits. Here, STS-95 crew member U.S. Senator John H. Glenn Jr. gets help with his partial-pressure launch and entry suit from experts Jean Alexander (left) and Carlous Gillis prior to a training session at the Johnson Space Center. Note that the astronaut is not a young man: age is no barrier to entering the high frontier ... Senator Glenn was 77 at the time of his flight aboard the Space Shuttle Discovery. Source: NASA Johnson Space Center.

6.3.2 Stage Two: Orientation

Once would-be spacefarers have been carefully evaluated and selected, the next step is to effectively orient them before departure to space living, safety, and culture through both self-learning and training. In preparing and educating space explorers and settlers, the curriculum and its scope would vary depending on the intended orbit, length of stay, particular mission, and previous spaceflight experience. Certainly, it would be based on similar training programs of global space agencies, but go beyond that with civilian space pioneers. Since the selection and training of NASA astronauts at the Johnson Space Center and the Russian cosmonauts in Star City are well documented [39], we feature here the European program as outlined in Exhibit 83. The basic training takes one year, while the whole program is four years in length. It begins at the European Astronaut Center, Cologne, Germany, where the principal crew training and medical facilities are located. Four branches specialize in additional preparation of spacefarers from that continent: in France, Marseilles hosts the External Servicing Facilities where EVA training takes place, while Toulouse is

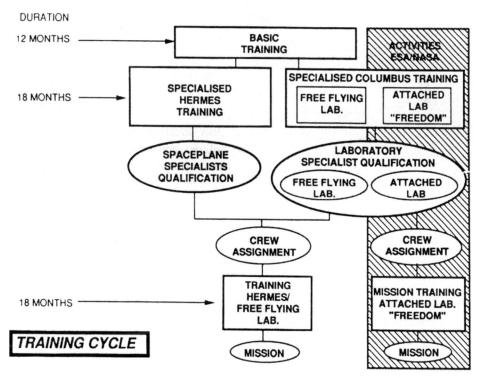

Exhibit 83. ESA Astronaut Training Cycle. This chart relates to 1996, hence the reference to "Freedom Lab" on what today is called International Space Station. As new missions are added, this chart obviously changes. For the current cycle, see [26]. (Source: ESA Fact File #3, European Astronaut Center, Cologne, Germany.)

the site of the Hermes Subsystem Training Facilities; in Belgium, Brussels offers the Pilot Training Facility for the Hermes flight simulator and trainer aircraft; in the Netherlands, the Robotic Training Center provides training on the Hermes robotic arm. The German Aerospace Research Establishment (DLR) has been the leader in the psychological training of scientists for spaceflight [40].

As the European Space Agency expands spaceflight agreements with the American, Canadian, Japanese, and Russian space agencies, their astronauts/cosmonauts undergo further crew training for specific multinational missions, such as the Johnson Space Center in Texas or Star City near Moscow. ESA is involved with 16 other nations in preparations for final construction and staffing of the International Space Station as previously explained, so that multinational crew training is essential for that macroproject.

COSMONAUT TRAINING

Cosmonaut training began with survival skills and expanded to include other competencies. Like NASA, the training takes places in various bases around the country, but is largely in the Gagarin Training Center at Star City located near Moscow. Like the Johnson Space Center in Houston, it has facilities for both training and mission control. When Valentina Tereshkova was chosen to be the first woman in space in 1963, there were also four other women selected with her for training; though they never got into orbit, they worked in ground support jobs. It would be 19 years before the Soviet system would permit the next Russian woman to fly in space—Svetlana Savitskaya in 1982. (Source: Dennis Newkirk, *Almanac of Soviet Manned Space Flight*. Houston, TX: Gulf Publishing, 1990.)

Next, we continue with a prognosis on the future orientation of spacefarers for long-duration missions and settlement:

- **Who?** A short-time visitor, for example, might not be required to undergo as extensive an indoctrination as one who is expected to stay months or years aloft. Novice spacefarers, on the other hand, might undertake a longer training course than veterans of spaceflight. Families of the space voyagers should be involved, but their program would vary, depending on whether they will remain on the ground or go aloft. Those who are to become space settlers on the Moon or Mars, for instance, might be exposed to the most rigorous and lengthy preparation, possibly as family teams.
- **What?** The pre-departure training should focus on the challenges in extraterrestrial living, from survival and safety to cultural and interpersonal skills [43]. The basic indoctrination should be on orbital survival skills, meaning the physical and psychological health care for those living offworld (see Appendix E). For example, in orbit a physical exercise regimen is required to offset the lack of or

diminishment of gravity forces. Another area of learning has to do with achievement of mission objectives, including the technical and operational aspects of transport, equipment, ecosystems, and coping in a zero or low-gravity environment. The next training component features the behavioral sciences so that crewmembers and future settlers function effectively as a team within a space community. The latter preparation might include human behavior issues and skills related to motivation, communication, conflict resolution, negotiation, leadership, team management, and family relations. Eventually, as colonies evolve, the preparation in advance of lift-off will have to include matter of governance and administration. Consideration needs to be given to appropriate and relevant mission instruction for spacefarers in the following subject areas:

(a) space sciences, such as astronomy, resource observation (including mapping and photography);
(b) space technologies, such as automation and robotics, communications and computers, orbital construction and mining;
(c) space migration skills, such as wellness and creative use of leisure aloft, and eventually community development;
(d) foreign languages/cultural understanding for multicultural crews and missions, such as basic English, Russian, Japanese, or Chinese as required.

The human resource development (HRD) program designed for 21st-century spacefarers must be cross-functional and enriched beyond pragmatic training of today's astronauts and cosmonauts. This was explained in Chapter 5. Dr. Gloria Leon, a psychologist at the University of Minnesota, recommends that preflight sensitivity training should place greater emphasis on gender and cultural factors, especially for enhancing crew teamwork and productivity.[3] Tomorrow, entirely new learning systems will have to be created, such as in *space ergonomics and ecology*, and *biocultural courses* in long-term human adaptation to microgravity. Such innovations will emerge eventually from long-term orbital experience of *living and working in space*.

Creative input will have to be developed for planning on new space careers which are only beginning [42]. As missions expand in size and scope, diverse multicultural workforces will require group dynamics and intercultural training to become synergistic teams. Furthermore, crews must receive special education to undertake expanded roles. Dr. David Baker reported that cosmonauts, for instance, have been given instruction in orbital observation of resource deposits, both gas and oil, within their country. Thus, spacefarers contribute to the development of maps for remote areas as well as to photography of civil-engineering projects and national environment needs [43]. Similarly, Shuttle astronauts engaged in "Mission to the Planet Earth" had to acquire new technical competencies to fulfill that goal.

---

[3] Leon, G. R. "Men and Women in Space," *Aviation, Space, and Environmental Medicine*, **76**(6), June 2005, B84–B88.

LUNAR DEPLOYMENT SCENARIO

In his science-based novel on lunar industrialization and settlement, your author proposes a Unispace Academy operated by a Global Space Trust, made up of representatives from private space enterprise and various space agencies. In this scenario that facility would be located in Hawaii and built to simulate a lunar base. Adjoining would be the Hawaii Space & Science Park which provides facilities for high-tech companies serving space enterprises. Another building would house the Matsunaga Ground Control Center that provides support services to lunar travelers and settlers; it is named in honor of the late Senator who originated the plans for a nearby *Spaceport Hawaii*. A supercomputer would manage a Human Assets Registry of all those going offworld, particularly those permanent colonists. HAR would not only keep personnel records, but data on all types of human resources necessary to support space development and its extraterrestrial human population. Beside this talent bank of knowledge workers, a complimentary unit would be Robotic Assets Registry. RAR would not only match automated capabilities with human partners, but maintain records of all robots and automated systems operating in orbit, including their maintenance and service needs.

The Academy, affiliated with the International Space University and The Challenger Center for Space Science Education, would have both a core and visiting faculty, made up of experienced astronauts and cosmonauts; behavioral, biological, and space scientists; space entrepreneurs and disabled instructors, such as Stephen Hawking. The latter would assist future *Selenians* to live and work on the Moon as "disabled" persons with their life support systems. The regular faculty would be supplemented by world-class space experts from the alumni of space agencies; universities and research institutes; corporations and non-profit agencies. The instructional methods would include group dynamics and team training; media and computer technologies; simulations and field trips; plus specialized workshops and seminars on-site or abroad. The educational focus would be on experiential action learning and research. The curriculum would be designed to meet individual and team needs based on length and type of mission undertaken. It would range from the scientific and technical to cross-cultural and family quality of life aloft. The overall goal would be to prepare people for creating a lunar culture quite different from that of Earth. Thus, the goal would be to educate a new breed of spacefarers, including children, who demonstrate

- *adaptability*, ability to cope with differences and demands of long-duration missions, particularly in isolated, confined environments;
- *self-management*, ability to practice human relations skills, both interpersonal and intergroup;
- *wellness*, ability to maintain a personality pattern that includes personal fitness, outgoingness, and a positive mental attitude;
- *competence*, ability to contribute expertise in one or more vocational fields or disciplines; a multi-talented learner capable of creative improvising.

Source: Philip R. Harris, *Launch Out*. West Conshohocken, PA: Infinity Publishing, 2003. Available from *www.buybooksontheweb.com* or *www.univelt.com*

To prepare the first lunar dwellers for private enterprise, your author has proposed a curriculum and training methodology [44]:

- **How?** Innovation in the pre-departure orientation of spacefarers must extend from the design of self-learning and intensive group learning, to new educational methodology and technology. The instructional possibilities for conveying the above subject content might include interactive video, video case studies, simulations (live and through computerized graphics), expert systems, programmed learning, instrumentation and tests, plus a variety of group processes, as well as field trips. Partnerships should be formed among government, corporations, and institutions of higher education not only to produce HRD instructional programs and materials for the deployment of spacefarers, but perhaps to plan for the establishment of space *academies and universities*, first ground-based and later in orbit. A *Global Unispace Academy* would seem desirable, possibly in conjunction with the International Space University for degree-granting purposes, and with the Challenger Center to meet children's needs for space science education and experiences. Such an international educational institution would facilitate research and education, providing for standardized training and resources for spacefaring candidates.
- **Why?** The costs, complexities, and risks presently inherent in long-duration space missions demand that, initially, access be limited to the competent, educated, and multi-skilled, as well as those who are well-balanced "copers" and "can-doers". In a space-based culture, knowledge and information ensure survival and success. As anthropologist Edward Hall reminds us: any culture is primarily a system for creating, sending, storing, and processing information with communication underlying everything, whether in words, behavior, or material things. Thus, the preparation for space culture should be comprehensive [45].

Space orientation research issues

Pre-launch training today for relatively short missions concentrates on the technical skills needed and the biomedical demands of the orbital environment, including the housekeeping procedures that facilitate life on the Shuttle or Space Station. Science writer Alcestis Oberg has described the typical preparation of an astronaut, with its emphasis on interdisciplinary familiarization in case of emergencies. Currently, classroom work focuses on basic sciences, particularly as this relates to the Shuttle system: electronics, computers, aerodynamics, orbital mechanics, spaceflight physiology, Earth observations, astronomy, planetary science and star identification, spacecraft design and operations, ground support activities, and mission control procedures. The current methodology for this includes real and computer simulations, technical work assignments, flying experience, and public relations activities. The training differs somewhat relative to role as either pilot or mission specialist, but exceptional performance and enthusiasm in applying what is learned is a factor in mission assignment. NASA's head of flight operations makes a final crew selection based on this training process, plus demonstration of competence, dedication, willingness to sacrifice and take on disagreeable tasks, and cooperation. The need for further research centers on how this orientation is suitable not just for future astronauts—but "technauts", technical specialists and contractors who will go aloft to build and maintain space infrastructures, as well as possibly settle there [46].

Because the International Space Station creates a different living situation aloft, Douglas, in the previously cited THURIS study, did ask experienced astronauts for recommendations relative to changes in the traditional NASA pre-launch training. Their pragmatic feedback included

- paramedic training, so all crewmembers could assist flight surgeons on long-duration missions;
- group dynamics and interactive training, so the crew may work together more effectively as a team;
- family training or information programs as to what to expect and do, which would be supportive and helpful while their spacefarers are away on long missions;
- operational training within a mock-up of the actual Space Station with simulations of actual control room situations;
- safety training on established procedures, dual warnings of problems, and how to cope with system failures and learn from them;
- potential problem-solving training on what to do if a certain event or problem occurs.

The group interviewed had faced the hazards of spaceflight, so they advocated cross-training, procedure conditioning, and continuing familiarization with specialized activities (e.g., EVA underwater training). We should remind ourselves that even ground training can be risky, for both astronauts and cosmonauts have died during their pre-launch preparation.

Clearwater and Harrison also identified some of the differences in training for long-duration missions, such as Mars [47]:

- length of spaceflight precludes full mission simulations, and forces partial training built around critical activities; it is more important to learn how to acquire and apply general knowledge to specific situations not covered in simulations;
- more emphasis on imparting maintenance skills, so critical on prolonged missions;
- onboard training will be required to shorten mission preparation time, reducing time between acquiring and using skills;
- expand training in interpersonal dynamics, human relations, and teamwork, so crewmembers understand each other better.

Some of the recommendations in the above studies have already been implemented, but additional research would be useful into those not yet acted on. Further investigations need to address such questions as:

(1) Is contemporary pre-departure training adequate in terms of subject matter, methodology, and final selection of such personnel?
(2) What should be dropped, what should be added, especially with reference to long-duration spaceflight?
(3) What can be learned and adapted for spacefarer orientation from the pre-departure training practices of ground-based organizations for professionals on expatriate global assignments?
(4) What innovative educational technology can be utilized in preparing people to go aloft (e.g., laptop computers, virtual reality, expert systems, animation)?

With regard to advanced HRD research with spacefarers, studies should be directed toward diversification both in the population going aloft and in organizational sponsorship. The trend would seem to be toward less theoretical and more experimental education, with greater emphasis on psychosocial learning for group living and space culture. Also the private sector sponsors may wish preparation to be different from that of the public sector in the past. Behavioral science research to date on such matters has been inadequate. For example, the Canadian Space Agency and DARA in Germany have integrated psychological components into their human factor training. But, have there also been follow-up studies on its effectiveness after their astronauts have returned from orbit? Let us next consider some other space education analogs.

Other orbital education strategies

Numerous other ground-based, space education analogs exist for developing a more broad-based program for spacefarers at all levels of education and for all ages. In the U.S.A., for example, these are wide-ranging from the NASA Astronaut Training Corps in Houston to the U.S. Space Camp in Huntsville, from the NASA National Space Grants and Fellowship Program to the Challenger Centers for Space Science Education, to consortia like the University Space Research Association and the University Corporation for Atmospheric Research.

Within worldwide universities and colleges, there are many aerospace courses and majors being offered without much central coordination or information sharing. In the United States, for instance, various university space centers and institutes offer majors in space sciences, astrophysics, and astronomy (e.g., University of California-San Diego, University of Arizona, and MIT). There are numerous schools of aerospace study (e.g., University of North Dakota). The space courses in higher education today may be the curriculum for tomorrow's space academy. For example, in the social sciences, at the University of Hawaii there were special courses offered in space anthropology; at the University of California-Davis in space psychology; at California State University-Northridge in space sociology and at Niagara University in space settlements; at the University of Houston-Clear Lake in space architecture, habitats, and food services, as well as currently an institute in space systems operations. Space commercialization and policy courses were available at Stanford University, Harvard University Business School, and George Washington University in the District of Columbia. For a time, National University San Diego, California, had a Master's degree in space commerce. Many North American universities with engineering degrees offer majors in aerospace engineering, such as MIT and San Diego State University... Canada has graduate degrees in space law at McGill University in Montreal... In Europe, the best graduate program in space sciences, as noted below, is offered at the International Space University in France. In Britain, Italy, and Germany, there are various aerospace technical institutes. Both the United Nations in Vienna (U.N. Office of Outer Space Affairs) and the International Astronomical Association in Paris sponsor an International Institute of Space Law ... In Russia, there is the Moscow Aviation Institute, a technical university; the Space Research Institute of the Russian Academy of Science, the Russian Institute of Space Device Engineering, and other such specialized institutes; and the National Youth League which is oriented to space activities. In many countries, there are professional societies which offer in-service education to members, such as the American Institute of Aeronautics and Astronautics, or the Association for the Advancement of Space Sciences and Technology in Russia. The need is for a common Internet directory of all these institutions engaged in some form of space education, especially within Australia, Canada, China, Europe, Japan, India, and the U.S.A.

The problem is that there is little coordination or exchanges of information among these resources on designing a learning program for spacefarers. There have been proposals discussed within the U.S. Congress to use the "land grant university" model for legislation, which so successfully promoted public higher education, to fund university space education on the ground and in orbit. For example, in 1987, Senator Lloyd Bentsen and Rep. Michael Andrews introduced a bill into both houses of Congress to establish and fund a system of space grant colleges and fellowships modeled on the land grant agricultural system of the past. That proposal has yet to be translated into national legislation and funded.

Relative to specialized educational institutions for the preparation of space travelers, three prototypes are emerging which are worthy of further investigation by scholars:

(1) The Japanese are developing plans for a *Moon Park* that would combine research and education on space settlements (*Space World*, June 1988, p. 22). The primary focus of the facility will be to study space crew relationships and psychological problems caused by isolating people who are dependent on closed ecological life support systems. At a site yet to be determined, the Earth-bound base would use off-the-shelf equipment and closed-loop systems in water and air reclamation. The project sponsors were the Japanese Institute of Space and Aeronautical Science, and the Obayashi Corporation.
(2) The *International Space University* would seem the ideal entity to lead the global orientation and education for space citizens of the future. Presently, ISU coordinates a multinational, multi-institutional, mobile educational program in space sciences and management at the graduate level. In different nations, it offers its instruction during the summer with a distinguished visiting faculty from many countries and institutions. It also has a permanent campus in Strasbourg, while

developing satellite affiliate branches linked together by an electronic network. Eventually, ISU expects to develop campus extensions on the high frontier (*www.isu.net*).

(3) The *California Space Institute* is a state research institute for all campuses of the University of California, headquartered in San Diego (*www.calspace.ucsd. edu*). It conducts space research for both NASA and private corporations, and sponsors seminars and symposia. It hosted a NASA summer study on "Strategic Planning for a Lunar Base", as well as a workshop on space robotics. It has yet to reach its potential by coordinating all space education throughout the state, as well as with aerospace corporations and space entrepreneurs.

There may come a time when national governments or the international community will consider funding *space academies* for the preparation of a more diverse spacefarer in everything from astrolaw to astrobusiness! In December 1985, a Japanese parliamentarian who headed up his nation's space committee, Tetsuo Kondos, was credited with circulating a proposal for the establishment of an international space center/university to be located in the Pacific area. Within the U.S.A., the late Senator Spark Matsunaga suggested the state of Hawaii for a space academy during a 1986 public forum held by the National Commission on Space [48]. Later that same year when the NRC issued its famous report on *Pioneering the High Frontier*, acknowledgement was given to the plan for a "Pacific Space Center", a potential international spaceport that would promote space-related educational and business activities for the Pacific Basin during the 21st century.

Dr. Bruce Cordell, a California space consultant and astronomer, has done extensive studies on all aspects of manned Mars missions, offering this profound observation on the role of education for such an accomplishment [49]:

> "Education and an elevated level of human consciousness are essential to human survival and prospects ... In reality, education should involve not merely the acquisition of skills, but an increase in perspective and an elevation of the human spirit. The prospect of the evolution of civilization into the cosmos offers humankind a superb opportunity to achieve continuance, expansion, prosperity and wisdom."

6.3.3 Stage Three: In-space services

In-space services are the on-site support and monitoring of people in orbit, both physical and psychological, which is necessary for survival and mission accomplishment. Here is one current example of in-space services. On Father's Day (June 19, 2004), astronaut Michael Fincke was sitting 225 miles in orbit on the International Space Station. Down on Earth, his spouse Renita had just given birth to daughter Tarali in Clear Lake, Texas. NASA arranged a telephone and video link so the new Dad-to-be became the first man in space actually to hear and see his wife give birth to their child. Michael found the experience "awesome", commenting: "My wife has already given me the Moon, and now she's given me a star!" Fincke had been launched into space the previous April 18th from the Cosmodrome in Kazakhstan for Expedition 9, along with Russia's Gennady Padalka.

To facilitate human adaptation into emerging space culture and communities, here are some of the issues to be considered:

- **Who?** Apart from space tourists, those who depart the Earth to live and work aloft deserve support from their sponsoring organizations until such time as a space settlement or colony can function somewhat independently. In addition to the families of the space voyagers, a host of professional personnel on the ground will be necessary to provide in-space services. Once at a space station or base, some specialists there should be made responsible for the management, distribution, and maintenance of such services. One proposal, for example, calls for contract corporation employees for the International Space Station to render separate services ranging from hotel accommodation and food, to heat and fuel (for both the station and spacecraft). Currently, the record for humans in space at the same time is only 13. Imagine the challenges for support services when these numbers jump to a 100, 1,000, 1,000,000! Consideration should be given to the leadership titles and roles, changing from a military model to civilian terminology.
- **What?** The support services aloft generally are likely to be of three types. The first would be *operational*, ensuring the technical or mechanical aspects of a station or base, so that it remains in orbit, or is functional and safe for human habitation... Another dimension is *human factors*, or the physical and biological support needed to maintain the quality of life for people in outer space. Obviously, this involves transport, habitat, communication, nourishment, equipment, and supplies to accomplish the mission goals. More specifically, it involves provision of "biospheres" for a total Earth-like environment, a closed ecological system for recycled air, water, food, and waste management (CELSS). Maurice Averner, who once managed this program at NASA headquarters, said the trend is toward examining *regenerable* life support for Moon and Mars' outposts, possibly a physical/chemical system based on growing plants to provide food, water, and oxygen (*Space News*, March 4–10, 1991, p. 6). NASA-Ames contracted for a *Life Science Centrifuge Facility* and *Support System* for the space station, which evaluates the effects of different levels of gravity on living organisms, including plants and small animals. Life support systems can be built not only into habitat spacesuits, but also in trailer-like vehicles that can be transported around the surface of another planet. On the Moon, for example, "life pods" that supply astronauts with emergency shelter, oxygen, water, food, first-aid supplies, and space suit repair kits could be strategically located around the lunar surface. Human engineering is improving the orbital work environment by new designs for workstations, habitability systems, human–computer interfaces, and extravehicular activities.

    The final dimension of in-space services is *psychosocial*, helping to delimit "space culture shock" on long-term flights and missions, while furthering learning in space safety, survival, and knowledge. Such activities contribute to the morale, productivity, and high performance of the spacefarers, as well as their integration into the space environment and culture. An example of one such educational/recreational service has been proposed by Nadja and Michael O'Hagan of Orincon Corp. in San Diego, California. These researchers seek to develop an interactive expert system for space personnel to assess and maintain psychological wellbeing onboard a spacecraft. The strategy would involve self-reporting interactive evaluation and supportive treatment of stress, coupled with computerized game playing based on adaptations of fuzzy set logic. Two other proposals were engendered by NETROLOGIC Inc. under Dr. James Grier Miller as principal investigator. One was a comparison of an expert system and neural network as a decision-aid for optimizing human performance in space. The objective was development of intellectual systems that would enhance the human/automation interface, group interactions and choices, while assisting other cognitive processes. The other proposal for the Department of Defense involved creation of portable infrared transponders and wall-mounted receivers to locate and monitor spacecraft personnel, permitting commanders to maintain real-time assessment of the whereabouts and status of key personnel, such as on an orbital station or base where this onboard system would enhance mission safety and efficiency. The whole field of artificial intelligence and expert systems has vast potential for space frontier applications as a means for improving human productivity aloft.

Throughout spaceflight, there have to be mechanisms and procedures in place for those on the ground and in orbit to evaluate human needs, progress, and performance. Ground training can never quite prepare spacefarers for the unknowns confronted aloft, especially when the equipment fails in some spectacular way. For example, in 1992, the rescue of Intelsat 6 required extensive improvization before Shuttle astronauts could salvage that wayward satellite. Similarly, innovation and risk-taking became necessary for repairs of the Hubble telescope.

- **How?** To illustrate the enormities of these human support challenges, consider these two reports. Since 1961 the Russians have been preparing for a $2\frac{1}{2}$ year

---

GROUND-BASED MANUFACTURING FOR EVA

To construct an orbiting station or colony requires extravehicular activity. Astronauts will have to work 300 or more kilometers above the Earth for long hours in temperatures as low as $-90°C$. For that purpose ground manufacturers must design and produce space suits that will protect humans aloft from the cold and support their other needs. Recall that during an EVA in February 1995, two astronauts from Shuttle *Discovery* realized that their hands and feet were getting too cold, so their 5-hour test outside had to be cut short. Dr. Bernard Harris, an African American physician, and Dr. Michael Foale, the first British-born spacewalker, performed well, but the insulated suit and other special clothing provided by NASA proved to be inadequate.

manned mission to Mars within decades. Their Academy of Sciences have, therefore, conducted extensive research for that purpose on long periods of confinement, and self-contained or "closed-loop" ecosystems; such findings were applied to the Mir station, enabling crewmembers to remain aloft for over a year. To place humans on Mars, the Russians estimate that a crew of three would need 31.5 tonnes of logistics, whereas a ten-person expedition would require 105 tons. Thus, to transport such weight from the Earth's gravity well into orbit, the Russians, in addition to their large Energia, are experimenting with a heavy-lift rocket (capacity 110 tons) and cryogenic rocket propellants. Support services to Mars will require a multinational joint venture, and possibly a nuclear powered spacecraft, similar to today's submarines of that type, for interplanetary travel.

In-space support services for Russian space stations have included morale boosts for psychological support, such as

(a) special radio and video programs, plus two-way television communication every Sunday between cosmonauts aloft and their families on the ground, and on special occasions (e.g., birthdays);
(b) visiting cosmonauts who stay one week, bringing mail and personal packages, as well as official supplies;
(c) partnerships with ground-based scientists through participation in their orbital experiments, especially with plants, gardening, or even animals;
(d) periodic television broadcasts to the homeland on their work and accomplishments for the Moscow evening news;
(e) opportunities for spacewalks and experiencing the majestic panorama outside, such as time for looking out windows and imagining;
(f) maintaining an "Earthlike" work schedule, including at least a day or two off weekly to rest and absorb the space experience.

To support a person in space with air, food, water, propellants, replacement parts, and other supplies, the Russians developed an unmanned cargo ship called Progress that visited their orbiting station at least every three months. They are currently repeating that same procedure with the International Space Station.

In the Bluth and Helppie NASA analysis of Soviet space station analogs [16], these sociologists examined the cosmonauts' environment, technology, organization systems, personality systems, and physical condition. Regarding support services aloft, these findings are appropriate to note here:

- Salyut/Mir stations were designed for a 7-year to 10-year lifespan with improving life support systems, windows, air/temperature/noise controls, washable leather walls, easy maintenance with replaceable parts, brighter lighting, light color orientation, computer and automated systems, adequate living quarters with private crew quarters;
- combinations of light and music to relax and stimulate, as well as to give a sense of Earth's seasons;
- disposable clothing designed for ventilation and avoiding wrinkles, as well as provisions for individual preferences to colors and pockets;

- variety in food selection, resupply, and eating arrangements, as well as use of vitamins, artificial stimulants, tranquilizers, and sleeping pills;
- structured and surprise leisure activities and exercises; a 5-day working week (typically 8 AM to 11 PM) with no weekend working or split work shifts;
- psychological support programs through a ground-based team, a monitoring system of psychological reactions and stress, and periodic visits of psychologists, sociologists, and physicians.

This report underscored broader human factors in the Russian program which contributed to higher performance on extended missions. It noted details about crew quarters, personal hygiene, living and social environment that contributed to crew morale and productivity. This remarkable research recorded matters of cosmonaut safety training, physical conditioning, sleep habits, role relationships, group management skills, and other station strategies to ensure alertness and performance. The learnings from this Russian experience are being applied to life aboard the International Space Station, and eventually to a lunar outpost. Other support service insights may be gained from the anecdote at the top of the next page.

Now, NASA is expanding its life sciences and human factor efforts on behalf of astronauts. Among its plans for offterrestrial manned missions, the Agency has been engaged in long-term research to circumvent the high transportation costs of taking food to space, a critical factor for future colonies in orbit. Under the "Controlled Ecology Life Support System" (CELSS), the Agency is experimenting with "astrofarming" so that food can be grown by spacefarers. At their Ames Research Center, biochemists are experimenting with "astrocrops" which are sealed in lucite cylinders used for growing wheat, corn, tomatoes, and soybeans on porous platforms at ten times the density of ordinary farming! The light, air, and wind are controlled artificially, while automatic spraying of roots takes place with nutrient mist; sensors monitor everything and a computer adjusts the mixes of the variables. Also planned is a Variable Gravity Research Facility of canisters tethered together and to the Space Station which will grow crops in a completely controlled orbiting garden. Other scientists at universities and private corporations have been contracted to test astrofarming on a real-life scale, ranging from fish farming, animal raising, and robotic farmers.

In-space services begin to function with lift-off, as experiences with Skylab and the Shuttle so well demonstrate [50]. On reaching a space station or base, more life support systems are utilized in unique ways. Perhaps the new arrival might be paired off with an experienced "buddy" and participate in some form of indoctrination briefings on the local situation aloft. Communication procedures are established to link the individual not just with "mission control"—but with families and friends on Earth. Provision for use of the latest communication technologies facilitate the ability of spacefarers to exchange, especially visually, with family, friends, and colleagues on the home planet! Medical and mental health services made available aloft will help those in orbit to cope with stressors related to gravity-free living and space sickness, circadian rhythms and sleep patterns, lack of privacy, and anxieties. Telemedicine will be critical to the maintenance of good health aloft ... The work

> **THE CASE OF THE PSYCHOLOGICALLY UNPREPARED ASTRONAUT ON MIR**
>
> Norm Thagard was a Marine aviator who flew 163 combat missions in Vietnam ... a graduate engineer and medical doctor ... a teacher and researcher ... a husband and father of two children. He also was a NASA astronaut who successfully completed a year of evaluation and training and became a mission specialist. But, despite four short-duration missions on the Shuttle, at 52 years of age he was not prepared for the isolation, confinement, and cultural differences of 115 days of weightlessness on the Mir station. As the first American to come aboard their orbiting home, his four cosmonaut colleagues gave him a welcoming celebration in March 1995. In fact, Elena Kondakova gave him a big hug and a kiss, before giving Norm the customary newcomer gift of bread and salt. At the start Thagard found the station nice and roomy, though he was disappointed that his experimental equipment was delayed by Russian customs; it was weeks into the mission before it finally arrived. Occasionally, he reported to ground control that he felt isolated, unable to speak enough with family and colleagues in Houston. He found the food strange and complained about not being able to speak English enough. Yet, in a conversation with the White House, he told the President, "The Russians took great care of me. We're great friends, so I think if what we did on the interpersonal level is any indication, there won't be any problems with this on the intergovernmental level as well."
>
> And, after $3\frac{1}{2}$ months aloft, the Shuttle *Atlantis* arrived to deliver a fresh Russian crew and to take Norm and two cosmonaut comrades back to Earth. On arrival home, Thagard was by far the hardiest of the three—the Russians were carried off on stretchers. Norm walked off without help, hugged his wife of 30 years and their three sons. All three spacefarers then underwent the medical tests and debriefing. Norm observed that if the mission had stretched another six months, he might not have made it ... Then to his credit, new NASA administrator Dan Goldin apologized to Norm, who had been the longest in space among his countrymen, compared with the Russian record of 438 days aloft. Dan Goldin admitted, "We put all our emphasis on the physical well-being of the astronauts and the success of the mission. We neglected the psychological well-being, and Dr. Thagard made it clear to us" (see Exhibit 53).

schedule, as well as social and recreational opportunities afforded spacefarers, is important for enhancing the quality of space life. Education and training aloft can be continued by means of team development and technology-assisted learning.

- **Why?** Until space settlements become self-sufficient well into this century, space voyagers will be dependent on the home planet. In their seminal book, *Envoys of Mankind*, Robinson and White (cited on pp. 23–24, 57) suggest that two mutually dependent types of people and societies are emerging: *earthkind* and *spacekind*. In the synergistic relations created between these two kinds, the former provides

Sec. 6.3] SPDS stages 289

support and service to those who live and work in space, so that its resources may be utilized to the benefit of all (see Appendices A and E). The situation is somewhat analogous to that of the 17th and 18th-century colonists to the New World who depended upon Old World institutions and people until they could eventually function on their own, contributing in time much more back to Europe than they initially received. So too will spacekind eventually return more than they receive from humankind on our orbiting planet.

Perhaps the scope of ground support for human spaceflight can best be appreciated in the *macromanagement* required for the manned Apollo missions: it took at one point 300,000 people in 20,000 companies to enable a lift-off of the Saturn V rocket and spacecraft (2,867 tons) for 13 missions (10 of them manned) which put 12 astronauts on the Moon! As the third element in a deployment system, in-space services requires the building of an infrastructure on Earth to support space-based activities. This necessitates more than the construction and transportation of space vehicles, as well as the providing of life support. It includes on this planet, for instance, regional bases, spaceports, space-training centers, communication and tracking systems, etc. Similarly, other infrastructures have to be financed and erected in orbit so that humans can live and work there, as well as explore and settle the Solar System. For example, Exhibit 84 illustrates one type of hardware that can provide human support both on the

Exhibit 84. Lunar/Mars Excursion Vehicle. Pat Rawlings of SAIC provides an artistic rendition of an exploration vehicle that could be first tested on the Moon, along with other relevant systems and technologies, with a view to later use on Mars. Operating on the lunar surface in this manner requires a whole new type of long-duration mission training and support system.

Moon, and on Mars. The International Academy of Astronauts proposed in 1990 that a multinational lunar base inhabited by 12 persons be established by 2010. Co-author of that proposal, H. H. Koelle of the Technical University of Berlin, said the base would cost $600 billion to build, $1.5 billion to operate each year, envisioning an expenditure of $75 billion over a 15-year period. Previously, in an *Ad Astra* feature column (January 1989, p. 48), Timothy Morgan made the case for such a base by reminding readers that the Moon's resources, just 3 days away, are essential to further human expansion. He recommended that a Lunar Settlement Simulator be constructed on Earth for applicants to spend a year in preparation before being permitted to settle on the high ground. Currently, U.S. national policy has rescheduled the permanent lunar landing and outpost to 2020, and expects costs to be much higher, making international participation critical so that investments and risks can be shared (see Chapter 10 and Appendix B).

In-space service research issues

Homo sapiens has only accumulated a little over 47 years' experience on how to sustain human life in space. The references cited in this book confirm the accomplishments and the setbacks. especially in terms of the U.S. Skylab station and the Shuttle fleet, the Russian Salyut and Soyuz/Mir missions, and ongoing habitat research at the International Space Station. But this experience and subsequent studies should provide more support and comfort for their inhabitants, as well as improve prospects for Moon and Mars dwellers. R&D must be more aggressively directed toward the infrastructure essential to support space outposts, then settlements.

In the U.S.A., the Shuttle Orbiter is the current prototype of not only future transportation—but also habitats. It is a model of "biospherics" [50]. Presently, it provides a home and work laboratory 300 kilometers aloft featuring a closed ecological system and complete life support for seven people for up to six days, or possibly more. Its food services offer a galley equipped with a microwave that can supply a diet programmed for 3,000 Calories per day to compensate for the body loss in orbit of potassium, calcium, nitrogen, and other minerals. Onboard, fitness equipment can slow the negative effects of weightlessness: loss of muscle tone, bone mass, concentration, and disposition. Apart from ground analogs and simulations, until it actually docks at a space station, its Orbiter is a real laboratory for expanded human factor and life science research related to in-space support services. Mission length there has been extended up to 16 days aloft for behavioral and biological, as well as other material science, investigations on the remaining fleet of Orbiters. Within them, there are unique research capabilities emerging that hold promise for researchers. One proven example is ESA's Spacelab, a $4\,m \times 7\,m$ cylindrical laboratory that is also fitted into the Orbiter's cargo bay (see Exhibit 85). This pressurized lab permits ESA researchers to work with a number of experiments, including using external pallets that can be exposed to the vacuum of space. The Orbiter and ISS are presently our only orbiting laboratories for studies up to six months or more. These are opportunities for joint human factor research through space agency agreements by NASA,

Sec. 6.3] SPDS stages 291

Exhibit 85. Shuttle Orbiter carrying Spacelab. Artistic rendering of a high-angle front view of the Shuttle Orbiter carrying Spacelab as its primary cargo. Flying since 1983 under management of NASA Marshall Space Flight Center in Huntsville, Alabama, the facility has hosted American, Russian, European, Saudi Arabian, and Japanese astronauts engaged in utilizing a variety of technologies. See *Space Shuttle: The First 20 Years*, Washington, DC: Smithsonian Institution/DK Publishing, 2002. Source: NASA Johnson Space Center.

CSA, RKA, ESA, and JAXA. The technologically complex Shuttle system is expected to be retired in 2010, and we await final decisions on its replacement. Hopefully, the private sector will not only utilize ISS more effectively, but provide the innovations for new space transportation systems to there and the Moon!

Private enterprise has many proposals for orbiting research facilities, such as that of *SPACEHAB, Inc.*, in conjunction with the McDonnell-Douglas Astronautics Co. and international partners in Europe and Japan. This manned experimental module fits into the Orbiter's payload area and is connected with the maindeck crew compartment. Originally conceived to carry researchers as passengers, currently it is used primarily for industrial microgravity experiments. It offers opportunities for relatively low-cost access to a manned space environment for the conduct of short-term experiments with reduced documentation and protection of proprietary rights. Its 21 lockers also provide possibilities for Space Station testbed experiments; even in life sciences and human factors, schemes exist to take advantage of the pressurized volume of space for research ... Another promising commercial prospect was described by John Cassanto of Instrumentation Technology Associates, Malvern, Pennsylvania. In an article entitled, "A University among the Stars" (*International Space Business Review*, Spring 1986), he proposed a *Material Dispersion Apparatus*. MDA was envisoned as a minilab that weighs less than 2.3 kg and occupies a volume less than $0.01 \, m^3$. Because of its small size, sample experiments can be sent aloft not only on the Shuttle and Space Station, but aboard expendable launch vehicles (ELVs). Those seeking to do basic research on materials behavior in space could use the MDA on NASA's *Hitchhiker* program. If only such devices could be adapted for life science and human factor experiments! Bigelow Aerospace is proposing inflatables that can be used as space stations, laboratories, or habitats.

Although much attention has been paid to the psycho-physiological aspects of spaceflight, much research yet remains to be undertaken relative to the psychosocial dimensions, particularly with regard to long-duration missions. On-site support will have to satisfy psychological and sociological needs contributing to effective group and community life in the orbital environment. Let us review four ground-based analogs which provide insights for further investigations related to life aloft:

(1) In examining studies of group living in exotic environments, Harrison and Connors [6] hypothesized that people have to be provided with latitude to respond to changing conditions in order to satisfy human needs; sharply defining roles can discourage functional, as well as dysfunctional, behaviors. What is the optimal balance aloft between prescribed and discretionary roles? How can effective leadership and management practice avoid role overload?

(2) Military personnel serving in NATO's northern flanks operate during the winter in near total darkness. Sources in Oslo, Norway, have reported a suicide epidemic that took four times as many young soldiers lives as during the previous ten years; some killed themselves while on duty; others on home leave did so as a result of suffering from severe depression and alcoholism. Apart from the impact of prolonged darkness and long separation from girlfriends, a fundamental factor contributing to the problem was the attitude of the defense establishment, which regarded this posting to the far north as punishment (*Insight*, November 25, 1985) ... A lunar personnel deployment system must consider the effects of two weeks of darkness after two weeks of light. What type of support services will Moon dwellers require to counteract this natural condition there? How can the latest research in human circadian rhythm and chronobiology (study of periodic changes in biological patterns) be utilized for the benefit of *Selenians*, the lunar settlers? For example, could a spacefarer's adjustment be facilitated by resetting their biological clocks during dark lunar periods through increased exposure to artificial light?

(3) Research should also be directed to sleeping problems in orbit. While one-third of terrestrial adults are estimated to have such difficulties, approximately 75% of astronauts on Shuttle missions experienced trouble with sleep. Former astronaut

Bruce McCandless II of Martin Marietta Astronautics admits that he had the problem, and attributes it to a combination of intense emotional and environmental problems, such as lack of gravity, odd sleeping times, noise, and frequent sunrises. Again, the matter is related to disruption in circadian rhythm, the body's daily cycle of sleep and wakefulness. To deal with this issue of concern, Dr. Jeffrey Davis, director of Preventative Medicine at the University of Texas, Galveston, reports that NASA is experimenting with astronaut behavior modification, such as bright light therapy, adjustment of sleeping schedules, sleeping masks, restraining devices, and opaque windows. Half the astronauts are utilizing sleep medication that works quickly with minimal side-effects to impair performance aloft. One of the newest is Ambien which is manufactured by G. D. Searle & Company of Skokie, Illinois, but may have undesirable side-effects (for information call the 24-hour hotline, 1-800-SHUTEYE).

(4) Research literature on living overseas abounds with descriptions of the phenomenon of culture shock and how to minimize it. Such transitional experiences can cause an identity crisis in an individual:

"Culture shock is a trauma experience in a new or different culture or environment because of having to learn and cope with a vast array of new cultural cues and expectations, while discovering that old ones probably do not fit or work. It is precipitated by anxiety that results from losing all our familiar signs and symbols of social intercourse which orient one to the situations of daily life, contributing to peace of mind and performance efficiency, even when not consciously aware of them" (Kalervo Oberg in Moran, R. T.; Harris, P. R.; and Moran, S. V. (eds.) *Managing Cultural Differences*, Seventh Edition. Burlington, MA: Elsevier/Butterworth-Heinemann, 2007, ch. 10).

For long-duration missions aloft, how can on-site support services counter the dysfunctional effects of *space culture shock*? The latter is psychological disorientation, exacerbated by lack of understanding and knowledge of an alien orbital environment. How can research lead to programs, provisions, and facilities that lessen personal rigidity in responding to the challenges of this new offterrestrial living situation? The answers are psychological counseling for individuals and groups, as well as possible drug therapy to reduce

- excessive fears, hypochondria, psychosomatic illness;
- severe confusion, loneliness, and depression;
- withdrawal, hostility, and paranoia;
- substance abuse or excessive interpersonal conflict.

Space living, culture, and environment are decidedly different, but people can be educated and helped to cope positively with such challenges. With their unusual orbital circumstances, cultural differences aloft can be improved when spacefarers and their sponsors create cultural synergy.

Workload and stress

Astronauts on short missions described space as a place without stress, but cosmonauts on longer flights (140 or more days aloft) showed evident signs of stress and fatigue. One hypothesis about long space voyages is that the body's immune system will respond more to psychological, rather than physical, conditions. Proper attention to habitability factors, such as esthetics, proxemics, and privacy, can contribute to lessening maladjustment, while positive programs can foster stress reduction [55]. Thoughtful scheduling can also counter excessive workload stress. Excessive workload aloft has been shown to affect cognitive performance by narrowing the attention space and interfering with concentration leading to forgetting the proper task sequence, incorrect evaluations, slowness in decisions, or failure to carry out decisions made. Related to this, further investigations are desirable in the emerging field of *cognitive space ergonomics* with reference to mental workload, human–machine interfaces, and adaptation of ground-based stress training to weightless environs. Pre-departure orientation that deals with issues of emergencies, illness, accidents, and death in orbit may facilitate stress management aloft when such events do occur.

Flight surgeon Douglas (cited on pp. 208–209, 229) confronted astronauts with the potential problem of dealing with death on the Space Station. For example, does one preserve the body, or dispose of it on-site? He also found that astronauts thought this would be less stressful if the crew received psychological support aloft, and knew that their families had access to similar services on the ground.

Dr. Peter Hancock, when at the University of Southern California's Institute of Safety Science & Systems, engaged in research for modeling and measuring the interaction between physical and psychological stress. His unobtrusive measure for evaluating physiological stress levels might be adapted for a station or base aloft. As part of the Living Systems Applications to Space Habitation project, Hancock proposed to collect and analyze data on an "envelope of life" in orbit, so as to determine a point beyond which stressors from the interacting physical, biological, and social variables may not go without jeopardizing survival.

Monitoring performance

Another area for study was identified by scientists providing input to the National Commission Space report (1986). The issue was described in *The Space Settlement Papers* as:

> "The experimental and demanding nature of human settlements in space will require constant, non-intrusive monitoring of experience, as well as analysis of the social forces in an isolated environment that is without precedent. Progress will depend heavily on learning from this process of feedback."

The Living Systems proposal of Drs. James Grier Miller and Philip R. Harris described in Chapter 3 envisioned a computer/sensor technology for such non-intrusive monitoring of inhabitants of a space station or base. Besides the improved management and safety which could result from such procedures, a continuing learning cycle could be established for the benefit of future spacefarers. Such experiments and data collection is possible now aboard the International Space Station and could be extended eventually to a lunar base!

Just as NASA today monitors its astronauts aloft as to vital functions and physical wellbeing, someday organizational sponsors may wish to have means for evaluating the wellness and acculturation of their space travelers. In addition to the above-mentioned methods for physical evaluation, psychological monitoring might also be accomplished aloft through computerized need adjustment surveys, individual and group counseling for the development of case studies, as well as forms of performance data analysis and camera/video observations. The purpose of any monitoring and data collection with spacefarers should not be to invade their privacy—but to ensure their wellbeing—while improving the performance and productivity of those who follow them in orbit [52].

Play and performance in space

Long-duration space living must have provisions for relaxation and recreation among spacefarers. Interviews with astronauts by Douglas (1984) revealed the issue of "humanness": that is, what a crew might sacrifice during a 10-day mission, but unwilling to renounce on a 90-day stay at a space station! These space veterans warned of overscheduling and the need for compensations for demands made on crewmembers. For the sake of mental health, they recommended opportunities for private time to contemplate the grandeur of the space experience, to socialize and foster interpersonal relations [53].

It is worth repeating here that when the Skylab IV crew worked an average of 16 hours a day with no day off, a sense of malaise set in. The crew found that they performed best when they worked and dined together, even going to bed at the same time. Both the American and the Russian experience aloft confirms these favorite pastimes: looking out the window, gardening, visitations from Earth and mail deliveries, packages, even grab bags. Among those who kept diaries or logs of their space experience, Jeffrey Hoffman [53] detailed what a play day on the Shuttle *Discovery* was like: running on the treadmill, zero-g games, night observation out flight deck windows, last-meal-in-orbit feasting, listening to music with headphones while floating free, and sleep trances. With new video/computer games, it should be possible for participants aloft to play not only with one another—but possibly with partners on the home planet!

In their seminal *Living Aloft* study, Connors, Harrison, and Akins (see Chapter 5) did examine leisure time activities aloft and discovered that they ranged from passive (media watching, reading, and meditation) to active (group singing and dancing, playing musical instruments, games and contests, and letter writing). On Soyuz 9, for instance, the cosmonauts watched concerts on video, listened to news and music piped up from their homeland, held two-way communication with family and friends in Russia, as well as with scientists and celebrities. One Indian cosmonaut practiced yoga.

The narrow human factors research to date has generally ignored the play phenomenon aloft, as well as interpersonal issues relating to sex, drugs, alcohol, and robots. We have yet to focus investigations on advanced communication technologies in orbit which could facilitate education and recreation, such as better use of multimedia and expert systems. With more permanent habitats aloft, planners will have to expand the range of in-space studies and services implied in settlements. The focus should be on development of positive, "space wellness" strategies and programs.

Human biological and psychosocial evolution will be accelerated in the centuries ahead by extraterrestrial living! Survival adaptations in orbit may eventually lead to species mutations. No wonder that the head of Soviet biomedicine, Dr. Oleg Gazenko, believed that the limitation to our living in space is psychological—not physiological; the Russian space program has always given psychological support high priority. Woman cosmonaut Svetlana Savitskaya goes further, suggesting that psychologists be included on long-duration missions. Certainly, future colonies may find that caring nurses and psychiatric social workers could prove very useful. In engaging in space human factor or life science studies, researchers would be well advised to consult NASA's *Man–Systems Integration Standards*, which contains much useful information, such as on human performance, orbital environment and safety, health management, workstations, and extravehicular activities [54].

To close the space deployment loop, we now consider issues related to the return of the spacefarer to this planet, especially after long duration aloft. As in all people migrations, eventually some colonists may prefer to remain offworld for the remainder of their lifetimes, unwilling or unable to come home again to Earth! Space tourists, possibly, may have less re-entry problems.

6.3.4 Stage Four: Re-entry policies and programs

These are required to deal with spacefarers' relocation back to Earth, including re-acculturation problems [19]. Living in a low or gravity-free environment, especially for lengthy stays, obviously affects body, mind, and spirit. Information should be systematically gathered and analyzed from space "expatriates" to improve future recruitment, selection, and training of spacefarers as is currently done with foreign deployment of workers on this planet. Re-entry preparation, counseling, and guidance, as well as debriefings, complete the space personnel deployment system. In this way, "re-entry" shock or "the overview effect" can be delimited to manageable proportions.

- **Who?** Just as with pre-departure orientation, the space voyager and his or her family should be involved, whether the latter accompanies that individual into space or remains back on Earth. The extraterrestrial experience, especially one that may involve many months or years aloft, impacts family and close friends, affecting both personal and organizational relations. In this 21st century when space settlers volunteer permanently for the new orbiting colonies, re-entry procedures will be necessary for their possible visits back to the home planet. Perhaps astronauts and cosmonauts who have been in space are the best persons to act as consultants in this and other aspects of the deployment system. Who better understands the "culture shock, overview affect and re-entry shock"?

- **How?** Policies need to be established on pragmatic matters as to
    (a) the length of space assignments and sabbaticals from such outposts, such as at the Space Station or Lunar Base;
    (b) when, where, and how reconditioning is to take place;
    (c) what the length and process of debriefing should be before a person leaves to begin or is re-assigned to orbit;
    (d) what type of data-gathering and follow-up procedures are to be undertaken on return to Earth, and how long after the last orbital experience should this be done;
    (e) what rights, benefits, and assignments the spacefarers may expect on return to the sponsoring organization, especially in matters of compensation, job assignments, and support services;
    (f) what kind of assistance may be expected if the spacefarer comes back with a disability, be it physical or psychological?

Before any rotation back to this planet from long-duration missions, the re-entry program should likely be inaugurated while still in orbit; extensive conditioning will be required to ready people physically for the gravity environment back on Earth.

- **What?** The re-entry process must deal with physical, psychological, social, and organizational aspects of return to gravity living. Research in this area will have to deal with programs that recondition the body, foster re- acculturation to home planet conditions and societies, as well as guidance for further career development or retirement. These programs will differ for those who were in orbit for a limited duration, in contrast to those who are away for years.
- **Why?** Spaceflight for most people will involve a *transitional experience*: from a new perspective of seeing our planet as an interconnected ecosystem to ground-based views that are quite myopic and provincial. On return, some may complain of severe adjustment problems or "re-entry shock", In his book, *The Overview Effect*, Frank White reported on this homegoing phenomenon among astronauts brought on by the fiery re-entry, adulation by well wishers, sense of gratitude for safe return, and even feeling let down. Constructs and perceptions may be so altered as to require psychological counseling so as to lead a normal Earthlife again. Some space expatriates bring back a global consciousness that inspires them to a leadership role; others get so depressed on the ground that may verge toward a nervous breakdown. For example, among ex-Apollo astronauts there was a high incidence of divorce from spouses, and life-changing trauma that led to substance abuse and even "born-again" religious experiences. Perhaps the case below best underscores the problems that may result.

The whole space experience, even for those who never go aloft but learn of it through media (be it photographs, paintings, television, or films), changes our imagery and imagination, our field of perception of life in space. For millions worldwide, the live television broadcasts from the Moon in 1969 had an impact comparable with viewing the 1995 film, *Apollo 13*, which was but a re-enactment. Think what the actual experience meant for the 12 men who actually landed on the lunar surface, the first among billions of their species!

Space re-entry research issues

Ideally, under the leadership of global space agencies, a data or knowledge bank should be developed with a pattern recognition and retrieval system relative to the broad human role, performance, and problems in space habitation. In addition to recording findings of terrestrial analogs as described previously, this information system would be a comprehensive storehouse on the ongoing human experience aloft [55]. Perhaps the Living Systems template of the late Dr. James Grier Miller described in Chapter 3 would serve as a framework for this cross-disciplinary analysis. Although NASA has been forthright about medical and biological insights gained from previous spaceflights, the agency has been circumspect on studying or releasing information on the psychosocial experience of its personnel in space. In the past, NASA has limited access to astronauts by social science researchers, even by its own psychiatrists and psychologists. According to psychologist Dr. Albert A. Harrison, the agency has failed to capitalize on the data it has collected which could improve spaceflight and living for others to follow. For example, there are indications that transcripts of crew communications going back as far as the Mercury flights have not been analyzed from a behavioral science context, so as to obtain clues to improve future missions and avoid tragedies.

There does not appear to have been any global, systematic data-gathering surveys of past and present astronauts/cosmonauts who have actually flown in space. This expert group could provide valuable input to better human performance in space if behavioral scientists were given the opportunity to construct an appropriate questionnaire, and analyze the results. For Neil Armstrong, Buzz Aldrin, and the other ten who visited the Moon, their lives and viewpoints were altered forever. A limited number of astronauts and cosmonauts have reported after their return (sometimes in books, poetry, and paintings) about their offworld lives. They experienced everything from vertigo, claustrophobia, and depression, to a sense of renewal, euphoria, and ecstasy. Veterans of numerous spaceflights seem to learn from prior missions and get better at their jobs aloft. Documenting the space explorer and settler experience on the high frontier should be a scholarly priority. Further consideration should be given on utilizing astronauts and cosmonauts who have returned from outer space. This valuable resource has been used in space agency administration and mission control, but they would be helpful both to train spacefarers and then assist them when they are aloft. For example, on September 1, 2006, veteran astronauts Ken Bowersox and Kent Rominger announced their retirement from NASA. Both had served as managers in the Agency's Flight Crews Operations Directorate. Both had experiences as commanders on Shuttle flights to ISS, including the longest and second longest Shuttle missions. Think of the contribution this pair could make to some institute or educational program that is sponsoring private enterprise in space!

Early in the space program, the Space Science Board (1966) became concerned about long-duration mission effects, decrying NASA's neglect of psychological and behavioral factors in such spaceflights. In an interview on his Skylab and Spacelab experiences, former astronaut Owen Garriott observed that it was human presence which ensured the success of those missions, and that there would be no functional limits to high performance on longer flights, if radiation levels are contained and a daily exercise routine of one hour is maintained. Then he significantly added, "Neuromuscular weakness is not evident in space; it is only noticed after return to Earth" [55].

Bluth and Helppie in recording Soviet space station experience, include this pithy quotation from O. G. Gazenko: "All people who have been in orbit experience certain difficulties, often significant ones after returning to Earth" [16]. These researchers described what cosmonauts noticed about themselves after landing: weakness, fatigue, feelings of increased body and object weight, gravitation shift in vector and vertigo, skin paleness, face puffiness, limited locomotor function, orthostatic stability decrease, perspiration increase, faintness, and other bodily difficulties. Consider this diary account of cosmonauts [57].

One Russian, Valentin Lebedev, wrote in his diary of his 1982 Salyut 7 space station experience that the reality is a story of loneliness, of fear, and homesickness, but the sheer routine of operating their Salyut 7 (meaning "salute") diverted these feelings somewhat. But he and his crew member, Berezovoi, were getting so fatigued, they required 12 hours of sleep each night. When they expressed a strong wish to be with their families by the New Year, the doctors and flight director approved of an early return ahead of schedule. After 211 days aloft, they came back to their families by New Year's Eve ... Yet, they could be satisfied with their high performance when in orbit: they produced 20,000 photographic plates, helped to locate oil and gas fields, guided the route for a transcontinental pipeline and railway, and saved a billion roubles a year by giving their countrymen advance warnings of adverse weather.

This is the type of anecdotal data that researchers need to be collecting worldwide on the spacefarer experience, both in orbit and on return! Since the Russians parachute their capsule on re-entry to land—rather than water—their cosmonauts have had many adventures: from crash landings in remote areas to encounters with blizzards and wolves. Again we draw from Brian Harvey's first edition of that wonderful volume now extensively revised and updated as *The New Russian Space Programme* [56]:

- When cosmonauts Vladimir Kovalyonok and Aleksandr Ivanchenkov plunged through the atmosphere and floated down to Earth after almost 140 days aloft aboard Salyut 6 in 1978, the former climbed out unaided, picked up a handful of soil and promptly fell back into the doctor's arms. Ivanchenkov said they just loved the intoxicating fresh air of mother Earth and the breeze of wind in their faces! Within two days they were up and walking to the canteen, trying to make up for weight loss: Kovalyonok had lost 1 kg, while Ivanchenkov lost 1.8 kg.

The same source reported that the former Soviets often sent their cosmonauts to a health resort in the Caucasus to recover and readapt after prolonged spaceflight. No small wonder when we consider how their endurance was tested as is evident in this 1987 report of a return from the Mir/Kvant complex:

- Yuri Romanenko achieved a space endurance record spending 326 days aboard Mir, completing 5,149 orbits, and covering 2,164 million kilometers. After the search team found him and his companions, Aleksandr Alexandrov and Anatoli Levchenko, in their snow-covered capsule amidst a ferocious blizzard, they were whisked away by helicopter to their Baikonour base. Once there, Levchenko's work was far from over; he was put behind the controls of a waiting TU-134 airliner and required to fly it to Moscow and back, so as to simulate the ability of Shuttle pilots to fly airplanes in level flight after a period of weightlessness!

Aboard Mir, the Russians discovered that cosmonauts start getting on one another's nerves after 8 or 9 months aloft, but recover dramatically within two weeks when told they are coming home. The anecdotal accounts given here illustrate the importance of such records to researchers (particularly if they are on videotape), since they could be utilized in the orientation of future space voyagers.

Research on what has been called "the overview effect" needs to be expanded. What are its pros and cons. For example, spaceflight also brings humans a sense of grandness and oneness with the universe.

Since the longer the duration of the space experience, the more likely the spacefarer will be subject to "re-entry-shock", it is thus necessary also to conduct studies on the physical and mental debilitation of returns which require rehabilitation through therapy or other treatment. As masses begin to move beyond terrestrial boundaries, there are questions to be answered that will further human absorption back into the gravity environment, such as

(1) How much has the space experience raised consciousness, altered perception, and actualized potential?
(2) Are top performers in space equally successful on the ground after their return?
(3) What can be learned from spacefarers who have been high performers aloft or well adjusted since their re-entry?
(4) What is the role in the readjustment process of specialized services ranging from social work to occupational placement?
(5) Does post-flight psychological counseling facilitate re-integration into family, society, and organizations?

If we do not get answers to such questions, how will we ever successfully achieve missions of several years and back to Mars and its moons? Exhibit 86 illustrates the challenges ahead for humans on the Red Planet.

Exhibit 86. Mars mission challenges. A painting by artist Robert McCall of a Mars Lander arriving and departing from a Mars Spaceport later in the 21st century. Source: National Commission on Space, *Pioneering the Space Frontier*. New York: Bantam Books, 1986, p. 69.

Long-term re-entry surveys and support

After a spacefarer has returned from orbit for a year or more, there should be additional studies of their adjustment to the home planet. For example, an annual survey of their acculturation needs would seem to be in order. When these data are analyzed and summarized, they should be utilized in two ways: (a) to provide continuing physical or psychosocial support to those space traveler respondents, and (b) to improve re-entry programs for future voyagers coming home from orbit. Perhaps the case below will highlight some of the problems and needs faced after a transitional experience in space, a turning point in most lives. The author is personally aware of these re-adjust "challenges" after he returned from a year abroad as a Fulbright professor in India [57]. The Russians have a name *asthenization* to indicate space mental health problems for persons who have flown in orbit for six weeks or more. The symptoms are fatigue, irritability, lack of appetite, sleep problems, and emotional instability.

Part of the re-entry struggle can be traced to the astronaut and cosmonaut cultures with their emphasis on achievement. As science writer Seth Borenstein put it (see [59]):

"From the dawn of the space program, American astronauts have been treated like stars, saluted as red-white-and-blue heroes, and indoctrinated into NASA can-do, failure-is-not-an- option ethos. Could that explain the downfall of Lisa Nowak, the astronaut accused of attempted murder? Were the expectations too high? Were the pressures too great?"

In the history of the Astronaut Corps, there never had been a major scandal about one of its members, so the public and the Agency were very shocked when the media reported the incidents in this next case. But, first, a few expert opinions obtained by reporter Borenstein about this "affair-of-the-heart gone wrong":

- Dr. **Patricia Santy**, former NASA psychiatrist and author of the book, *Choosing the Right Stuff*: "I really think NASA goes overboard in promoting how heroic and super all these people are. They themselves have forgotten these are ordinary people and in that celebrity culture, there is a sense of entitlement. The Astronaut Corps is like family, but it's almost like a dysfunctional family when it comes to understanding that interpersonal issues have profound impact."
- **Jerry Linenger**, former astronaut featured in the popular book, *Dragonfly*: "Astronauts take pride in their self-discipline, and set a goal and it's then just going, going, going, and you let nothing get in the way." Linenger speculated whether this singlemindedness would somewhat explain Nowak's waywardness after her spaceflight experience. He thought her actions on this occasion were contrary to her formation: "To not make a midcourse correction is scary. It's just off her training and everything else."
- **Jay Barbree**, NBC correspondent who has covered many space launches and author of the book, *Live from Cape Canaveral*: "The biggest problem astronauts have had after they obtained their goals is, what do I do next? It was more commonplace than people think because overachievers had no more major goals in front of them. Among earlier astronauts, most every one of them came back with adjustment issues."
- Like many of her colleagues, **Lisa Nowak** pursued a career in spaceflight since childhood. After her last Shuttle trip with her goal achieved, her prospects for another mission were dim. The astronauts had been told that further space assignment would be hard to come by with the planned retirement of the Shuttle in 2010, and a replacement vehicle several years away.
- **Buzz Aldrin**, second man on the Moon who had his own depression issues on return, commented that the space agency "can deal with physics and engineering, and the science of things. They behave according to the science of predictability. But you can't predict human reactions and responses and how complex it is. It is not easy for NASA."

Ah, yes, it is hard for an engineering culture to deal with human needs and failures. Robots are much more predictable, so they have a great future in space developments. Although studies have confirmed a low rate of mental illness among astronauts, Howard McCurdy, an American University professor who has written about space policy and stress, maintains that the program puts pressure on personal relationships, resulting in a high divorce rate. "The astronaut culture is a carry over from *The Right Stuff* days. It is very high intensive and competitive" [58]. The next case underscores this cultural reality.

> **THE CASE OF THE EMOTIONALLY UNSTABLE ASTRONAUT** [59]
>
> Capt. Lisa Nowak, USN, had successfully served as a NASA astronaut. Having performed well on an STS-121 Shuttle round-trip flight to and from the International Space Station in July 2006, she had been working since at the Johnson Space Center near Houston. In February 2007 at 43, she was preparing for a new assignment as lead Mission Control specialist; she would have been in charge of communications between ground control and the crew of the next Shuttle flight. Lisa was also "apparently" happily married to Richard, a NASA contractor, and the mother of three children. Although she had given birth to twin daughters in 2002, she separated five years later from her husband. At the time of this incident, the Center director, Col. Robert Cabana, reported that Nowak had been a "vibrant, hardworking, and energetic person who did her job well." Apparently, none of her co-workers had noted any emotional or personality problems. Then her world, marriage, and career blew apart because of an "affair of the heart".
>
> Capt. Nowak fancied herself in love with another divorced astronaut, Commander William Ofelein, USAF, who piloted the Shuttle *Discovery* on mission STS-166 to ISS in December 2006. And she perceived as her rival for his affections, Capt. Colleen Shipman, USAF. In a moment of extreme agitation and jealously, she drove 900 miles non-stop from Texas to Florida to confront Shipman after learning of the latter's flight plans on Ofelein's computer. Police said that Lisa wearing a wig and frustrated, pepper-sprayed Colleen in the Orlando Airport parking lot on the morning of February 7, 2007. Driving away Shipman had called police who found (in a nearby trash can) Lisa's bag, disguise, compressed air pistol, steel mallet, knife, rubber tubing latex gloves, and garbage bags. In a rage about "her boyfriend", she had worn diapers on the long drive so she would not have to stop and relieve herself. Police arrested Lisa charging her with attempted kidnapping and murder. In a request for protective custody, Shipman reported that Nowak had been stalking her for two months though she had never met Lisa before the violent encounter. After a court appearance, the accused astronaut was returned to Clear Lake, Texas, in shackles, while her lawyer argued for a reduction in charges and bail, pleading "not guilty" to the charges leveled. Her trial is still pending.
>
> Meanwhile, until these criminal charges are ultimately resolved, NASA, the space community, and the public are trying to understand Nowak's bizarre behavior. A *New York Times* editorial (February 8, 2007) observed that "astronauts are human, subject to the same material strains and mental aberrations that afflict ordinary mortals. The space agency may need to revise its procedures for psychological health of astronauts—particularly those who might embark on long space travels ... Currently, astronauts undergo a rigorous psychological and psychiatric screening to gain admission to the corps. But there is no

formal evaluation during their service thereafter, unless a medical doctor senses the need for one, or an astronaut is scheduled for an extended flight on the space station."

While there is no evidence of any psychological problems in Capt. Nowak's well-regarded performance during the seven months since her return from outer space, many questions are being raised. Was her breakdown due to re-entry shock, or simply a female, mid-life crisis? Was she unable to cope with her "transitional experience" after being in orbit? Or was it simply a matter of hormone imbalance or an emotional love affair? Did she only wish to frighten a perceived rival, or was she really intent on harming her? Whatever the outcome, the space agency has been shaken by the event, and is reviewing its personnel policies.

Capt. Nowak was attached to two government entities, and it is interesting to note how their organizational cultures reacted to this problem. Within a month, NASA had fired the offending officer from the Astronaut Corps. But, when she returned to duty in the U.S. Navy, pending the court judgement, the Captain was assigned to curriculum development at the Naval Training Command in Corpus Christi, Texas.

Meanwhile, NASA is changing its policies and procedures relative to astronaut mental health, before, during, and after spaceflight. These have been clear for instability demonstrated in space: a detailed, written process exists for dealing with offworld psychotic or suicidal behavior. It includes restraints for disturbed spacefarers, tranquilizers, and consultant flight surgeons. In fact, the flight commander has a 1,051-page manual for dealing with all types of health emergencies in orbit. But, there are no instructions for dealing with unhinged astronauts when back on the ground. The psychiatric studies of stress focus on astronauts aloft—not after returning back home. Agency officials insisted that if an astronaut asked for help with an emotional problem, it would not affect his or her flight status. But, in the self-reliant astronaut culture they are not likely to seek personal assistance.

Furthermore, the organization has no rules governing astronaut fraternization and infidelity that correspond to those in the military. Now, NASA administrator Michael Griffin has ordered a review of psychological testing, oversight, and counseling of astronauts at the Johnson Space Center. Among those who empathize with the plight of Lisa were her parents and estranged husband, who offered their support. The legal case goes on. In court, Capt. Nowak apologized to Capt. Shipman for her behavior on the day in question. Stay tuned to the media for the outcome of this unfortunate case or call 1-703/614-2000, Naval Personnel Public Affairs.

6.4 TRANSFORMING SPACE DEVELOPMENT THROUGH TOURISM

For more than a half a century, members of the global space community have gathered regularly for professional development and sharing, as illustrated in Exhibit 87. Their meetings and conferences around the world have often been sponsored

Sec. 6.4] **Transforming space development through tourism** 305

Exhibit 87. Innovations in space development. Worldwide conferences and meetings like the one pictured above bring experts and advocates together to promote technical space progress, innovation, and entrepreneurism, as well as valuable networking. Thus, on May 4, 2006, the Space Tourism Society celebrated its tenth anniversary by co-sponsoring a similar gathering in Los Angeles called the *Orbit Award Program*. Source: Derek Webber, director of Spaceport Associates and Transformational Space Corporation (*www.spaceportassociates.org* or email *Derek@TransformSpace*). Derek is pictured second from the left seated at the dais in the above photograph and is the contributor of Case Study 7.2.1.

by space societies of both experts and activists, as well as by space agencies. For example, the National Space Society conducts an annual International Space Development Conference at various cities within the U.S.A. (*www.nss.org*). More often than not the varied continuing education sessions are focused on technical issues of spaceflight and settlement, strategic space planning, or human factors in orbit. Then in 1996, when tourism and eco-tourism were big business, John Spenser, a space architect, founded with other colleagues, the Space Tourism Society (*www.spacetourismsociety.org*). Many laughed at the very idea of privately sponsored tourism in space, but now it may become, after satellites, the second major space industry [60]. STS's goals are

> "to conduct research, build public desire, and acquire the financial and political power to make space tourism experiences available to as many people as possible and as soon as possible."

The growing Society defines space tourism as experiences in Earth-based simulations; cyber/web world space tourism; tours and entertainment; in-Earth orbit; and beyond Earth orbit. Such orbital travels may be in LEO and GEO, or eventually to the

Moon, Mars, and beyond. With distinguished boards of governors and directors, the organization sponsors an annual conference, exhibits, and publications, as well as promoting the Orbit awards with the National Space Society for leaders and pioneers in this emerging industry. For example, the Space Tourism Society promoted a *Space Tourism Expo 2001* in Pasadena, California, which announced the founding of a Space Tourism Society and the Space Pioneers Awards Show (*www.space-tourism-society.org*).

Aerospace engineer Michael van Pelt wrote the second book ever published on space tourism [61]. This author believes that leaving Earth's atmosphere with human passengers who pay for their own travel will soon become as routine and safe as flying airlines today. Looking ahead with clarity, he has created in this work likely scenarios that cover the selection and preparation of ordinary space tourists; launching into orbit to space stations and orbital hotels; the realities and possibilities in such microgravity experiences; and the return and future prospects for this private enterprise. Van Pelt reminds us that

> "we are all actually space travelers, dashing at 30 kilometers (18 miles) per second around the Sun onboard a big blue spacecraft, while the whole solar system circles around the center of the Milky Way at an incredible 217 kilometers (135 miles) per second. Our ship carries everything we need to survive: water, a breathable atmosphere, climate control, radiation protection, living space, and a huge crew. Spaceship Earth is an amazing place."

He urges us to explore it and beyond!

Lest one think this is all "pie in the sky", people are already paying millions to go into space. And others are paying large deposits to Virgin Galactic owned by Sir Richard Branson. He has ordered a fleet of suborbital spaceships and is taking reservations from 100 "founder passengers" who will pay $200,000 per ticket; 60,000 or more people have already registered for their future flights and are known as "voyagers". Founder candidates are screened carefully for geographical and occupational diversity. In 2006, Namira Salim, 35, an artist who is a VG founder and would-be passenger, hosted an exhibition on space tourism in her native Pakistan; she even had handcrafted replicas of SpaceShipOne. After test flights of his newly manufactured SpaceShipTwo, the innovative design engineer Burt Rutan hopes to deliver the first one for paying customers possibly by 2008. Virgin Galactic's goal is to send 500+ persons into space during their first year of regular operations, roughly the same number of orbital travelers for the past 45 years [62]!

The history of human spaceflight for the past 50 years has been dominated by government-funded programs which have only allowed fewer than 500 people access to orbit. That number could double and triple by 2015 because private companies are competing feverishly to provide tickets to space at a six-figure price. Literally, the pioneering tourists opening the space frontier for the masses to follow are paying a higher price: at the time of writing as much as $25 million to be trained in Star City, then fly to and from the International Space Station on a Russian rocket! Since the death of a teacher (as a crewmember on the ill-fated Shuttle *Challenger*), NASA has been reluctant to fly "civilians" on an American space vehicle, though a few politicians, members of the U.S. Congress, actually made it into orbit. Until now the closest the average person could get to a space experience was through movies, television, video games, and simulations.[4]

Simulated space travel on the ground is also available. Incredible Adventures of Sarasota, Florida, utilizes the NASTAR Center in Southampton, Pennsyvania, to present the thrill of spaceflight experiences while staying on our home planet (*www.nastarcenter.com*) ... Other entrepreneurs plan short orbital experiences for the public, some of which are now being implemented. For example, Pat Kelley of Vela Technologies took reservations for a vacation package costing $98,000 which includes six days of training, two and a half minutes of weightlessness, two flight suits, and an otherworldly view of Earth. To do this, his Space Cruiser rocket plane would fly from a conventional runway on a two-stage trip: first, on an adapted aircraft to 50,000 feet high; then, blasts off under rocket power cruising at 2,300 miles per hour; and, finally, retrorockets slow the cruiser for re-entry into the atmosphere ... For $3,750, Zero Gravity Corporation of Fort Lauderdale, Florida, is another space tourism and entertainment firm actually flying passengers 32,000 feet then plunging downward 8,000 feet for 25-second snippets of zero gravity during descent. In 2007, its CEO Peter Diamandis, M.D., waved that fee so that one of the greatest physicists of all time, the paralyzed Stephen Hawking, could experience the advantages of low gravity ... In addition, the International Symposium for Personal Spaceflight brings together veteran astronauts from the Association of Space Explorers with future space tourists to discuss such out-of-this-world experience. One Shuttle flyer, Tom Jones, described space travel as a "life changing experience—a strange but wonderful occurrence in microgravity, so make the most of the opportunity."

The first private explorers were a group of venturesome, very wealthy space enthusiasts who "boldly went where no paying customers had gone before." Their travel arrangements were made by Space Adventures, a firm in Arlington, Virginia, who has an agreement with the Russian Federal Space Agency for training and transport of their clients at an agreed price which offsets the cost of that flight. The company has over 200 reservations (including pop star Lance Bass, 28) who have paid deposits to go aloft. *Rocketplane Kistler* also have undisclosed plans for space tourists (see Exhibit 151). Here are the initial group of *pioneer space tourists*:

- **Dennis Tito**, a 61-year-old investment management executive and former JPL aerospace engineer, took the original six-day jaunt on the Russian Soyuz rocket from Kazakhstan with two cosmonauts. Despite NASA objections to his presence and barring him from the U.S. station section, he received a hearty reception and hug on arrival at ISS from Cosmonaut Kazarbayev. The latter remarked: "I offer an especially warm welcome to Tito. Until recently you would only read in science fiction that an ordinary man could go into space. You have paved the way for space tourism!"

---

[4] Public Broadcasting Services offers a CD-ROM recording these programs via satellites: *The Space Race, Apollo 13, Astronauts*, and *The New Cosmos: Eyes of the Universe, Mysteries of Deep Space*. Inquire at *www.pbs.org/als* or telephone 1-800/257-2578.

After returning from his 8-day adventure, Dennis, smiling but wobbly, said: "My personal experience was well beyond my dreams... up there I felt the best in my entire life." Later on a helicopter ride back to Moscow, he added: "It's an experience that cannot be described. It is being in a different world, being in a different life. I loved living outside gravity. I slept better than I did in years!" (*San Diego Union-Tribune*, May 7, 2001, pp. A1 and A3).

- **Mark Shuttleworth**, a 28-year-old South African Internet millionaire, was prepared at the Gagarin Cosmonaut Training Center, flown on the Russian Soyuz, and enjoyed a 10-day space voyage to and from the space station in 2002. This second space tourist returned on a 3.3-ton spacecraft that became a fireball at temperatures of 18,000 degrees as it plunged at 10 times the speed of sound back into the Earth's atmosphere. While the vehicle landed by parachute on the Kazakh steppes, his parents and brother watched the touchdown at Moscow's mission control. Adjusting to Earth's gravity pull after his soft landing, Mark bubbled over, saying: "It was the most wonderful experience ever. It was fantastic." Next Shuttleworth plans to tour his country's schools promoting space studies (Mara Bellaby, "Space Tourist Returns to Earth," Associated Press via *America On-Line*, May 17, 2002. For more infomation, call 1-888/85-SPACE).

- **Gregory Olsen**, 60, another U.S. citizen, got a week-long stay on the International Space Station in 2005. With NASA's relaxed stance on space travel by "amateurs", he received basic training on station procedures at the Johndon Space Center. The flight director there commented: "He's, as you can imagine, a very smart person and in a phrase 'he gets it'." With Russian assistance, Olsen arrived safely at ISS, roamed the entire station, conducted experiments, and emailed or telephoned home from orbit. At a news conference from the station, Gregory commented, "I feel welcome." Cosmonaut Tokarev, a crewmate, observed: "He is a real space researcher. We can no longer simply call him a tourist" (Tad Watson, "Space Growing More Tourist Friendly," *USA Today*, October 2005).

- **Anousheh Ansari**, 40-year-old chairman and founder of Prodea Systems, a digital home technology company, was the first female space tourist. The Iranian-born, U.S. citizen trained for six months at the cosmonaut center near Moscow. Officially she was a crewmember on Soyuz TMA-9 which flew to the ISS on September 18, 2006. The lady entrepreneur was accompanied by NASA Expedition 14 crew commander Michael Lopez-Alegria and Russian flight engineer Mikhail Tyurin who docked the craft smoothly with the station's aft port Zvezda module. Together they joined astronauts and cosmonauts for a total of 12 aboard ISS. During her ten-day adventure, she called herself "the space cadet" on an Internet blog informing many friends on Earth of her challenges with motion sickness and weightlessness, as well as the hazards of personal hygiene and the value of Velcro. Having earned a Master's degree in engineering and a fortune in the telecommunications industry, Ansari was both optimistic and enthusiastic aloft, while her spouse followed her antics on video monitors in Korolev. In orbit Anousheh did a few experiments for the European Space Agency, studied microbial lifeforms onboard and how they spread, and videotaped educational programs on different laws of physics. Like her counterparts above, this fourth space tourist has been fascinated by outer space since childhood. Now Anousheh hopes her life and career will be a role model for other young women, so they too will have an out-of-this-world experience. After her safe return to the home planet, she set to work launching a new company with the help of her spouse. Through the X-Prize Foundation, she promotes educational activities about the infinite resources of space, and the need for new technologies to utilize them for humanity's benefit (Reuters, "U.S. Woman to Become First Female Space Tourist," September 1–7, 2006; Sara Gooudarzi, "Interview with the First Female Space Tourist," *www.space.com/missionlaunches*, September 17, 2006; Ker Than, "Fourth Space Tourist, Expedition 14 Crew Docks at ISS," *www.space.com/mission launches*, September 24, 2006; Associated Press, "Space Tourist Blog Was out of this World," *San Diego Union-Tribune*, September 29, 2006; see website *www.anoushehansari.com*).

- **Charles Simonyi**, at 58 this Hungarian-born billionaire and American entrepreneur was the fifth tourist to be launched from the Baikonour Cosmodrome. With two Russian cosmonauts aboard TMA-10, the spacefarers lifted off from Central Asia on April 7, 2007. In his mission blog, Simonyi wrote: "The sight of the booster is staggering. I had no basis for what to expect and went into this project with a lot of hope." The former Microsoft software developer brought the Station crew a six-course meal selected by his close friend, Martha Stewart. During his 13 days in orbit, he conducted experiments for ESA, JAXA, and the Hungarian Space Office; for the latter, he mapped the radiation environment within the ISS. Charles hoped that his research would contribute to future human permanence in space. He also had HAM radio sessions with students while onboard. After re-entry, Simonyi engaged via website, videos, and lectures with children around the world who seek to learn more about space and space travel (*www.charlesinspace.com*). An accomplished pilot, he was so motivated by his early childhood as a Junior Cosmonaut in then-Communist Hungary, traveling to Moscow to learn about the Soviet space program (Tariq Malik, "The Fifth Space Tourist: American Entrepreneur Charles Simonyi Prepares for Liftoff," *www.space.com/mission launches*, April 5, 2007).

An interesting group profile emerges on these five space tourist pioneers. First, they represent a wide age range from 28 to 61. Second, since childhood, they have had a long-standing interest in outer space and travel there. Third, they are well adjusted and educated; successful career professionals who carried out useful activities on the space station. Fourth, they are entrepreneurs in high technology. Fifth, these first space tourists are affluent, able to pay for the high cost of early access to orbit. Finally, they all flew aloft on Russian spacecraft and their fees helped underwrite the launch costs for an agency whose finances are quite limited.

A recent poll by Zogby International, a market research firm, found that 19% of well-off Americans would be prepared to spend $100,000 on short suborbital flights of some 30–90 minutes. Eric Anderson, CEO of Space Adventures, maintains there are thousands of potential customers willing to pay that for such an adventure and experience of weightlessness (*The Economist*, June 15th, 2002, p. 64). In any event, health and safety issues within the emerging space tourism industry are critical issues, as Appendix D will expand on.

6.5 CONCLUSIONS ON SPACE PERSONNEL DEPLOYMENT AND TOURISM

If space habitation is to advance rapidly, then a more comprehensive system approach to personnel deployment would seem to be appropriate. The model described in this presentation offers components which could be utilized in this decade for analysis of the "wintering-over" experience in Antarctica. Furthermore, knowledge from such an Antarctic Personnel Deployment System would be directly transferable in the next decades for creation of comparable deployment systems for orbiting stations, as well as for future Moon and Mars bases. At this preliminary stage in the human "passover" from Earth's boundaries, it is my contention that ground-based research of this type helps to further our transition to a space-based culture.

There is increasing American/European/Russian exchange on the human experience in spaceflight. Certainly, such cross-cultural communication and research should be expanded, especially with other spacefaring nations, such as China, India, and Japan. In fact, all international interchanges about human behavior in isolated, confined, and extreme environments can improve survivability, habitats, and life style in orbit. The range of data shared, whether about living in Arctic or desert regions, or in space, should be multidimensional (e.g., the requirements for food and shelter, health and wellness, group dynamics and interpersonal skills, transport and ecosystems, as well as play, performance, and productivity). Coping successfully with alien environments contributes to self-discovery and the actualization of human potential.

Among the deployment issues, there are three worthy of immediate study which can be undertaken equally in Antarctica or in orbit. The *first* has been mentioned above: the creation of *new cultures* by scientists and the military, by construction workers and colonists, as they explore and develop unknown territories. Space exploration represents an unusual opportunity to employ methodologies of cultural anthropology, so as to plan consciously for such cultural community experiments on the Moon or Mars.

The *second* research opportunity has to do with the creation of *new space careers*, whether on the ground or aloft. The high frontier will need people with a variety of occupational skills (technauts), to build and service spacecraft and facilities, in addition to the inhabitants. A recent U.S. Presidential commission anticipated such emerging vocational endeavors:

"Many people from non-aerospace fields are turning to space as a career. In the 21st century, professionals will view space as a new arena in which to develop their careers. Space doctors and medical researchers will be challenged by the physiological effects of prolonged weightlessness on the human body. Researchers in the interactive human sciences will study human adaptation in alien worlds and environments. Space architects, environmental engineers, and human factor engineers will join together to design remote living and working quarters. Virtually every trade and discipline will be involved in space endeavors, from obstetrics to insurance" (*Pioneering the Space Frontier*, National Commission on Space, 1986, p. 78).

With such a diverse, multiskilled, multicultural workforce leaving the home planet, broader personnel deployment systems, such as discussed in this chapter, will be essential!

The *third* opportunity for personnel deployment research is for *behavioral scientists in orbit*, to undertake such studies as discussed in this chapter, whether formal or informal. Psychologists, psychiatrists, sociologists, and other social scientists can be cross-trained to perform other functions at space settlements, in addition to applications of their own professional expertise.

During the next 50 years, the initial stations and bases aloft will be the "laboratories" for making later interstellar migration of our species possible. These outposts will permit us to experiment with offworld transfer of our human and material resources. The next few thousand space pioneers could make a remarkable contribution to a fund of knowledge for the establishment of spacefaring civilizations. But, as William MacDaniel, Niagara University professor emeritus of sociology, wrote to me in 1986, the real migration challenges are downrange: "selecting permanent human populations in orbit or on celestial bodies." Maj. MacDaniel believes that systems research will then be directed toward selection of the most appropriate groups (optimum size, age, sex, occupation, and social distribution) to support necessary social institutions on the new frontier.

Among the many spinoffs from space technologies and explorations to benefit humanity, the greatest so far is the *satellite communications industry*. The next may be the *space transportation industry* which lowers the costs which facilitate orbital access. But, most promising in the near future may be the *space tourism industry* which increases the possibilities for personal space travel. But, whether short or long-term space missions, people going aloft will have to be carefully selected, trained, and supported both in orbit and on re-entry.

The late James Fletcher, a former NASA administrator, stated that one of the most intriguing options ahead is the transforming of the environments of other worlds to make them habitable for human beings (*Space World*, July 1987, p. 36). The *terraforming of celestial bodies* will challenge us through this millennium. Presently, there is no convincing evidence of life in our galaxy other than here on Earth. Seemingly, the human mammals from this planet have the opportunity to extend life into our Solar System and beyond!

> **STEPPING OUT FOR A SPACE WALK**
>
> I was not in a strange or forbidding place at all, but in a place where I—"I" in the sense of being a member of the human race—was meant to be. I thought of those who claim that man should not fly because he was not given wings, and have conjured up, I am sure, similar analogies for space travel. As I looked at my space suit and the MMU, however, I knew that we are meant to travel away from the Earth because we have been given the curiosity, the intelligence, and the will to devise the means and the wonderful machines—such as those that now enclosed me—to permit such adventures. (Source: Astronaut Dale Garner in *The Jetpack: Flight without Wings*, by Nelson Lewis Olivio, 1995.)

6.6 REFERENCES

[1] Campbell, J. *Hero with a Thousand Faces*. Princeton, NJ: Princeton University Press, 1958.

[2] Larsen, K. M. *Cosmology 101*. Westport, CT: Greenwood Publishing Group, 2007.

[3] Goehlich, R. *Spaceships: A Reference Guide to International Reusable Launch Vehicle Concepts from 1944 to the Present*. Burlington, Ontario: Apogee Books/CGPublishing, 2007 (www.cgpublishing.com).

[4] Cordell, B. "Man Mars Overview," paper presented at 25th Joint Propulsion Conference, American Institute of Aeronautics and Astronautics Conference, Monterey, CA, July 10–11, 1987 (www.aiaa.org) ... Zubrin, Z. *The Case for Mars: The Plan to Settle the Red Planet and Why We Must*. New York: Free Press/Simon & Schuster, 1996, ch. 5, pp. 113–137.

[5] Finney, B. R.; and Jones, E. M. (eds.) *Interstellar Migration and the Human Experience*. Berkeley, CA: University of California Press, 1984 ... Finney, B. R. "Scientists and Seamen," in Harrison, A. A.; Clearwater, Y. A.; and McKay, C. P. (eds.) *From Antarctica to Outer Space*. New York: Springer-Verlag, pp. 89–102.

[6] Harrison, A. A.; and Connors, M. M. "Groups in Exotic Environments," in Berkowitz, L. (ed.) *Advances in Experimental Social Psychology*. New York: Academic Press, 1984 ... See back issues of *Journal of Human Performance in Extreme Environments*, 8 vols., 16 issues (www.society@hpee.org).

[7] Lugg, D. J. "Behavioral Health in Antarctica: Implications for Long-Duration Space Missions," in Harrison, A. A.; and Nunneley, S. A. (eds.) *Aviation, Space, and Environmental Medicine*, **76**(6), Section II Supplement, June 2005, B74–B77 (published by the Aerospace Medical Association, Alexandria, VA) ... Gunderson, E. K. "Emotional Symptoms in Extreme Isolated Groups," *Archives of General Psychiatry*, **9**, 1963, 362–368 ... Palinkas, L. A. "Group Adaptations and Individual Adjustment: A Summary of Research in 1948–1988," in Harrison, A. A.; Clearwater, Y. A.; and McKay, C. P. (eds.) *op. cit* [5, 9], 1991, ch. 21.

[8] Moran, R. T.; Harris, P. R.; and Moran, S. V. *Managing Cultural Differences*, Seventh Edition. Burlington, MA: Elsevier/Butterworth-Heinemann, 2007 (www.books@elsevier/business) ... Ritsher, J. B. "Cultural Factors and the International Space Station," *Aviation, Space, and Environmental Medicine*, **76**(6), June 2005, B135–B144.

[9] Harrison, A. A.; Clearwater, Y. A.; and McKay, C. P. (eds.) *From Antarctica to Outer Space: Life in Isolation and Confinement*. New York: Springer-Verlag, 1991 (see chs. 4, 5, 10, 11).

[10] Osland, J. S. *The Adventure of Working Abroad: Hero Tales from the Global Frontier*. San Francisco, CA: Jossey-Bass, 1992.
[11] Black, J. S.; Gregerson, H. B.; and Mendenhall, M. E. *Global Assignments: Successfully Expatriating and Repatriating International Managers*. San Francisco, CA: Jossey-Bass, 1992.
[12] Tung, R. L. *The New Expatriates: Managing Human Resources in an International Mix*. Cambridge, MA: Ballinger, 1987 ... Brewster, C. *The Management of Expatriates*. London: Kogan Page, 1991.
[13] Torbiorn, L. *Living Abroad: Personal Adjustment and Personnel Policy*. New York: Wiley, 1982 ... See *Mobility*, the magazine of the Employee Relocation Council (1720 N St. NW, Washington, D.C. 20036; *www.erc.org*).
[14] Harris, P. R. *Relocation Preparation Index: Twenty Reproducible Assessment Instruments*. Amherst, MA: Human Resource Development Press, 1995 (this is a source of other instruments, *www.hrdpress.com*) ... Kelley, C.; and Meyers, J. *The Cross-cultural Adaptability Inventory*. Boston, MA: Nicholas Brealey/Intercultural Press, 1993 (*www. intercultural press.com*) ... Dunbar, R. L.; Bird, A.; and Gudelis, R. *Expatriate Profile: Computer-Based Training Program*. New York: New York University/Stern School of Business, 1992.
[15] Palinkas, L. A.; Gunderson, E. K.; Johnson, J. C.; and Holland, A. W. "Behavior and Performance on Long-duration Spaceflights; Evidence from Analogue Environments," *Aviation, Space, and Environmental Medicine*, **71**(9), Supplement, 2000, A29–A36 ... Palinkas, L. A.; Allred, C. A.; and Landsverk, J. A. "Models of Research: Operational Collaboration for Behavioral Health in Space," *Aviation, Space and Environmental Medicine*, **76**(6), June 2005, B52–B60 ... Palinkas, L. A. *Health Performance of Antarctic Winter-over Personnel*. San Diego, CA: Naval Health Research Center/U.S. Naval Hospital, 1985 (Reports 85-18, 85-48, 85-40).
[16] Bluth, B. J.; and Helppie, M. *Soviet Space Station as Analogs*. Washington, D.C.: NASA Headquarters, 1986 (NAG-659) ... Johnson, S. L. *Handbook of Soviet Manned Spaceflight*. San Diego, CA: UNIVELT/AAS, Vol. 48.
[17] Furnham, A.; and Bochner, S. *Culture Shock: Psychological Reactions to Unfamiliar Environments*. New York: Methuen, 1987.
[18] White, F. "The Overview Effect and the Future of Humans" in Krone, M. L. and Cox, K. (eds.) *Beyond Earth*. Burlington, Ontario: Apogee Books. CGPublishing, 2006, pp. 38–40 ... *The Overview Effect: Space Exploration and Human Evolution*. New York: Houghton-Mifflin, 1987.
[19] Austin, C. N. (ed.) *Cross-cultural Re-entry: A Book of Readings*. Boston, MA: Nicholas Brealey/Intercultural Press, 1987.
[20] Harrision, A. A.; and Nunneley, S. A. (eds.) *New Directions in Spaceflight Behavioral Health*. Alexandria, VA: Aerospace Medical Association, 2005 (*Aviation, Space and Environmental Medicine*, **76**(6), Supplement, *www.asma.org*) ... Levine, A. S. "Psychological Effects of Long-Duration Space Missions and Stress Amelioration Techniques," in Harrison, A. A.; Clearwater, Y. A.; and McKay, C. P. (eds.) *From Antarctica to Outer Space*. New York: Springer-Verlag, 1991, ch. 23 ... Leon, G. R. "Men and Women in Space," *Aviation, Space and Environmental Medicine*, **76**(6), Section II Supplement, June 2005, B84–B88 (published by the Aerospace Medical Association, Alexandria, VA).
[21] Helmreich, R. L.; Wilhelm, J. A.; and Runge, T. E. "Psychological Considerations in Future Space Missions," in Cheston, S.; and Winters, D. (eds.) *Human Factors in Outer Space Production*. Boulder, CO: Westview Press, 1983 ... Flynn, C. F. "An Operational Approach to Long-Duration Behavioral Health and Performance," *Aviation, Space, and Environmental Medicine*, **76**(6), Section II Supplement, June 2005, B42–B51 (published by the Aerospace Medical Association, Alexandria, VA).

[22] Lane, H. W.; Sauer, R. L.; and Feeback, D. L. *Isolation: NASA Experiments in Closed Living*. San Diego, CA: UNIVELT/AAS, Vol. 104, 2002.
[23] Brady, J. V. "Small Groups in Confined Microsocieties," in Harrison, A. A.; Clearwater, Y. A.; and McKay, C. P. (eds.) *op. cit.* [20], 1991 ... Wood, J.; Schmidt, L.; Lugg, D.; Phillips, A. T.; and Shepanek, M. "Life, Survival and Behavioral Health in Small Closed Communities: 10 Years of Studying Isolated Antarctic Groups," *Aviation, Space, and Environmental Medicine*, **76**(6), Section II Supplement, June 2005, B89–B94 (published by the Aerospace Medical Association, Alexandria, VA).
[24] Stuster, J. *Bold Endeavors: Lessons from Polar and Space Exploration*. Annapolis, MD: Naval Institute Press, 1996 ... Barbabrasz, A. E. "A Review of Antarctica Behavioral Research," in Harrison, A. A.; Clearwater, Y. A.; and McKay, C. P. (eds.), *op. cit.* [20], 1991, ch. 3.
[25] Collins, P. "Choosing the Right Stuff," *Space Policy*, **6**(3), August 1990, 281–282.
[26] For further information on these findings, contact Long-Term Program Office, European Space Agency (8/10 rue Mario Nikis, 75738, Paris, Cédex 12), or ESA Washington Office (955 L'Enfant Plaza, SW, Ste. 7899, Washington, D.C. 20024; email: *Mary.Davis@ESA.int*).
[27] Poynter, J. *Biosphere 2: Lessons Learned*, 2000 (*www.amazon.com*) ... Singhama, L. "Be Sure to Pack Scientific Curiosity," *San Diego Union-Tribune* (Associated Press), July 21, 2002, p. D12 ... David, L. "Biosphere 2 Dwelling on the Future," *Final Frontier*, January/February 1994, 41–45 ... Shanks, C. "Genesis of a Space Colony: Biosphere 2, *Ad Astra*, **3**(1), 1991, 30–34 ... Crawford, B., "Inside the Bonsai World of Biosphere 2," *American Way*, March 1990, 34–36, 107 ... Allen, J.; and Nelson, M. *Space Biospheres*. Melbourne, FL: Orbit Books/Krieger Publications, 1987.
[28] For information on proceedings for IDEEA ONE/TWO contact Sasakawa International Center for Space Architecture, University of Houston (Houston, TX 77204), and for IDEEA THREE contact the UK's Oxford University School of Architecture (email: *dbibby@brookes.ac.uk*).
[29] Zubrin, R. *Mars on Earth: The Adventures of Space Pioneers in the High Arctic*. New York: Tarcher/Penguin, 2003.
[30] Clarke, J. D. (ed.) *Mars Analog Research*. San Diego, CA: UNIVELT/AAS, Vol. 111, 2006. From the same source, consult related publications on *The Case for Mars* in the AAS "Science & Technology Series": Vol. 57, 1981; Vol. 62, 1984; Vols. 74 and 75, 1987; Vol. 86, 1996 (*Strategies for Mars: Guide to Human Exploration*); Vols. 89 and 90, 1990; Vol. 97, 1993; Vol. 98, 1996; Vol. 107, 2003 (*Martian Expedition Planning*).
[31] Harris, P. R. "Personnel Deployment Systems for Space," *Living and Working in Space*. Chichester, U.K.: Wiley/Praxis, 1996, Second Edition ... "Personnel Deployment Systems: Managing People in Polar and Outer Space Settings," in Harrison, A. A.; Clearwater, Y. A.; and McKay, C. P. (eds.) *From Antarctica to Outer Space*. New York: Springer-Verlag, 1991, ch. 7.
[32] Piradov, A. S. "Creating a World Space Organization," *Space Policy*, **4**(2), May 1988, 112–114 ... O'Donnell, D. J. "Founding a Space Nation Using Living Systems," *Behavioral Science*, **39**(2), April 1994, 93–116 ... O'Donnell, D. J.; and Harris, P.R. "Legal Strategies for a Lunar Development Authority," *Annals of Air and Space Law*, **XXI–XXII**, 1996, 121–130 ... Schwab, M. "Views of Global Leadership: Government, Business, Academia, and Faith," in Krone, M. L. and Cox, K. (eds.) *Beyond Earth*. Burlington, Ontario: Apogee Books/CGPublishing, 2006, pp. 24–35 (see *www.martinschwab.com* or *www.homeplanet defense.org*).
[33] ESA. "The European Astronaut Centre," ESA Fact File No. 3, July 10, 1990; "ISEMI: The Human Factor," ESA Fact File No. 4, November 14, 1990 ... "Astronauts Wanted," *ESA Newsletter*, January 1991, pp. 1–4.

[34] Manzey, D.; Horman, H. J.; Fassbender, C.; and Schiewe, A. "Implementing Human Factors Training for Space Crews," *Earth Space Review*, **4**(1), January/March 1995.
[35] Moran. R. T.; Harris, P. R.; and Moran, S. V. *Managing Cultural Differences*, Seventh Edition. Burlington, MA: Elsevier/Butterworth-Heinemann, 2007, ch. 10, pp. 260–304 ... Also see Barclay, J. *Psychological Assessment*. Melbourne, FL: Krieger Publishing, 1991 and Black, J. S.; Gregersen, H. B.; and Mendenhall, M. E. *Global Assignments*. San Francisco, CA: Jossey-Bass, 1992, ch. 3.
[36] Harvey, B. *Russia in Space*. Chichester, U.K.: Springer/Praxis, 2001 ... Harland, D. M. *The Mir Space Station*. Chichester, U.K.: Wiley/Praxis, 1997.
[37] Williams, R. S.; and Davis, J. R., "Ensuring Behavioral Health during Extended-Duration Spaceflight Missions" ... Filbert, H. E.; and Keller, D. J. "Astronaut Interdisciplinary Training for Manned Mars Missions," in Stocker, C. (ed.) *The Case for Mars, III: Strategies for Exploration*. San Diego, CA: UNIVELT/AAS, Vols. 74/75, 1989, pp. 161–171.
[38] Swanson, G. E. (ed.) "STS-71 Preview: Shuttle/Mir Docking Mission," *Countdown*, **13**(3), May/June 1995, 44–47 (crew biographies) (published by CSPACE Press, 123 32nd St., Grand Rapids, MI 49509).
[39] Cooper, H. S. *Before Lift-Off: The Making of a Space Shuttle Crew*. Baltimore, MD: Johns Hopkins University Press, 1987. For Russian assessment of cosmonauts see Harvey citation [36].
[40] Manzey, D.; and Schiewe, A. "Psychological Training of German Science Astronauts," *Acta Astronautica*, **27**, 1992, 147–154 ... Manzey, D.; Schiewe, A.; and Fassbender, C. "Psychological Countermeasures for Extended Manned Spaceflight," *Acta Astronautica*, **30**, 1995.
[41] Nicholas, J. M. "Interpersonal and Group Behavior Skills for Crews on Space Station," *Aviation, Space, and Environmental Medicine*, **60**, 1989, 603–608.
[42] Sheffield, C.; and Rosen, C. *Space Careers*. New York: William Morrow, 1984.
[43] Harrison, A. A. "Behavioral Health: Integrating Research and Application in Support of Long-term Missions," *Aviation, Space, and Environmental Medicine*, **76**(6), Section II Supplement, June 2005, B3–B12 (published by the Aerospace Medical Association, Alexandria, VA) ... Baker, D. *The History of Manned Spaceflight*. New York: Crown Books, 1985.
[44] Harris, P. R. *Launch Out: A Science-based Novel about Space Enterprise, Lunar Industrialization, and Settlement*. West Conshohocken, PA: Infinity Publishing, ch. 14, pp. 353–403 (www.buybooksontheweb.org).
[45] Hall, E. T.; and Hall, M. E. *Hidden Differences*. Garden City, NY: Anchor Press/Doubleday, 1987.
[46] Oberg, A. R. *Spacefarers of the 80's and 90's*. New York: Columbia University Press, 1985.
[47] Clearwater, Y. A.; and Harrison, A. A. "Crew Support Systems for an Initial Mars Mission," *Journal of the British Interplanetary Society*, **43**, 1990, 513–518.
[48] National Commission on Space. *Pioneering the Space Frontier*. New York: Bantam Books, 1986.
[49] Cordell, B. "Manned Mars Overview," paper presented at 25th Joint Propulsion Conference, American Institute Aeronautics and Astronautics, Monterey, CA, July 10–11, 1989.
[50] Reichhardt, T. *Space Shuttle: The First 20 Years*. Washington, D.C.: Smithsonian Institution/DK Publishing, 2002 ... Harland, D. M. *The Space Shuttle: Roles, Missions, Accomplishments*. Chichester, U.K.: Wiley/Praxis, 1998.
[51] Jones, E. M. (ed.) *The Space Settlement Papers*. Los Alamos, NM: Los Alamos National Laboratory, 1985, p. 13 (Report LA-UR-85-3874).

[52] Harrison, A. A.; Sommers, R.; Struthers, N.; and Hoyt, K. *Privacy and Habitat Design*. Moffet Field, CA: NASA Ames Field Center, 1986 (Report NAG-2-357).
[53] Hoffman, J. A. *An Astronaut's Diary*. Montclair, NJ: Caliban Press, 1986.
[54] NASA. *Man–Systems Integration Standards*. Houston, TX: Johnson Space Center, July 1995 (NASA-STD-3000-B).
[55] Hall, S. B. (ed.) *The Human Role in Space*. Park Ridge, NJ: Noyes Publications, 1985.
[56] Harvey, B. *Race into Space: The Soviet Manned Space Programme*, 1988 ... The new edition is *Russia in Space*. Chichester, U.K.: Springer/Praxis, 2001.
[57] Harris, P. R. "Transitional Experiences as a Fulbright Professor in India," in Usmani, Z.; and Ghosh, N. K. (eds.) *Beyond Boundaries: Reflections of Indian and U.S. Scholars*. New York: iUniverse Inc., 2007, pp. 360–362 (*www.iuniverse.com*).
[58] Wolf, T. *The Right Stuff*. New York: Bantam Books, 1981 ... Santy, P. *Choosing the Right Stuff: The Psychological Selection of Astronauts and Cosmonauts*. Westport, CT: Praeger, 1994.
[59] The Nowak Case Study was based on these sources: "Astronaut Granted Bond on Attempted Murder Charge," *CNN.com*, February 6, 2007; Borenstein, S. "Astronaut Culture Stresses Achievement," Associated Press, February 8, 2007; Kaufman, M. "NASA to Review How It Assesses Astronauts," *San Diego Union-Tribune*, February 8, 2007; Schwartz, J. "Astronaut's Arrest Spurs Review of NASA Testing," *New York Times*, February 8, 2007; Schneider, M. "NASA Has Plan for Unstable Astronauts," Associated Press, February 23, 2007; and Schneider, M. "Ousted from NASA, Former Astronaut to Start New Job," Associated Press, March 21, 2007, *www.space.com*
[60] Spencer, J.; and Rugg, K. L. *Space Tourism: Do You Want to Go?* Burlington, Ontario: Apogee Books, 2004 (*www.cgpublishing.com*) ... Also see Stine, G. H. *Living in Space: A Handbook for Work and Exploration Stations beyond Earth's Atmosphere*. New York: M. Evans & Co., 1999.
[61] Van Pelt, M. *Space Tourism: Adventures in Earth Orbit and Beyond*. New York: bCopernicus Books/Springer (*www.copernicusbooks.com*) ... Also see Tarzwell, R. "The Medical Implications of Space Tourism," and Wichman, H. A. "Behavioral Health and Implications of Civilian Spaceflight," *Aviation, Space, and Environmental Medicine*, **71**(6), 49–61, B164–B171.
[62] Bernstein, P. *Making Space Happen: Private Space Ventures and the Visionaries behind Them*. Medford, PA: Plexus Publishing, 2002 ... McElyea, T. *Vision of Future Space Transportation: A Visual Guide to Spacecraft Concepts*. Burlington, Ontario: Apogee Books/CGPublishing, 2003 (*www.cgpublishing.com*) ... Also consult the following websites *www.spacevoyages.com* and *www.astronautix.com*

SPACE ENTERPRISE UPDATE

NASA is using a new vertical treadmill to improve astronaut health. It enables spacefarers to run while suspended horizontally, thus preparing for long-duration flights to the Moon. Engineers at Cleveland's Glenn Research Center built a Standalone Zero Gravity Simulator to imitate conditions for exercising in space, so essential to lessen the harmful effects of microgravity.

7

Macrothinking in strategic space planning

> *One of the major challenges facing leaders around the world is macroengineering. This is the process of marshalling money, materials, personnel, technology, and logistics on a huge scale to carry out complex projects—often international in nature that last over a long period of time.*
>
> Neil E. Carter [1]

To open up and develop the space frontier, a new type of *macrothinking* is required, as implied in the above quotation. Such reasoning and actions were evident during the Apollo mission period when the human family was "turned on" to the daring achievement of rocketing humans to the Moon! Unfortunately, since that decade (the 1960s), space programs throughout the world have been the exclusive concern of national space agencies, aerospace contractors, planetary scientists, and government legislators. Their somewhat myopic focus on special interest missions has garnered only the support of direct beneficiaries and committed space activists. As a result, public interest has waned over technical accomplishments and failures which do not capture the imagination of *Everyman*. Because would-be space planners fail to provide a 21st-century vision of humankind's future within our Solar System, they do not inspire the masses to financially back risk-taking endeavors in the orbital environment. Further, too many mission planners and managers neither control escalating costs nor provide adequate information about the return on investment, especially the spinoff value of space technologies. Instead they alienate suffering taxpayers, potential investors, and politicians with grandiose plans that are hugely costly to implement and manage effectively. Only in this first decade of the new millennium have space entrepreneurs recaptured the public's interest in orbital enterprises, as evident in less expensive spaceplanes, space inflatables, and space tourism. Whether offworld endeavors are undertaken by sponsors in the public or private sectors, or through global consortia, a different and innovative type of "macro" thinking, planning, and management is required, as will be discussed in this and Chapter 8.

7.1 UNDERSTANDING MACROTHINKING AND PLANNING

Now that some 40 years have passed since humanity's lunar landings, we need again to think *big*, think *ahead*, and think *systems*, if we are to permanently return to the Moon and eventually beyond! The case for investment in space enterprise has to reflect macrothinking if wide public support is to be obtained!

Macro, in its Greek origin, means "large, great, or long". In human endeavors, the macro concept has many applications. When intellectual disciplines or academic specialities devote themselves to issues and activities of society at large, they are termed macrosociology, macroeconomics, or even macromanagement (a topic discussed in Chapter 8). The late professor at the University of Alabama, Dalton McFarland, explained that *macrosocial* phenomena include a variety of forces that integrate the elements of a society, such as broad consensus among both the public and professionals [2]. Since society is itself a system, there must be linkages among subsystems that give legitimacy to the central institutional system, so that the larger community's purpose can be achieved and resources/rewards allocated. Macrothinkers (foresighted leaders and visionaries) give expression to goals and plans in which the citizenry concur, mobilizing themselves to achieve what is envisioned.[1] NASA, for example, has held numerous summer studies, published proceedings with bold plans, but then failed to implement them for one reason or another [3].

President John F. Kennedy demonstrated such thinking in 1961 when he set a goal for the United States within a decade to land a man on the Moon and safely return him. The macrothinking that led to that Presidential statement and policy came from a report of a combined Department of Defense/NASA Task Force under the leadership of Dr. Lawrence L. Kavanau. National conditions were such that the private and public systems supported and achieved that ambition. On July 20, 1989, while commemorating the 20th anniversary of that feat, another U.S. President, George Bush (senior), attempted to set forth the next major space goal for the nation. As explained in Chapter 1, that Space Exploration Initiative's objective included construction of a space station, a permanent return to the Moon, and a manned mission to Mars by 2019. While SEI showed elements of macrothinking, it failed to engender sufficient consensus, backing, and resources. So, the President's ambitions were aborted; the whole proposal lacked macroplanning, including a rationale that Congress would finance. Now his son, President George W. Bush, is presently trying to revive the essence of that plan with the *Vision for Space Exploration*.[2] NASA is

---

[1] At the Massachusetts Institute of Technology, there are two Macroengineering Collections worthy of contacting: *Books@DeweyLibrary*, 30 Wadsworth St., Bldg. E53-100; *Films@MIT Museum*, Bldg. N52-2nd Fl ... both at MIT, Cambridge, MA 02139.
[2] "Spaceflight: Objective Moon," *The Economist*, May 28, 2005, pp. 79–80 (*www.economist.org*).

expected to be doing the planning and funding out of its regular budget to put humans permanently back on the Moon by year 2020, and eventually to explore Mars and beyond. The problem within that space agency is the lack of macrothinking. Currently, NASA planners are working on transportation back to the Moon, and building an outpost there for about half a dozen astronauts. Macrothinking would develop strategic plans for building and maintaining a lunar base, but in conjunction with other space agencies (such as those of Russia, Europe, China, India, and Japan) it would also involve private enterprise in this venture that would in time create a twin-planet economy by utilizing lunar resources.

The United States Congress and Presidential Commissions are also not noted for their "macro" thinking or planning. But consider this account from a Commission appointed by the President in 2004 on how best to carry out the new VSE policy for permanently returning to the Moon and later exploring Mars. The *New York Times* headline "Half-Baked Proposal for Space," was followed by this commentary:

> "The nine-member presidential commission led by Edward Aldridge, former secretary of the Air Force, was breathtakingly bold in some of its recommendations. It proposed a radical change in the way NASA operates, even a $1 billion prize for the organization that first puts human on the Moon and sustains them there for a fixed period before returning them to Earth. The glaring defect in the planners' report was its failure to provide any thoughtful analysis or detailed justification for its proposals. The 60-page document amounts to a pep talk that is too skimpy to be persuasive ... Congress will need to take these half-baked proposals and cook them more thoroughly" (*www.nytimes.com/2004/06/21/opinion/21MON1html*).

7.1.1 Large-scale space enterprises

Enterprise has been defined as a project of some purpose and importance which is undertaken with energy and boldness by resolute participants. People are said to be enterprising when they engage in ventures of great difficulty with imagination and initiative. When individual inventiveness combines with collective action, then big thinking and things happen! Burt Rutan building SpaceShipOne and Sir Richard Branson ordering a fleet of them to launch his Virgin Galactic project for LEO tourism are indeed enterprising!

Macroenterprises, whether on the ground or aloft, are by their very nature macroengineering programs which pose a combination of technical, managerial, legal, and financial challenges. Macroengineering requires macrothinking to study, plan, manage, and incorporate comprehensive technological ventures which integrate people, hardware, and software for major accomplishments. While macroengineering builds on present knowledge and technology, it strains the outer limits of the "state of the art", stretching the competencies of the macroengineer. Currently, there is some macrothinking going on in terms of environmental change and restoration. For example, John Darabaris has written on macroengineering with regard to environmental restoration macroprojects, such as those involved in major oil spills or environmental cleanups caused by nuclear accidents and radiation.[3]

A *macroproject*, for instance, has been described as a large, complex, long-term undertaking, usually involving expenditures of $100 million plus; this is the case with most space enterprises, such as the Space Shuttle or Hubble Space Telescope. Macromanagement refers to a new type of 21st-century interdisciplinary management for such large-scale and complex programs, as in constructing and maintaining an orbiting space station, a lunar base, or other proposals for utilizing space resources. In macroengineering projects, significant quantities of public and private resources must be committed and put at risk within extended timeframes. In marshaling resources for such projects, they tend to have profound sociocultural impacts and environmental implications [4]. This expertise has taken on new relevance because of global efforts to cope with climate change and planetary pollution.

Unfortunately, most professional schools, whether in engineering, science, business, or whatever, do not prepare their graduates to engage in either macroprojects or macromanagement where practitioners must move beyond narrow academic disciplines and specialities to conceive "holistically" in terms of interacting systems. Two European Union exceptions are the Center for Macro-Projects and Diplomacy at Roger Williams University in Bristol, U.K., and the Candida Oancea Institute in Bucharest, Romania.

Complex space enterprises today, and more so tomorrow, are both multinational and multidisciplinary, requiring multicultural skills. Macrothinkers manifest multifaceted capacities to effectively coordinate human, material, and technical resources, as well as to deal with sociocultural, political, financial, and legal dimensions, plus ecological and environmental concerns. Too much contemporary space planning is largely concerned with hardware and technical matters, neglecting broader human factors, so essential to long-duration missions and space settlement (see Exhibit 20, Chapter 2). Most space planning is narrow and mission-oriented, rather than placing each project in a larger macrothinking context that may involve decades.

As the "prophet of macroengineering", MIT's Frank Davidson, has so aptly stated in a score of books and articles, "we need a clear vision of how technology should shape our future." In addition, this farsighted lawyer argues that to benefit from modern science and technology, we have to transform our thinking, behavior, institutions and vocabulary, so that

"we abandon the sterile political rhetoric of public versus private sector, and learn instead the neglected art of intersectoral harmony and collaboration" [5].

Exhibits 88A–88C offer insights into the concept of macrothinking and the process of macroplanning. In Exhibit 88A, macroengineers suggest eight aspects to include in any macroplanning. To conclude this introduction, Exhibit 88B synthe-

---

[3] Darabaris, J. *Macroengineering: An Environmental Restoration.* Boca Raton, FL: Taylor & Francis/CRC Press, 2006.

Exhibit 88A. THE EMPHASIS OF MACROPLANNING CHARACTERISTICS.

- Problems critical to large-scale developments
- Database management systems and development tools
- Analytical tools
- Integration of available program management facilities
- Computer graphics and report generation
- Exploratory data analysis
- Natural language and knowledge representation

Source: Meader, C. L.; and Parthe, A. C. "Managing Macro-Development: Policy, Planning and Control Implications," in Davidson, F. P.; Meader, C. L.; and Salkeld, R. (eds.) *How Big and Still Beautiful? Macroengineering Revisited.* AIAA Selected Symposium, Vol. 40, 1980, p. 90.

Exhibit 88B. REQUIREMENTS FOR MACROTHINKING AND MACROPLANNING.

A convergence in thinking and planning for large-scale enterprise which is

- Multidimensional
- Transnational in scope and leadership
- Interdisciplinary
- Market-oiented
- Engaging in preliminary studies, demonstration models, and prototype testing
- Able to mobilize resources (technical, material, and human)
- Sensitive to the ecology and environment
- Synergistic and able to promote collaboration

Source: P. R. Harris, "Macrothinking in Global Space Planning," *Earth Space Review*, 3(1), 1994, 5–8.

Exhibit 88C. ATTRIBUTES OF MACROTHINKING AND MACROPLANNING.

- Demonstrates an imaginative mindset that practices the art of the long view
- Promotes coalitions and consortia which match purpose to means
- Advances integrative education and practice of advanced technologies
- Facilitates joint technological venturing across sectors, both public and private
- Conceptualizes holistically in terms of systems and human needs
- Ensures multicultural participation while considering broad impacts
- Transforms attitudes and behavior to stimulate innovation

Source: Inspired by Frank B. Davidson's *MACRO: A Clear Vision of How Science and Technology Will Shape Our Future.* New York: William Morrow & Co., 1993.

sizes the requirements for *macrothinking and planning*, while Exhibit 88C offers an overview of the attributes for successful macroprojects.

From the vantage point of 45 years of system consulting experience, your author has concluded that those in the scientific and aerospace communities who offer proposals for space investment would benefit by studying the field of strategic planning and management. One lesson I learned from this speciality is to "think big and think ahead", at least a decade or more. As a consultant in NASA headquarters before the Apollo 11 landing, I experienced the opposite view. When I asked what plans these top administrators had for space exploration and development after putting the first persons on the Moon's surface, they expressed uncertainty. Finally, reflective of an engineering mindset, someone replied, "We're waiting for Congress to tell us." Perhaps such thinking may help to explain why the U.S. Congress and Administration were unwilling for four decades to capitalize on the momentous feat of placing American astronauts on the Moon. During this hiatus NASA leaders were often criticized for lacking vision. For a time, the Agency emphasis was on the "cheaper, faster, approach", such as in the Mars Pathfinder mission. A global macro space plan has the capacity to capture humanity's sense of destiny and spirit of exploration. Even when construction of an offterrestrial infrastructure may be described in incremental steps and features miniaturization, it can get support over costly "big science" programs [6].

Some futurists, like Milan Cirkovic of the Astronomical Institute of Belgrade. are even thinking "Macroengineering in a Galactic Context" (*www.arxiv.org/astro-ph/ 0606102*). Such thinking usually occurs in the fields of astrobiology and astrophysics. Other science and science fiction writers, like Gregory Benford, look back over a millennium and then project ahead a thousand years (see *http://baens-universe.com/ articles*/A Pocket History of MacroEngineering: The First Millennium). The Foundation for the Future also takes a millennium perspective in its symposiums, videos, and publications, such as the *Humanity 3000* 2005 Symposium Proceedings (*www. futurefoundation.org*).

7.1.2 Space strategic planning and management

Strategic planning and management (SPM) today is both an art and science. Space organizations of any type, big or small, private or public, are well advised to employ experts in this field, either as an internal or external consultant. In my experience with over 200 human systems, I discovered too many of these institutions were unaware of the SPM resources available, whether in terms of management literature or trained professionals. Remember that SPM is an important tool that executives and administrators cannot afford to ignore.

One such expert in the field of SPM is Professor Marios Katsioloudes of the American University in Sharjah, United Arab Emitates. In his recent book on the subject, he defines strategic planning as:

"the process by which a system maintains its competitiveness within its work

Sec. 7.1]   **Understanding macrothinking and planning**   323

environment by determining where the organization is, where it wants to go, and how it wishes to get there" [7].

In Exhibit 89, Dr. Katsioloudes elaborates further on the key elements to be considered in a strategic management process.

Using the above model, a space enterprise begins strategic planning with a clear *mission statement* that incorporates both its vision and reason for existence. Then it needs to articulate its *goals*, whether for the whole organization or a specific macroproject. Usually, these are open-ended statements of planned accomplishments. Finally, a plan states the *objectives* that will enable its personnel and contractors to achieve these goals: specific *targets* or milestones with end results for the planned accomplishments. Here is where the term *strategy* must be clarified: it is the methods for achievement of goals, a planned operation for execution of the project. The evolving plan is an attempt to match internal capabilities with external relationships, at all levels of the enterprise. And the plan should be continuously updated and even revised.

There are three other key elements in the above paradigm:

(1) Strategy formulation requires the defining of responsibilities for this strategic planning and management. With this clarification of roles and relationship, the leaders can better assess their strengths, weaknesses, opportunities, and threats relative to implementation. In this SWOT analysis and data gathering, allow for people's *intuitions*, that individual power for individuals to make mental leaps, to

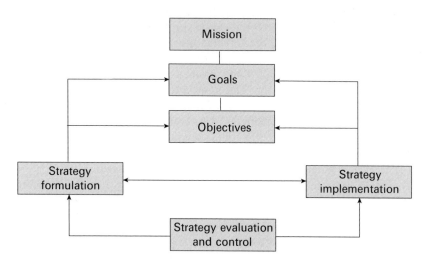

Exhibit 89. Dynamic management strategic process. Source: M. I. Katsioloudes, *Strategic Management: Global Cultural Perspective for Profit and Non-Profit Organizations*. Burlington, MA: Elsevier/Butterworth-Heinemann, 2006, p. 8. This volume is part of the Managing Cultural Differences Series (*www.books@elsevier/business*).

grasp facts and knowledge without recourse to reason. Thus, both intuitive and quantitative analysis is included in the process.
(2) Strategy implementation ensures that things happen as planned, rather than remaining fantasy. It involves attaining consensus and commitments on not only the broader goals and objectives—but specific targets and time tables for achievement. Here is when the refinement of strategies transforms such into specific programs, budgets, and procedure.
(3) Strategy evaluation and control means monitoring project progress, so as to be flexible in changing or altering strategies.

Strategic planning and management also identify and manage the various stakeholders in any macroproject. These range internally from stockholders, directors/trustees, managers, and personnel, to external partners, unions, customers, contractors, suppliers, distributors. Other possible stakeholders may be government, community, venture capitalists, creditors, trade associations, and even competitors. Depending on their stake in the enterprise, strategic leaders seek appropriate stakeholder involvement, support, and collaboration. Space strategic planners have to consider the influence of both global and interplanetary environments, as well as culture, on their plans! Finally, they have to be flexible and adaptive relative to their plans, especially when experiencing failures or disasters. One can only speculate what strategic plan Sir Richard Branson has for his Virgin Galactic enterprise (see *www.virgingalactic.com*)!

> *Make no small plans for they have not the power to stir men's blood*
>
> Niccolo Machiavelli

> Man is an engineering animal. Since prehistoric times, we have used our skills at engineering to adapt our environment to suit ourselves. We have levelled mountains, engineered man-made lakes and drained swamps, reversed the course of rivers, and made deserts bloom. The stone monoliths of Stonehenge were an engineering feat unrivalled for its time. Later great pyramids boasted in stone to the gods of Egypt that we humans, too, could make mountains. In modern times, Gustav Eiffel's tower—at the time, the tallest building in the world—was a marvel of its age, a poem in steel. For better or worse, as our technology becomes more powerful, our engineering feats are moving to planetary scale. (Source: Geoffrey A. Landis in the Foreword to R. B. Cathcart and R. D. Schuiling (eds.) *Macroengineering for the Future*. New York: Springer-Verlag, 2006.

7.2 MACROTHINKING EXAMPLES FROM THE PAST

Macrothinking has been carried out by *Homo sapiens* since our ancestors climbed down from the trees and walked upright. Even hunters and gatherers demonstrated macrothinking in the way they migrated across the planet. The story of civilization contains numerous examples of large-scale engineering projects, beginning with the agricultural period of human development and the need for infrastructure. Over millennia, history records macrothinking feats in building the walls of Jericho, beautiful cities like Babylon, the aqueducts of the Roman Empire, the Great Wall of China, medieval castles and cathedrals, Mayan temples and metropolises. But, it has only been in the past few centuries that humankind has distinguished itself by macrothinking which resulted in huge system engineering of roadways, dams, canals, bridges, waterways, pipelines, transcontinental railways, ship and air ports, and many a megalopolis—the very infrastructure of a modern, functioning society. Thinking big happens everywhere, and is most evident on Earth in the Suez and Panama Canals, the Manhattan Project, the St. Lawrence Seaway and Power Project, the Aswan High Dam, the Osaka island redevelopment, the Saudi Arabian Jubal city, and more recently the Anglo-French Channel tunnel or Chunnel. As Davidson and Meader indicate, macrothinking demonstrates that a society has "both the competence and compassion to address human needs on a scale commensurate with our problems" [6].

Examples of individual macrothinkers abound in the past and present, from ancient builders of the Seven Wonders of the World and Egyptian Pharaoh Sesosteris I, who connected the River Nile to the Red Sea in 2000 BC, to the 16th-century Dutch dike builders such as engineer-scientist Andries Vierlingh. The results of "thinking big and ahead" are found in every century, country, culture, and career. History honors their vision, as among artists like the Italian Michelangelo; printers and statesmen like the American Ben Franklin; architects like the Englishman John Rennie, who built the original London Bridge; soldier-builders like French emperor Napoleon Bonaparte who united Europe under his leadership; writers like the Frenchman Jules Verne; engineer-managers like the Anglo-French Isambard Kingdom Brunel; pragmatic philosophers like the American William James; theoretical scientists like the German–Swiss American Albert Einstein; political leaders like the U.S. President Franklin D. Roosevelt, who launched massive social programs and projects. Many entrepreneurs illustrate macrothinking at its best, from Pierre-Paul Ricquet in the 17th century who designed French canals, to Ted Turner in the 20th century, who created a global television network. Then in the 20th century, Olaf Stapledon, John B. S. Haldane, and Freeman Dyson who promoted macro-enginering projects as focal points to extrapolate humanity and, by analogy, other intelligent species into the Milky Way in the 21st Century. You also see macrothinking at work in the United Arab Emirates, especially at Dubai. There in the middle of the desert, amazing large-scale engineering projects are under way. The same thinking is evident in the building of high-speed railroad systems in France and Japan, or immense skyscraper buildings throughout Asia. Modern macroengineering emerged in the 1960s, and today Japan and China lead in its advancement (*www.daviddarling. info/encyclopedia/M/Macroengineering_Cathcart.html*). Throughout the ages there have been "big thinkers" with visions of what humanity needed to accomplish.

However, it is in the Space Age that macrothinking became both a science and a necessity. It is evident in the thinking of Russians: such as mathematician Konstantin

Tsiolkovski, who wrote philosophical and scientific space books in the 1880s, while designing pre-revolutionary atomic spaceships; Yuri Kondratyuk, who described orbiting space stations in the 1920s; Friedrich Tsander, who outlined space shuttles in the 1930s. Despite the secrecy and mystery in the former Soviet Union, the names of its brilliant 20th-century space macrothinkers emerged: such as Valentin Glushko, Vladimir Chelomei, and especially chief grand designer Sergei Korolev, a Ukrainian. With their Soviet colleagues, they launched a spectacular series of macroengineering "firsts", both unmanned and manned, including the first orbiting satellite Sputnik 1 (1957); the first man in space Yuri Gagarin aboard Vostok 1 (1961); the first woman in space Valentina Tereshkova (1963); the first spacewalker, cosmonaut Alexei Leonov in Voskhod 2 (1965); the first soft-landing on the Moon, by Luna 9 (1966); the first automated docking of spacecraft with Cosmos 186 (1967); the first unmanned lunar flyby/return with Zond 5 (1968); the first unmanned lunar "crawler", Lunokhod 1 (1970); and the first manned space station with Salyut 1 (1971).

Of course, there were many manifestations elsewhere of space macrothinkers, such as Romanian-born Hermann Oberth, an inspiring space writer; Germany's rocket engineers Wernher von Braun and Krafft A. Ehricke; the United States' Robert Goddard, pioneering rocket scientist, and James E. Webb, second NASA administrator; British scientists Fred Hoyle, the astronomer, and Arthur C. Clarke, proposer of global communication satellites and science fiction writer. During the Cold War, the Americans joined the space race with their own demonstrations of macrothinking at its best, including: the first successful planetary probe with Mariner 2 to Venus (1962); first successful communications satellite with Telstar (1962); first manned lunar landing with Apollo 11 (1969); first successfully manned space station of long duration with Skylab (1973); first successful unmanned landing on Mars with Viking (1975); the world's first planetary Grand Tour, Voyager 2 (1977–1989); the world's first reusable Space Shuttle, *Columbia* (1981). Both the Luna and the Apollo mission series not only reflected exceptional macrothinking, but gave impetus to the whole field of macroengineering [8]. In both instances, space planners coordinated and managed cumulative complexity. Throughout this opening stage of space exploration in the last quarter of the 20th century, there are numerous examples of macrothinking and macroengineering by leading spacefaring nations [9]. Often we turn to artists to illustrate our futuristic, visionary thinking, as Exhibit 90 demonstrates.

But, humanity has yet to reveal sufficient macrothinking in terms of offworld *settlement and industrialization*. Here are some seminal illustrations, arranged chronologically, which point toward that direction within the U.S.A. Each shows prospects for long-term payback which would engender broader public and industry support, while gaining investment by entrepreneurs, venture capitalists, and consortia.

- In 1969, Astronauts Neil Armstrong and Buzz Aldrin deployed reflectors in the lunar soil. Ever since scientists have been firing lasers at these same reflectors so as to measure the distance between Earth and the Moon down to 1 millimeter. Why? Such lunar ranging enables them to (a) gain insights into the orbital

Sec. 7.2] **Macrothinking examples from the past** 327

Exhibit 90. Macrothinking: orbiting spaceport. Macrothinking would be necessary to construct an L1 spaceport as a jumping off point for spaceships cycling between Earth and Mars. Transfer vehicles, like those pictured in this painting by artist Robert McCall, would be serviced by astronauts through EVAs. Source: National Commission on Space, *Pioneering the Space Frontier*. New York: Bantam Books, 1986, p. 135.

  mechanics of both bodies; (b) better understand the spiraling of the Moon away from the Earth due to tidal forces (about 4 centimeters per year).
- The 1976 publication of physics professor and macrothinker Gerard K. O'Neill's *The High Frontier: Human Colonies in Space*, and the founding the next year of his Space Studies Institute to promote and research his ideas. In 1977, O'Neill directed a NASA summer study and its proceedings on *Space Resources and Space Settlements* (NASA SP-428, 1979).
- The 1979 General Dynamics Space Systems study on *Lunar Resources for Space Construction* that envisioned putting 1,600 people in space, including up to 400 on the Moon.
- The 1981–1999 conferences on *Mars Underground* for space experts and activists interested in developing Mars as a place for travel, science, and settlement. Co-sponsored by various professional space organizations, it has produced 12 volumes to date on *The Case for Mars*, published by the American Astronautical Society (available from UNIVELT, PO Box 28130, San Diego, CA 92128; www.univelt.com).

- The 1984 formation of The American Society for Macro Engineering and its publication *TASME News & Views* (4701 Kenwick Rd., Baltimore, MD 21210; email: *tasme@ecostructure.com*; *www.tasme.org*).
- The 1984 scholarly summer study and resulting proceedings by NASA with the California Space Institute: *Space Resources* (NASA SP-509, 1992); and with the National Academy of Science: *Lunar Bases and Space Activities of the 21st Century* (Lunar Planetary Institute, 1985).
- The 1986 report and recommendations of the National Space Commission for *Pioneering the Space Frontier*, chaired by macrothinker Dr. Thomas O. Paine
- The 1986–1987 Smithsonian conferences at the National Air and Space Museum, leading to a *Declaration of First Principles for the Governance of Outer Space Societies* (see Appendix A), and publication of George S. Robinson and Harold M. White's *Envoys of Mankind 9*. Washington, D.C.: Smithsonian Institution Press, 1986.
- The 1988 amendment of Rep. George E. Brown, Jr. to NASA's charter through a Space Settlement Act, subsequently passed in part by the 100th U.S. Congress as an attachment to the Agency's appropriations; this legislation calls for strategies and programs that promote colonization and permanent presence aloft.

There is evidence of macrothinking among the many space proposals and plans of the past 25 years, such as

- manifested in Russian/American space synergy with 16 other national sponsors, involving combined missions for construction, maintenance, and operation of the orbiting International Space Station. (If only comparable agreements were formulated for a lunar base, or shared research on utilizing space-based energy; automated and manned Mars exploration!)
- The 1996 first gathering of scientists at the NASA Ames Research Center to plan for the emerging field of *astrobiology*. Their efforts focus on the origins of life, where and how other habitable worlds were formed, how the Earth and its biosphere influence each other over time, whether terrestrial life can be sustained on our planet, and how can the human presence on Mars be expanded?
- The program for Value-Added Transfer of Space Technology to NASA's commercial field and incubation centers under the leadership of macrothinker Dr. George Kozmetsky at the University of Texas' Institute of Constructive Capitalism in Austin. Kozmetsky, a founder of Teledyne-Ryan was convinced that schools of business and engineering have to include education on macrothinking and macroplanning [10].
- The restructuring of the aerospace industry worldwide to changing global market conditions, particularly by mergers and acquisitions, such as in the synergistic consolidation of Lockheed Corporation, purchaser of General Dynamics fighter

airplane division, with Martin Marietta, acquirer of General Electric Aerospace and General Dynamics Space Systems.
- The research in both the public and private sectors for affordable space transportation, ranging from the studies of the Space Propulsion Synergy Teams of three U.S. agencies (NASA, DOD, and DOT), to current investigations for a Crew Exploration Vehicle that will take us back to the Moon. One innovative scheme in this regard has been put forth by Dr. Robert Forward who advocates a spacecraft with a vast "thin sail" that is propelled by pure light: lasers and particle beams. His *Starship*, a lightweight sail, would channel the power of laser light and reflect microwaves for interstellar travel ("Science and Technology: The Crossing," *The Economist*, September 10, 1994, pp. 99–100).
- In October 1998, NASA has launched Deep Space 1 which has already traveled more than 3.5 million miles from the Earth at more than a mile per second. The 1,000-pound spacecraft is testing numerous devices, techniques, and advanced technologies. Managed by JPL, the experimental craft has an unusual solar-electric propulsion system with a small ion-driving engine. It was programmed to fly by the asteroid 9969 Braille as well as periodic comet Borrelly. The vehicle also has a new type of solar power array with panels spanning 38 feet across that produce 20% more power than conventional cells.

In the near future, macrothinking will be further evident when *space-based energy* is utilized. That will occur when the U.S. Departments of Energy and Defense, plus energy corporations or global consortia: (a) fund R&D demonstration power-beaming studies either for solar power satellites, such as proposed by Dr. Peter Glaser, or a lunar solar power system as advocated by Drs. David Criswell and Robert Waldron (see Appendix B) [11]; and (b) the DOE designates one national laboratory as the research center for space solar energy. Exhibit 91 shows an artist's concept of a future moonbase. One aspect of such a base might be the collection of solar power and its transmission back to Earth, thus providing for rising global energy needs.

7.3 MINI CASE STUDIES OF MACROTHINKING

The following cases illustrate contemporary space macrothinking that will impact our future: one from both private/public sector collaboration, another from a space agency in the public sector, and two from non-profit enterprises. If these proposals as envisioned for the 21st century are to be implemented, it would require the practice of macroplanning, macroengineering, and macromanagement. Each demonstrates vision and innovation toward space development, but the latter two failed due to a lack of strategic planning and adequate resources.

Exhibit 91. Macrothinking: space-based energy from the Moon. An artist's depiction of a future major base on the surface of the Moon. One aspect of such a base might be the construction of large solar energy collectors on the lunar surface, to convert the collected solar power into usable energy, and then transmit the energy as low intensity microwave beams to receivers on Earth, thereby helping to satisfy rising global energy needs. The system could be built on the Moon from lunar materials and operated on the Moon and on Earth using existing technologies. Source: Original painting by Dennis M. Davidson for *Space Resources* (NASA SP-509), Washington D.C., U.S. Government Printing office, 1992.

7.3.1 Spaceport case study by Derek Webber[4]

7.3.1.1 Global spaceport planning

We cannot imagine space flights that do not start from a tower of flame at a launch pad. We have become familiar with the sights of Kennedy Space Center with its line of gantries marching along the Atlantic Ocean coastline at Cape Canaveral. But this is about to change. In 2004, the only astronauts to enter space from the United States departed from Mojave Spaceport. Mojave is now managed by the American Air Force. It has no gantries, no high-energy LOX and propellant bunkers, or block-

---

[4] Derek Webber, Director, Spaceport Associates (5909 Rolston Road, Bethesda, MD 20817; *www.SpaceportAssociates.com*). For further information, see About the contributors section (p. xxxii).

houses. It did not even have a launch pad! Mojave represents the prototype of a new kind of spaceport that will take over from the earlier types as we look forward to the next stages of humanity's journey into space. Let us consider why we must strategically plan for these changes in the way spaceports are designed, built, and operated.

A. Geographical factors

Underneath the trajectory of a rocket heading into space is a ground track where you do not want to have people. They would be at risk from debris resulting from a possible explosion caused by the typically high-energy oxidizer and propellants onboard the rocket. The best zones on the Earth's surface which meet this objective are deserts and oceans. During the Cold War, the Soviet Union developed a series of desert spaceports and the U.S.A. used a combination of desert and coastal spaceports. The main driving and organizing force behind the design, building, and operation of these spaceports was the military.

B. Early launch sites

The early coastal launch sites of the U.S.A. were at Wallops and Cape Canaveral (KSC) on the East Coast for equatorial launches; and Vandenberg on the West Coast seven years later, firing southerly for polar launches. The United States also had a desert site at White Sands, New Mexico, for early suborbital northerly firings. The former Soviet Union had desert sites at Kapustin-Yar, at Baikonour, and ten years later at Plesetsk. The only other active launch sites at that time were two contrasting desert sites: one in the cold Northern desert of Churchill in Canada, used for sounding rockets, and the other being in the decidedly hot desert of Woomera, Australia, where the British tested their Blue Streak ICBM in northerly suborbital firings.

C. Present available global spaceports

Since that Cold War era, much has changed. There are now some 16 countries with launch sites, many with more than one spaceport, so that there are currently about 35 active spaceports in the world [see Exhibit 92 developed by the contributor of this case study, Derek Webber]. Although there has been some gradual adaptation toward the needs of commercial users of launch services, nevertheless generally the operations still take place within a military operating regime [12].

D. Predicting future launch demand

The first step in the planning process for new spaceports is to look at demand for their services. How many orbital launches a year are provided from these existing 35 global spaceports? What kinds of customers and payloads are served? And how do we see this changing over the next few decades? The historical picture has evidenced a rather severe decline in global launches. From 1965 through 1985, the global launch rate for orbital missions was around 120 per year. Now it hovers around 60 launches per year, and this figure includes all countries, and all military, all civilian, and all commercial launches.

Exhibit 92. LAUNCH DETAILS.

Country	Spaceport	Degrees latitude	Easterly azimuths	N/S azimuths	Years operational	Mission types*	Launch vehicles
U.S.A.	KSC	28.5°N	35–120	None	58	L, G, H	Atlas, Delta, Titan, STS
	Edwards	35.0°N	N/A	N/A	61	A	Pegasus, X-Planes
	Vandenberg	34.7°N	None	140–201	51	L, P, A	Atlas, Delta, Titan, Pegasus, Taurus
	Wallops	37.5°N	36–60	None	63	L, S, A	Black Brant, Pegasus, Scout
	White Sands	32.5°N	None	Yes (N/K)	63	S	Starchaser
	California Spaceport	34.7°N	None	147–220	13	S, P	Taurus, Minotaur
	Kodiak	57.0°N	None	125–235	10	S, P, L, M	Athena
	Florida SA	28.5°N	35–120	None	51	L, G	Athena, Microstar
	MARS	37.5°N	38–60	None	11	S, L, A	Minotaur
	Mojave	35.0°N	N/A	N/A	3	H	SpaceShipOne, X-Cor
	Oklahoma	35.5°N	N/A	N/A	1		Rocketplane
Australia	Woomera	31.0°S	None	350–155	9	P, S	Blue Streak, Skylark, Kistler
Brazil	Alcantara	2.2°S	10–100	10–100	15	S, L, G	Sonda, VLS, Cyclone 4
	Barr do Inf	5.5°S	10–100	10–100	40	S	Sonda, Nike-Apache
Canada	Churchill	57.7°N	None	Yes (N/K)	48	S, P	Aerobee, Nike, Black Brant
China	Jiuquan	40.6°N	135–153	135–153	41	L, P, H	Long March 1, 2 and CZ2F
	Taijuan	37.8°N	90–190	90–190	17	L, P	Long March 2, 4
	Xichang	28.2°N	94–105	N/K	27	G	Long March 2, 3
	Hainan Island	18.0°N	N/A	N/A	N/K	S	Sounding rockets
Fr. Guiana	Kourou	5.2°N	350–933	50–933	7	S, N, G, H, P	Ariane, Soyuz, Vega, Cyclone

India	Sriharikota	13.7°N	18–50	18–50	25	L, G, P	PSLV, GSLV
International	Odyssey	0.0°N	Any	Any	10	G, P	Zenit 3SL
Israel	Palmachim	32.0°N	Yes (N/K)	None	N/K	L	Shavit
Japan	Kagoshima	31.2°N	31–100	None	43	L	M5
	Tanegashima	0.4°N	Yes (N/K)	None	38	L, G	H2
	Takesaki	3.0°N	N/A	N/A	37	S	Sounding rockets
Kazakhstan	Baikonour	45.6°N	25–62	193	48	L, M, G, H	Cosmos, Dnepr, Rocketplane, Soyuz, Proton
Marshall Is.	Kwajalein	8.0°N	Yes (N/K)	N/K	N/K	G, L, A, S	Falcon, Pegasus
Norway	Andøya	69.3°N	N/A	N/A	43	S	Skylark, Black Brant
Pakistan	Suparco	25.0°N	None	220–310	45	S	N/K
Russia	Kapustin Yar	48.4°N	51–107	None	48	S, L, P	Cosmos
	Plesetsk	62.8°N	90	14	39	M, L, P	Cosmos, Molniya, Rocketplane, Soyuz, Angara
	Novosibirsk	54.0°N	N/K	N/K	N/K	L	Shtil
Sweden	Kiruna	68.0°N	N/A	N/A	41	S	Skylark, Black Brant
Taiwan	Ping Tung	22.5°N	N/K	N/K	N/K	N/K	N/K

*G = GEO; L = LEO; P = polar; MARS = Mid-Atlantic Regional Spaceport, M = Molniya; S = suborbital; H = human; A = air-launched.

And what of the future? The ASCENT Study, organized by NASA's Marshall Space Flight Center and issued in 2003, made it clear that the 20-year forecast would remain at about the same annual level of launches for these same payload types (i.e., only between about 50 to 70 launches per year in total globally) [13]. The study, however, identified the dramatic impact that would be made by the emergence of a successful space tourism industry, with as many as 50 additional launches a year for orbital space tourism, even at today's $20-million prices. And clearly, the number would be much higher at lower ticket prices.

These are transforming numbers compared with today's annual launch rates worldwide. These are the kinds of numbers that lead to a new kind of spaceflight. They indicate the beginnings of airline-like operations, and better operability and reliability, and lower costs per pound for payload into orbit. Starting in low Earth orbit, space tourism can be seen as an enabling and transforming technology and business opportunity.

E. Emerging space tourism

Chapter 6 had a section on the emergence of the space tourism industry. Briefly, following the success of Burt Rutan with his SpaceShipOne winning the X-Prize in 2004, it has become a certainty that at least suborbital space tourism will soon be operational. Sir Richard Branson of Virgin Galactic in Mojave, and the Rocketplane team flying from Oklahoma, will probably provide the first space tourism offerings. A survey of millionaires indicated that up to 15,000 passengers a year are interested in suborbital spaceflight at prices of around $100,000 [13].

What are the prospects for orbital space tourism? The Russians already offer the service via Soyuz from Baikonour. It may be safely assumed that the Chinese will eventually offer tourist rides in their Shenzhou spacecraft, once they have obtained some more operating experience. In the U.S.A., other entrepreneurial companies have announced their intention to eventually provide an orbital space tourism experience, among them the companies being run by the billionaires Jeff Bezos and Elton Musk.

How does all of this affect the spaceport scene? If all of the growth in the launch industry is going to come from the tourism sector, then attention needs to be given to the needs of this new marketplace, and its implications for the design, build, and operations of the new spaceports. Mojave, and subsequently Oklahoma, have become the first of a "new breed" of licensed U.S. non-Federal spaceports. Throughout America, various states are exploring the potential economic benefits of creating spaceports to serve the new space tourism markets, and some of them have begun the process of obtaining FAA operating licenses. These states are anticipating potential benefits in terms of employment, building contracts, tax base, and terrestrial tourism as a result.

SpaceShipTwo flight testing and early passenger operations by Virgin Galactic will take place at Mojave. The *Rocketplane* venture is doing its work based at the Oklahoma spaceport (see Exhibit 151). In Upham, New Mexico, Spaceport America is being developed, and is the venue for a series of annual X-Prize Cup races into space. Virgin Galactic plans to move its base of operations there from Mojave once the custom-made spaceport is ready. Texas is proposing several different possible spaceports, and the following states propose one each: Nevada, Alabama, Washington, Wisconsin, Utah. In all, about a dozen U.S. spaceports are currently in various stages of the process of evaluation and regulatory approval, and others are planned elsewhere in the world (e.g., Singapore and Dubai).

F. Spaceport transformation

For a spaceport to succeed as a provider of space tourism opportunities, the management must focus on satisfying the needs of the new sector. These needs can be almost directly in contrast to the needs of existing users of traditional spaceports, which have served us in the first half century of human spaceflight. It may well prove to be the case, because of this, that existing spaceports will be constrained to serve only their traditional markets, with their limited growth potential, while the new high-growth sector in this century is likely to be innovative, specially designed, *space tourism spaceports*.

So, what are the kinds of transformation that the new spaceports need to embody? The most obvious, and in a way the most emblematic, is the fact that existing Federal spaceports are generally military establishments. Among their payloads are military cargoes and traditional rocket fuels that are highly dangerous mixes of propellants and oxidants, from which the public needs to be protected. By contrast, the SpaceShipOne flights of 2004 employed a new kind of hybrid rocket engine that used HTBP (an ingredient of tire rubber) and nitrous oxide (laughing gas). This mixture of fuel and oxidizer is so inherently safe that it presented no danger to the crowds who waved at the passing carrier aircraft and suspended space plane as it went by *en route* to space.

How do we plan for this new kind of customer-of-the-future spaceport? For space tourism to succeed, the public needs to be encouraged—not discouraged—from entering a spaceport to watch launches. Open access is an important feature, even in such traditionally sensitive areas as launch control and training facilities. Some space tourism flights will take off horizontally, and some vertically, and there will need to be facilities for each type of launch. The Webber/Reifert *Adventures Survey* offers some insight into the preferences of future tourists for each of these approaches [14]. There needs to be new facilities for training and medical care of space tourists. Entertainment such as IMAX movie theaters and space theme parks will be important, and central to the experience, especially for the friends and families of tourists who are going off into space. New residential facilities and restaurants and shops will need to be built for staff, tourists, families, and terrestrial tourism customers. This will decidedly not be your father's spaceport experience! Exhibit 93 provides a schematic for a potential Dubai spaceport in the U.A.E. as proposed by Space Adventures.

It is self-evident that the design, construction, and management of such spaceports cannot follow the former military paradigm. The entrepreneurs who are creating the space tourism industry are very focused on the needs of paying customers, and their families and friends, so therefore insist that the spaceports that they use as their bases reflect this perspective.

336 Macrothinking in strategic space planning [Ch. 7

Exhibit 93. Tourist spaceport of the future. Artist rendering of what a potential spaceport for tourism purposes might look like. This is an example of macrothinking and macroplanning. Source: Space Adventures.

In the United States, for example, spaceport management must deal with a mixture of funding from local, State, Federal, and commercial sources. Nowadays, the Federal government interface is much less likely to be the Department of Defense, and is more likely to be the Department of Transport or Commerce. The new breed of managers must be able to comply with the developing regulatory requirements that the FAA is providing for the new industry. These new regulations include elements that describe the spaceport itself (e.g., its ability to meet environmental and public safety guidelines); the design of the vehicles; the training of the spacefarer crews and passengers; and the medical guidelines for space tourism (see Appendix E).

G. Conclusions on the future of spaceports

Space tourism is likely to transform spaceflight and spaceports to a 21st-century model. It probably will recreate the global aerospace industry with its renewed growth prospects and a reinvigorated future for the younger generation. The spaceport is an essential element of the whole space business experience, but must undergo a dramatic transformation in order to enable this new industry to develop. It remains to be seen how many of the traditional spaceports will be able to adapt to this changing role, or how many totally new spaceports will emerge that will provide the space tourists of the future the kind of experience that they will demand for the price of their ticket into space. Obviously, creating spaceports is a costly and risky business that requires macrothinking, macroplanning, and macromanagement!

7.3.2 The European Moon program

The European Space Agency (ESA) has many studies under way that also demonstrate macrothinking, such as a four-phase lunar program early in this century. Roger M. Bonnet, when director of scientific programs, indicated that Europeans envision the Moon as a natural space platform or station (38 million square kilometers) which only requires infrastructure to exploit its scientific potential and its resources. Although aware of problems inherent in a lunar base, ESA sees it as a macroarena for developing high technology and undertaking challenging projects in exobiology, radiation biology, ecology, and human physiology. Further, ESA planners envision this Earth satellite as a test bed for learning more about our universe by astronomical observations from the lunar surface, and by developing skills there for interplanetary travel and exploration. As a part of the European Union, the Agency is well aware of a range of lunar resources to be utilized from energy and water to mining and manufacturing.

Such lunar development provides European member states with a common long-term vision and an independent peaceful goal for scientific and cultural knowledge that have political and sociological benefits. This European Moon program has four steps that build on initial experience through gradual progress and interrelated missions [15]:

(a) *Unmanned, automated investigations*, including possibly MORO (Moon Orbiting Observatory) to characterize the lunar surface in terms of geology, morphology, geochemistry, mineralogy, topography, and heat flow, as well as to study the interior in terms of geodesy and gravimetry (launched on Ariane 5).
(b) *Permanent robotic presences*, including possibly LEDA (Lunar European Demonstration Approach), a lunar lander carrying a payload of robotic and scientific instruments (a rover, soil-processing facility, and a robotic arm) which will land close to the lunar south pole.
(c) *Robotic deployment of scientific instruments* to gather information for establishing a human outpost; some instruments will determine micrometeorite flux, seismic noise, soil mechanics, thermal properties of the surface, soil characterization, stereo imaging of the surface, measurements (gases, dust particles, sky background, and radiation doses), and others will search for water.
(d) *Establishing a human outpost on the Moon*, including building facilities and an oxygen production plant (see Exhibit 94).

In a letter to the author (January 7, 1994), Dr. Bonnet indicated that a second-phase report is under way to define a European global view for both lunar exploration and exploitation. However, he wrote that ESA sees lunar industrialization as a longer term goal which they are not addressing at this stage. This writer

Exhibit 94. ESA utilization of lunar resources. The European Space Agency envision humans at a lunar outpost engaged in power generation by thermonuclear fusion. Source: ESA-SP 1150, *Mission to the Moon*. Paris: ESA, 1992, p. 130.

believes that true *macrothinking* would necessitate that aspect now if European taxpayers and investors are to support such a large-scale, long-term lunar enterprise: on that continent, people expect to see some return on the huge expenditures required to carry out these lunar goals!

Wisely, the ESA proposes international involvement, a worldwide initiative aimed at creating consensus among scientists, politicians, and space organizations. The assumption here is that the Moon is already part of human territory, and should be made accessible to all humanity through telepresence and virtual reality. Although Europeans have obviously done some impressive macroplanning, they seem slow on implementation, something NASA is now engaged in with the U.S. National Policy, *Vision for Space Exploration*. While the American VSE program moves ahead to return permanently to the Moon and establish a lunar outpost by 2020, the Europeans are considering joining in this venture under some type of global agreement. Such an approach is necessary for a lunar macroproject that should also include China, India, and Japan, all of whom have near-term plans for Moon missions (for further information on ESA, see Case Study 1.4).

For further information about their current and future planning for planetary exploration, astrophysics, and heliophysics missions, visit *www.esa.int.esaSC/index.html* and for the Moon/Mars Aurora program visit *www.esa.int/Specials/Aurora/index/html*

7.4 MACROTHINKING PERILS

There are many challenges and hurdles to be overcome before macrothinking can be translated into actions and accomplishments that lead to the achievement of big goals. The next example tells the story of macrothinking that was ahead of its time. Following this are two more cases of macrothinking that were too grandiose without the needed strategic planning and adequate resources.

7.4.1 Orion: the original proposal for a nuclear-powered rocket

At the beginning of the Space Age in the late 1950s, a group of innovative professionals did some commendable macrothinking and macroplanning about utilizing nuclear power for space travel [16]. *Project Orion* brought together a group of distinguished scientists and engineers under the leadership of Theodore Taylor, Los Alamos veteran then at General Atomics, and Freeman Dyson, a physicist at Princeton University. Their team proposed to create a massive 40-person rocket, which would be nuclear-powered and capable of missions to Mars and Saturn within a decade. To prove the technical challenges were surmountable, they built several demonstration models to test the propulsion concept: one succeeded in a 100-meter stable flight, confirming the possibilities in the proposal.

But, the thinking was too advanced, and political forces brought it to a halt. Struggles within the Presidential administration, and between Congress, NASA, and the military doomed this macroproject. It could not compete with Kennedy's goal to put a man on the Moon, or go forward because of the signing in 1963 of a nuclear test ban treaty. Today, General Atomics, a company founded by Frederick de Hoffman who championed Orion, is a subsidiary of General Dynamics, so this plan is still classified. Someday we may yet build interplanetary spaceships that are nuclear-powered, just like those atomic submarines that revolutionized ocean travel! ... The next two illustrations are also indicative of interesting marcrothinking, both of which lacked effective strategic planning and sufficient resources, so never achieved their intriguing goals.

7.4.2 United Societies in Space, Inc.

In harmony with the U.N. Outer Space Treaty (1967), this private initiative sought to confirm that outer space is indeed the "common heritage of humankind". Under the leadership of Colorado attorney, Declan J. O'Donnell, its primary goal is to settle

Exhibit 95. Space peace symbol. The logo of United Societies in Space, Inc. (*www.usis.org*). The dove, symbol of peace, flies over our Solar System. Source: USIS designer, Royal Publishing Company of Denver, Colorado.

and populate the high frontier as an "interplanetary commons" for the human species. For that purpose, *United Societies in Space (USIS)*, was founded as a non-profit corporation recognized by the U.S. Internal Revenue Service. USIS also fosters *space* legal, financial, and governance systems which encourage investment, industrialization, and settlement aloft. USIS operates on the concept of *metaspace*, which views space as humanity's common territory. It furthers individual and institutional efforts promoting freedom and peaceful enterprise throughout our Solar System. Its motto is *space is a place for synergy*, its official journal is entitled *Space Governance* (your author was the founding editor), whose content emphasized space governance, law, and treaties. A Council of Regents was established for the future space metanation that included many of the contributors to this book. USIS also sponsored an annual Countdown Conference, a Global Space Essay Contest to encourage solutions to these challenges, and the Buzz Aldrin Space Library. In retrospect, USIS's greatest contribution to macrothinking may be its plans for lunar, Mars, and orbital economic development authorities. Today, USIS has diminished to a special interest chapter within the National Space Society (*www.nss.org*). However, the organization's founders did demonstrate macrothinking in these ways [17]:

- *Space metanation*. They proposed a strategy to establish this entity on behalf of all humanity, possibly under the U.N. Charter provision of the Trusteeship Council for New Territories. With the existing spacefaring nations as trustees, this innovative proposal would then have the legal, financial, and governance capability to further exploration, utilization, and colonization of the space frontier. With the adoption of a new Declaration of Interdependence between *earthkind* and *spacekind*, as well as a Constitution for the space nation, a transportation system and infrastructure can be undertaken to promote offworld development.
- *Space authorities*. These are seen as the mechanism by which a metanation can encourage orbital investment and development. These quasi-governmental bodies could coordinate transnational endeavors for specific bodies, such as

the Moon or Mars. Such multi-jurisdictional authorities would be modeled on the legislation and administration of the Tennessee Valley Authority and the New York Port Authority. Initially, the plan calls for creation within ten years of a *Lunar Economic Development Authority (LEDA)* to issue bonds, leases, and insurance, as well as provide public information, regulation, and management for development projects and industrial parks on the Moon, or its vicinity (see Section 10.2.2). LEDA might be founded privately, under an Act of the U.S. Congress, or under the United Nations, or a combination thereof. Certainly, it could provide the legal framework for the projects of the above European Moon program. In 1995, the second National Space Society's *International Lunar Exploration Conference* endorsed the formation of an Advisory Consortium to the LEDA. Subsequently, O'Donnell, a tax/space lawyer, prepared LEDA articles of incorporation for the state of Colorado. In 1996 while preparing the *Omnibus Space Commercialization* bill for the U.S. House of Representatives, Congressman Robert Walker, chairman of the House Science Committee, requested a proposal from USIS on the LEDA strategy. The organization responded with these recommendations for national legislation:

- ○ *Lunar venue*, extending the legal scope of space resource management to the Moon based on the precedent of Intelsat, as a means for worldwide participation.
- ○ *Triple tax exemption*, legislative provision for three income tax exemptions to investors in Authority bonds intended to promote space commerce and lunar infrastructure.
- ○ *International participation*, provisions for both private and public-sector involvement (e.g., advisory consortia, nation states, and national space agencies as reported in *Space Governance*, **3**(1), June 1996, the USIS journal).

• *Space macroprojects* could be underwritten through creative funding alternatives, such as the above bonds, or through a metabank that would issue its own currency. In addition to funding lower cost, mass access to space, beginning with a single-stage-to-orbit vehicle, other relevant income-producing proposals might be sponsored, such as
  (a) space-based energy via power-beaming and fusion energy;
  (b) lunar industrial park and base;
  (c) lunar campus of the International Space University;
  (d) space transportation system capable of interplanetary travel, beginning with Mars;
  (e) research and development for constructing orbital "city ships" comparable with Gerard O'Neill's "Stanford torus".

In these ways, the metanation and the space authorities would contract with global consortia of entrepreneurs, agencies, corporations, and universities to advance free enterprise aloft, with the purpose of facilitating mass human migration and colonization of the Cosmos (for further information consult *www.usis.org*).

7.4.3 First Millennial Foundation

Macrothinking at its best is evident in the work of another Colorado visionary, Marshall T. Savage, and his colleagues. This non-profit foundation believes that there is a cosmic imperative to colonize space which will lead to a dramatic explosion in economic growth and creation of unlimited sources of new wealth. Founded in 1994, the *First Millennial Foundation (FMF)* has core members who have evolved a step-by-step strategy for this purpose, publishing *First Foundation News* and *New Millennium Magazine*, as well as holding annual core conclaves [18]. Among their programs is the *Millennial Project*, a remarkable 515-page, well-documented volume which spells out in detail the FMF strategy for eight stages of colonizing the Galaxy during this millennium. These proposals include

(a) *Aquarius*, a prototype space colony under the sea;
(b) *Bi-frost*, a 21st-century space launch system;
(c) *Asgard*, the first space colony in orbit for 100,000 residents;
(d) *Avallon*, ecospheres on the Moon;
(e) *Elysium*, the terraforming of Mars;
(f) *Solaria*, the colonizing of the Solar System;
(g) *Galactia*, the colonizing of the Galaxy;
(h) *Foundation*, the Millennial Movement which has already been started.

The value of organizations like United Societies in Space and the First Millennial Foundation is that they "push the envelope" in our macrothinking. Although their ideas are not presently implemented fully, these visionaries plant seeds that someday in the future may blossom, though not as their founders may have envisioned. Long before the Shuttle was built, macrothinkers imagined spacecraft that could fly to orbit and land back on Earth, just like Burt Rutan is doing today. Similarly, advanced thinkers like Wernher von Braun and Gerard O'Neill designed space platforms and stations decades before the International Space Station was actually built in orbit. Such creative and futuristic thinking now enables astronauts to live and work in microgravity, as shown in Exhibit 96.

7.5 CURRENT MACROTHINKING ILLUSTRATIONS

The media has been reporting some "big space thinking" since the turn of this century. Some of it is good, and can be placed within the context of both macrothinking and macroplanning, while others are lacking in sufficient research and resources. Perhaps the following excerpts will help our readers to do a little "mindstretching" on this chapter's themes:

- **NASA and Google** have announced a collaboration to develop open source software for the processing of Earth remote-sensing data and planetary data sets. Essentially, the endeavor will meld NASA data sets with Google's

Sec. 7.5] Current macrothinking illustrations 343

Exhibit 96. Living and working in microgravity. On July 9, 2006, the European Space Agency astronaut from Germany, Thomas Reiter, and NASA mission specialist, Stephanie Walker, of the U.S.A., go over their checklists in the Destiny laboratory on the International Space Station (Expedition ISS013-E-49432). Source: NASA.

geolocation products, and may revolutionize science data access and presentation. One aim is to produce software that will enable viewers to experience a virtual flyover of the Moon, something the Lunar Explorers company has already been working on (email: *many@lunarexplorers.com*). (Source: Google Press Release, December 2006. The moral here is that before entering into a macroproject, check out what others in the field are doing.)

- **Sea Launch** utilized a converted oil-drilling platform, Odyssey, to launch a television broadcasting satellite on a 200-foot rocket. Odyssey is one of the world's largest semi-submersible, self-propelled vehicles, and sailed from California to a remote Pacific location for liftoff on September 25, 1999. This site 1,400 miles south of Hawaii on the Earth's equator. There are several benefits of an equatorial launch platform: first, the rotational speed of the Earth is greatest at the equator, providing an extra launch boost. Even more important, the need for a plane change to the zero degree inclination of geostationary orbit is eliminated, providing a major additional launch boost. Any orbital inclination can be reached, combining in one launch site the inclinations attainable from both Cape Canaveral and Vandenberg, for example. Sea Launch which owns the Odyssey Launch Platform is part of an international consortium which also includes Boeing Commercial Space Company of the U.S.A., RSC Energia of the Russian Federation, Kvaerner Maritime of Sweden, and KB Yuzhnoye/PO Yuzhmash of the Ukraine. The ship *Sea Launch Commander* is a customized mission control center with many amenities for a crew of 310. The launch system is fully automated, so the Odyssey crew can cross over to Commander; the kerosene/liquid fueled Zenit 3SL rocket can carry commercial payloads up to 6,000 kg to geostationary transfer orbit. Sea Launch plans eight launches per year, and has 19 orders for its services. The biggest problems faced by the company are not technological—but cultural and political. Despite a rocket that exploded on the launch pad in January 2007, the company is still forging ahead with its macroproject. (Source: Dawn Stover, *Popular Science*, December 1999.)

- **Utilizing near-Earth resources** have only recently attracted serious research, especially the economic tools for this analysis, such as offworld mineral assets. As humanity advances into outer space, more creativity in engineering and science is necessary to transcend the limits of a one-planet biosphere, so as to harvest orbital resources. Further, spacecraft built on the Moon or near asteroids can help us overcome the current constraints in space travel and payloads. For this to happen, major private capital investment is required, but that usually follows demand. Space planners need to go beyond traditional assessment of costs, market conditions, and risk factors. Dr. Brad Blair of the Colorado School of Mines is investigating the evolving economic tools needed for analysis of financial prospects for using space resources. (Source: Brad R. Blair, Ph.D., *The Evolution of Demand for Space Mineral Resources*, EB 695 Research Paper, April 29, 1999 (email: *bblair@mines.edu*).)
- **Space tourism enterprises** were discussed in Chapter 6 with such endeavors as Sir Richard Branson's Virgin Galactic and its "Spaceport America" in Las Cruces, New Mexico. This company also has an agreement with the Swedish government for suborbital flight of its SpaceShipTwo from Esrange's Kiruna launch site; Esrange proposes launching a sounding rocket in March 2008 so SpaceShipTwo passengers can view the Northern Lights ... Today the public space travel business is further expanding with the advent of Blue Origin, LLC, which plans to use a West Texas site for its New Shepard Reusable Launch System. Founded by billionaire Jeff Bezos who started Amazon.com, Blue Origin proposes to launch its RLVs on a suborbital, ballistic trajectory in excess of 325,000 feet aloft at the privately owned spaceport in Culberston County, 25 miles north of Van Horn, Texas. This macroproject calls for a vehicle-processing center, a launch complex, landing, and recovery area, and other amenities. The intention is to build and test New Shepard RLVs before starting commercial operations. The system would have a propulsion module with a crew capsule on top for three or more spaceflight participants; the spaceship is patterned after the Delta Clipper experiments of DOD and NASA. The reusable propulsion module with its own avionics would operate autonomously under control of onboard

Sec. 7.5] Current macrothinking illustrations 345

computers. The fuel would be high-test peroxide and a kerosene propellant. The spacecraft would ascend vertically, disengage the crew capsule into orbit, and then steer back to the landing pad. Following construction of the necessary infrastructure, including a manufacturing facility, and completion of testing, the strategy calls for the macroproject to be operational by 2010, with 52 launches per year for paying passengers. (Source: Leonard David. *Tourism Update: Jeff Bezos' Spaceship Plans Revealed* (www.space.com/businesstechnology, July 5, 2006). For information, visit www.livescience.com/blogs/author/leonarddavid)

- **California Spaceport Authority** has issued *A California Space Enterprise Strategic Plan 2007–2010*. CASA in Santa Maria is a non-profit corporation representing the state's commercial, civil, and national interests in defense and homeland security. Its activities already raise $22 billion in space revenues, which is 19% of the world's space market, not including another $52 billion in classified programs. Through its California Space Economic Advisory Council, the stakeholders in this macroplanning include 200 individuals representing 120 state organizations. SEAC offers 100 specific recommendations to expand space enterprise in this state, including policy/legislative monitoring of the aging aerospace workforce and the education of replacements. It established four strategic goals in these areas: space business development, retention, and growth; science and research technology development; education and workforce development; public and policymaker awareness efforts. One of its aims is to ensure that California, where transformational aerospace was born and thrives, remains in the forefront of the space enterprise/tourism business. (Source: Leonard David, *Report: Strategic Space Plan for California* (www.space.com/news, December 2, 2006).)

- **NASA Institute for Advanced Concepts** seeks to identify innovative undergraduates who have shown exceptional creativity and promise for success in building visions of the future. Thus, in conjunction with its affiliate, the Universities Space Research Association, the Institute inaugurated the *NIAC Student Fellows Prize* in 2005. To attract and facilitate high-performing applicants. individual students or groups may develop proposals around NIAC topics related to projects 10–40 years into the future. The $9,000 prize for winners is intended to foster student mentoring, networking, and creativity, especially through responsibility in project management. Applicants must be in an institution of U.S. higher education, apply no later than their junior year, and be able to work in the United States (see www.niac.usra.edu/students/winners.html). The annual award to promote marcrothinking usually has an April submission deadline for September funding. (Source: Dr. Robert Cassanova, Director, and Dr. Diana Jennings, Associate Director, NASA Institute for Advanced Concepts, September 16, 2007 (visit www.niac.usra.edu/student/call.html or email niacstudents@niac.usra.edu).)

- **Challenger Center for Space Science Education** was founded in April 1986 by the grieving families of the 51-L Shuttle crew, three months after the loss of their beloved seven astronauts (see Dedication, p. xii). With unusual macrothinking, they wanted this Center not only to be a memorial—but an educational science center that would continue *Challenger*'s teaching mission. Twenty-two years later there are 61 Challenger Learning Centers across the United States, Canada, and the U.K. Furthermore, over 400,000 students, teachers, and members of the general public are served yearly with their programs which follow National Science Education Standards. Their projects are sponsored by aerospace corporations, such as *Soar* which is a clearinghouse for space-teaching activities and professional development of teachers in science, mathematics, and technologies. These learning centers "keep the space dream alive" among youth and educators, especially through simulated Moon, Mars, and Comet mission scenarios that also include studies of planet Earth. These undertakings foster attitudes of inspiring, exploring, and learning among the next generation of space workers and travelers. (Source: William A. Gutsch, Ph.D., CEO of CCSSE, *Challenger Log*, 2006. For information, call 888/683-9740, email *dcoates@challenger.org* or visit *www.challenger.org*)

- **International Space University** is a constant source of futuristic macrothinking and macroplanning. Its mission is to provide graduate-level training to future leaders in the emerging global space community. Its Masters in Space Studies Program may be pursued at its main campus in Strasbourg, France, or in Summer Session held at various universities abroad. ISU envisions someday offering its education in space, possibly on the Moon. Its multidisciplinary curriculum covers space science and engineering, systems engineering, space policy and law, space business and management, space and society. Publications result from student research on a wide range of topics from a lunar base to multicultural space missions. Since its founding in 1987, ISU has graduated over 2,300 students from 87 countries. Along with a faculty and lecturers, these alumni form an effective network for individual career connections, professional development, and a powerful lobby for global space cooperation. ISU graduates are on the leading edge of the $90 billion worldwide space business. Besides tuition and scholarships, ISU's annual budget of some $5 million is partially from a mix of some 19 national space agencies and corporate donors, like Boeing and Lockheed Martin [19]. The co-founders of ISU were three macrothinking MIT and Harvard Medical School students, Peter H. Diamandis, Todd B. Hawley, and Robert D. Richards who issued an inspiring credo in 1995 on the university's purpose. For information, visit *www.isu.edu*

- **International Lunar Observatory Association** is a macroproject under way as a result of the leadership of Steve Durst, founder of Space Age Publishing. ILOA is positioning itself to be the first scientific, low-cost commercial enterprise for interplanetary astronomy and communications. The Hawaii-based non-profit organization aims to construct a multifunctional observatory with telescopes on the Moon. The site proposed is at Malapert Mountain, near the lunar south pole. In addition to that facility, the strategy calls for a power station using solar energy, as well as a commercial communication center with broadcasting, imaging, and advertising capacities. The latter would transmit astrophysical data, lunar video and imaging to varied institutions back on Earth. Follow-on human service missions are planned to enhance robotic capabilities there. To achieve this, beginning in 2009 the Association is seeking endorsement from the major terrestrial observatories, such as the Optical Hawaiian Array for Nanoradian Astronomy on Mauna Kea. The sponsors seek to advance Hawaii as their Earthbase for support services, putting that state on the leading edge of 21st-century astronomy. Astronomers and investors in Canada, China, Europe, India, Japan, Russia, and the U.S.A. met in 2007 at Hawaii and Hyderabad to promote the venture. For information, visit *www.ILOA.org* or email *info@iloa.org* (see Appendix E).

- **Lunar agricultural engineers** may be operating on the Moon within two decades or less [20]. Through their corporate and professional association research, especially at the Kennedy Space Center, these food specialists are thinking big and ahead. As part of the life support systems for a Moonbase, they contribute to the survival of future lunar dwellers by ensuring sufficient food and water. Their participation in NASA's CELSS program fosters investigations using plants as the means to remove carbon dioxide, while providing food, water, and oxygen. Their experiments have already been successful on spacecraft and stations through controlled environment agriculture. Now they are planning to grow plants continually on the Moon, either hydroponically with recirculating nutrient solution, or in mixtures of lunar soil. Light pipes of tiny fiber optics supplement the light of the lunar day within biomass production chambers. Astroagriculture staggers crop harvesting automatically and benefits from sensors and other instrumentation. Robots will be doing the harvesting within the lunar greenhouses, as well as food packaging. Genetic engineering protects the varied crops from solar radiation, using small digesters to break down cellulose. Food and water are recycled with plants acting as filters for the later, plus as a backup source of oxygen. Automated computer networks throughout the base will enable these systems to function effectively. For information, visit *www.A-SAE.org*

- **Futuristic space nurses**, already members of the Space Nursing Society, are planning a vital role for the nursing profession on the high frontier (see Appendix D) [21]. Following the pioneering space-nursing vision of the late Dr. Martha E. Rogers, some SNS founders and others have formed *Space Medicine Associates*, an entrepreneurial private corporation offering space-related health care services, aimed initially at space tourists. The SMA mission statement is

    "to provide multidisciplinary space medicine and bioastronautic consultation, training, and oversight in support of all aspects of world travel, serving both the commercial launch industry and the personal space traveler. Space Medicine Associates endeavors to safely open human spaceflight to as many individuals as possible."

    Its macrothinking founders include Eleanor O'Rangers, Pharm.D.; Linda Plush, MSN, FNP; Petra Illeg, MD; and Jim Logan, MD, MS. If only other professionals would follow this example and seriously plan for their astrocareers! For information, visit *www.spacemedicineassociates.com* or email *plushsn@IX.netcom.com* (see Appendix D).

- **Micro/nanotechnology (MNT)** is tiny in size, but big in thinking! MNT represents the confluence of diverse disciplines in proprietary partnerships that integrate mass production of microinstruments. This innovation has wide applications from medicine and communications, to transportation and spacecraft. Prognostications have already been realized for the use of this technology in spacecraft inertial navigation units and propulsion systems. MNT with its smart sensors has great implications for offworld construction and low-power wireless telemetry, utilizing condition-based maintenance (CBM). The Eshellon company in Palo Alto, California, has a neural network transceiver chip for MNT/CBM applications, such as lunar robotic workers ... Another example of such innovations is *SuperBot*, a modular, multifunctional, and reconfigurable robot. Designed by Dr. Wei-Men Shen and his team at the University of Southern California, the project has received $8 million in grants from the National Science Foundation and NASA/DARPA. For affordable space exploration, this project seeks to develop small robots capable of diverse functionalities, such as their SuperBot, a robot that can act in its own right with microprocessors, sensors, communications, three degrees of freedom, and six connecting interfaces to other modules (images). For launch and landing, the robotic package takes a minimum amount of space, but when landed on the Moon, SuperBot can unpack itself, taking a wide variety of forms, such as an exploration rover. For space agencies and private enterprises the advantage is cost reduction through reusable parts, enhancement of reliability and safety, and requiring less active direction from astronauts or ground crews. In 2008, a hundred self-replicating modular robots will be tested in desert conditions. Each ten-centimeter cube is an autonomous unit with microprocessor and instructions on how to link up with the bot, creating a nanotechnology swarm of workers that is capable of altering shape to move smoothly over rocky lunar terrains ... Further, automated micromachines have a big future in space macroprojects! Thus, over ten years ago, Japan established a government-financed Micromachines Center, in anticipation of a $32 billion market by 2010 that emulates the existing $100 million plus market in microelectronics!

7.6 FUTURE MACROTHINKING

Thinking and planning in a macro context about humanity's future is currently under way, especially among professional futurists. For example, the Foundation for the Future has been conducting a series of seminars and publishing proceedings for this millennium (*www.futurefoundation.org*). Under the leadership of their founder, Walter Kistler, who also established Kistler Aerospace, the Foundation conducted a Humanity 3000 workshop on *Humans and Space: The Next Thousand Years* (*www.futurefoundation.org*). Amother seminar with UNESCO was held in Paris (September 20–22, 2006) on *Humanity and the Biosphere: The Next Thousand Years*. The 246-page volume of proceedings which resulted cover aspects of the biosphere related to information and expert knowledge, climate change and threats, conservation and extinction. Among the many worldwide publications forecasting the future of our species, your own author has written a book (published in 2008) on *Toward*

Human Emergence: A Human Resource Philosophy for the Future (*www.hrdpress. com*).

With reference to humanity's future in outer space among the stars, let us examine two positive possibilities that would require macrothinking.

- **Terraforming celestial bodies** may be the ultimate test for macrothinking and macroplanning [22]. Richard Taylor of London's Centre for Extra Mural Studies, described *terraforming* as a process of planetary transformation from a naturally existing abiogenic state to one that is Earth-like in all important respects. Ideally, a completely terraformed planet will possess a stable, self-regulating ecosphere in which all the prevailing conditions fall within the range of human physiological tolerance. The transformed world should be automated with regard to air conditioning, climate control, and food production. However, such macroengineering should aim to protect the environmental integrity of the celestial body under human transformation, while avoiding pollution of its soil and atmosphere, especially with debris. Taylor advocates *paraterraforming* which he explains as projecting beyond present limits of available technology, engineering, and scientific knowledge. This process differs in that it does not seek to achieve the total transformation of a planet to fully Earth-like conditions. The aim on a chosen planet is an almost immediately habitable environment, but within an ecosphere that is delimited. This restricted approach does not achieve total homeostasis, allowing for quite different biomes to exist and function in separate locations and equilibria. In this strategy, timescales can be significantly shorter and elastic.
- **The Global Exploration Strategy** provides a framework for future international coordination on space development. It manifests the principle described in Chapter 2 that *space is the place for synergy*. The high risks, costs, and complexity of macroprojects in space demand that humanity work together as a family in space exploration and settlement. This is exactly what 14 space agencies did in May 2007 when they pledged joint cooperative efforts to extend human presence beyond Earth's orbit, physically and culturally. GES nations recommend a voluntary forum and coordination mechanism for collaboration on individual projects and collective efforts, beginning with the Moon as our nearest and closest first goal.[5] Perhaps this quotation from their excellent report on the GES planning strategy sums up their intentions, and our concept of macrothinking [22]! Exhibit 97 sets forth what the GES participants envision as their principles for cooperation [23].

    "Space exploration enriches and strengthens humanity's future. Searching for answers to fundamental questions like 'Where did we come from? What is our place in the universe? And what is our destiny?' can bring nations

---

[5] In alphabetical order, the 14 GES government space organizations are ASI (Italy), BNSC (U.K.), CNES (France), CNSA (China), CSA (Canada), CSIRO (Australia), DLR (Germany), ESA (European Space Agency), ISRO (India), JAXA (Japan), KARI (Republic of Korea), NASA (U.S.A.), NSAU (Ukraine), and Roscosmos (Russia).

Exhibit 97. PRINCIPLES OF INTERNATIONAL COORDINATION FOR SPACE DEVELOPMENT.

Principles	*Resulting requirements*
Open and inclusive	• Receives input from all interested agency participants that invest in and perform activities related to space exploration. • Provides consultation among all interested agencies with a vested interest in space exploration, and also space agencies* or national government agencies without specific related capabilities.
Flexible and evolutionary	• Takes into account and may integrate existing consultations and coordination mechanisms. • Allows consultation and coordination structures and mechanism(s) to gradually build and evolve as requirements for these activities grow. • Allows for assigned representatives and clear stake in space exploration. • Provides for different levels of consultation and coordination.
Effective	• Encourages participating agencies to accept the role of the coordination process and act on anticipated results of the coordination mechanism.
Mutual interest	• Contributes to common peaceful goals and benefits all participants. • Respects national prerogatives of participating agencies. • Allows for optional participation based on the level of each agency's interest.

Source: Global Exploration Strategy. *The Framework for Coordination*, May 2007, ch. 6, pp. 23–24 (*www.BNSC.gov.uk*, press release #31-05.07).
*The words "space agency" refer to government space agencies, science organizations, and groups of space agencies that have been designated by their governments to represent them, such as ESA and BNSC.

together in a common cause, reveal new knowledge, inspire young people, and stimulate technical and commercial innovations on Earth. The Global Exploration Strategy is the key to delivering these benefits!"

The GES framework and principles are synergistic, and a positive, long-overdue step toward macrothinking about our common future in outer space. However, it is all public-sector oriented, and the strategy could easily degenerate into bureaucracy at its worse. If humanity is truly to move forward together, then the private sector has to be included in any entity that might result. Government agencies cannot continue to dominate the space undertaking, and private enterprise is likely to become a prime force in this venture. That is why we have previously suggested a Global Space Trust be founded that represents our joint movement offworld.[6]

Your author considers two pioneering rocket scientists, Wernher von Braun and Krafft Ehricke, to be ideal examples of macrothinkers. As previously noted, Dr. Ehricke, for example, set forth a detailed plan for creating a lunar civilization and city named *Selenopolis* [24]. This Space Age philosopher also proposed three laws to guide us in carrying out our *extraterrestrial imperative*, which I have slightly edited below [25]:

— Law One   Nobody and nothing under natural law of this Universe should impose any limitations on our thinking, planning, and actions—but humans themselves.
— Law Two   Not only the Earth, but the entire Solar System, and as much of the universe as we can reach under the laws of nature, are humanity's rightful field of activity.
— Law Three By expanding through the universe, humans fulfill their destiny as an element of life, endowed with powers of reason and the wisdom of moral law within themselves!

And so this futurist concludes that these laws of astronautics challenge our species to write its own declaration of independence from *a priori* thinking, from uncritically accepted conditions; in other words, from a past, principally different, pre-technological world that still clings to our mindset. In our thinking and planning, rather than placing constraints on them—we should be open to endless possibilities.

7.7 MACROTHINKING CONCLUSIONS

The type of macrothinking and macroplanning described in this chapter has the following advantages if applied to development of the space frontier:

(a) it captures public imagination by its boldness, scope, and timeframe, for it provides supportable goals;
(b) it permits the incorporation of incremental stages for accomplishment, putting individual space missions into a larger, longer context;
(c) it encourages synergy and interdependent actions, ideal for international cooperation and financing;
(d) it gives the younger generation a vision of future possibilities and even orbital career opportunities.

Macrothinking is in consonance with exploring and pioneering the high frontier, offering solutions to the challenge of creating a cosmic civilization, for macrothinking

---

[6] The strategy of the Global Space Trust was explained in the author's novel *Launch Out* (*www.buybooksontheweb.com*).

opens up the frontiers of the mind, stretching our capacities through long-term planning. In his book, *The Pale Blue Dot*, astronomer Carl Sagan places our planet in the context of the universe [26]. That is the context of macrothinking! The best example in modern times of such judgment was manifested by those involved in planning and executing the Apollo mission program that sent 12 humans to the Moon and back! It took some 400,000 workers and some $36 million to do that four decades ago. And at the first lunar landing in 1969, a billion people worldwide watched on television the results of such macrothinking personified in the three envoys of humankind: Astronauts Armstrong, Aldrin, and Collins who represented us all.

It is our *imagination and thinking* which may delimit us from developing the richness and resources of outer space, just as it did with those inadequate politicians in 1972 who curtailed the last Apollo missions at #17, failing to follow up with a vigorous program of lunar science and industrialization. Since space exploration and exploitation are imperatives for our species, only macrothinking can adequately deal with future mass migration aloft, providing the rationale for human spaceflight, settlement, and interstellar travel. Only macrothinking can translate the excitement of that journey *en masse*, so that we utilize offworld resources to protect this planet, and counter poverty among its inhabitants.

In his classic work, *MACRO* [5], Frank Davidson concludes with a discussion of the neo-industrial paradigm that links technology to articulate purposes for improving the human condition. As we celebrate the 50th anniversary of spaceflight, remember that macrothinking provides space technology with such purpose in our voyage through the stars. Futurist Alvin Toffler said it best [27]:

"Space will not only shape our descendants' view of our time; it also shapes our view of the future. It increases the gravity of tomorrow, the consciousness of the future, in our culture today and in a world of exploding change."

Such macrothinking suggests that space exploration and space-derived knowledge will define humanity's place in the universe!

SCENARIO FOR MASS SPACE MIGRATION—2050

Humanity's access to outer space is presently constrained by high transportation costs and limited spacecraft capacity. Mass migration to low Earth or geosynchronous orbits will become feasible later in this 21st century when transportation costs of space travel are expected to drop as low as $9 to $26 per kilogram of payload (compared with current estimates of $4,000 to $9,000 per kilogram via today's Shuttle and expendable launch vehicles).

Presently, laser propulsion scientists have research under way that may dramatically lower such investments, while boosting the number of people able to go aloft. Arthur Kantrowitz of Dartmouth College, New Hampshire, U.S.A. and Avco Everett Research Laboratories, as well as Leik Myrabo of Rensselaer Polytechnic Institute, forecast that by the middle of this century, air breathing/

laser propulsion technologies may result in small private spacecraft (5 meters in diameter) for three to five persons, capable of a round-trip to and from LEO at a transportation cost of $8.40 per kilogram!

Before powerful lasers began to lift expensive payloads into orbit, including tens of thousands of people, more study is also needed on personnel deployment and habitation systems for long-duration missions ... The media fans of Gene Roddenberry's *Star Trek* already know that the ultimate answer to transmigration and space travel on Starships will be, "Beam me up, Mr Spock"! Now, that is macrothinking at its best.

7.8 REFERENCES

[1] Carter, N. E. "The Challenge of Macroengineering," *Battelle Today*, No. 43, June 1985, p. 1 (Battelle, 506 King Avenue, Columbus, OH 43201) ... Cathcart. R. B. *Macroengineering: Its History and Future*, 2005 (available from GEOGRAPHICS, 1300 West Olive Ave., Ste. M, Burbank, CA 91508).

[2] McFarland, D. E. *The Management Imperative: The Age of Macromanagement*. Cambridge, MA: Ballinger/Harper & Row, 1985.

[3] McKay, M. F.; McKay, D. S.; and Duke, M. B. (eds.) *Space Resources*. Washington, D.C.: U.S. Government Printing Office, NASA SP-509, 1992, 5 vols. (*www.univelt.com*). See Vol. 4, *Social Concerns*, "New Space Management and Structure: NASA in the 21st Century" (K. J. Murphy, pp. 16–34); "The Future Management" (P. R. Harris, pp. 120–141).

[4] Badescu, V.; Cathcart, R. B.; and Schuiling, R. D. *Macroengineering: A Challenge for the Future*. New York: Springer-Verlag, 2006 ... See R. B. Cathcart's chapter in Badescu, V.; and Rudhakrishnan, N. (eds.) *Macroengineers' Dreams* (to download textbook, go to *http://textbookrevolution,org/*).

[5] Davidson, F. P.; and Brooke, L. B. *Building the World: An Encyclopedia of Great Engineering Projects in History*. Oxford, UK: Greenwood Publishing, 2000, 2 vols. ... Davidson, F. P.; and Meader, C. L. (eds.) *Macroengineering: Global Infrastructure Solutions*. Chichester, U.K.: Ellis Horwood, 1988 ... Davidson, F. P.; and J. B. Cox, *MACRO*. New York: William Morrow & Co., 1983 (see *www.tasme.org* or email *tasme@ecostructure.com*).

[6] Pritchett, P.; and Muirhead, B. *The Mars Pathfinder Approach to Faster, Better, Cheaper*. Dallas, TX: Pritchett Associates, 1998 ... See *www.nasawatch.com/fbc.html*

[7] Katsioloudes, M. I. *Strategic Management: Global Cultural Perspectives for Profit and Non-Profit Organizations*. Burlington, MA: Elsevier/Butterworth-Heinemann, 2006, ch. 1.

[8] Salkeld, R. "Space Macro-arena for Macro-engineering," in Davidson, F. P.; Giacoletto, J.; and Salkeld, R. (eds.) *Macro-Engineering and the Infrastructure of Tomorrow*. Washington, D.C.: American Association for the Advancement of Science, Vol. 23, 1978, pp. 131–138 ... See *en.wikipedia.org/wiki/Macro-engineering*

[9] Surkov, Y. *Exploration of Terrestial Planets from Spacecraft*. Chichester, U.K.: Ellis Horwood, 1990 ... Curtis, A. R. *Space Satellite Handbook*. Houston, TX: Gulf Publishing, Third Edition, 1992–1994.

[10] Kozmetsky, G. "Education for Large-scale and Complex Systems Based on Technology Venturing," *Technology in Society*, **6**, 1984, 173–176. This recommended macrothinking

journal is published by the Institute for Technology Management and Policy at Pratt Polytechnic Institute of New York (333 Jay Street, Brooklyn, NY 11201).

[11] Glaser, P. E.; Davidson, F. P.; and Csigi, K. I. (eds.) *Solar Power Satellites: A Space Energy System for Earth.* Chichester, UK: Wiley/Praxis, Second Edition, 1998.

[12] Webber, D., "Horses for Courses: Spaceport Types," address at the International Space Development Conference, Washington, D.C., May 2005.

[13] Space Tourism Market Study. *The Futron/Zogby Survey.* Futron Corporation, October 2002 (Futron Corporation, 7315 Wisconsin Ave., Ste. 900W, Bethesda, MD 20814; www.futron.com).

[14] Webber, D.; and Reifert, J. *The Adventurers' Survey: Full Report.* Rockville, MD: Spaceport Associates, November 2006 (www.spaceportassociates.com) ... Space Adventures, 8000 Towers Crescent Road, Ste. 1000, Vienna, VA 22182; www.SpaceAdventures.com).

[15] Bonnet, R. M. "Taking the Next Steps in the European Moon Program," *Planetary Review,* **XV,** January 1995, 8–11 ... ESA. *Mission to the Moon: Europe's Priorities for Scientific Exploration and Utilization of the Moon.* Paris: European Space Agency, ESA SP-1150, June 1992, 190 pp. ... For further information on current lunar planning, contact ESA Headquarters (8–10 Rue Mario Nikis, Paris, Cédex 15, or 955 L'Enfant Plaza, SW, Ste. 7800, Washington, D.C. 20023; tel.: 202/488-4148; fax.: 488-4930).

[16] Dyson, G. *Project Orion: The True Story of the Atomic Spaceship.* New York: Henry Holt & Co., 2002.

[17] O'Donnell, D. J. "Founding a Space Nation Utilizing Living Systems Theory," *Behavioral Science,* **39**(2), April 1994, 93–117 ... "Overcoming Barriers to Space Travel," *Space Policy,* **10**(4), November 1944, 252–255 ... For further information consult back volumes of the journal *Space Governance,* from 1994 to the present (published by United Societies in Space, Inc., 499 Larkspur Rd., Castle Rock, CO 80104; tel.: 303//688-2432; email: *djope@quest.net*; *www.usis.org*).

[18] Savage, M. T. *The Millennial Project: Colonizing the Galaxy in Eight Easy Steps.* New York: Little Brown & Co., 1994 ... "The Millennial Project Strategy," *Space Governance,* **2**(1), June 1005, 26–31 ... For further information, contact First Millennial Foundation (PO Box 347, Rife, CO 81650; tel.: 303/625-2815; email: *mtsavage@pipeline.com*).

[19] Schoenberger, C. R. "The Space Mafia: If You Haven't Heard of International Space University, You Will Soon: Rocket Science Goes Commercial," *Forbes,* April 17, 2000, pp. 104–108 ... As an example of the type of project or thesis research produced at ISU, consider Rei Uda's report on *Influence of Different Language, Culture, and Nationality among Team Members on Space Missions and Projects,* May 1996. Direct inquiries to International Space University (Communuaté Urbaine de Strasbourg, Blvd. Gonthier d'Andernach, 67400 Illkirch, France; North American office: 3400 International Dr., NW, 4M Ste. 400, Washington, D.C. 20008).

[20] Forston, R., "Cultivating the High Frontier," *Agricultural Engineering: Technology for Food and Agriculture,* November 1992, 20–23 (reprints available from Agricultural Engineering, 2950 Niles Rd., St. Joseph, MI 49085; Russ Forston's contact details are The Bionetics Corp., Code BIO = 3, Kennedy Space Center, FL 32899).

[21] Harris, P. R. "Future Role of Space Nurses in a 21st Century Lunar Colony," June 22, 2001, 10 pp. in Proceedings of *2001, A Nursing Odyssey: A Tribute to Martha Rogers,* American Holistic Nurses Association Conference (www.ahna.org or www.sns.org) ... "The Space Nurse's Role in Moon/Mars Personnel Deployment Systems," September 27, 1997, pp. 4–78, in the program *Pushing the Envelope II: Medicine on Mars,* University of Texas Medical Branch at Galveston (www.UTMB.edu or www.sns.org) ... Malinski, V. M.; and Barrett, E. A. (eds.) *Martha E. Rogers: Her Life and Her Works.* Philadelphia,

PA: F. A. Davis Co., 1994 ... O'Rangers, E.; and Plush, L. "A Proposal for the Oversight of Medical Care for Travelers and Crew Flying in Virgin Galactic Spacecraft at Mojave Spaceport," April 2005, 10 pp. (*www.spacemedicineassociates.com*).

[22] Taylor, R. L. "Paraterraforming: The Worldhouse Concept," *Journal of the British Interplanetary Society*, August 1991, 341–352 ... The author, Richard Taylor, works at the Centre for Extra Mural Studies, Birkbeck College (26 Russell Square, London WC1B 5DQ).

[23] Global Exploration Strategy. *The Framework for Coordination*, May 2007. Press release of British National Space Centre, #31.05.07, "New Era for Space Exploration as 14 Space Agencies Take Historic Step" (*www.bnsc.gov.uk/home*).

[24] Ehricke, K. "Selenopolis," in Mendell, W. W. (ed.) *Lunar Bases and Space Activities of the 21st Century*. Houston, TX: Lunar and Planetary Institute, pp. 731–736.

[25] Freeman, M. *Challenges of Human Space Exploration*. Chichester, U.K.: Springer/Praxis, 2000, p. 234 ... "Krafft Ehricke's Extraterrestrial Imperative," *21st Century Science & Technology*, **7**(4), Winter 1994 ... Ehricke, K., "The Anthropology of Astronautics," *Astronautics*, **2**(4), 1957, 26–27.

[26] Sagan, C. *The Pale Blue Dot*. New York: Random House, 1994.

[27] Toffler, A. "The Space Program's Impact on Society," in Korn, P. (ed.) *Humans and Machines in Space*. San Diego, CA: UNIVELT/AAS, 1992 (*www.univelt.com*).

SPACE ENTERPRISE UPDATES

Two macrothinkers are evident in Burt Rutan, founder of Scaled Composites in California's Mojave Desert, and Richard Branson, founder of Britain's Virgin Group. With the investment of Paul Allen, co-founder of Microsoft, Rutan built and tested SpaceShipOne, his winning entry in the $10 million Ansari X-Prize. But Branson made this new spaceplane commercially viable by licensing the whole technology for use in offering suborbital flights to paying passengers. He and his executive, Will Whitehorn, created a milestone in space tourism when they contracted with Rutan to design their SpaceShipTwo, including its WhiteKnightTwo, the redesigned conventional aicraft that gets the former into orbit at less cost than two-stage rockets. The latter would provide enought lift to get eight people into orbit and re-entry to Earth. Virgin Galactic is marketing a $200,000 spaceflight, and already has collected $30 million in deposits as Rutan was building his second factory. The spacecraft is revolutionary in that it can take passengers and payloads offworld. Thinking big, Rutan and Branson are planning their next step: supersonic vehicles that fly across the world, and eventually to the Moon itself. NASA already has a signed agreement with Virgin Galactic, while Northrop-Grumman has increased its 40% stake in Scaled Composites. Rutan is hoping to built up to 50 of this new generation of launch craft (adapted from *The Economist*, January 26th, 2008, pp. 66–68).

8

Macromanagement of space enterprise

> *Global scale space exploration represents the sum of many projects undertaken nationally and internationally. But it also signifies the collective will to find answers to profound scientific questions, to create new economic opportunity, and to expand the boundaries of human life beyond Earth. These goals of space exploration in service of society are embodied in the recurring themes of the Global Space Exploration Strategy* [1].

The prototypes of 21st-century management, particularly with reference to managing large-scale enterprises, may be found partially in the last half century within the global space program [2]. The Space Age inaugurated a number of macroprojects by both the U.S.A. and the former U.S.S.R. of such scope and magnitude that another type of management had to be created to ensure successful achievement [3]. The opening of the high frontier is a powerful catalyst not only to the development of new technologies, but also to the advancement of numerous fields of knowledge, such as macroengineering and *macromanagement*. The premier space writer, Leonard David, said it best:

> "Cynics like to say the future of space exploration isn't what it used to be. Or even worse, that the future has been cancelled. But you have only to look skyward on any clear night to see the future is alive and doing well. Silently sweeping across the backdrop of stars, humankind's genius parades itself in the form of artificial satellites, space shuttles, and orbiting outposts. Each is a foothold into the future!" [4].

8.1 MANAGEMENT CHALLENGES FOR THE SPACE ERA

Further extension of human presence into outer space during the decades ahead offers opportunities for management innovations by those with responsibilities for projects aloft, whether in the public or private sector, whether manned or unmanned missions. In the past, NASA spaceflights, such as Mercury, Gemini, Apollo and Skylab, as well as building and operating the Space Shuttle Transportation System, the Hubble Space Telescope, or now the International Space Station (ISS), are examples of large-scale enterprises requiring an inventive type of administration and contracting. While a seminal study of *Living Aloft* anticipated the expanded human requirements for long-duration missions, it devoted only a chapter to organization and management, limiting discussion to spacecrew structures, motivation, and external relations. Even an updating of that work entitled *Spacefaring* fails to mention the art and science of management as applied to space development [5]. In fact, space and management literature is sparse relative to the *how of space management*, the exception being books and journals that deal with the subject of macroengineering [6]. Yet, currently, as well as in the immediate future, post-industrial technical and management challenges exist relative to satellite expansion, planetary scientific exploration, spaceplanes and heavy-lift launch vehicles, maintaining orbiting stations and platforms in both LEO and GEO, returning to the Moon and constructing a lunar outpost, then possible human missions to Mars.

Today, reduced funding and rising costs compel the world's space agencies and companies to collaborate and share, especially with reference to information and management systems. Two examples from the last century confirm how the space superpowers have been financially constricted: (1) the American *Space Exploration Initiative* proposed by the first President George Bush was postponed because the U.S. Congress failed to finance the program; (2) cash-short Russia has several times kept cosmonauts aloft for extra stays on station Mir because of financial problems. In October 1995, the Russian Space Agency (RKA) reported that construction of a booster rocket for crew replacement was delayed; this was caused by a shortage of funds that caused two Russians and one German to stay in orbit for an extra month. Finally, age and costs forced the Russians to discontinue that orbiting platform.

Thus, it is obvious that for both NASA and RKA, there will be no manned Moon or Mars missions without multinational partnerships and financing. Whether for scientific, commerce, settlement, or defense purposes, international space activities feature more complex missions, involving multiple locations, plus greater numbers and varieties of personnel and nationalities. Furthermore, such major global space projects require the integration and synthesis of differing technologies and management systems/styles. Even within one nation, such as the U.S.A., the sponsorship of most space operations increasingly will come from multiple sources—not just NASA, the DOD or military branches and other governmental agencies, such as the Department of Transportation or Commerce—but increasingly via the world's private sector, from the aerospace industry to new commercial space corporations.

But my point is that 21st-century space exploration requires more than collaborative management: what is needed is a new type of *macromanagement for large-scale enterprises*! Perhaps this may be best appreciated by reviewing Exhibit 98 with its prototype of a future micromanager.

Exhibit 98. PROTOTYPE OF FUTURE MACROMANAGERS: JOHN PIERCE.

When John Robinson Pierce died in 2002 at age 92, his obituary recalled more than his design and launch of Telstar 1, the world's first commercial, communication satellite. He was also remembered for his contribution of writing 20 books and 300 technical papers. While few may have read his "Theory and Design of Electron Beams," millions have devoured his science fiction stories and books published under the name "J. J. Coupling" (meaning a process in physics for determining wave functions). For this author realized that most leaps forward in knowledge and technology are anticipated by writers from Leonardo daVinci to Arthur C. Clarke, to the latest science fiction scribes, like David Brin. Pierce confessed that he only tried to make real what he and others dream about by making calculated guesses about the future. For him fiction was creative thought that leads to solutions, yet observes scientific laws. His writings helped pay for his tuition and expenses at California Institute of Technology where he received a doctorate in electronics.

By the mid-1930s, John was employed by Bell Telephone Laboratories where he worked for the next 35 years. There he and colleagues engaged in advanced scientific research and development. His team invented what this wordsmith called a *transistor*, so indispensable to radios and computers. During WWII, he worked on radar where signals directed at an object are bounced back to the transmitter. As the Russian and American governments launched artificial satellites, Pierce calculated that 25 of them, suitably placed, could provide continuous communication around the world. In the course of his career, he was granted 90 patents, won many awards, and received dozens of honorary doctoral degrees. To maintain balance in his life, he had hobbies in gliding and musical composition by computer. For 12 years at Stanford University, he was a professor of music, but never accepted a salary. One theme in his science fiction concerns human-looking *cyborgs*, about which he raised issues of whether these artificial creations of ours should have emotions and sexual feelings. John Pierce did think big and far ahead, so fits our description of a macrothinker and macromanager. May his example encourage our readers to *think out of the box*!

Source: Adapted from "Obituary: John Pierce," *The Economist*, April 13th, 2002 (*www.economist.com/*).

8.1.1 Macromanagement prototypes

The management of space enterprises is in transition, from the way short-duration spaceflights have been *managed* for the past 40 years, to high-technology, computer-based project management and leadership, which is quite flexible and multinational in scope. The new global space approach may possibly create an *ad hoc* consortium, possibly with the United Nations or UNESCO as sponsor, comparable with the coalition of allies in the Gulf War. Or it may require the creation of a world space administration [7]. One harbinger of the future may be the contemporary Space Agency Forum, which presently meets annually in conjunction with the International Astronautical Federation. There, SAF brings together the heads of 30 space agencies from around the world with representatives of numerous international space organizations. At the moment, it is simply a useful exchange of information and views on issues facing them in undertaking upcoming space activities. In 2007, another positive step was the formation of the Global Space Exploration Strategy by 14 space agencies, explained in Chapter 7.

There is also potential for tomorrow's multinational space management in today's European Space Agency (see Chapters 1 and 7). Managed by an ESA Council of Ministers, originally they represented 14 nations (13 member states and the cooperating state of Canada). As the governing body, the Council approves or disapproves the Agency's space plans for Europe. The Agency then manages ESA's research, operations, and facilities within the European Union. Exhibit 99 shows the locations of some of the latter, which are spread across that continent and in countries around the world. However, eventually, the private sector may lead in confronting the interwoven issues of space governance and management, as was suggested in Chapter 7 and will be enlarged on in the next chapters, especially relative to establishment of a global space trust or even space authorities.

The inherent orbital challenges go far beyond space technology and management, encompassing human resource and cultural dimensions. Over 20 years ago, William MacDaniel, professor emeritus of sociology at Niagara University, forewarned of such multiple difficulties in terms of just one project being planned: the International Space Station (ISS) [8]:

"Any way that we look at it ... NASA will be confronted with management problems that will be totally unique. Space station management is going to be an entirely new ball game, requiring new and imaginative approaches if serious problems are to be resolved and conflict avoided."

MacDaniel, a retired Air Force pilot who co-founded the Space Settlement Studies Project (3SP) at Niagara University, then analyzed one people-management dimension that results from the sociocultural mix of international scientific/engineering teams and onboard space crews. The multicultural inhabitants of the Space Station are coping not only with technical challenges, but the many practical aspects of their cultural differences which alter their perceptions and ways of functioning relative to everything from communication and problem-solving, to spatial needs and diet. In June 2007, the ISS computers on the Russian component shut down threatening control of the station's orientation and oxygen production. Fortunately, the Shuttle *Atlantis* was docked there to help with the problem, while technical managers onboard and in Moscow and Houston got the six processors online again. Because of such management crises, a special case study on the International Space Station is provided later in this chapter.

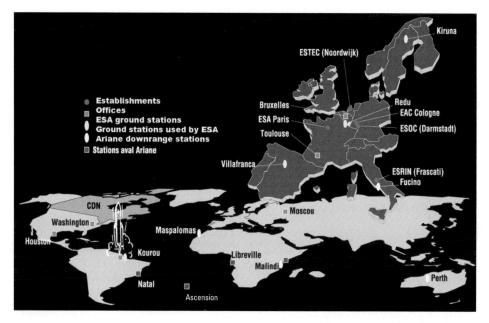

Exhibit 99. ESA ground-based global establishment. This illustration of Europe's space facilities and offices indicates the global scope of the European Space Agency in carrying out its mission: *to provide for and to promote, for exclusively peaceful purposes, cooperation among the European states in space research and technology, and their space applications.* Source: ESA.

Whether the orbiting of greater numbers of people for longer periods of time is to be done by a combined consortium from both the world's private and public sectors, project leaders will also have to set the management of cultural synergy as a priority, while creating new system capabilities. Unmanned probes of spacecraft into the far corners of the universe will also demand increased inter-institutional and international cooperative management. In any event, futurists, students of management, and those concerned with technological administration would do well to review the limited literature and experience of emerging space management, for its wider implications. Despite current difficulties associated with the management of NASA programs, that Agency does offer scholars both experimental history and a possible demonstration model of future management practice.

8.2 THE APOLLO HERITAGE OF INNOVATIVE MANAGEMENT

In worldwide management, a transformation is under way from industrial organization designs and leadership styles to those suitable for a new knowledge work culture [9]. In an AT&T report on emerging issues, the term metaindustrial was used to designate the evolving management approach to human systems. One driving force for this transition may very well have been the innovation within the space program by NASA in the 1950s and 1960s. That agency in conjunction with its partners in the aerospace business contributed to the emergence of high-technology industry and its flexible management practices. Because of the very complexity of the Apollo lunar mission, space administrators and contractors also invented new ways of organizing and managing in order to achieve agency goals [10]. As Seamans and Ordway observed [11]:

> "The Apollo project which landed a team of American astronauts on the Moon is generally considered as one of the greatest technological endeavors in the history of mankind. But in order to achieve this, a managerial effort, no less prodigious than the technological one, was required."

It is this writer's contention that much of what is currently being characterized as the *new management* is partially the heritage of that space effort. This is especially pertinent with the building of large-scale technological undertakings, whether on this planet or in orbit. Those engaged in macro enterprises that involve many systems, disciplines, institutions, and even nations will have to apply in even more creative ways the legacy which the Apollo program gave to management. And the forerunner of space macromanagers to come was James E. Webb, NASA administrator during that momentous period [12]. Not only is more research needed in this regard by academics, but studies and conferences should be directed to what constitutes *macromanagement*, which the late Dalton McFarland described as the managerial imperative [13]. Currently, the term has a double, but complementary meaning: macromanagement represents a post- or metaindustrial management approach, but it can also be employed to describe the complex management of macroprojects (such as required for the building of the Anglo-French Channel Tunnel or in a manned mission to Mars).

In the 1983 inaugural issue of the periodical *New Management*, the editor, James O'Toole, provided ten guidelines of what contributes today toward organizational greatness or excellence:

(1) oriented toward tomorrow and attuned to long-term future;
(2) oriented toward people and development of human resources;
(3) oriented toward product and commitment to consumer market;
(4) oriented toward technology and employment of most advanced tools and technical process;
(5) oriented toward quality and emphasis on excellence, service, and competence;
(6) oriented toward external environment and concern for all stakeholders;
(7) oriented toward free-market competition and imbued with the spirit of risk-taking capitalism;
(8) oriented toward continuing examination and revision of organizational values, compensation, rewards, and incentives;

(9) oriented toward basic management concerns and making, selling, or providing services;
(10) oriented toward innovation/openness to new ideas and nurtures and encourages those who question organizational assumptions and propose bold changes.

Dr. O'Toole was later to elaborate on this theme through a book entitled *Vanguard Management* [14].

An examination of the Apollo program history will confirm that leaders then in NASA and among its aerospace contractors followed such principles, with the possible exception of the third item, which does not quite apply to public enterprise. But their aerospace civilian counterparts had this concern for the customer: in their case, the Agency and its personnel. Otherwise, the Moon mission would not have been so successful. Almost four decades ago, NASA anticipated post-industrial management. The very scope and complexity of putting humans on the lunar surface forced such creative alterations in administration. Even now NASA engineers building spacecraft to put humans back on the Moon by 2020 are studying the technology of the 1969 lunar transportation system. Utilizing the Apollo legacy, they have gone to museums to study the Saturn S48 rocket, analyzing and "cannibalizing" parts from the Apollo program! Using space design heritage, the new Ares solid rocket booster will also feature aspects of the Shuttle system, and its capsule will carry six astronauts, while its lunar lander will replicate the original used by Neil Armstrong and Buzz Aldrin. Over the next 15 years, $125 billion is expected to be spent on this lunar macroproject: another contribution of the American taxpayer to humanity!

8.2.1 Matrix management

Among the primary management innovations coming out of the Apollo space program some 40 years ago was the *matrix organization*, with its emphasis on dual reporting and team management. The adoption of this management approach was necessitated because of the complexity of mission tasks, and the inadequacy of traditional industrial age management. Then space contractor, TRW Systems of Redondo Beach, California, was a leader in a matrix-type management that two decades later would be a common feature of "new" management operations. Their vice president, Sheldon Davis, pioneered team building as a means of helping technical people work together better with those from other disciplines and divisions. Other aerospace contractors used project and team management as a form of *ad hoc* organization with new start-up activities. General Dynamics, for instance, could quickly assemble experienced team members for its later Shuttle–Centaur project because of experienced work groups who had developed the Atlas–Centaur rocket, still functioning in a different form 50 years later! Among these contractors, Hughes Aircraft was a principal exponent of the new matrix management. One of its executives, Jack Baugh, did a doctoral dissertation on how their decision-making was accomplished through this means [15]. His thesis was that matrix management is essential in some aerospace projects *when*

- simultaneous dual decisions are needed in situations of great uncertainty generated by high information-processing requirements;
- strong financial and human resource constraints are present;
- decisions must be speeded up;
- quantity of data, products, and services to be managed demand it.

Matrix management is useful in the management of projects that involve different departments of an organizations, and different disciplines. It is also a way to implement team management, when individuals have to report to one or more supervisors, so as to accomplish an objective or goal.

Today, and more frequently tomorrow, these are the conditions demanding macromanagement and high-performance teams. A contemporary profile of a metaindustrial organization would include these characteristics [16]:

- use of state-of-the-art technology, ranging from microcomputers to robotics;
- flexibility in management policies, procedures, and priorities, continuously adapting to the market—a norm of ultrastability (continuous change built into the system);
- autonomy and decentralization, so that people have more control over their own work space and are responsible for decisions, but with integrating controls present;
- open, circular communication with emphasis on rapid feedback, relevant information exchange at all locations, networking, and multimedia use;
- participation and involvement of personnel encouraged, especially through team, project, or matrix management;
- work relations that are informal and interdependent, cooperative and mutually respectful, adaptive and cross-functional;
- organizational norms that support competence, high performance, professionalism, innovation, and risk-taking, even to making allowance for learning from occasional failure;
- creative work environment that energizes people and enhances the quality of work life, so that it is more meaningful;
- research and development orientation that continually seeks to identify the best people, processes, products, markets, services, so as to better achieve mission objectives.

Currently, high-tech and software firms operate on such principles. It is interesting that many of these qualities were also identified some 40 years ago as essential to the interdisciplinary management of large-scale endeavors [17]. These were also the characteristics, to a great extent, practiced by NASA management in the Apollo era. They are considered essential for organizational excellence now and in the future, particularly for large-scale undertakings such as renewing American infrastructure, or developing a permanent infrastructure for the space frontier. Space agencies and aerospace corporations which have developed huge bureaucratic managements must also transform or reengineer themselves into such entities to prosper in the next century [18].

TECHNOLOGY INNOVATION

This the name of a useful publication by NASA's Exploration Systems Directorate (Washington, D.C. 20546; *www.ipp.nasa.gov*). In this complimentary magazine, the 12th volume (2005) focused on "The Role of the Innovative Partnership Program in Space Exploration." IPP plays a major role in implementing the goals of the *Vision for Space Exploration* strategy; namely, to implement a sustained and affordable program; extend human presence across the solar system; develop innovative technologies, knowledge, and infrastructures; promote international and commercial partitipation. The Innovative Partnership Network operates out of ten NASA field centers, fosters eight business facilitator or incubator start-up enterprises, and eight technology transfer organizations for public/private sector collaboration (RTTC, tel.: 1-800/642-2872). Further, at NASA headquarters there is a Small Business Innovation Research Program (email: *cray@hq.nasa.gov*), in addition to a Small Business Technology Transfer Program at the Goddard Space Flight Center (email: *paul.mexcur@pop700.gsfc.nasa.gov*). In the magazine's 13th volume (2006), the theme was "Nanotechnology Paves the Way for New Business Venture." All these Agency outreach programs in technology R&D impact business, medicine, education, and exploration aloft, while underwriting the application of space technology for the benefit of humanity on Earth! All such innovations advance the practice of space management. Such ventures are in line with these thoughts of Albert Einstein: "The most beautiful thing we can experience is the mysterious. It is the source of all true art and science!"

8.2.2 Knowledge management

In this Information Age, *knowledge management* is the facilitation and support of processes for creating, sustaining, sharing, and renewing organizational knowledge in order to generate economic wealth, create value, and improve performance ... Within the global workforce there is much diversity from the perspective of both cultural backgrounds and academic disciplines. Those in technical professions especially need to recognize the impact of cultural differences on both their research and development [19].

Since those in R&D management, especially those coming from engineering and technological fields, may have some differing perceptions about the management process, Exhibit 100 is included here. This classic model by Professor R. Alec MacKenzie illustrates the comprehensive art and science of managing both human and material resources effectively. Its central activities are managing ideas/ information, materials, and people through conceptual thinking, administration, and leadership. This involves analyzing problems, making decisions, and communicating. Management's task is to plan, organize, staff, direct, and control. The

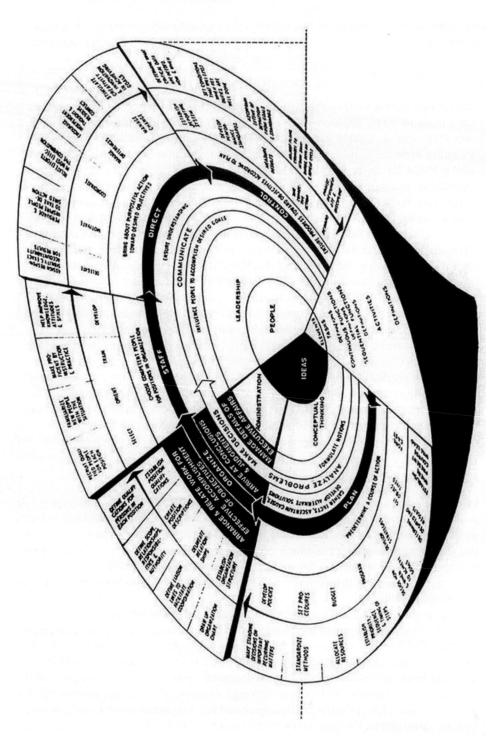

Exhibit 100. The management process. Source: R. Alec MacKenzie, "The Management Process in 3-D," *Harvard Business Review*, November/December 1969. Reproduced with permission.

paradigm highlights among its central facets, the management of change (stimulation of innovation and creativity in achieving goals), as well as managing differences effectively (encouraging interdependent thought and conflict resolution) [20]. It still seems a relevant conceptualization for managing large-scale undertakings, whether on the ground or in orbit to promote the formation of a network or collaborative organization. From the viewpoint of this researcher, a management psychologist who has served as a NASA consultant, it would appear that the main difficulties facing space leaders in the future may be found on the right side of the paradigm illustrated below: the people dimension calls for expert behavioral science management solutions.

The former U.N. Chief of the Outer Space Affairs Division, Lubos Perek, suggests we think of outer space as a resource to be managed [21]. Thus, in that context, "management is everything that improves safety, efficiency, and economy of space activities for our own and future generations." In this conceptualization of Perek, space management preserves the offworld environment; prevents interference in space communications; controls spacecraft traffic and debris aloft; delimits human error in spacecraft manufacture and operation; prevents and controls technical malfunctions; and creates organizational structures to ensure that space is used for reasonable and beneficial purposes. In the 21st century, the emphasis is on the management of knowledge and knowledge workers!

8.3 THE IMPACT OF ORGANIZATIONAL CULTURE

The work culture affects organizational planning, decisions, and behavior. Edgar Schein, an MIT professor, maintains that this is the mechanism for conveying, explicitly, ambiguously, or implicitly, the values, norms, and assumptions of the institution. Organizational culture is embedded and transmitted in human systems through [22]

- formal statements of philosophy or mission, charters, creeds, published materials for recruitment or personnel;
- design of physical spaces, facades, buildings;
- leader role modeling, training, coaching, or assessing;
- explicit reward and status system, promotion criteria;
- organizational fit (recruitment, selections, career development, retirements, or "excommunication");
- stories, legends, myths, parables about key people and events;
- leader reactions or coping, especially with organizational crises and critical situations;
- organizational design, structure, and systems;
- organizational policies, procedures, and processes.

In Chapter 4, the case was made for the influence of *culture* on space developments, including the organizational culture of space agencies and the aerospace industry (see Exhibit 52 on Aerospace Organizational Culture, which illustrates the many other dimensions of system expression of identity). Since research indicates that excellent organizations manifest strong functional cultures, leaders today give much attention to that dimension of macromanagment!

Global space agencies rightly focus on technology and mission accomplishment. However, it is advisable that some of their contracting be directed toward improvement and performance in internal management leadership. NASA, for example, frequently consults with university experts in the field of aerospace engineering, biotechnology, and medicine. As a matter of practice, it rarely connects with schools of business and world-class consulting firms to assist with management renewal or updating. Your author became a NASA consultant as a result of one such endeavor through a unique management development contract with Leadership Resources Inc. Unfortunately, engineers and technicians do not normally receive training in the art and science of management, even when they get promoted into positions of supervision and leadership There is an urgent need for space agencies to improve their *mission management*. That could happen if these government entities subsidized such research by professors and students in graduate business programs, or utilized outstanding international consulting corporations.

8.3.1 Learning from organizational setbacks

All human systems are subject to failures, so it is important to *learn* from such setbacks, so as to avoid their replication. Culture is a helpful coping tool in this regard, a means for adaptation and improvements. As a case for learning, consider the flaws in NASA's culture highlighted in two monumental reports of external experts analyzing the causes of the horrific Shuttle losses: *Challenger* and *Columbia*. Both investigations underscored the failure of mission managers to recognize dangers developing in these spacecraft and their designs, including many missed opportunities for discovering damage and instituting repairs. At every juncture, the organization's structure, processes, and management resisted new information or feedback from previous and present flights. Based on these findings, NASA continually has striven to alter its culture and re-invent itself.

In the aftermath of the demise of *Challenger* and its crew, the Presidential Commission (1986) investigating the tragedy revealed that its source lay not just in technological flaws, but also in management breakdown by NASA and its contractors. One of the positive outcomes within that agency was improved space management and scheduling, plus introduction of new supercomputer and management information system. In 1990, The Advisory Committee on the Future of the U.S. Space Program also provided an objective analysis of NASA's management turbulence, institutional aging, personnel, and technology base. Its report to both the agency and the National Space Council included a section on management with detailed recommendations for internal reorganizations, such as

- that a Systems Concepts and Analysis Group be formed in the NASA headquarters to serve the Administrators;

- that NASA adopt as standard for the management of multi-center programs a headquarters project manager and staff located at or near the "Primary Center" in the undertaking;
- that NASA management review the mission of each (field) center, consolidating and refocusing centers of excellence in currently relevant fields of science and technology with minimum overlap between centers;
- that an appropriate balance between in-house and external activity should also be developed;
- that NASA should concentrate its "hands-on" expertise in those areas unique to its mission ... Contract monitoring is best accomplished by a cadre of professional systems managers with appropriate experience. Increased use of performance requirements, rather than design specifications, will increase the effectiveness of this approach.

When external investigations or studies are commissioned, the issues become how well that institution analyzes and implements their recommendations, if feasible? For example, the above Advisory Committee recommendations, if adopted, would have profoundly changed the management of NASA's nine field centers. The proposal called for the agency to selectively convert some of these centers into a university-operated and government-owned organization, managed like the successful prototype model, the Jet Propulsion Laboratory in Pasadena, California, which operates in conjunction with the California Institute of Technology. Such changes have yet to be initiated ... Other space agencies are also in the process of organizational reforms and renewal for varying reasons. With the European Space Agency, the alterations are being propelled by creation of the European Union and launch failures with their Ariane rocket. For the Russian Space Agency, the reforms have been prompted by the collapse of the U.S.S.R. and the shift toward a market-economy, as well as the need for international agreements to finance existing space programs. For Japan, it was the excessive competition between two government space entities and their mission failures.

Testifying before the House Appropriations Subcommittee in 1995, former NASA administrator Dan Goldin outlined then plans for restructuring NASA management [23]:

- Cut out duplications/overlaps by consolidation.
- Transfer functions that need not be done by NASA to universities and/or the private sector.
- Change relationships with prime contractors, so NASA does less and contractors do more.
- Privatize and commercialize where appropriate.
- Shift the operations activities to the private sector, while focusing on "cutting edge" research and development.
- Emphasize objective contracting that enables rather than directs industry and academia.

The second Shuttle loss, of *Columbia* on February 1, 2003, led to significant investigations as to the cause and the failures. A monumental report by external experts in August 2003, revealed that mission managers failed to recognize the danger when foam was lost on these spacecraft. All told, there were eight such missed opportunities to prevent such accidents, while safety warnings were ignored. The investigating board's chairman, retired admiral Harold Gehman, rightly observed that it is not enough to identify the widget that failed or fire the managers to blame—but to examine the whole NASA culture and organization that led to this tragedy which resulted in the loss of not only a costly Shuttle, but the lives of seven valuable astronauts! Many organizational problems were identified and rectified relative to this very complex, risky, and experimental space transportation system, proving that learning from failure is vital. Thus, second-generation space plane designs are being improved by private enterprise, such as XCOR Aerospace demonstrates with its experimental rockets in Mojave, California: it has a record of reliable, capable, and less expensive rockets. And other space agencies have had their rocket failures, but are they gaining knowledge and benefit from those setbacks? For example, Brazil has an excellent launch site in its northeast at Alcantara which is 2° south of the equator. Yet on August 22nd, 2003, one of its rockets blew up during launch, killing 21 people on the ground. Since 1979, its management has had three failures of VLS 1 satellite launch rockets, so one wonders if they are truly learning from their failures?

But getting back to NASA, many question the whole way it does business with its overblown workforce and field centers. At $500 million per Shuttle launch, the Agency should be contracting out more routine tasks for lifting humans and materials into "near space" (anything closer than the Moon). The Space Frontier Foundation maintains that NASA should become a spaceflight customer, not a provider, so it may concentrate on its mandate to explore the unknown. That would really open the high-frontier market to private enterprise as will be suggested in Chapter 9. In 2008, NASA continues to be a case in point about changing organizational culture. It is still attempting to renew itself, to excel again in its Shuttle/Space Station/lunar-planning phase of its organizational development, but it is being done in a political and economic climate that demands downsizing and cutbacks in expenditures, especially by a country trying to fight a "war on terrorism".

Space management was examined at the California Space Institute when your author was a faculty fellow during a NASA Summer Study [24]. To ensure success with future space planning, a team of scholars and space experts at that time recommended that a survey and analysis of NASA organizational culture be conducted from its headquarters to the various field centers. Although this would facilitate planned change and renewal within the Agency, the proposal has yet to be fully implemented. Furthermore, these professionals concluded that if post–Space Station plans for a lunar base are to be effectively carried out, then it is likely that NASA must literally transform its management attitudes, styles, strategies, and operations. Whether it is NASA as an organizational system, or one of its top aerospace contracting partners, the aerospace work culture must shift from industrial or bureaucratic mode to that of enterprises characterized as *metaindustrial*. It may even require some structural changes in this Agency to give that organization more autonomy and mission focus. Then, NASA would be positioned to live up to its Apollo management heritage, enabling it to remain a principal actor in global space business, so as to take advantage of the vast resources on the high frontier well through the 21st century.

Management consultants envision organizations as energy exchange systems. Institutional culture can energize the use of the psychic and physical energies of its people in achieving organizational goals, or it can undermine and dissipate those efforts. This is the lesson of the Apollo Moon program for all engaged in space enterprises. In the private sector, space management is being transformed by economic forces which have caused many aerospace companies to disappear, such as General Dynamics Space Systems, when acquired and merged with a more visible industry leader, Lockheed-Martin. But it is space entrepreneurialism, discussed in the Chapter 9 and Appendix C, which may have the greatest positive impact in the decades ahead on the management of offworld enterprise. Start-up companies like Burt Rutan's Scaled Composites have more management flexibility than the traditional, large aerospace corporation. The new space entrepreneurs are more willing to draw on resources outside the aerospace field, as well as to take risks on smaller budgets.

"In changing NASA's organizational culture, there have been attempts to improve relationships with its external environment, both domestic and global as the following quotation underscores: "The international interactions of NASA have changed more slowly, representing a work in progress, rather than a deliberately designed change. In their own way, the changes being forced by the ISS program are just as profound as the 'faster, better, cheaper' initiative and the internal efficiency reforms. NASA is being forced of necessity to begin learning how to engage in truly international projects. Truly international meaning that the Agency can no longer unilaterally change the program, and then consult with its partners" (Roger Handberg, *Reinventing NASA: Human Spaceflight, Bureaucracy, and Politics*. Westport, CT: Praeger, 2003).

8.3.2 Space macromangement culture

To achieve comparable success in large-scale technological ventures, here is a sampling of cultural issues in space management to be confronted and altered:

(1) The mindset of the engineer and technologist requires expansion to more of a generalist; too often, present approaches exclude consideration of human issues, and the contributions of the management and behavioral sciences to planning and decision-making are downplayed.
(2) The need for more synergistic relationships in space endeavors, so as to replace obsolete, competitive postures by individual companies. Tasks related to exploitation of space resources are so immense that national space entities and efforts require both collaboration and partnerships. Perhaps the time has come for some type of global space administration or trust to be formulated for better

coordination of the space development of various countries, including both the private and public sectors?

(3) In spacefaring nations, legal mechanisms must be created that enable the aerospace industry and start-up space enterprises to work together in solving common problems, be it matters of quality control on launch pads and space vehicles, or greater sharing of research and development knowledge, such as in the creation of next-generation spacecraft. The great international aerospace corporations benefit the public more from joint venturing and sharing than by competitive duplication. Furthermore, new ways for cooperative inclusion of university and government research laboratories should be fostered. Perhaps the regional space consortium model being currently developed by the European Space Agency in this regard is worthy of increased emulation elsewhere in the world by cooperative agreements, such as proposed by the Global Exploration Strategy previously discussed?

(4) As space endeavors reach out to include more business participation beyond the traditional aerospace companies, attitudes and regulations regarding contractors and international technology sharing deserve revision. Perhaps the tradition of partnership with suppliers and other stakeholders is more appropriate than the mentality of seeing such simply as "users". Space enterprises would benefit from marketplace concerns for satisfying clients and customers. The commercial challenges in space management go beyond the opening of an office for this purpose in a government Department of Transportation or Commerce. To build a spacefaring civilization, nation-states have to facilitate, rather than over-regulate, space exploration and enterprise.

(5) Technology development timespans often have been increased, rather than condensed, because those in the space arena have become more bureaucratic, and less entrepreneurial and innovative. From goal-setting to implementation, the Apollo Moon missions were accomplished in a dozen years. Now, many space agency planners use a 15-year to 30-year timeframe from inception to completion of a new technology, while just the opposite experience occurs in the growing high-technology industry. Perhaps the time has come to re-examine the cultural assumptions on which practices of redundancy, over-design, over-preparation, over-study, and excessive timidity become imbedded habits and traditions? Such matters are worsened when these approaches go beyond aerospace design and get applied even to non-technical areas, like conference management and reporting. The high frontier needs ventures that not only cut costs, but the planning time before the implementation stage.

(6) Organizational renewal implies a continuing process for clarification of roles, relationships, and missions. It requires tradeoffs from the ways we always did it, to the adaptations and innovations necessary to remain in the emerging 21st-century "space game". Perhaps space stations and lunar outposts are better designed as habitat modules by architects and hotel chains than by traditional aerospace vehicle designers? Possibly the functions of such space activities should be privatized, so that space agency personnel can be given a more supervisory rather than operational role, thus freeing them for more basic space research and

development! Or maybe heath services aloft are better administered by space nurses and paramedics than physicians? Thinking "out of the box" has to be institutionalized!

These are some issues and questions that deserve consideration by management leaders in the global space community seeking to revitalize their organizational cultures and design a management strategy attuned to future demands [25]. Some aspects of terrestrial management are transferable offworld; others are inappropriate, requiring new management processes and approaches aloft. An example of the former may be found within the petroleum industry relative to managing effectively offshore facilities, as well as managing risk. The latter occurred as a result of large oil spills which led Exxon Mobil to study and adopt the Operations Integrity Management System which very well may have applications with lunar operations. Exhibit 101 summarizes this management innovation which has offworld implications.

Perhaps the global ground operations of the aerospace industries, space entrepreneurs, and space agencies is the place to install a similar risk management system! *Then apply OIMS to orbital operations, like those illustrated in Exhibit 102.*

8.4 EMERGING SPACE ROLES OF EARTH-BASED MANAGERS

Many management issues identified in the previous section are basically cultural and point up the need for planned changes within organizations which manage space projects. At the above-mentioned California Space Institute/NASA study on "Technological Springboard to the 21st Century," resource speakers addressed such issues and provided numerous insights related to the strategic planning for a lunar base. Five significant volumes of proceedings were edited by NASA's Mary Fae and David McKay with Michael Duke, and have been previously cited under the title *Space Resources* (SP-509). Washington, D.C.: U.S. Government Printing Office, 1992.

Several of the more telling comments from that futuristic report are relevant here to our management topic:

> "The aerospace industry culture is extraordinarily conservative. It suffers from a syndrome, "If it has not been done for the last twenty years, forget it." The industry and NASA are not bold enough in their planning and requests for funding. Major programs come into being because someone champions them (puts his "butt" on the line and helps bring it into being)" (William E. Wright, Defence Advanced Research Projects Agency).

> "Shared a document in the form of stock prospects for the establishment of a fictional corporation, "Consolidated Space Enterprises". It envisioned nine companies that could profit by serving customer needs and functions on the space station. Four were in the category of space service providers engaged in space transport, repair, research, and products; three were housekeeping companies

Exhibit 101. MANAGING RISK OFFWORLD.

Operations Integrity Management System is a system of management systems designed to identify potential hazards and manage the associated risks, while promoting safety, health, and environmental consciousness within the workforce. To foster the integrity of operations, OIMS provides discipline and structure to ensure improved performance, especially in exploration operations. It comprises the following 11 elements for every operation to meet; these offer a framework for meeting 64 expectations that address issues of safety, security, health, and environmental management. These key OIMS elements are

(1) *Management, leadership, commitment, and accountability* by management and workers relative to the above objectives.
(2) *Risk assessment and management* that include systematic reviews to help prevent incidents or accidents.
(3) *Facility design and construction* are evaluated early in the design for safety, security, health, and environmental impact.
(4) *Information and documentation* that is accurate, complete, accessible, and essential for safe and reliable operations.
(5) *Personnel and training* so that workers can meet high standards of performance.
(6) *Operations and maintenance* procedures subject to frequent assessment and modification, so as to improve safety and environmental performance by all involved.
(7) *Management of change* so that all alterations and innovations are evaluated relative to improved safety, health, and environmental impact.
(8) *Third-party services* are utilized to ensure responsible, safe, secure, and environmentally-sound operations.
(9) *Incident investigation and analysis* so that all incidents and "near misses" that do occur are thoroughly investigated.
(10) *Community awareness and emergence preparedness* that contribute to community involvement which may reduce the impact of incidents and accidents if they occur.
(11) *Operations integrity assessment and improvement* is a process for measuring performance relative to expectations regarding improved operations.

Source: "Managing Risk in a Challenging Business," *The Lamp*, June 2007, pp. 26–27. Published by Exxon Mobil Corporation (5959 Las Colinas Blvd., Irving, TX 75039; email: *exxonmobil.com*).

engaged in providing at the station, hotel, power, communications services; two were support companies providing special space services and fuel. The concept of commercial operations on the station, each "feeding" upon the other's needs, is not only stimulating to thought, but changes the role and relationship between public and private sector participation in space undertakings ... Vajk, now an independent space consultant, also cited examples of new, more sophisticated management information systems that can alter the role of space project

Sec. 8.4] Emerging space roles of Earth-based managers 375

Exhibit 102. Managing unmanned space missions: Galileo. This unmanned spacecraft was managed by teams of JPL scientists of different generations during its six-year journey across the Solar System until July 1995, when Galileo automatically released a probe to descend into Jupiter's atmosphere at a speed of 170,600 kilometers per hour. The probe, pictured going through the Jovian atmosphere, collected the first sample of that planet's chemical make-up, winds, and clouds, while radioing back data to Earth for 57 minutes. The automated Galileo Orbiter mission studied the Jovian system's moons, rings, and magnetic fields. Source: artwork, JPL/NASA in *Ad Astra*, September/October 1995, p. 17; text, a *Los Angeles Times* editorial (December 7, 1995). Tom Harris, a Canadian space writer, observed: "Today's Galileo spacecraft encounter with Jupiter provides exactly the kind of cultural influence society needs today ... Although Galileo is one of the last of the large planetary explorers, a new series of smaller, less costly spacecraft will soon be flying, giving us the thrill of extraterrestrial exploration and discovery on a regular basis.

managers. New computer tools, such as relational database management systems, give managers better capability for literature search" (Peter Vajk, SAI, and Michael Simon, then of General Dynamics).

"Pointed out that management issues related to space station and lunar base represent a departure from traditional NASA management practices. First, there is the matter of managing the development of such projects, and precursor missions; then there is the issue of operational management of a space facility when it is functioning. There are precedents with meteorological and communication satellites gained by NOAA and commercial operators. There are new challenges relative to man–machine interactions, operational cost containment, and private participation in such space activities" (Ronald Maehl, RCA).

These three samples of inputs from the above experts are indications for tomorrow's management of space enterprises that warrant innovative study and policy changes by space agencies and their contractors. The above observations also have implications for start-up space enterprises.

University schools of business and management have yet to produce significant research on the challenges in managing offworld macroprojects. Analysis needs to be devoted to the expertise and skills necessary for *Earth-based managers* of projects occurring hundreds or thousands of miles away from them. We have much to learn in this regard from previous project managers of unmanned probes by spacecraft, such as the Mariner, Viking, Pioneer, Voyager and Galileo missions, as well as from those who managed the Luna, Soyuz, Mir, and Energia programs. A book on space project management has yet to be written: the issues range from limited controls to teleoperations (the control on Earth by an operator of a machine that is at a remote location in space). New management problems are likely to be experienced related to "queuing time" (signal delays between operator and command and machine response and back). Some of the management changes relate to the use of automation and robotics in space. The role of ground-based managers of space enterprises will be further altered by other emerging technologies, particularly in the field of information and communications. This quotation illustrates the challenges for the new Mission Control Center at Houston's Johnson Space Center:

"Replacing the huge custom-built mainframe computer that drove monochrome monitors is a high speed fiber optic cable network of 200 Digital Equipment Corporation workstations with full-color graphic capability. The upgrade will enable the center to control a space shuttle and the proposed space station simultaneously, but at $30 million less each year than the old mission control" (*Air & Space*, Smithsonian Institution, October 1995, p. 14).

The many information technology innovations of the past decade and those to come are transforming the role of ground-based managers. One recent example is how Moscow mission controllers in June 2007 had to quickly problem-solve the restoration of power on Russian parts of the International Space Station. Space managers always have to deal with the unknown and the unexpected. It took a combination of three teams of managers (two on the gound, and one in orbit) to solve this problem.

As more manned space operations occur at multiple locations and for longer durations, new infrastructure will be needed on this planet to support such activities, often international and eventually interplanetary in scope. In addition to integrated worldwide mission control systems, there may be regional support centers, some under government or military auspices, and some under private corporations. When spaceports both on the ground and in orbit proliferate, unique management synergies will be invented which bring both the private and public sectors together in global joint ventures. For the next 50 years, we are likely to experiment with a variety of Earth-based management practices for orbital undertakings. As spacefaring becomes more complex and permanent, one trend seems evident: less control by those on the ground, and more control by those aloft. Ground management's role is likely to become more supportive and consultative, while orbital managers take charge of programs that involve human survival and performance in the microgravity environment.

Unique partnerships and synergistic relationships will be essential between managers and consultants on Earth, and their counterparts in orbit. More research is essential on the emerging role of *astromanagement* or manage in orbit. The learning laboratory for this is the International Space Station, and soon a lunar base. Whole new management systems and practices will evolve aloft. Innovative *management practices offworld* will occur through creative human responses to challenges posed by that unique environment, as illustrated in Exhibit 103.

Exhibit 103. Managing a new world: Mars. The exploration and eventual human ssettlement on other planets raises significant management challenges and hazards! How can we seek out life elsewhere in the solar system without harming it? Can robotic probes built on Earth be made clean enough to search for life on other planets without contaminating it? If we bring samples of alien life back to Earth, how do we prevent them from contaminating Earth's biosphere? "Planetary protection" is the prevention of "cross-contamination". That is, preventing life from getting from one planet to another and causing harm. NASA cares about planetary protection and that's why the agency has spent over 30 years and countless dollars trying to prevent such cross contamination. Source: NASA.

> **ENVOYS OF MANKIND**
>
> Thus, we stand in the late 20th Century, on the threshold of extending old civilizations into space, perhaps even creating new ones, in which our own sons and daughters may become extraterrestrials from every point of view ... Not only are our own sons and daughters pioneers in the firmament, they could also become biologically, if not taxonomically, different.
>
> What is the evidence that our original human space envoys are in any way different from their Earth-sitting counterparts, or that both earthkind and spacekind might become different in the course of generations? Are we really suggesting the possibility of a new subspecies in space ...? (Source: George S. Robinson and Harold M. White, Jr. in their seminal book, *Envoys of Mankind*, published in 1986 by the Smithsonian Institution Press, Washington, D.C. (see Appendix A).)

8.5 MANAGEMENT IN ORBIT

Diverse people on-site at the International Space Station, and soon at a Moon or Mars outpost, will require more freedom for decision-making and creative problem-solving than the astronauts and cosmonauts currently enjoy wherever a mission control center exists on Earth. Decentralized space and team management aloft will come into prominence with the erection of 21st-century space infrastructures: the creation of a spacefaring civilization. Now is the time to begin planning for the practical matters to be faced by orbiting stations and outpost managers, especially when the personnel and organizational components come from diverse sources and nationalities. With international space partners, there are bound to be matters of cross-cultural management and leadership differences arising, as already happens on this planet [25].

Relative to immediate macroprojects (e.g., an operational lunar base with advanced transportation vehicles), Earth/Moon systems research is needed now. Such studies by universities, space agencies, and corporations should focus on management concerns, such as communications, habitation, life support, emergencies, autonomy, leadership, and other such issues contributing to survival and quality of life on the lunar surface. If a manned mission to Mars were undertaken by 2050 under the auspices of a global consortium, then consider such crucial matters, as the above, magnified by the cultural differences of the sponsors as to the nature of management and leadership. Let us use Exhibit 104 to visualize how this "New Martian World" will be managed: the conditions, the processes, the products, the type of people required, all something quite different from terrestrial management.

Mixed crews aloft (men and women, military and civilian, private sector and public service workers, scientists and other professionals, diverse nationalities and cultures) will pose more complex management challenges and responses. The people in increasing numbers who visit a space station or lunar outpost by 2025 will include more than astronauts/cosmonauts, or even "technauts" (contractor technicians);

they will involve a broader segment of Earth's society from politicians to tourists. In past colonial explorations, trading companies were formed to recruit and sustain colonists in new, remote environs; perhaps some of these previous exploration strategies could be replicated by a Space Trading Company? The commercialization of the high frontier will be a profound force in altering the management of space projects.

As tomorrow's populations aloft increase in size and heterogeneity, as well as in length of their stay away from this planet, Ben Finney, when a University of Hawaii anthropologist, warned planners to expect more stress and strain among space personnel, requiring that inhabitants be given more autonomy in coping with their unique orbital environment. To maximize safe, congenial, effective performance by such pioneers in space living and work, behavioral science research and applications should be instituted related to orbital team development and group dynamics, new leadership training and responsibilities, and even wellness programs in outer space communities (see Chapter 3). Space managers on the ground and in orbit will have to concern themselves with such deployment issues, so as to facilitate a spacefarer's acculturation in an alien, isolated, sometime hostile, environment as was discussed in Chapter 6. Managing multicultural crews in space on long rotation involves participants who can deal with isolation and remoteness, are self-sufficient and multi-skilled, are sensitive to human relations issues, and operate by the norm of competence. It will call for new applications of leadership and followership. Whether on an orbiting platform in either LEO and GEO, or residing at a Moon or Martian base, the realities require costly, risky, and long-term programs involving new management procedures which provide continuity and consistency, quite apart from changes in personnel and annual budget constraints.

Another management concern to be addressed more vigorously is that of multipurpose missions, which combine civilian and military personnel and payloads. Economies of scale, piggybacking to constrain costs are arguments for such endeavors. Technical and management complexity, foreign policy and international cooperation considerations may provide stronger cases for not keeping separate commercial and defense space activities. Perhaps the biggest challenge may be in the education and training of scientists and technologists in space management, or the reverse, educating business managers in science and technology. Further, as space populations increase, peacekeepers and criminal justice systems will have to be introduced [26]. In this 21st century, diverse management competencies will be required in managing large-scale enterprises, as Exhibit 104 portrays.

8.6 MACROMANAGEMENT FOR/IN OUTER SPACE

As already indicated, large-scale and complex technical programs require a new type of *macromanagement*, whether to rebuild this planet's infrastructure or to create a space infrastructure. Exhibit 101 depicts my conceptual model in this regard, defining the scope of this term from a management perspective. Long-term projects costing $100 million or more require administrative skills across a range of activities that

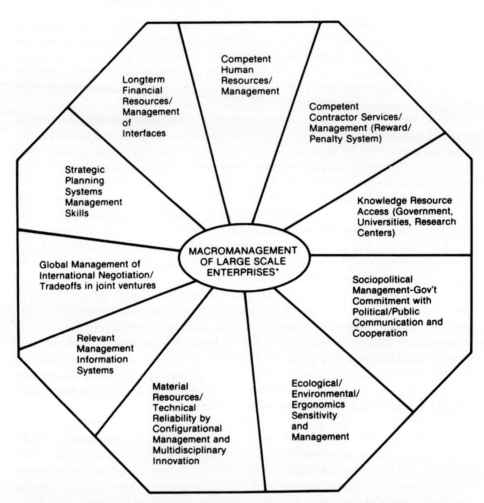

Exhibit 104. Macromanagement for space enterprises. Source: Astronautical artist Dennis M. Davidson illustrates the author's conceptualization for ten dimensions of *macromanagement*, whether it is practiced on the ground or in orbit. Most space macroprojects require this type of new, multidisciplinary, integrated management. *Long-term projects of $100m plus, such as rebuilding American infrastructure and building a space infrastructure.

begin with strategic planning and extend to global or interplanetary management of material and human resources.

Macroengineering projects have occurred for centuries, and may well shape our future. Space programs, like Apollo and the MIR/ISS space stations, advanced this field. They, along with the complex Shuttle technology, were the laboratories for the emerging discipline: space macromanagement. Most space programs are macroengineering in scope because they usually share these characteristics [27]:

(1) involve difficult, complex engineering and management problems which must be resolved before program completion;
(2) require significant public and private-sector resources that must be committed over long timeframes;
(3) include scientific and technical problems of complexity, unusual size or circumstances, and often unknown technologies or resources;
(4) have profound sociocultural impacts on societies which develop them relative to environmental, legal, and regulatory factors; economic and political effects, etc.;
(5) require massive funding from varied sources;
(6) have dimensions of large scale;
(7) extend technologies's states of the art;
(8) involve long-construction periods before completion;
(9) take place in difficult and sometimes hostile environments;
(10) have multinational and multicultural implications;
(11) require sophisticated project management techniques.

In the past few decades, project management of large-scale enterprises has benefited from creative tools, such as PERT (Program Evaluation and Review Technique), CPM (Critical Path Method), PMSS (Project Management State Space), Operations Management, systems modeling, and other such techniques. Similarly, developments in the supercomputer and management information systems have made macro-projects more feasible and manageable. Many of these management innovations owe their origins and refinement to the U.S. Department of Defense and NASA, as well as to high-tech companies.

8.6.1 Emerging macromanagement

Macromanagement may very well become a dominant theme in management theory and practice of the 21st century. According to the late Dalton McFarland, it represents post-industrial management. This distinguished professor maintained [28]:

> "Macromanagement theory will not only serve to guide and interpret research on management processes in society at large, but will also play a vital role in the determination of social, political, and economic policy."

By 2014, NASA and its partners seek to build the new crew exploration vehicle Orion to replace the Shuttle, or carry out its *VSE* plans to return permanently to the Moon by 2020. Not only will macromanagement strategies be needed within a culture of innovation, but managers aloft may also pioneer the process. With more corporations outside the aerospace industry participating in space ventures, all space agencies will face a unique set of interface challenges with these new stakeholders. Already space entrepreneurs expect to launch small space satellites and a variety of other commercial ventures that require creating synergy with governmental entities that control space access now. Burgeoning space technology enterprises will necessitate the adoption of macromanagement methods, as will the involvement of multiple countries in a single space macroproject.

Research funding should be expanded into matters of macromanagement by a global consortium made up of space agencies, world corporations, universities, and others. Such R&D influences the future of management, demanding a new type of managerial thinking, style, and skills. Presently, only the World Development Council is addressing issues of planning/financing/managing global superprojects [29]. Much must be done beyond the present endeavors of the Global Infrastructure Fund and the American Society of Macroengineering. Such organizations are in a position to develop a curriculum to educate macromanagers. Another positive step was taken when IBM cooperated with Cornell University in the Large Scale Scientific Computer Project, an essential tool for macromanging!

Exhibit 101 depicts macromanagement, whether on this planet or in space, as necessitating leadership capable of

- *synergy*, facilitating cooperation and collaboration in bringing together diverse elements so as to produce more than the sum of the parts;
- *intercultural and interpersonal skill*, managing diversity by overcoming differences between and among peoples, groups, and nations, particularly through effective cross-cultural communication, negotiation, and consensus-building;
- *political savvy*, gaining agreement and support for project goals from the various political or governmental actors/entities, as well as from the public if their support is essential;
- *financial competence*, understanding the economic realities of a long-term project, and capable of putting together the necessary business plan and funding to complete the undertaking, while containing excessive expenditures;
- *managing interfaces*, leadership in bringing together and on time the various resources required to achieve project goals (human, informational, technical, material).
- *cosmopolitan mindset*, sensitivity to global and interplanetary issues affecting the project, such as legal, ecological, environmental, and human, as well as the ability to cope with such international over national concerns.

8.6.2 Educating tomorrow's macromanagers

The above are but a sampling of qualities desirable in tomorrow's macromanagers, found in part today among global executives. In fact, no one person may possess all of these, but a management team may together exercise such competencies. If NASA engineers and administrators had possessed such competencies, their planned Space Station might have been in orbit in 1995. Certainly, a traditionally educated engineer is not likely to possess such skills. At MIT, research has been under way by Frank Davidson and others on a more comprehensive, multidisciplinary education of macroengineers. The American Society of Macro-Engineering, especially through its conferences and publications, such as *Technology in Review*, is also beginning to address these concerns [30]. Dr. Bob Krone, professor emeritus at the University of

> **EMERGENCE OF MACROMANAGEMENT**
>
> The emergence of macromanagement with the onset of World War II, followed by the Space Race between the U.S.A. and the U.S.S.R., eclipsed the era of individual Big Thinkers, more or less. New management was needed for macro or mega projects! Some of their well-known defining characteristics are: (1) complex and trying multi-scale engineering, geographical, and other management problems that must be solved even before the megaproject is begun; (2) significant public and private commitments are necessary, both in terms of financial and societal dedication; (3) sometimes solving scientific and operational problems of inordinate complexity are essential if favourable outcomes are to be realized; (4) macro-projects usually impact the planetary biosphere or the Earth's human-inhabited volume. As Geoffrey Dobbs noted in his book in the 1950s, planning implies control; and the stealing of the best choices. (Source: Richard Brooke Cathcart, *Macroengineering: Its History and Future* on *The Internet Encyclopedia of Science*, December 10, 2007.)

Southern California, believes that macromanagers capture the best of global brainpower. However, schools of businesses, as well as the Academy of Management, are lagging behind in synergistic education of engineering managers. Technology or R&D managers need to become more management generalists, more open to new ideas outside their own fields and industry, more competent in emerging skills of strategic planning and management. For this to happen, schools of engineering and business will have to design comprehensive, integrated curricula which include instruction in law, political science, finance, entrepreneurialism, or technological venturing. It may be one of the central issues of 21st-century management that explains why some scholar/practitioners, like the late Dr. George Kozmetsky when at the University of Texas-Austin, called for *transformational* management strategies [30]. There, his Institute for Constructive Capitalism fostered two progams akin to this chapter's theme:

(1) It co-sponsored with the Society for Design and Process Science a 1995 conference and publication on *Integrated Design and Process Technology*. IDPT seeks to use both technologies to create a symbiotic relationship among scientists, engineers, decision-makers, and other critical thinkers.
(2) It developed through two NASA Technology Commercialization Centers at the Johnson Space Center in Houston and the Ames Research Center in Sunnyvale, California entrepreneurial, incubator programs to transfer space research and technology into commercial enterprises.

During the previously mentioned NASA summer study at the California Space Institute [24], two resource speakers pointed out existing management models worthy of further analysis by space planners. To create the necessary infrastructures for tomorrow's space programs, New York consultant Kathleen Murphy proposed that we could learn from macrodevelopment projects around the world as another terrestrial analog for space:

- Major "greensite" projects in the Third World have already resolved problems of owners and contractors, conflict resolution and negotiations, reward and penalty provisions, as well as tested arrangement practices that might prove feasible for space development [31]. These include new financing models, joint ventures, consortia, shared R&D between government and industry, venture capital and national bank syndicate investments, and other macromanagement strategies.

    Further, consulting engineer Peter Vajk observed that many global projects relate to new terrestrial materials (NTMs) that offer insights for the exploration and exploitation of space resources.

- New terrestrial material enterprises are high risk, capital intensive, and involve very large R&D, start-up, and operating costs. Companies developing NTMs are beginning to use a macromanagement approach through a global corporate headquarters that sets general policy, negotiates major contracts, and keeps accounting/systems records, but follows distributed or semi-autonomous management with regard to subsidiary operations or facilities. NTM projects are also technology/multiskill-intensive, and involve macroengineering. Such project sponsors own, lease, or hire their transportation; they operate distribution centers, retail outlets, and sales/marketing offices. Their programs are extended in time and space with reference to R&D, deployment, and operations phases. Their activities are transnational and utilize sophisticated computer information networks involving high-rate data transfer. Vajk believes that macroprojects involving extraterrestrial resources (ETMs) can operate like these Earth-based analogs: "space is just a different place to do the same kinds of things we do on this planet." Exhibit 105 is an example of an ETM.

For eons humans have been "terraforming" their home planet and now we are planning to do it on other celestial bodies. As we engage in the latter, micromanagement will be essential.

8.7 SPACE MANAGEMENT IN THE FUTURE

But space is also a place for technological venturing and large-scale endeavors of a peaceful and commercial nature, such as envisioned in Exhibit 102 [32]. It opens opportunities on the part of human institutions and governments to produce synergy—not war. It not only requires a change in mindsets, but new management practices, multilateral action, and collaborative research. Thirty-seven years ago, a classic work provided us with the charter for that purpose. In *Managing Large Systems: Organization for the Future*, the authors reminded us that such enterprises are interdisciplinary in character, and integrate an array of scientific, technological, social, political, and other personalities and resources [33]. That describes the large-scale programs of most national space agencies, and was well understood by the key administrators of the Apollo program. Leonard Sayles and Margaret

Sec. 8.7] Space management in the future 385

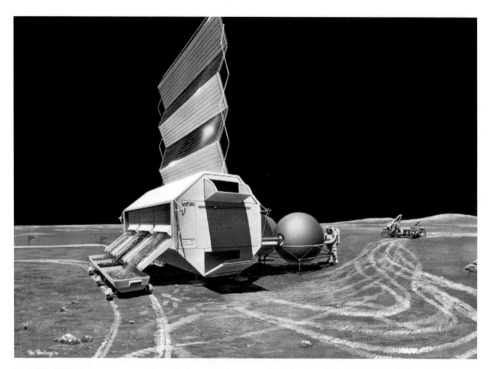

Exhibit 105. Lunar power plant. Macroprojects on the Moon might include building a compact oxygen plant like this one illustrated by Pat Rawlings of SAIC. It will process *lunar regolith*, using a hydrogen reduction treatment technique to refine oxygen for propulsion. Source: NASA Johnson Space Center.

Chandler's book is worthy of further restudy and application relative to the macromanagement of space enterprises, such as the Lunar Solar Power System (Appendix B).

Over 20 years ago, the National Commission on Space appointed by the President of the United States recommended investing $700 billion on the space frontier up to year 2036. The macro space undertakings proposed in their 1986 seminal report on *Pioneering the Space Frontier* are of such scope as to need more than bold vision—they require a new systems approach for managing continuity over long time periods, despite fluctuations in personnel, policies, governmental administrations, and finances. Three such macroproject examples are presented in Exhibit 106. Imagine where we would now be in space, if humanity had invested that kind of funding for offworld development, instead of wasting limited resources on military weaponry and wars! Today, the global community is only just talking about getting back to the Moon permanently in a dozen years, and initial planning activities are getting under way!

The struggle for humanity in this century is whether space will be used for peaceful, commercial, and settlement enterprises, or for nationalistic military and "defense" of the high ground, such as with space-based ballistic missiles and spy

Exhibit 106. Managing Moon macroprojects. Artist Robert McCall envisions activities of the next century on the Moon, which may include astronomy with both optical and radio waves; use of lunar materials to manufacture products in space; and a mass driver (foreground) that propels baseball-sized lunar materials into space. Source: National Commission on Space, *Pioneering the Space Frontier*. New York: Bantam Books, 1986, p. 64.

satellites! Yet, the United Nations treaty that most countries signed about outer space made it clear that space was humanity's heritage and should be utilized for peaceful purposes.

8.7.1 Re-education of humanity

Today *earthkind* is still terrestrially oriented, and only a minority of the world's population comprehend why and where we should be going beyond this planet. Gaining national and international consensus to support such space visions and ventures as pictured above is a cultural problem. Implementing plans for the purpose implies innovative approaches to space management, such as discussed here. Why we are so timid in carrying out the Commission's bold recommendations has been well expressed by *Washington Post* writer, Charles Krauthammer (April 9, 1995):

> "Ours was the generation that first escaped gravity, walked on the Moon, visited Saturn—and then, overtaken by inexpiable lassitude and narrowness of vision, turned its cathedrals of flight into wind tunnels."

For existing space organizations, whether global space agencies, their aerospace partners, or space activists, the mass re-education of humanity about offworld exploration and development is essential, particularly by use of modern media. But their own personnel also have to be prepared for the challenges inherent in space management in general, and macromanagement in particular. Whether for oldtimers or newcomers in space enterprises, executive/management development programs should be redesigned to deal with these considerations. In transnational partnerships, corporate HRD (human resource development) specialists need to cooperate with R&D professionals to create more appropriate in-house training of macromanagement skills. Then, universities and colleges have to design appropriate space management curricula and majors, as well as to encourage writers to produce textbook material on this new subject matter.

Space management in the future will necessitate crossing traditional academic disciplines and industrial differentiations, as the quotation of Frank Davidson's book *MACRO* so aptly describes [34]:

"Space development is a critical case-in-point, because it will test the ability of our diverse, rather relaxed society to set long-range goals, to hue the line despite disappointments and setbacks, and to devise institutional arrangements that will assure continuity ... Low-cost approaches are indispensable, because an increasingly educated public will rightly insist on return on investment ... Now is the time, for the aerospace community to reach out to the mining industry, the heavy construction industry, and the ground transportation industry, so that joint ventures on land and sea, as well as 'up there,' may set a pattern of partnership and a network of personal relationships which will benefit all systems engineering programs that are so necessary for the future health, safety, and prosperity of the Republic."

A prototype of future *macromanagement education* may be in the making at the International Space University in France in their Master's Degree Program in Space Science (*www.isu.edu*). This graduate program is available during the regular academic year, as well as in a ten-week summer session at different campuses abroad. ISU departments are multidisciplinary in Space Systems Architecture and Mission Design; Space Business and Management; Space Engineering; Space Life Sciences; Space Policy and Law; Space Resources, Robotics and Manufacturing; Satellite Applications; Space Physical Sciences; Space and Society. In the second department mentioned, the business and management aspects of space projects are examined. The core lectures cover a broad range of topics: management techniques, financing, costing, business planning and structure. Principles are illustrated with examples from space enterprises, while management tools are used in terms of the managing and economics of space activities. Special lectures deal with space failures and quality management, space project management, and space business case studies. The core curriculum is taught by an international faculty, supplemented by lecturers from around the world who are experts in their subject matter. This author sees this emerging model as most promising for the preparation of tomorrow's macromanagers. My hope someday is that this innovative approach will be spread to other universities electronically, and eventually in orbit!

8.8 SPACE STATION CASE STUDY

Orbiting platforms and stations are the predecessors of tomorrow's orbiting resorts and cities! Back in 1952, Wernher von Braun proposed building an orbiting wheel which artist Chesley Bonestell so beautifully illustrated. It inspired Princeton physicist Gerard O'Neill to research and speculate about orbiting space colonies on the high frontier. Then in the late 20th century, such dreams were translated into realities with the launching of Skylab, Mir, and finally, the International Space Station.

To appreciate why macromanagement is so vital to space development, the following mini case study is presented. As suggested in Chapter 7 on *macrothinking*, this summarizes why project managers must utilize a combination of political, financial, technological,, communication, and managerial skills to achieve their goals. In this instance, the macroproject is the construction of an orbital platform and laboratory as depicted next in Exhibit 107. Its proponents have had a difficult time putting together a case for investment that sufficiently convinces both the public, legislators, and scientists to support this venture whose future is still uncertain as of this writing [35]. The global public is generally unaware of the technological and cultural feats being accomplished at the orbiting International Space Station, except when there is a crisis or mishap up there. This account contains many learnings about trying to manage a space macroproject, including some good and some bad approaches.

8.8.1 Mini case study: evolution of the International Space Station

Visionaries have dreamed and written of housing people on islands in the sky for centuries. Wernher von Braun, who began the Space Age from Peenemünde on the Baltic coast of Germany on October 3, 1942 with the launching of the first rocket to reach space (an early V-2), wrote over 50 years ago in *Colliers* magazine about future orbiting stations and platforms. This great rocket scientist actually designed a demonstration model, a 77-meter diameter wheel which was to become immortalized in the film, *2001: A Space Odyssey*, based on Arthur C. Clarke's book. Von Braun envisioned the station as the starting point for expeditions to the Moon—instead the U.S. went there directly, postponing the orbiting station. Others, like your author, argue for using the Moon as an orbiting space station [36].

8.8.1.1 Background briefing

After many abortive proposals and debates, NASA actually launched the first Earth-orbiting station in Skylab, flying three missions there in 1973. The astronaut crews stayed aboard it for up to three months, while performing 270 multidisciplinary

Exhibit 107. International Space Station cooperation. As we end the first decade in the New Millennium, this macroproject will finally be completed, representing the most challenging attempt at synergy in space among the world community. The often-changed name and design for this orbiting joint venture by the U.S.A., Europe, Canada, Japan and Russia may yet face further alterations by the sixteen nations in this allied partnership [see the next case study for the evolving story]. Source: NASA Johnson Space Center.

investigations. The spacecraft contained a house-sized workshop, airlock, docking adapter, solar observatory, and command service module. Weighing 100 tons, the orbiting facility provided 350 m^3 of workspace and was 54 meters in length. A walk through the Skylab mock-up in the Smithsonian Air and Space Museum will confirm the size and amenities which its astronauts enjoyed while aloft. Much was learned from this project about human capacity to live and work in the microgravity environment. With insufficient public support and devastating budget cuts, its lifetime was limited until Skylab plunged back into the atmosphere six years later [37].

After a year of experimenting with an orbiting space station, the former Soviets began to establish semi-permanent manned facilities in 1977 with the launch of Salyut 6. The Russian station evolved as modules were added to subsequent Salyut and Cosmos craft until the Mir superstation appeared in 1986. Political and social turmoil within the then U.S.S.R. in the 1990s postponed their plans for Mir 2 and Kosmograd, a space city in the sky. Within the new Commonwealth of Independent States, the Russian Federation then developed agreements with NASA, ESA, CSA, and NASDA for their personnel to conduct life and material science experiments aboard Mir, and both astronauts and cosmonauts have used each other's spacecraft for this purpose. In 1995, the U.S. Shuttle docked for the first of several such dockings with the aging Russian station which hosted space visitors from many nations. Finally, deterioration of its systems and costs of maintenance forced the Russians to decommission this great "orbital laboratory for learning". In 1999, on the 13th anniversary of Mir's original launching, the last three-person crew blasted off the station which had enabled 100 multinational men and women to live and work there in orbit! By the next year, Moscow mission control directed the 130-ton unmanned station through a fiery re-entry and destruction, with its remnants falling into a desolate ocean location. With this symbol of space glory and national pride gone, the Russians were able to direct their talent and limited resources to finishing, with other national partners, the International Space Station [38]. Helpful results were a 1994 RSC Energia hazard report to NASA, an analysis of 1994 and 1997 fires aboard Mir, plus reports of problems with Soyuz flights to their station. The design of ISS was improved by considering the way Mir systems wore out or failed over time. The whole experience with the Russian station brought about improvements in the new international station, such as their crises with air leaks, spacewalk emergencies, life support collapse, loss of control, major medical and psychological crises, and many others. The Russians have proven to be a very valuable station partner in many ways, especially since they signed up for the construction of ISS in 1993, when a major redesign of the project took place.

When fully assembled by 2010, the International Space Station will be five times as large as Skylab, and four times as large as the Mir station!

8.8.1.2 NASA's station saga

In the 1970s, NASA began the search for a permanently manned space station design, the configuration of which had been subject to numerous debates among politicians, space scientists, engineers, and the public. After President Ronald Reagan in 1981 requested from the U.S. Congress, $8 billion to actually build it, Hans Mark attempted to record the history of these furious debates which are still going on today [39]. An NCB network broadcaster reported on April 6, 1991 that NASA has now *spen*t $5 billion on a space station that is not likely to be permanently occupied until 1999, and then by only four astronauts. This commentator said the price tag for the finished product is likely to be $30 billion, yet some called the station an "orbiting pork barrel". By 1995, some insiders claimed that the Agency had actually spent $12 billion on station *planning*, estimating that 70 cents of every $1 so expended went on "paperwork" to complete the project; critics then estimated $40 billion more would be required to complete this macroproject ... Each year when NASA seeks its funding authorization for this program, the debate heats up again, instead of getting on with multi-year authorization for the financing of the oft-redesigned station. Meanwhile, NASA engineers and bureaucrats have been severely criticized for over-promising what might be done elsewhere more cheaply. No one has a convincing proposal about how an orbiting facility might be funded through the sale of bonds

Apart from failing to manage effectively its station budget needs, the basic problem with this macroproject is that the Agency did not make a substantial and convincing *case for investment* to the public, the Congress, the scientific and industrial communities. Over the years, the cost estimates and purposes of the program have altered many times. In effect, the Agency did not properly *macromanage* the program! Despite the efforts of well-meaning space activists to promote this large-scale technological venture, a political base and consensus for the station has yet to be developed. Many rationales have been offered by NASA and its aerospace partners as to why the station needs to be constructed at such huge cost, plus even larger sums that will be required to keep it in orbit in terms of maintenance and human occupancy. Typical justifications range from a place for the Shuttle fleet to dock and a platform for exploration of the universe, to a laboratory for microgravity manufacturing, for general scientific experimentation and observations, for life science research on the effects of weightlessness, and even for defense purposes. The arguments in favor maintain the station is needed for learning about orbital living and mechanics; for development of new space-based vehicles, and technologies; for missions to return to the Moon and lunar industrialization, for going to Mars and beyond. Some expert study panels and commissions agreed, such as the National Commission on Space (1986). Others did not, especially among planetary scientists whose astronomers proposed putting the money into unmanned space explorations. Once, the University Space Research Association sponsored a study by 57 universities that recommended using the Shuttle's external tank, now discarded, to build the station; NASA showed no interest in the proposal. Given that the design life of the station is only projected for 15 or so years, some observers, like your author, wonder if it would not have been better to invest in the Moon itself as a permanent space station and launch pad into the universe? By the mid-1990s, the project had been scaled backed in scope, with a life science laboratory then given as the principal reason for building it. International cooperation with Russian participation was another thrust in advocating the program.

NASA did try to win support for the enterprise; it produced beautiful brochures as to why the station was necessary as an orbital foothold to the future. Proponents argued the station was needed as a new national laboratory for science and technology, for exploration, for U.S. competitiveness and technology leadership, and for humankind. The agency did enlist *Spacecause* to lobby for the project; it did sign up four major contractors and numerous subcontractors who would also seek to "win friends" for the program. But its designs for habitation, laboratory, and logistics modules were considered by some experts to be grandiose. What started as the U.S. Space Station Freedom, for a time called Alpha, had to be transformed today into an "international partnership" to reduce expenses, plus obtaining funds and technical assistance from the space agencies of Europe, Japan, Canada, and principally Russia. However, these same partners became increasingly disillusioned at: (a) the slow pace and scope of negotiations, even after agreements were signed; (b) the numerous redesigns and budget reductions which caused scheduling delays caused by U.S. politicians, which impacted the programs of their countries to provide separate laboratory/habitat modules. By November 2001, the then NASA administrator, Sean O'Keefe, a budget technology expert, was testifying before the U.S. Congress that *the ISS was a program at the crossroads*, and that he was dismayed at the out-of-control finances of that project.

The U.S.A., which provides the primary funding, is distracted from underwriting such a massive space undertaking because of the tremendous national budget deficit, and the unexpected costs of two Middle Eastern wars, compounded by ever-growing demands for domestic spending to remedy problems in the homeland (e.g., from decaying infrastructure, to homelessness and substance abuse). In the minds of many Americans, these concerns seemed to outweigh the scientific, economic, social, and political advantages claimed by NASA for the station. Thus, budget cuts consistently erode original plans to house a crew of eight permanently in a low-inclination orbit of $28.5°$ to the equator. A dual-keel station weighing well over 200 tonnes, with 5×5 meter truss structure elements, had to be scaled back on orders from the U.S. Congress (a $6 billion curtailment over five years). The complicated management support system had to be reorganized: the headquarters' Space Station Program Office supposedly coordinates station activities in nine of the agency's field centers, with 15 major international partners, as well as numerous private contractors.

For 26 years, this station macroproject has produced enormous management problems: (a) difficult negotiations to obtain agreement from all station partners to keep the facility design and equipment compatible; (b) counteracting criticism of the station designs from engineers who boasted of superior, less expensive designs; (c) critical reports from other governmental agencies, such as the White House's Office of Science and Technology. Meanwhile *the incredibly shrinking station* has been redesigned from originally 122 meters long, with one-sixth the transmitting capability originally conceived; the $1.8 billion annually operating cost has been scaled down by 80% because of Congressional mandate permitting only $2.1 billion a year to be spent on the whole endeavor. Yet, in 1995, NASA was still promising to build the station by June 2002 for a mere $17.4 billion! By 1998, $20 billion had already been spent over the 14 years from drawing-board to launch, and the station was far from complete. Frustration was evident in a newspaper feature, "Scrap the Space Station" which appeared in *The Economist* (November 14th, 1998, p. 19), predicting the macroproject would cost $100 billion before it was finished. That estimate is likely to be proven close to total program costs which are now forecast at $130 billion by year 2010, the completion goal.

8.8.1.3 *Emerging adaptations*

Finally named the *International Space Station (ISS)*, the project is coordinated by the NASA Office of Space Flight with a total now of some 17 international partners. The NASA Office of Space Access and Technology has been designated to spearhead new ways for the Agency to do business, transferring space technology to commercial applications, including what is learned from the ISS operations. OSAT developed a Payload Traffic Model, a long-range operations plan determining which experiments will be sent to the station when it is in orbit. NASA states that the ISS goal now is to provide, at 445 km above the Earth, a "long-term research platform in a near zero gravity environment to conduct basic and applied science, technology, and commercial driven research, as well as provide an observation platform for Earth and space science." Since 1984 when NASA started to actually build a space station, there have been many meetings among the Agency and its partners as to the station's rationale and its future use. For example, in 2002, a conference was held in Montreal, Canada, in which NASA program managers led discussions on preparing a business plan for privatization of the station over the next 15 years! In 2007, that strategy is still being pursued by ISS sponsors. Proposals for using this orbital facility have extended from life science research and commercial manufacturing, to a destination for tourists and other non-conventional activities.

The principal contractors then, Boeing, Rocketdyne, North American Rockwell, and McDonnald-Douglas, built hardware for launching in 1997–1998. Beside well-proven space technology, the contractors innovated with new technologies, such as a computer-simulated system called Integrated Equipment Assembly that was used to test various elements on the ground, instead of in microgravity. Then the ISS configuration called for eight pressurized modules, a long metal connecting truss, a Canadian robot arm for exterior work, a solar power system, and numerous smaller systems to maintain the "laboratory". The latest is that it will be 88 meters long, with a 110-meter wingspan, a volume of $1,230 \, m^3$, and a total of 110 kW of power. The revised orbital inclination of ISS is 51.6°, permitting launches from Russia. The loss of two shuttles, *Challenger* and *Columbia*, severely handicapped the whole station construction schedule. When the Shuttle fleet was grounded for safety studies for a total of 975 days, transportation to and from the station was totally dependent on Russian spacecraft. The original plan called for a November 1997 start-up, that was to inaugurate some 44 spaceflights to assemble 100 components until June 2002, when basic assembly was to be completed and the crew installed. The Technical and Management Information System (TMIS) developed the computerized means for assembling and distributing data about the project. But the elaborate expenditures for ground support facilities and programs had to be reduced.

It would seem to this writer that the ISS construction plan was based on four activity phases requiring 73 launches to take place in a sequence over 55 months, involving possibly 888 hours of both American/Russian EVAs at great risk. The evolution of this station over the past 17-year period, achieved a series of milestones for this orbital macroproject.[1] These began in November 1998 when the first Zarya element was launched on a Proton rocket from Russia's Baikonour Cosmodrone in Kazakhstan (though Russian-built, NASA paid $240 million for it). In December 1998, the crew of Shuttle *Endeavor* snatched the 22-ton, unmanned Russian service module in orbit, mating it with an American-built module released from the Shuttle's

---

[1] See "Milestones" in *Creating the International Space Station* by D. M. Harland and J. E. Catchpole. Chichester, UK: Springer/Praxis, 2002, pp. 339–346. For further clarification of this term, see M. J. Fogg, *Terraforming: Engineering Projects and the Environment*. Warrendale, PA: SAE, 1995.

payload deck. The formidable 240-mile high construction job included three EVAs using special tools to attach electrical connectors and cables. Three days of spacewalks completed the rigging, but two days inside were required to install computers, communication systems, and supplies. With the help of both Houston and Moscow controllers, the station was operational, with power provided by the Russian Zarya (meaning "Sunrise") module, and communications by the American-made chamber called Unity (36 foot and 25,000 pounds). At the cost of another $300 million, this U.S. component also serves as a connecting vestibule or passageway between the units. This feat marked the beginning of human presence, made possible by a Soyuz flight the following year. The term "Expedition" was designated for crew missions, from Expedition 1 in 1999 (STS-960 on Shuttle *Discovery*), to Expedition 17 in June 2008, as this was being completed. With each station mission, whether on American or Russian spacecraft, humans have been accomplishing spectacular feats as more modules are added through this monumental construction in orbit.

Since 1998, 14 or more ISS components have actually been assembled in space. Among the milestones of installation was Russia's Zvezda service module in 2000 brought by a Proton rocket; the U.S. Spacehab single-cargo module in 2006 on Shuttle *Discovery*; Europe's Columbus laboratory module in 2007 brought on the *Discovery* Orbiter; Japanese Kibo experimental logistics module in 2007 arriving on Shuttle *Endeavor*. Incremental missions and assemblies continue to build ISS at this writing... Furthermore, the station has been continuously staffed since November 2, 2000, engaging in both assembly and research activities. Much of the latter is done in a Boeing-built laboratory called "Destiny" which has 13 telephone booth–sized racks for science experiments, plus the control room for the robotic arm.

The whole docking maneuver, whether by a Russian or American spacecraft, is also something of a technological wonder. With only three Shuttles left in NASA's fleet, this 122-foot long vehicle weighing 240,000 pounds can deliver up to 35,000 pounds of cargo. Docking requires a gradual approach to ISS. This occurs at a foot per second while both spacecraft race through orbit at 17,000 miles an hour. Just before docking, the Shuttle now does a graceful backflip, so station occupants can photograph details of its bottom surface to inspect for any tile damage or loss, which is repaired after docking and before re-entry. Following a successful docking, there is a transfer of equipment and supplies to the orbiting lab. These are often stored in the Italian-made reusable cargo carrier *Leonardo* which can hold up to 5,000 pounds. Some of the ISS activities can be watched by those on the home planet by means of television, videos, or the IMAX film entitled, *Space Station*. Thus, millions of people are aware of problems, delays, and accomplishments with both the Shuttles and the station that often require creative solutions, repairs, and replacements. NASA's rebuttal is that every dollar spent on ISS will produce $2 in direct and indirect benefit to humanity.

8.8.1.4 Station management

Based on the Intergovernmental Agreement (IGA) previously signed and revised in 1995 by Canada, ESA members, Japan, and Russia, the *partners'* roles and responsibilities have been spelled out, along with their financial contributions and allocated onboard space. *Users* are of several types: facility-class payloads for microgravity/life science research; technology and commercial payloads; express and external payloads (the latter are attached to four outside ports which may be plugged into station resources). The user community has a greater voice in program preparation through a special representative; users now sit directly on various product teams developing the station. The seven science facilities available aboard the station are: laboratory; 2.5 m centrifuge; a furnace; facilities for gravitational biology; human research; fluids and convection; and biotechnology.

At NASA, the station chief is now the Deputy Administrator of the Office of Space Flight, who heads up a streamlined management team, most of whose members are located at JSC in Houston, Texas. The design team includes both Americans and Russians (e.g., chief engineers from NPO *Energia*). The management approach of Total Quality Management has been incorporated into the *Integrated Product Team*. IPT is organized along product lines—instead of function and disciplines. Hundreds of these multidisciplinary teams will have resident product experts, including contractors, business managers, and customers. McDonnell-Douglas, for instance, had five IPTs for each segment of the main truss it delivered to NASA under its contract. Within that company there is a Vice President for the Space Station, so that the IPT process encourages management to get its act together and plan, while making sure roles and responsibilities are well defined. The international partners agreed that three resources are essential for research at the station: crew time, onboard laboratory volume, and available power.

All the ISS contractors and partners, whether in the U.S.A., Russia, Europe, Japan, and Canada, have learned much about micromanagement while working on this large-scale joint venture. Boeing Integrated Defense System, for instance, was the prime contractor for the design, development, integration, testing, and delivery of the American-built elements. Principally, these consisted of three connecting modules or nodes: a laboratory module; truss segments; four solar array modules; three mating adapters; a cupola; and an unpressurized carrier. This aerospace corporation's contract for $13.3 billion provided thermal control, life support, guidance, navigation/control, data handling, power systems, communications, and tracking. About 1,850 employees across the U.S. work on this ISS program in four different locations. Obviously, among such partners and contractors are numerous macroproject experts whose knowledge should be collected, published, and become the basis for micromanagement education and training.

Once aloft, either a NASA astronaut or an RKA cosmonaut may be designated as Commander during a set time period to direct onboard operations. Interestingly, in November 1998, NASA issued an *ISS Commercial Development Plan* which called for a non-governmental organization to manage the station when completed. This NGO would act as landlord and station manager, handling operations and leases by negotiating and contracting with governmental and commercial entities for space and services. This strategy for using private-sector management which has efficiently and profitably run similar terrestrial projects makes sense. It is not unlike the Orbital Space Authority proposal put forward by United Societies in Space, Inc.

8.8.1.5 Current situation

Now, there are 17 space agencies involved in the ISS macroproject all of which have coped with many obstacles in building this orbital platform incrementally. Apart from the loss of two Shuttles, they have dealt with multiple problems, such as inadequate budget, eroding performance schedules, power demands on the station that exceed generating capacity, etc. But gradually these have been overcome; Russia in particular proved to be a very reliable and creative partner. By November 2000, the first crew of three were aboard to facilitate the station's further construction. As of this writing in 2007, the high-performing spacefarers are successful in their trouble shooting and problem solving. For example, when a new addition caused a power failure in the Russian segment's central computer, the ISS station commander followed an established trouble-shooting procedure for repair and replacement of computer systems. In the 15 Expeditions to date, the station has grown bigger, better, and more powerful. The Shuttle and Soyuz craft regularly dock at ISS for personnel exchanges and to bring equipment, while the Russian unmanned Progress freighters deliver food and supplies, up to 2.5 tons. Diverse people of mixed genders, nationalities, and skills work well together in this orbital laboratory. Some have worked over 5–6 hours on EVAs to install new equipment and make adjustments, or do external maintenance on both the Shuttle and the station itself. A crane-like Strela telescoping boom is used to move both spacewalkers and cargo. There is also a mobile transporter; it moves a platform containing a robotic arm along the truss of the complex.

Within this expensive orbital observatory, there is commodious living, dining, and exercise quarters: the three-level Transhab module houses up to six inhabitants providing an $84\,\text{ft}^3$ bedroom for each; common kitchen with dining table for up to 12; pantry, toilet, bath, small gym, and storage areas. This module is crafted from fabric and composite materials. Inflated in space, the walls are a foot thick to provide better protection from space debris, but have thinner radiation shields ... In ISS laboratory work stations, the PROMISS experiment in station communications uses diagnostic equipment to monitor the exact growth conditions of protein crystals, which may contribute to improved medicines ... All U.S. science activities on the station are coordinated by a payload operations team at NASA's Marshall Space Flight Center in Huntsville, Alabama. The scope of changing onboard experiments ranges from time lapse photography to studying the physics of Earth's surface crystallization and fluids. Astronaut Barbara Morgan has even conducted educational sessions with students back on Earth! Some orbital efforts include producing videos on the weightless environment for use in education. The major modules attached to ISS as scientific laboratories involve personnel from NASA, RKA, ESA, JAXA, and other nations (e.g., a South Korean woman astronaut).

8.8.1.6 ISS finished product

In 2008, JAXA's Experimental Logistics Pressurized Module was installed on ISS. This Kibo (meaning "hope" in Japanese) laboratory and storage unit is part of the Japan Experiment Module or JEM, that will permit external research experiments at the station. RSC Energia is constructing a multipurpose research laboratory for Russia's Federal Space Agency; it is to be launched on a Proton rocket for ISS installation by the end of 2008.

Because of the multiple hazards this orbiting platform could encounter in space, NASA has an ISS Independent Safety Task Force ordered by the U.S. Congress. Their 2007 report considered potential disastrous risks posed by the impact on the station of space debris and micrometeorites, collisions from visiting spacecraft or the ISS robotic arm, onboard fires and toxic spills, undermining of station systems by mistakes made by Mission Control or computer failures. Recommendations were made to cope with such vulnerabilities, including changes that reduce risks. Furthermore, astronaut health aboard the station is now monitored more closely because *operating in space is, and will be for the foreseeable future, inherently risky, requiring continuing discipline and diligence to maintain safe operations*!

By 2010, ISS will be complete with re-designed air-conditioning and electrical systems, new Shuttle docking port, four additional solar arrays. The latter give the station a rotating space an acre in size to draw power from the Sun. Thus, ISS has become the first human-made object in orbit that can be seen with the naked eye from Earth during daylight. The six assembly flights during the previous 18 months have doubled the station's size. Shuttle flights have been trimmed to 17, so as to retire the orbiters by the end of the decade. Thereafter, station re-supply will depend mainly on cargo vehicles from Russia, Europe, and possibly Japan. However, NASA has under development the Orion Crew Exploration Vehicle (CEV) which is intended to take crews both to the ISS and the Moon. Details of the finished station product are given in the Case Study Exhibit.

To sustain a crew of six at ISS until year 2025 will likely cost up to $60 billion. Will it be worth it? Yes, the station is expected to open biological research frontiers, so as to unlock the secrets of proteins through space crystallization experiments. NASA also created a Bioreactor device for tissue culture research aloft, enabling long-term cancer research and treatment testing. Investigations are under way to study fluids, fire, molten metal, and other materials within the low-gravity environment. There are also 14 different sites outside the ISS that can be used to attach exterior experiments. The purpose is to advance both Space Science and Fundamental Physics. Finally, the orbiting platform is used for Earth observation, studying long-term environmental changes on the home planet. The aim is to advance Earth Science related to geology, oceanography, and ecology.

As a macroproject, the International Space Station is so far the greatest construction project of humankind, rivaling the pyramids of Egypt, the Great Wall of China, and the grand canals built in Suez and Panama. ISS is the largest, most complex international scientific project in history, as well as the largest ever undertaken offworld. When ISS becomes fully functional, it remains to be seen whether the facility will justify the investment and effort put into the venture. And the humans aloft on ISS continue to astound us with their high-performance feats. For example, on June 19, 2002, the Space Shuttle *Endeavor* returned from the station and its 5.8-million-mile journey with two astronauts who broke the U.S. orbital endurance record: Daniel Bursch and Carl Walz passed Astronaut Sharon Lucid's record at that time of 188 days on a station. Again, on October 21, 2007, two women for the first time were simultaneously in change of spacecraft at the same time: Pamela Melroy commanded the *Discovery* mission, while Peggy Whitson was in charge of the ISS mission ... For ISS updates, refer to NASA Johnson Space Center electronic Internet reports on Shuttle Missions/Space Station by contacting listserv@jsc.nasa.gov (type in "subscribe hsfnews"). Fore moer information on ISS activities offworld, visit www.nasa.gov/station or tune into the NASA TV channel. Additional data may be obtained from http://nim.nih.gov and choose SPACELINE; www.discovery.com/stories/science/iss/iss.html; en.wikipedia.org/wiki/international_Space_Station;

Case Study Exhibit. ISS AT COMPLETION IN YEAR 2010.

Solar array wingspan	356 ft or 108.5 m (port to starboard, but 239 ft or 72.8 m tip to tip)
Length (pressurized section)	167 ft or 51.0 m
Total length	192 ft or 58.5 m
Total height	100 ft or 30.5 m
Integrated truss length	310 ft or 94.5 m
Mass weight (with 2 Soyuz vehicles docked)	927,316 lb or 420,623 kg
Operating altitude	220 nautical miles (average) or 407 km
Inclination	51.6° to the equator
Atmosphere inside	14.7 psi or 101.36 kPa
Pressured volume (with two Soyuz docked)	34,700 ft^3 or habitable volume 14,400 ft^3
Onboard computers to control ISS	52
Robotic arms	3 ft–55 ft Canadian RA for assembly lift of 220,000 lb ... 30 ft European RA for external experiments on Russian laboratory ... 30 ft Japanese RA based on Kibo, to move and deploy experiments on JEM
Power generation	84 kW
Human crew	6 (holds up to 12 when Shuttle docks)
Total components	100 via 88 orbital assembly spaceflights

Source: Boeing, *Space Exploration: International Space Station Backgrounder*, 2006, 6 pp. Available from IDS Business Support, Communications and Community Affairs (PO Box 516, St. Louis, MO 63166; www.boeing.com/spacedenfense/space/spacestation).

www.boeing.com/defense-space/space/spacestation and choose "International Space Station" PDF overview.

8.9 ISS LEARNINGS IN MACROMANAGEMENT

Building, orbiting, maintaining. and staffing the two orbital laboratories called Mir and the International Space Station were momentous achievements, as well as awesome engineering feats that confirm the need for improved space management synergy [40]. However, given the histories of both these space stations, there must be a better way to manage space macroprojects. Building on experience in high-technology management, it will require macromanagement skills if numerous spaceports are to be constructed on the ground or in orbit, along with a lunar base in the decades ahead. That means creating a macroproject with a business focus, financial controls, adaptability, organizational cohesion, entrepreneurial culture, and a sense of integrity (as Professor Sean O'Keefe once lectured to his students at Syracuse University's Maxwell School of Citizenship and Public Affairs). Also space agencies and corporations, as well as schools of business and engineering, will have to provide education in managing large-scale projects. Even the whole approach of project management will have to be expanded for larger, long-term undertakings. Similarly, our present manner of engaging in strategic planning and financing has to be revised in the context of macroprojects.

Dr. Michael Griffin, NASA administrator at this writing, described the International Space Station as *one of the most amazing construction projects that human beings have ever undertaken*. That statement appeared in a *New York Times* editorial (September 22, 2006, p. A20), which rightly added that the greatest value of the space station enterprise is simply the experience of building such a large, complex structure in space. This venture makes the case for further study of the whole concept of macromanagement. It's 27 years since a U.S. President mandated the construction of this orbiting laboratory, and there is much to learn from both the failures and successes in achieving that goal. Since ISS has only a projected life of 15 years, the planning and assembly period from start to finish has to be drastically reduced for future ventures offworld!

The ongoing knowledge obtained from this endeavor underscores these lessons about planning and carrying out macroprojects:

- international cooperation and participation is essential;
- human resources have to be developed for macrothinking, macroplanning, and macromanagement;
- funding requires large budgets, possibly in 5-year increments, while innovative investment is sought in the private sector, as well as from governments;
- creative multidisciplinary input and team management are critical in macroprojects;
- technical excellence must be matched by effective management of costs and schedules;

- careful monitoring of milestone achievements and fiscal controls are necessary;
- revising the rationale for using the ISS beyond that of science, technology, and commercial operations to include paying tourists and a way-station for lunar activities.

The station will also provide much learning about managing people in orbit. Once ISS is fully operational, perhaps the biggest challenge will be ascertaining the ability of humans to live and work together effectively on the facility for long durations. As JoAnna Wood, a psychologist at Baylor University put it: "one of the biggest showstoppers is psychological" [41]. Literally, the station will also be a laboratory as to how long our species can productively and comfortably stay aloft: six months to three years or more? Johnson Space Center architects have done their utmost to make the facility hospitable: the habitation module assures spacefarers of privacy with cubicles which they can decorate with photos of family and friends. A dining area permits the crew to gather together for at least one meal, for such social interaction is vital in remote environments. NASA's human support technologies will continue to provide for a balanced diet of both American and Russian foods until the day when fresh food can be grown on the station. Communication provisions have been made so astronauts and cosmonauts may keep in touch with friends and family on Earth. But there are many questions regarding human relations (including sexual) which have to be resolved. Candidates for missions to the ISS have to demonstrate their capacity for getting along well with others. The cultural backgrounds of participants will expand, as more Europeans, Japanese, Chinese, and even Brazilians go aloft! Expect diversity as the number of tourists to ISS increases.

Some planners hope to see the ISS become a site for building a lunar transportation system, consisting of robotic Lunar Transfer Vehicles to ferry cargo to the Moon, and Lunar Exploration Vehicles that could take both humans and materials to the Moon. Only at the orbiting station can the very large orbital frames for LTV aerobrakes be constructed with the required welding and fittings. Aerospace engineers envision the rocket-propelled LTVs and LEVs eventually providing a ferry service with ISS as a way-station to and from the Moon, and then possibly evolving into an interplanetary spaceport.

8.10 CONCLUSIONS ON SPACE MACROMANAGEMENT

The plans which global space agencies have for space developments during the next 50 years will have to include private enterprise if they are to be realized. Even at the minimum level of maintaining the International Space Station, building new space transportation systems and a lunar outpost, construction and operating costs will be billions of dollars, requiring a new generation of technological and management advances. To accomplish such minimal offworld goals demands: (1) more international technological partnerships involving private-sector participation with the public sector; (2) a different type of space management and leadership from that exercised during the past 50 years which inaugurated the Space Age; (3) a new method of financing for space infrastructure that may involve strategies from the selling of space bonds to global space lotteries; (4) enlisting the aid of the global media to convince the public as to why humanity needs to explore space and utilize its resources.

Building on the Apollo heritage of technological/administrative innovation, macromanagement has been proposed here for 21st-century space macroprojects. These begin by facilitating lunar enterprises, and might range from actually building space settlements aloft, then manned missions to Mars, to mining the asteroids. However, the technicians and professionals involved must first have mastered the basic skills of managing [42]. Furthermore, this chapter has highlighted some of the issues that need to be addressed in the strategic planning/management of offworld enterprises, such as

(1) large-scale technological undertakings involving humans dictate innovative approaches to their financing and management;
(2) increases in orbital human populations for longer periods of time will necessitate a decrease in the influence and management of ground control from Earth;
(3) expanding space-based programs, whether manned or unmanned, entail more effective management by spacefarers on-site;
(4) space management, both on the ground and in orbit, will be the laboratory for advancing the art and science of all macromanagement in this third millennium.

For any of the above to happen, governments with developed economies need to allocate more than 1% of their national budgets to the next stage of space exploration and exploitation. The latter will transform our species, while opening up space resources at a magnitude greater than anything experienced until now by humankind. Here are some of the offworld natural resources that, through macrothinking, macroplanning, and macromanagement, will benefit humanity:

- solar energy could solve Earth's energy problems while being utilized at space settlements;
- materials on the Moon or near-Earth asteroids that could be put to industrial use;
- microgravity environment and vacuum that offers advantages for stable orbits and transmissions, as well as for food and drug production.

Those with foresight and competence have already tapped this bonanza by means of communication satellites, remote sensing, and surveying for mapping and imaging of Earth resources. Now big-thinking planners are working on (1) space telescopes to be placed on the Moon and at the L2 libration point; (2) mining of asteroids; and (3) even building space business parks. Chapter 9 will hopefully further reader understanding of the potential beyond the home planet.

> **DIVERSITY IN SPACE**
>
> With the globalization of the space program, managers, trainers, and commanders in orbit will have to possess multiple management and diversity skills. Cross-cultural management competencies will be essential for missions that are more international in scope, in terms both of operations and personnel composition. Space macroprojects will bring together people from many nations and cultures, all conditioned in a particular way of managing and operating. Crew diversity will take on added significance in terms of the International Space Station, a lunar base, or manned mission to Mars. This new millennium will be the focus of space settlement and industrialization, pointing up the interdependence of both earth-kind and spacekind. Diversity begins as terrestrials migrate offworld, but real diversity will evolve when our species settles and adapts to the varied environments of our Solar System. And, if we should meet life from other worlds that may be the ultimate challenge in coping with diversity! (Source: Philip R. Harris, Ph.D., space psychologist and author.)

8.11 REFERENCES

[1] The Global Exploration Strategy, *The Framework for Coordination*. London: British National Space Centre, 2007-06-18 (BNSC Press Release 31.05.07 entitled "New Era for Space Exploration as 14 Space Agencies Take Historic Step"; *www.bnsc.gov.uk/home*) ... For multicultural challenges, see Simons, G. F. *Eurodiversity: A Business Guide to Managing Differences*. Burlington, MA: Elsevier/Butterworth-Heinemann, 2002.

[2] Elder, D. C.; and James, G. S. *History of Rocketry and Astronautics*, Vol. 26. San Diego, CA: UNIVELT, 2005 (this is part of the American Astronautics Society's History Series of some 15 volumes with the same title, going from Vol. 7 in 1986 to this present release; inquire *www.univelt.com*) ... NASA, *Orders of Magnitude: A History of NACA and NASA, 1915–1990*; NASA, *A History of the U.S. Civil Space Program*. Washington, D.C.: U.S. Government Printing Office, 1989; 1995 ... Levine, A. S. *Managing NASA in the Apollo Era*. Washington, D.C.: NASA Scientific and Technical Information Branch (NASA SP-4102), 1984 ... Mark, H.; and Levine, A. S. *The Management of Research Institutions: A Look at Government Laboratories*. Washington, D.C.: U.S. Government Printing Office (NASA SP-481), 1984 ... Johnson, W. L. (ed.) *The Management of Aerospace Programs*. San Diego, CA: UNIVELT/AAS, Vol. 12, 1967.

[3] Harvey, B. *The New Russian Space Programme: From Competition to Collaboration*. Chichester, U.K.: Wiley/Praxis, 1996 ... Johnson, N. *The Soviet Space Program: 1980–1985*. San Diego, CA: UNIVELT/AAS, Vol. 66, 1985.

[4] David, L. "More than a Global Market," in *Space and Industry*, the program booklet for the International Space Year issued at the World Space Congress, 1992 (*www.aiaa.org*).

[5] Harrison, A. A. *Spacefaring: The Human Dimension*. Berkeley, CA: University of California Press, 2001 ... Connors, M. M.; Harrison, A. A.; and Atkins, E. E. (eds.) *Living Aloft: Human Requirements for Extended Space Flight*. Washington, D.C.: U.S. Government Printing Office (NASA SP-483), 1985.

[6] Davidson, F. P.; and Meador, C. L. (eds.) *Macroengineering: Global Infrastructure Solutions*. Chichester, U.K.: Ellis Horwood, 1992. See also *Technology in Society* (an international journal published by the Polytechnic University, 333 Jay St., Brooklyn, NY 11201).

[7] Cordell, B. "Interspace: Design for an International Space Agency," *Space Policy*, November 1992, pp. 287–294 ... Contact ANSER, Center for International Aerospace Cooperation, a non-profit public service research institute (1215 Jefferson Davis Highway, Ste. 800, Arlington, VA 22202; www.anser.org).

[8] MacDaniel, W. E. "Intellectual Stimulant," *Extraterrestrial Society Newsletter*, **7**(2), 1985 (then published by the Sociology Dept., Niagara University, New York, NY 14109).

[9] Harris, P. R. *Managing the Knowledge Culture; The New Work Culture and HRD Transformational Management*. Amherst, MA: HRD Press, 2005; 1996 ... Moran, R. T.; Harris, P. R.; and Moran, S. V. *Managing Cultural Differences: Global Leadership Strategies for the 21st Century*. Burlington, MA: Elsevier/Butterworth-Heinemann, 2007, Seventh Edition.

[10] Reynolds, D. W. *Apollo: The Epic Journey to the Moon*. New York: Harcourt/Tehabi Books, 2002 (www.harcourtbooks.com) ... Chaikin, A. A. *Man on the Moon: The Voyages of the Apollo Astronauts*. New York: Viking, 1994.

[11] Seamans, R. C.; and Ordway, F. I. "The Apollo Tradition: An Object Lesson for Management of Large-scale Technological Endeavors," *Interdisciplinary Science Review*, **2**(4), 207–303.

[12] Bizony, P. *The Man Who Ran the Moon: James E. Webb, NASA, and the Secret History of Project Apollo*. New York: Thunder's Mouth Press, 2006 ... Levine, A. J. *Managing NASA in the Apollo Era*. Washington, D.C.: U.S. Government Printing Office.

[13] MacFarland, D. E. *Managerial Imperative: Age of Macromanagement*. Cambridge, MA: Ballinger/Harper & Row, 1985.

[14] O'Toole, J. *Vanguard Management*. New York: Berkley Publishing Group, 1987.

[15] Bauch, J. A *Study of Decision-Making within Matrix Organizations*. San Diego, CA: United States (now Alliant) International University, 1981 (available through University Microfilms International, 300 N. Zeeb Rd., Ann Arbor, MI 48106).

[16] Harris, P. R. *High Performance Leadership*. Amherst, MA: HRD Press, 1994 (www.drphilipharris.com/).

[17] Sayles, L. R.; and Chandler, M. K. *Managing Large Systems: Organizations for the Future*. New York: Harper & Row, 1971.

[18] Charnes, A.; and Cooper, W. W. (eds.) *Creative and Innovative Management*. Cambridge, MA: Ballinger/Harper & Row, 1984. See also Amidon, D. M. *The Innovation Superhighway: Harnessing Intellectual Capital for Sustainable Collaborative Advantage*. Burlington, MA: Elsevier/Butterworth-Heinemann, 2003.

[19] McElroy, M. W. *The New Knowledge Management: Complexity, Learning and Sustainable Innovation*. Burlington, MA: Elsevier/Butterworth-Heinemann, 2003 ... LaRoche, L. *Managing Cultural Diversity in Technical Professions*. Burlington, MA: Elsevier, 2002 (www.books@elsevier/business/).

[20] Nohria, N.; and Eccles, E. G. *Networking and Organizations: Structure, Form, Action*. Boston, MA: Harvard University Press, 1994. See also Skyrme, D. *Knowledge Networking: Creating Collaborative Enterprise*. Burlington, MA: Elsevier/Butterworth-Heinemann, 1999.

[21] Lubos, P. "Management of Outer Space," *Space Policy*, **10**(3), August 1994, 189–198 ... See also McCurdy, H. E. *Faster, Better, Cheaper: Low Cost Innovation in the U.S. Space Program*. Baltimore, MD: Johns Hopkins Press, 2001. See same author and publisher for

titles: *Inside NASA: High Technology and Organizational Change in the U.S. Space Program*, 1993 ... *The Space Station Decision: Incremental Politics and Technological Choice*, 1993.

[22] Schein, E. H. *Organizational Culture and Leadership*. San Francisco, CA: Jossey-Bass, 1985. See also Jones, E. *Innovating at the Edge: How Organizations Evolve and Embed Innovation Capability*. Burlington, MA: Elsevier/Butterworth-Heinemann, 2000.

[23] Handberg, R. *Reinventing NASA: Human Spaceflights, Bureaucracy, and Politics*. Westport, CT: Praeger, 2003 ... "The Future of NASA: Lost in Space," *The Economist*, August 30th, 2003, pp. 57–58 ... Launius, R. D.; and McCurdy, H. E. *Imagining Space: Achievements, Predictions, and Possibilities*. San Francisco, CA: Chronicle Books, 2001.

[24] Harris, P. R.; and Murphy, K. J. "New Space Management and Structure," *Social Concerns*, Vol. 4, pp. 1–141, in McKay, M. F.; McKay, D. S.; and Duke, M. B. (eds.) *Space Resources*. Washington, D.C.: U.S. Government Printing Office, 1992, 4 vols. (NASA SP-509; *www.univelt.com*).

[25] Wibbeke, E. S. *Global Management Leadership*. Burlington, MA: Elsevier/Butterworth-Heinemann, 2008 (*www.books@elsevier.com/business*).

[26] Shusta, R. M.; Levine, D. R.; Wong, H. Z.; Olson, A. T.; and Harris, P. R. *Multicultural Law Enforcement: Strategies for Peacekeeping in a Diverse Society*. Upper Saddle River, NJ: Prentice Hall, 2007, Fourth Edition (*www.prenhall.com/criminaljustice*).

[27] Badescu, V.; Cathcart, R. B.; and Schuiling, R. D. *Macro-engineering: A Challenge for the Future*. New York: Springer-Verlag, 2006 ... Davidson, F. P.; Meador, C. L.; and Salkeld, R. (eds.) *How Big Is Beautiful: Macroengineering Revisited*. Boulder, CO: Westview Press, 1980 ... Carter, N. E. "The Challenge of Macro-engineering," *Battelle Today*, 1985, Reprint #24 (published by Battelle Institute, 505 King Ave., Cleveland, OH 43201).

[28] McFarland, D. E. "Management, Humanism, and Society: The Case for Micromanagement," *Academy of Management Review*, October 1977.

[29] For information on the conferences and proceedings of the World Development Council (40 Technology Park, Ste. 200, Norcross, GA 30092) consult the American Society of Macroengineering's *Technology in Society* (an international journal published by the Institute for Technology Management and Policy at the Polytechnic Institute (333 Jay St., Brooklyn, NY 11201).

[30] Kozmetsky, G. *Transformational Management*. Cambridge, MA: Ballinger/Harper & Row, 1985; "Education for Large-scale and Complex Systems Based on Technology Venturing," *Technology in Society*, **6**, 1984, 173–176 ... Krone, B.; and Krone, S. *Ideas Unlimited: Capturing Global Business Brainpower*. West Conshohocken, PA: Infinity Publishing, 2007 (*www.buybooksontheweb.com*) ... Goudie, A. S.; and Cuffs, D. J. *Encyclopedia of Change*. Oxford, U.K.: Oxford University Press, 2002, pp. 495–502.

[31] Murphy, K. J. *Macroproject Development in the Third World: An Analysis of Transnational Partnerships*. Boulder, CO: Westview Press, 1983.

[32] Konecci, E. B.; and Kuhn, R. L. (eds.) *Technology Venturing: American Innovation and Risk Taking*. New York: Praeger, 1985 ... Mendell, W. W. (ed.) *Lunar Bases and Space Activities of the 21st Century*. Houston, TX: Lunar and Planetary Institute, 1985.

[33] Sayles, L. R.; and Chandler, M. K. *Managing Large Systems: Organization for the Future*. New York: Harper & Row, 1971.

[34] Davidson, F. P.; and Brooke, K. L. *Building the World: An Encyclopedia of Great Engineering Projects in History*. Oxford, U.K.: Greenwood Publishing Group, 2006 ... Davidson, F. P.; and Box, J. B. *MACRO*. New York: William Morrow & Co., 1983.

[35] Harland, D. M.; and Catchpole, J. E. *Creating the International Space Station*. Chichester, U.K.: Springer/Praxis, 2002 ... Heppenheimer, T. A. *Development of the Space Shuttle: 1972–1981*. Washington, D.C.: Smithsonian Institution Press, 2001.
[36] Harris, P. R. "Why Not Use the Moon as a Space Station?" *Earth Space Review*, **4**(4), December 1995, 7–10.
[37] Beckey, I.; and Herman, D. (eds.) *Space Stations and Platforms: Concepts, Designs, Infrastructures, and Use*. Washington, D.C.: American Institute of Aeronautics and Astronautics, 1986 ... McCurdy, H. E. *The Space Station: Incremental Politics and Technological Choice*. Baltimore, MD: Johns Hopkins University Press, 1990.
[38] Harland, D. M. *The Mir Space Station: A Precursor to Space Colonies*. Chichester, U.K.: Wiley/Praxis, 1997 ... Burrough, B. *Dragonfly: NASA and the Crisis aboard Mir*. New York: Harper/Collins Publishers, 1998 ... Kanasa, N. *et al*., "Crewmembers' Interactions during Joint U.S./Russian Mir Missions," *Journal of Human Performance in Extreme Environments*, June 1999 (email: *performance@HPEE.org*).
[39] Marks, H. *The Space Station: A Personal Journey*. Durham, NC: Duke University Press, 1990 ... See also the cover story on "The International Space Station," in *Launchspace*, official magazine of the space industry, June/July, 1997, pp. 30–42 (published by Launchspace Publications, 7777 Leesburg Pike, Falls Church, VA 22034).
[40] Beattie, D. A. *ISScapades: The Crippling of America's Space Program ... Reference Guide to the International Space Station*. Burlington, Ontario: Apogee Books/CGPublishing, 2007; 2006 (*www.cgpublishing.com*).
[41] McFarling, M. E. "Space Station Will Be Test of Endurance," *Los Angeles Times*, October 20, 2000, pp. A10–A14. See also Marsha Freeman, "Space Station Opens New Biomedical Frontiers," *21st Century Science & Technology*, Fall 1998, 72–79.
[42] Sears, W. H. *Front Line Guide to Mastering the Manager's Job ... Creating a Winning Management Style ... Thinking Clearly ... Communicating with Employees ... Building High Performance Teams*. Amherst, MA: Human Resource Development Press, 2007, 5 vols. (short, pragmatic books; *www.hrdpress.com*).

SPACE ENTERPRISE UPDATE

Macrolearning can come from the careful analysis of previous efforts with large-scale space undertakings. This is especially evident in the above space station case study. One can speculate whether the Russians have done this relative to their experience with orbiting stations, especially Mir. Similarly, American macroplanners would gain much from an assessment of the International Space Station. Was it worth spending the expected $130 billion during twelve years of construction, given the limited lifespan of ISS?

Some will argue that such a vast investment should have been directed to using the Moon as a natural space station. What if the Nixon presidential administration had not made the decision to discontinue the Apollo program after its 17th mission? What if the Reagan administration had not decided to build the present space station? Would we have been much farther ahead in space development if we had developed an infrastructure and utilized the resources on the Moon during the past 35 years? On the other hand, one could also argue that insights and skills acquired from building these space stations might someday enable humanity to macroplan for orbiting cities in space.

Yes, macroprojects beyond Earth call for a new type of innovative thinking, planning, and management, preferably on an international basis. Right now, Europe has such a macroproject under way. Called Galileo, this will be a global network of 30 global positioning satellites providing precise timing and location information to users in the air and on the ground. It will be able to pinpoint locations on Earth to a meter or so. The development of Galileo has already cost more than $1.5 billion, and will probably need at least a further $3.4 billion to get it operational by 2014. This level of investment makes Galileo the biggest space project initiated in Europe. How much macrothinking and macroplanning will go into this expensive venture offworld?

9

Challenges in offworld private enterprise

Because of the changes that were happening in Russia, there was a unique opportunity in changing the way people think about space. This change is still underway, and the growth of space tourism means that in coming years, we'll all have members of our family who will have been there and had their thinking changed by that experience.

Mark Shuttleworth, Software Entrepreneur, First African in Space,
World's Second Space Tourist (2002) [1]

Although the human species required over three and a half billion years to evolve, George Robinson, a space philosopher and attorney, observed that in the past 50 years we have moved beyond Earth to penetrate near space, deep space, and other planets [2]. In the process of transforming our perceptions of humanity, space law scholars speculate that *Homo spatialis* or *spacekind* will develop as a new species, altered in time from *Homo sapiens*, physically, psychologically, and socially (see Appendix A). In contemplating the human occupation of outer space, issues related to its industrialization and settlement may be viewed as problems or challenges [3]. Preferring the latter approach, there are indeed numerous multidimensional challenges: the first are technological, biological, and financial. However, in this chapter, we will delimit analysis to just three dimensions: commercial, legal, and political, ending with specific action plans to further space enterprise.[1] We will also revisit and amplify some of the themes discussed in Chapter 1.

---

[1] The author acknowledges the special contribution to this chapter of his esteemed colleague, Dr. Nathan C. Goldman, space attorney and professor. The research reported here was initially undertaken when we were Faculty Fellows together in a NASA Summer Study on utilizing space resources [4].

9.1 CHALLENGES AND REALITIES IN SPACE ENTERPRISES

In the 21st century, the most exciting developments are the internationalization of space endeavors, and the emerging role of private enterprise in offworld activities! Let us examine these trends within the following context.

Globally, all of the existing national space programs are moving forward but with more economic constraints in the midst of varying degrees of chaos within their countries. Yet, even government space agencies have recognized the need for fostering space commerce: NASA, for instance, set up special offices and programs to encourage it. A U.S. Congressional Budget Office study observed that the civilian space program has been justified as a means for realizing human destiny, an investment in the international standing of the country, and a provider of economic benefits [5]. The latter can be demonstrated in terms of public goods and services, stimulation of private sector R&D, and the creation of new industries, technologies, products, and services as a result of space activities. NASA has tried to make a case for space investment by highlighting the spinoff applications from its research, and underwriting commercial technology transfer [6]. The U.S.A. is still reaping technological returns from the Apollo legacy, ranging from laser heart surgery, magnetic resonance imaging, and voice-controlled wheelchairs for the handicapped, to cued speech devices for the deaf, reading machines for the blind, and "aquaculture" techniques for recycled water.

When president of the National Space Society, Ben Bova said it best in a *Space World* editorial (September 1987, p. 6):

> "Space is not a luxury. The space program is not intended merely for exploration and adventure. Space is an economic necessity ... and offers important economic paybacks ... Since NASA's creation, Washington has appropriated roughly $130 billion for the space agency—less than half of one year's Defense budget ... Yet the money we taxpayers have invested in space comes back to us magnified 20, 30, 50 times each year. Space-derived technology is responsible for $500 billion per year in the U.S. economy ... As space begins to develop, as industrial plants take advantage of low gravity, high vacuum, and endless solar energy, space technologies will become the dominant force in 21st Century economies."

9.1.1 Space entrepreneurialism

Many nations through their space agencies, such as ESA and RKA, have done studies and reports on the importance of involving the private sector in their space programs. The major point was to convince taxpayers and business leaders that what happens "up there" has not only direct benefits "down here"—but that there is also commercial opportunity offworld! One such attempt to influence national space policy and the public's mindset occurred when NASA gave a grant for this purpose to the National Academy of Public Administration. As a result an expert panel issued a 53-page publication, *Encouraging Business Ventures in Space Technologies* (see Exhibit 108 for excerpts from the summary).

Exhibit 108. SPACE BUSINESS VENTURES.

Within the past quarter century, the United States has penetrated the frontiers of space exploring even distant reaches of the Solar System. The space program, born of national resolve and financed by the American people, has opened up the space environment to the scrutiny of mankind ... Now, moreover, we begin to see emerging possibilities for private industry to use space technology for new commercial ventures. The resulting business could strengthen the economy, expand employment, and improve the nation's posture in the global competition for high-technology markets.

The extent to which past investment in space technology contributes to our future economic wellbeing and national growth will depend in large measure on policies and actions taken in the spirit of collaboration by the Federal government and industry. Unless the public and private sector join to develop the opportunities presented by new space technologies, and unless entrepreneurial forces are engaged more fully, the United States will fall behind in the contest for leadership in space and economic rewards associated with that position.

The National Academy of Public Administration Panel recommends the following policies and initiatives for adoption by NASA to encourage business ventures in space technologies:

(1) Declare and institutionalize a major commitment to the commercialization of space technology.
(2) Assist industry in pursuing opportunities for profitable investment in space.
(3) Offer NASA facilities and services for use by private companies under conditions that encourage commercial development.
(4) Continue R&D including study of long-range space opportunities.
(5) Reduce the risks and restrictions that impede commercial exploitation of space technologies.

Source: Kloman, E. H. (ed.) *Encouraging Business Ventures in Space Technologies*. National Academy for Public Administration (1120 G Street N.W., Washington, D.C. 20005; tel.: 202/347-3190), 1982, p. 53.

To NASA's credit, the Agency did try to implement many of the recommendations of this distinguished panel of ten. Unfortunately, the bureaucracy in many Federal departments still provides private firms with many confusing hurdles to overcome in order to engage in space business: these center primarily around regulations and licenses related to launches. In 2006, NASA finally selected Lockheed-Martin to be its prime contractor of the next-generation Orion Crew Exploration Vehicle. The noted magazine *Aviation Week & Space Technology* claims this CEV proposal is said to contain numerous technical enhancements, plus effective and realistic concepts for avionics and software development, as well as proven operations and innovative technologies. But critics maintain that this move to replace the Shuttle through aerospace giant corporations only cripples innovation, creativity, and the chance for breakthroughs in spacecraft design by entrepreneurs.

Yet, the Agency has inaugurated its *Centennial Challenges* competition which is sponsoring a contest to win contracts for a Lunar Lander and Vertical Rocket designs; it also sponsors Space Elevator games and the Wirefly X-Prize Cup ... Another milestone was achieved in December 2006 when the Federal Aviation Administration issued its *Final Rule on Human Space Flight Requirements for Crew and Space Flight Participants*. The FAA concern is primarily on space crew qualifications while acknowledging the paramount importance of safety and training in the budding private launch industry.

After 30 years of research on the contents of this book, the author has concluded that it will be private enterprise that truly opens up the space frontier through commerce. The history of exploration confirms a pattern: a small number of explorers and traders move first into the new frontier; then governments take interest in the territorial acquisition prospects, so military outposts are established, often with the help of missionaries, and a basic infrastructure emerges; but, for settlement, it is large commercial trading companies that bring civilization in the form of colonists seeking to improve their life prospects. Examine the opening and development of the African and American frontiers by Europeans, and this pattern becomes evident. Similarly, with regard to outer space, it was the explorers in science fiction and the rocket enthusiasts who opened our minds to the possibilities beyond Earth. Then, it was governments, like those of the U.S.A. and the U.S.S.R., that got into a competitive political race to use offworld opportunities. In the former country, space leadership came from two government agencies, the Department of Defense and NASA, both of whom employed civilian contractors. Recall that pioneering astronauts and cosmonauts were usually from a military background, while the actual unmanned exploration resulted from civilian teams of scientists, engineers, and academics. Growing out of the birth and maturity of worldwide aviation, the great aerospace industry arose in the last century. And these big corporations innovated and succeeded in ventures to build rockets and spacecraft that could take humans to the Moon, or the far corners of the universe. Unfortunately, too many of these companies, despite advantages from mergers and acquisitions, became overly dependent on their government contracts. Thus, like those businesses in the defense/military industrial complex, they often are less creative and risk-taking, so end up on "government welfare". Their efforts are concentrated on lobbying and obtaining the next contract supported by public funds, rather than becoming more enterprising.

9.1.2 Rocket renaissance

Now is the *era of true space enterprise*! The personal space travel industry is making progress, thanks to improved regulatory decisions, availability of insurance, spaceport development, and spaceship design testing to ensure safe passage. This is presently apparent in the small start-up enterprises to build less expensive spacecraft and satellites, to provide services to the space station or lunar base, or to sign up space tourists (see Appendices C and D). Speaking before the Second International Symposium for Personal Spaceflight in October 2006, Dr. Peter Diamandis, chairman of the X-Prize Foundation, stated: "We're at the birth of a new industry ... but to advance this industry, a flourishing private market place is needed." At that same event, Michael Simpson, president of the International Space University, observed: "We're knocking at the door of the future ... and we are the privileged generation to see it crack open just enough!"

Relative to personal spaceflight, *Popular Science* magazine's editor said it best about the long-term possibilities when SpaceShipOne (see photograph in Exhibit 109) and its White Knight spacecraft won the Ansari X-Prize: "Some of the boldest, most mind-blowing innovations we've ever surveyed!" Having made history with the first manned private spaceflight, despite a plant fire in 2008, Burt Rutan's Scaled Composites still has a commitment of $1 billion to build a fleet of SpaceShipTwo vehicles, which are powered by a hybrid rocket, partially filled with solid fuel, but no oxidant (*www.scaled.com/contactus.php*). Because this may be safer for his passengers, Sir Richard Branson has ordered five of these vehicles at a cost of $240 million. The latter's Virgin Galactic spaceline has already banked $15.6m in passenger reservations even before that spaceliner has flown a suborbital flight. At a ticket price of possibly $1000,000–$200,000 each, the six people, plus two pilots, will get a few days of training for two hours of weightlessness in a large cabin with big windows while cruising high above the Earth. The craft will be launched into orbit from Eve, a "White Knight Two" specially built aircraft. After separation at some 50,000 feet aloft, SpaceShipTwo will continue to an orbital peak of 400,000 feet, so travelers connected to a tether may fly around the roomy interior. Cabin seats will ensure the passengers can be comfortable during the $5g$ launch and re-entry. Take-off and landing of experimental flights take place in California's Mojave desert, until regular operations start around 2008 at New Mexico's *Spaceport America*. Virgin Galactic hopes to have 450 people on its first-year orbital flights, including the initial 100 "Virgin Founders" from 18 countries who have already paid their deposits. Presently, Virgin's three airlines have the best safety record in the world, a feat which Will Whitehorn, president of Virgin Galactic seeks to emulate in outer space with their spaceliners! He also hopes to bring the ticket price down to $50,000 by the fifth year of operation, and $25,000 before their tenth year. The company's world headquarters will be the Upham, New Mexico spaceport, while pursuing their admirable goal of providing a foundation for actual space colonization (*www.virgingalactic.com*)![2] Chief designer Rutan wants to go from mass production of suborbital spaceships, to designing an orbital system that goes to the Moon and back. But the entrepreneur, now 62, who first designed Voyager for its round-the-world flight, admits that breakthroughs like his are also beset with overcoming obstacles and ridicule from people saying "that's impossible." Rutan forecasts that within five years there will be competing suborbital spacelines flying up to 3,000 people, and within the next decade there will be 80,000 astronauts.

The founder of Virgin Galactic, along with son Sam and staff members have completed their training for their flights on SpaceShipTwo at the National Aerospace

---

[2] To view video on the training of the Virgin Galactic founders, go to *http://video.msn.com/?mkt=en-US&brand=msnbc&vid=ddda6712-d1a3-46c4-87b7–b90eb163d814*

Training and Research Center (*www.nastarcenter.com*). Here their future passengers will also practice and prepare for spaceflight. About 60 of the first 100 VG customers, known as "founders", have also undergone the 2-day training. Dick Leland, NASTAR president, said: "This training is as much as creating mindsets, about anxiety reduction, as it is about physiological training." After the learning experience, these would-be space tourists exclaimed: "Absolutely fantastic ... Loved it ... Wow, I feel like one of the luckiest people in the world." Branson added: "It was an amazing experience. Due to the flight simulation combined with G forces created by the STS-400 centrifuge, I really felt like I was launching into space." As the CEO of the Center's parent company, Environmental Tectonics, commented: "This is the beginning of a new era in space activities!"

Further evidence of advancing manned space enterprise is the formation in 2007 of the Personal Spaceflight Federation (PSF), and the announcement of the Wirefly X-Prize Cup in Las Cruces, New Mexico near Spaceport America. After test flights of the new spaceplane and licensing by the Federal Aviation Administration, commercial operations hopefully will begin there between 2008 and 2010. Exhibit 109 highlights a few of the other creative rocket entrepreneurs, engineers, and designers leading the emerging private spaceflight industry.

Robert Bigelow, the hotel billionaire, who is innovating with orbital inflatables at Bigelow Aerospace in Los Angeles, is underwriting a $50m prize for a manned vehicle that is able to reach an altitude of 200 km and complete two orbits of the Earth (see Exhibit 111). The second feat must be accomplished within 60 days of the first orbit and before the end of year 2010! Among the commercial rewards for the winner is a chance to become a supplier for Bigelow inflatable space habitats, as small space stations or hotels ... Other encouraging trends in space commerce are the Space Frontier Foundation conferences, and even the aborted Space Entrepreneurs Trade Association. Still another example is the Space Stock Surfers Club which facilitates investment in aerospace ventures that promote space economic development (email: *alexho@aol.com*).

The quintessential space entrepreneur Steve Durst is also sponsoring periodic Lunar Commercial Communication Workshops, as well as Lunar Development Conferences (see *www.spaceagepub.com/ilo* or Appendix E) Ultimately, global consortia will be formed to utilize space resources through macroprojects connected with astronomy, space-based energy, the mining of the Moon and asteroids, as well as space tourism and resorts (see Appendices B, C, and D). Increasingly, it will be entrepreneurs who will take the lead in private spaceflight, industrialization, and settlement, individuals like those described in Exhibit 109.

9.1.3 Space entrepreneur analysis

Entrepreneurs, often working out of their garages, built the global, high-tech industries. And it appears now the same is happening with regard to space enterprise. *Entrepreneurialism* is a manifestation of the innate human sense of curiosity and discovery. In his book, *The Discovers*, historian Daniel Boorstin expressed admiration for the amateur or expert willing to try something new [7]: "Every true discoverer

Exhibit 109. PRIVATE LAUNCH ENTREPRENEURS.

- **Eric Anderson**, Space Adventures of Arlington, Virginia in partnership with Prodea (below) to use Prodea's Explorer rocket ... in addition to this suborbital spacecraft, Space Adventures is planning two spaceports in the United Arab Emirates and Singapore ... funding two other spacecraft development efforts: Aerospace's XCOR and the Russian MDB's Cosmopolis XX1 ... already in space tourism business with Russians sending wealthy spacefarers to ISS.
- **Ansari Family**, Prodea Corp. of Texas—Explorer rocket is still in the design stage for five passengers and supervised by the Russian Federal Space Agency ... funded the $10m X-Prize competition ... Anousheh Ansari became first female space tourist on the E14 crew to ISS.
- **Steve Bennett**, Starchasers Industries of the U.K.—Storm, Skybolt, and Thunderstar rockets.
- **Jeff Bezos**, Blue Origin of Seattle, Oregon—New Shepard 1 in development, but is a liquid rocket VTOL (vertical take-off/landing) that may be able to carry up to seven passengers.
- **Robert Bigelow**, Bigelow Aerospace of Las Vegas, Nevada—Genesis-1 spacecraft has already been launched on top of a Dnepr Booster from a Russian cosmodrome ... also investing half a billion dollars into development of space inflatable habitats (see Exhibit 108).
- **John Carmack**, Armadillo Aerospace of Mesquite, Texas—VTOL suborbital design development.
- **Brian Feeney**, Da Vinci Project of Canada—XF1, Excalibur, and Valkyne spacecraft designs.
- **Hokkaido Aerospace of Japan** and its HASTIC hybrid rocket, cooperating with Walter Kistler and Chuck Lauer's Rocketplane Kistler of Oklahoma City ... XP rocket scheduled to be launched on the ground from the new Hokkaido Spaceport (see Exhibit 151).
- **David W. Thompson**, Orbital Sciences Corporation of Dulles, Virginia—Pegasus XL rocket, plus Magellan satellite navigation and communications equipment.
- **Eric Knight**, UP Aerospace of Unionville, Connecticut—SpaceLoft XL vehicle testing to 70-mile high altitude, and plans for 30 launches per year by 2008.
- **Elon Musk**, SpaceX of El Segundo, California—Falcon 1 turbopump-fed Marlin engine being tested to launch non-human cargo (see Exhibit 107).
- **Dumitru Popescu**, Stabilo of Romania—ARCO European team design of a suborbital system with an unconventional carrier balloon.
- **Geoff Sheerin**, PlanetSpace of Chicago, Illinois—spacecraft programs under way include Silver Dart and Canadian Arrow ... with Canadian government, developing a Cape Breton spaceport in Nova Scotia.

Source: Expansion on "Private Spaceflight: Rocket Renaissance," *The Economist*, May 13th, 2006, pp. 90–91.

Exhibit 109a. Private launch entrepreneurs. The successful flights, in 2004, of Scaled Composites' SpaceShipOne, the brainchild of aerospace engineer Burt Rutan, generated enormous media and industry interest in space tourism. Since those epic flights, Sir Richard Branson's Virgin Group and Scaled Composites have signed an agreement to form Virgin Galactic, a company that will own and operate privately built spaceships, modelled on the remarkable SpaceShipOne. Here SpaceShipOne glides down for approach to Mojave airport during an early test flight. Source: Scaled Composites, LLC.

or inventor was doing something for the first time," as contrasted with bureaucrats and traditionalists who fear change and risk. The credo of the American Association of Entrepreneurs describes this spirit: "I do not choose to be common—it is my right to be uncommon. I seek opportunity, not security. I want to take calculated risk, to dream and to build, to fail and to succeed." In another book, your author listed some qualities of the promoter which somewhat describe the entrepreneur [8]:

> "Perceives opportunity ... Has short-term orientation ... Makes minimal commitment of resources when decisions are made to pursue opportunities ... Prefers minimal overhead, seeking to borrow, barter, lease, or obtain donated talent ... Is comfortable with a flat, lean organization that emphasizes team management and networking ... Creates high potential ventures to meet human needs, while using varied management styles and systems ... Seeks to build change into the enterprise, so values a norm of ultrastability."

One innovative example of the above qualities is embodied in Robert Bigelow (see Exhibit 111).

Exhibit 110. PROFILE OF A *NEWSPACE* ENTREPRENEUR.

They call him part playboy and part space cowboy. Elon Musk has no sense of occasion. He talks expansively about saving the planet and conquering space. Moreover, like other Silicon Valley "thrillionaires" who throw money earned in the Internet boom in voguish new hobbies, Musk is proving original in his thinking about his new pursuits.

On March 21, 2007, the Falcon, a two-stage rocket of his Space Exploration Technologies (SpaceX) lifted off from the Marshall Islands and climbed to an altitude of 200 miles. Although the second stage failed to reach its intended orbit, the Falcon can claim to be among the first rockets to be designed, developed, and financed by the private sector that is anywhere near carrying payloads into space. Mr. Musk founded SpaceX five years ago and designed much of the rocket himself.

Though he is only 35, Elon has already made surprising progress toward achieving three modest goals he set for himself when he was in college: to transform the internet; make a breakthrough in clean energy; and propel humankind toward interplanetary travel. As a graduate student, he arrived at Stanford University intending to do a doctorate dissertation on batteries for electric cars, but dropped out to jump on the Internet bandwagon. Musk struck gold when he sold his online payments firm Pay Pal to eBay for $1.5 billion in 2002. Rather than retire comfortably, he says he "doubled down" his proceeds into his two other passions: clean energy and space.

The first bet was Tesla Motors, an electric car company, which last July unveiled its first model: a sports car that is faster than a Ferrari, more environmentally friendly than a Toyota Prius, and can travel 200 miles after charging overnight in an ordinary household. Taking on Detroit's auto industry hardly counts as easy, compared with conquering space. Giant defense/aersospace contractors close to the Pentagon and NASA, have long dominated the business of launching satellites. Musk complains that "launch vehicles today are little changed from those of 40 years ago." So he set about redesigning rockets from the bottom up at his manufacturing facilities near Los Angeles ... A display there of "all the stuff that didn't work" re-inforces SpaceX's culture of experimentation. His firm is stocked with experts poached from Northrop Grumman, Boeing, and other aerospace giants. Elon observed, "These guys get frustrated at the old bureaucracies. Here it is more Google-ish."

Because it is designed from scratch, his Falcon is much simpler than most rockets and free of some of the risks and costs of complexity. The version launched last week cost under $7m, and the company's competitors charge four or five times as much. SpaceX rockets can be recovered and recycled. In time, Musk expects his rockets will cost a tenth of the competition, while lifting payloads larger than traditional rockets. This entrepreneur alleges that the aerospace industry has unfairly tried to keep him out. When the Air Force recently awarded two dozen future rocket launches to a consortium formed by Boeing and Lockheed-Martin, SpaceX cried "foul." It sued them on the grounds that they are colluding to keep low-cost competitors out of the business which they dominate ... Entrenched incumbents are the least of his worries, Elon believes because rocket science is like designing computer codes which can be tested "only in parts and must be bug free," Falcon's progress has convinced its founder that his dream of interplanetary travel may yet come true. "If normal humans ever travel beyond Earth, it will be because of SpaceX or companies like it." Musk reckons that Mars is amenable to civilization, and ought to be colonized as a "life insurance policy" that guarantees the continuity of humanity. (Source: adapted from "Face Value: Rocket Man," *The Economist*, March 24th, 2007, p. 78.)

9.1.4 *NewSpace* business

The above is almost a profile of space entrepreneurs in general. Today the entrepreneurial space industry is commonly called by its advocates *NewSpace* [9]. The emphasis at the moment is suborbital and orbital vehicles, space travel and tourism, and orbital services and structures (see Appendices for further information). The industry will address emerging concerns and needs aloft or on the ground, such as with building and improving spaceports. The concerns cover a wide range from safe spaceflight and accident prevention, to coping with varied amounts of radiation and government regulations. Public space agencies can inhibit or facilitate this growing industry as partner, customer, or investor. One template for *NewSpace* is NASA's competitive program, Commercial Orbital Transportation Services, which awards contracts to private firms. In 2006, COTS awarded nearly $500 million to Rocketplane Kistler (RpK) and Space Exploration Technologies (SpaceX) to aid in development vehicles for the International Space Station. In addition, the Agency negotiates big contracts with the aerospace giants, such as Lockheed-Martin and Boeing to redesign the Atlas 5 and Delta rockets as passenger vehicles. The Agency is funding other ventures through the Space Act agreements for ISS, such as resupply and commercial orbital missions. This approach is illustrated in the Bigelow Aerospace contracts related to inflatable orbital habitats.

Furthermore, private enterprise is trying to promote space enterprise. In 1997, the Cheap Access to Space prize was announced. It is funded by the Foundation for International Non-Govermental Development of Space (FINDS) and managed by the Space Frontier Foundation (SFF). The CATS prize of $250,000 will go to the first private team that launches a 2-kilogram payload to an altitude of 200 km or greater. SFF also holds annual conferences in Los Angeles for entrepreneurs, investors, media producers, and reporters (*www.space-frontier.org/EVENTS*). But it was the X-Prize, created by Dr. Peter Diamandis, co-founder of the International Space University, that really provided space entreprepreneurs with a serious financial incentive. Winning that $10 million prize with the SpaceShipOne design enabled Burt Rutan to fill Sir Richard Branson's order for five more spaceplanes required by his Virgin Galactic company. The latter seeks to make space tourism affordable before 2010; their $200,000 spaceflight will only require two days training for one day in LEO. Contrast that with Space Adventure's Russian trip to ISS which means six months of

Exhibit 111. PORTRAIT OF A *NEWSPACE* COMPANY.

Bigelow Aerospace, founded in 1999, is playing a high-stakes game to privately develop offworld lightweight but strong inflatables, or flexible space complex architecture. The company is testing Earth–space modules and subsystems in its 50-acre Las Vegas facilities. Its "skunk works" collaborates with NASA's Johnson Space Center in the hope that the Agency will someday use the Bigelow inflatable technology on the Moon or Mars, or even in orbit as stations or hotels. The firm's trip module metallic simulator demonstrates how three inflatables can be docked to a central node. One prospect is to sell or lease out small space stations or habitats parked in orbit. These Nautilus modules have $330\,\text{m}^2$ of volume; the Zarya and Zvezda modules on ISS have volumes of about $170\,\text{m}^2$.

Currently, one-scale Genesis modules are being tested, based on technology licensed originally from NASA, but redefined and advanced. The vessel is made of several layers of vectran, a strong artificial fiber, that has foam shielding to protect against micrometeorite impacts. In later missions, water blankets will be used for radiation shielding. Inflatables take up less room on a launch vehicle and are cheaper to put into orbit. To save costs, Genesis-1 Pathfinder was launched in July 2006 on an ISC Kosmotras Dnepr rocket, a converted Russian ICBM. To do that, Bigelow Aerospace had to resolve off-shore, third-party liability insurance and certified vehicle issues. Its Inflatable Module Orbiting Earth (R) was protected by its Inflatable Shell for Satellites Patented (L). The next milestone is Guardian inflatables, 45% scale modules. The company aims to launch an unmanned Nautilus on a Proton-class booster by 2008, then connect it with a manned module by 2010. The expectation is to launch ten inflatable missions in a decade.

Nautilus is designed for a crew of five to seven, with an environmental and life-support system that can provide air, water, and carbon dioxide removal. Multiple subcontracting provides components for this purpose, such as EADS Astrium. Subcontractors make more than 50% of the parts designed by Bigelow engineers. The interior with a longitudinal deck layout in the vessel improves habitability with its three-deck web flooring. Beside creative fabrication of materials, the business is pioneering inflatable "soft goods" to form modules, as well as expertise in folding and packaging those goods around the module's aluminum core. Thus, a unique technology mix is emerging by using multilayered "soft goods" composed of advanced materials with traditional aerospace aluminum structures. Some of the component testing is done with JPL under the Space Act agreement with NASA.

The enterprise envisions a diverse market from commercial to government to military. It anticipates the space tourism and spaceport industries will need Bigelow orbiting habitats, laboratories, shopping centers, and office buildings! Ultimately, the founder's vision is to bring terrestrial principles of real estate development and financing into space. Since tycoon Robert T. Bigelow made his fortune in the hotel and general contracting business, it would be natural to build on that experience offworld. The reclusive entrepreneur plans to spend half a billion of his dollars by 2015. As he stated: "This is not typical in aerospace, but one reason I am doing it after so many years as a general contractor is to bring philosophies from that realm into this business." Thus, he has put together a competent team with this pragmatic philosophy: "We are a 100% experimental program, and we have to prepare for failures and not be overly shocked if they happen. We realize that this is going to be done at significant risk."

To achieve his goals of making space travel available to the masses of human beings, Robert Bigelow is also funding the America's Space Prize of $50m to be awarded by 2010 to those who can build a low-cost commercial rocket that could hold up to seven spacefarers. In this space launch contest, the winning team must also produce a vehicle that can maneuver to dock in orbit at well over the 100-mile altitude and twice survive a 17,000 mph re-entry. In addition, as noted previously, the winner will receive first rights on a contract from Bigelow Aerospace for ongoing orbital servicing of its Nautilus modules. (Source: Graig Covault, "Bigelow's Gamble," *Aviation News & Space Technology*, September 27, 2004, 54–58 ... "Space Travel—From Russia with Love: A Private Space Building is Launched," *The Economist*, July 22nd, 2006, p. 78 ... Leonard David, "Genesis-1: Reaching Escape Velocity from Red Tape" (*www.space.com/businesstechnology*, July 28, 2006).)

training, plus $20 million for the privilege of being on the space station for up to seven days. And Rutan's new generation of private space vehicles was made possible by another entrepreneur's investment: Paul Allen, co-founder of Microsoft!

An early innovator in space enterprise is SPACEHAB Inc. which in December 2000 received a second NASA contract of $30.9 million for the company to conduct Space Shuttle research. This provider of commercial space services prepared another SPACEHAB Research Double Module (RDM) for flight on STS-112. The company then markets this facility to other national space agencies, such as those of Japan and Germany, for $8m payload services. Its CEO, Dan Bland, commented: "These Shuttle research missions are an excellent opportunity for cutting-edge researchers to explore the microgravity of space and to prepare for long-duration research on the ISS."

Another success story is SpaceDev of Poway, California, founded by Jim Benson. At first it hoped to capture asteroids, then got serious with NASA contracts, and proceeded with a number of small and effective high-tech projects. Now it is concentrating on building Dream Chaser, a six-passenger human space transport aiming for orbital flight testing by 2010. It is powered by a scaled down version of the hybrid rocket motor which it supplied to SpaceShipOne. But first it must build a large hybrid motor for an Air Force contract of $2.7m, nine times the thrust of the motor which helped toward Rutan winning the X-Prize. SpaceDev, listed on the stock exchange, is also working on the International Lunar Observatory project.

As space writer Alcestis Oberg observed, this second wave of space enterprise has brought numerous benefits to the public: global positioning systems in automobiles;

thriving orbital mortuary business; virtual space pioneers who send their photo and DNA on Celestis funeral voyages; the necklace of refrigerator-size satellites in orbit that make the whole global phone business prosperous; suborbital flights by small aerospace planes, some launched by an airliner 35,000 feet or more aloft. Midway in this first decade of the 21st century, it is estimated that over $100 billion is already being spent worldwide on space commerce. Today thousands of businesses are engaged in non-NASA space services, while dozens more use GPS for a variety of navigational products. The remote-sensing industry is prepared to sell pictures of houses, property, and even cities, taken from outer space. And entrepreneurs will be dreaming up innovative ways to participate, like Applied Space Resources Inc. in Long Island which wants to scoop up lunar ice to sell back on Earth! Some companies hope to make profits on ground-based enterprises, like Space Age Games of Houston. Phil Kopitake, the aerospace engineer who created the game, said that various modules enable players to build their own planet, space station, and orbital bases. His *Space Race* aims to promote colonization of the Solar System and utilization of fusion power. So beside fun, participants learn many lessons about the space environment. Some *NewSpace* activities will be funded by the public sector. For example, Wyle Laboratories of Houston, Texas, labors to support the Space Life Sciences Directorate at the Johnson Space Center. These *bioastronautics* contracts sustain the work of ISS, Space Shuttle, Constellation and Human Resource programs. The work is performed at JSC, as well as at the University of Texas Medical Branch in Galveston, plus subcontractors such as Barrios Technology, Bastion Technologies, Easi, Muniz Engineering, Lockheed-Martin, and Futron Corporation. Wyle Labs has had this agreement since 2003 with extension options until April 30, 2013; if both options are exercised by NASA, the total contract may amount to $973 million. And this is only the beginning of space health and orbital services to be rendered (see Appendix D).

Another application of the new entrepreneurial spirit is the Aerospace Corporation which is launching Orbiting Picosat Automated Launchers (OPAL), six microsatellites the size of a cigarette pack weighing only 250 grams. Launched less expensively from a mother ship, these are presently circling the Earth at an altitude of about 400 miles. They are testing micro-electronic mechanical systems (MEMS) which share particular tasks. The aim is to develop nanosats of less than 10 kg that could fly in formation to establish global navigation and communications systems; when re-arranged the swarm can transmit their results back to Earth. For interplanetary probes, these tiny sats could be used as relay stations to the home planet. The prototype OPAL was constructed by enterprising students at Stanford University and Santa Clara University. Ultimately, the plan calls for mass-produced manufacturing of picosat components.

Space News and Space.com have a feature writer, Leonard David, who is constantly producing mind-stretching articles on business opportunities offworld. For example, one such story dealt with the technology of rocketbelts; until now Textron Systems has been using high-pressure nitrogen gas for backpacking power. David reported on a Rocketbelt Convention held in Niagara Falls, New York, in September 2006, which discussed new technologies and possibilities for personal flying whether on this planet or in orbit. Manned maneuvering units have been used in the Gemini and Apollo missions, as well as in EVAs from the Shuttle. The future of jet belts, however, awaits a replacement for heavy hydrogen peroxide fuel tanks, such as by using ducted fans or small jet engines. Such flying without wings would be ideal for exploring the lunar surface (see *www.rocketbelt.nl*).

Of course, not all space commerce ventures succeed, often because they lack the resources to push ahead. But we can also learn from failures, such as General Astronautics of British Columbia—in which I invested; their ambitious plans for a Urania space transport system never even reached the launch pad! The reach for a single-stage-to-orbit (SSTO) goal has been unrealized for many entrepreneurs, like Rotary Rocket's Roton of Redwood City, California. Then there was the frustration of Kelly Space & Technology in never being able to complete their Eclipse astroliner. The San Bernardino, California company hoped to market reusable satellite launchers by means of a modified 747 airplane which towed their astroliner into orbit where it would release its payloads.

Post-Soviet Russia also has had disappointments trying to get into the space business for profit. For many of their advanced space technologies had little or no commercial value. Only Russian rockets, like Proton, interested Western buyers, especially their low-cost launch facilities. NASA contracts to build ISS components have provided necessary cash income. Russia's most famous aerospace corporation, Energia, has had difficulty adapting to a market orientation. Energia's rocket is presently too big for most customers, and nobody wants to buy its down-scaled version capable of placing up to 34,000 kilograms in LEO. Further, while Russia's Khrunichev firm was more open, accommodating, and interested in cooperative projects with the West, Energia was reluctant to adopt capitalistic business ways. Their leaders made unrealistic proposals for government funding and often were in conflict with Roscosmos, the nation's space agency. Even with the new leadership of Nikolai Sevastianov who makes bold announcements for new programs, the rift continues, especially concerning manned spaceflight. Roscosmos, lacking in visionary leadership which includes no cosmonauts in top administration, has refused to fund most Energia proposals, and even underfunded a Soyuz upgrade and a future Kliper spacecraft. Seemingly, Energia has yet to learn the basic lessons of a market economy: offering to build what both the public and private sectors in the global marketplace want to buy—rather than what it wants to sell! Although Russia's GNP has grown to $1 trillion (by 2006), the government is making insufficient investment in the country's space future. "To keep informed about their space program, read Russia's *Novosti Kosmonavtiki*," recommends Dwayne Day whose writings in *Space Review*, March 5, 2007, contributed to this paragraph (he is contactable at *zirconic@earthlink.net*).

The concept of the *business accelerator* might prove to be the stimulus that *NewSpace* start-up companies need. Since ordinary venture capitalist have been slow to invest in space enterprises, Rocky Persaud, chairman of the Canadian Space Commerce Association, heads up an entrepreneurial group interested in commercial space opportunities. He recommends the formation of entities that mentor *NewSpace* entrepreneurs from start-up to seed stage until their companies obtain initial funding.[3] He argues that such professional help would provide expertise in management, finance, accounting, product development, marketing, and research services. Then the start-up firm would be positioned to take full advantage of the newly founded *Space Angels Network*, which addresses the needs of such new undertakings for investment in the range of $100,000 to $2 million. The approach has been used successfully with high-tech start-ups by The Foundry, an accelerator in Menlo Park, California.

Those who invest in or invent better products and services for space commerce are positioning themselves as the leaders of tomorrow's offworld enterprises. Jeff Bezos, founder of Amazon.com, is just one such example. He envisions that, before the end of this century, millions may experience the free-floating environment of microgravity! By then, fleets of spacecraft may offer passenger opportunities to float, dance, swim, exercise, play, race, and make love in zero gravity. Expect industries to evolve that cater to space fashions and hair styles, as well as orbital hotels and resorts (visit *www.RedPlanetVentures.com*). For entrepreneurs with imagination and vision, endless prospects await beyond Earth as Exhibit 112 illustrates (see Appendix C for more information)!

> "In my opinion, there are two directions in which NASA must proceed for its future success and the benefit of the American people. The first is true commercialization of services; the second is to make the space program more participatory and engaging for the public" (Peter H. Diamandis, M.D.).

9.2 MACROCHALLENGES ON THE COMMERCIAL SPACE FRONTIER

Increasingly, global space macroprojects will have to involve international partnerships and consortia, so packaged as to attract participation by venture capitalists, investors, and entrepreneurs. Instead of competition in a narrow space market, national agencies and corporations need to collaborate in joint technological ventures for "do-able" enterprises on the high frontier that will provide return on investment. All the current spacefaring nations have plans for lunar missions, while the U.S.A. has a national policy, *Vision for Space Exploration*, which calls for permanent return to the Moon by 2020. As such, this represents a macrochallenge for space commerce to meet the needs of this emerging market!

Spacefarers are already taking their culture and society into orbit. But the microgravity environment alters humanity, our social organization, and institutions. It is not merely integration of outer space commercial activity within our global economy—we are in the process of creating a twin-planet economy between Earth and the Moon. The requirements of *astrobusiness* is but one factor influencing changes in our terrestrial ways [10]. The emerging space-based economy presents unique financial, technical, legal, transportation, construction, and political problems

---

[3] Persaud, R. "What Space Start-ups Really Need," *Space Review*, November 26, 2007 (*www.thespacereview.com/article/1007/1*).

Exhibit 112. Lunar market prospects. Relations of lunar base outputs and lunar base potential users.

Products	New knowledge	Services	Material goods and energy
Lunar enterprises	New information relevant to lunar operations	Maintenance and repair Social needs of lunar crew	Construction material, feedstock, consumables, and propellants Electrical and thermal energy
In-space enterprises	Research results relevant to space operations	Communication, maintenance, and repair Recreation	Propellants, construction material, and consumables Energy
Earth enterprises	Research results relevant to life on Earth	Environmental observation Adventure	Helium-3 Electrical energy

The International Astronautical Federation has a Sub-committee on Lunar Development which is doing forecasting studies on the Moon market. The above table is a typical model for commercial enterprise in space; it was developed by Professor H. H. Koelle of the Berlin Technical University. Source: "A Frame of Reference for Extraterrestrial Enterprises," *Acta Astronautica*, **29**(10/11), 1993, 735–741.

that range from risk management to transforming gross national products. Dr. Nathan Goldman maintains that the impact of orbital business will dwarf in comparison with the exploration and development of the New World in the 15th and 16th centuries. Space enterprise will gradually dominate our world economy throughout the 21st century!

Over the past three decades various forecasts have been made by experts as to what new products and services to expect from non-terrestrial industries. Beside the burgeoning launch and satellite businesses, predictions include

- orbital materials research, processing, and manufacturing, such as crystal growth, higher strength alloys, and pharmaceuticals;
- building large space structures for power systems, telecommunications, astronomy, navigation and positioning receivers (to improve ground and sea safety), remote-sensing and environmental controls (for weather research, environmental monitoring and land use, and resource management);
- space-related manufacturing of computer hardware and software for analyzing orbital data;
- mining the Moon and asteroids for valuable natural and mineral resources, including nuclear fusion or helium-3 mining;

- construction of space habitats and facilities, and platforms made of non-terrestrial materials;
- innovative applications of automation and robotics aloft, such as with picosatellites, made possible by miniaturization;
- defense systems, such as for intelligence gathering, and weather forecasting;
- affordable space transportation systems from launch vehicles to interplanetary spacecraft;
- space-based solar energy, such as described in Appendix B;
- ground-based support systems for space enterprises, such as providing funding, insurance, legal, communications, and other services, including spaceports.

Strangely enough, most of these prognosticators failed to identify the one business that may provide the most immediate payback: space tourism. The space entertainment industry was also overlooked; this big space-related, ground-based business now ranges from movies and television programs, to video games, simulations, and virtual reality. Expect more space museums, exhibits, and space theme parks that soon may permit manipulation of rovers on the Moon or Mars.

Government entities worldwide have sought to encourage their space agencies to promote the commercial aspects of public investment in space technology and exploration. NASA, for instance, has increasingly set up programs and schemes to encourage private space commerce. In 2001, the Agency circulated a report on *Enhanced Strategies for the Development of Space Commerce*.[4] The document suggests changes that might include income-producing strategies like

- limited tourism on the International Space Station;
- allowing family-friendly corporate sponsors to place emblems and logos on spacecraft beside NASA's, including a unique ISS logo;
- merchandizing that promotes the NASA brand;
- media and entertainment activities that boost the space program;
- hiring a non-governmental organization to manage the U.S. side of ISS.

Critics react that the plan is another public relations campaign, rather than the Agency creating a climate that promotes private space commerce. Obviously, the proposed strategy was a reaction to the cash-strapped Russian space agency's creative efforts to use its spacecraft for everything from advertising and merchandizing, to billionaire space tourists. Even before the death of the Soviet Union, enterprising aerospace leaders in that Eastern bloc empire sought to promote the capabilities of their launch vehicles and other space hardware. Elsewhere, in Europe, a consortium of companies decided to exploit commercial space travel, creating Arianespace SA, the world's first commercial space transportation company, a highly successful venture, as Exhibit 113 confirms.

Other Federal government departments in the United States are involved in the arena of space commerce. The Office of Air and Space Commerce in the U.S.

---

[4] Dunn, M. "NASA's Taking a Look at the Commercial Side of Space," *San Diego Union-Tribune*, October 2001, p. A29.

Exhibit 113. European space commerce. Following the first flight of the Ariane launch vehicle in 1979, a consortium of European firms decide to exploit commercial space travel, creating Arianespace SA, the world's first commercial space transportation company. By the end of 1999 the five different versions of the Ariane launch vehicle had carried over 70 satellites for research and communication into space. Arianespace is now the commercial launch services leader, holding more than 50 percent of the world market for satellites to geostationary transfer orbit. This series of images shows how the Ariane launch vehicle evolved from the Ariane 1 L01 in 1979 (far left) to the current Ariane 5 ECA (far right). Source: ESA.

Department of Commerce has done studies and reports on *Commercial Space Ventures*; Exhibit 114 describes the thinking of the Office of Commercial Space Transportation in the U.S. Department of Transportation. OCST has a public affairs program to provide the public, especially private industry, with information on their regulation of commercial launch enterprises. However, too often these various government space offices are too bureaucratic, failing to work in unison, while often promoting policies, regulations, and licenses that inhibit space business and entrepreneurialism!

To be as successful as the satellite communication industry, space commerce needs less domestic competition, and more collaboration. Indeed, *space is the place for synergy*, where partnerships and consortia develop together cases for investment that will attract venture capitalists and private equity funds.

9.2.1 Model for astrobusiness analysis

One model for analyzing the future of space industrialization or commerce has been proposed by Dr. Nathan Goldman and adapted by your author. Exhibit 115 offers ten sectors for consideration that are somewhat comparable with the categories explained above.

The above model was developed almost 25 years ago, but it did leave a box with a question mark to indicate other sectors would emerge for offworld business. Within this framework, Goldman originally described six sectors of space business as major economic arenas of the future:

- ***Space infrastructure sector.*** Beginning in LEO (low Earth orbit), future requirements range from growing transportation systems and space stations or

Sec. 9.2] Macrochallenges on the commercial space frontier 425

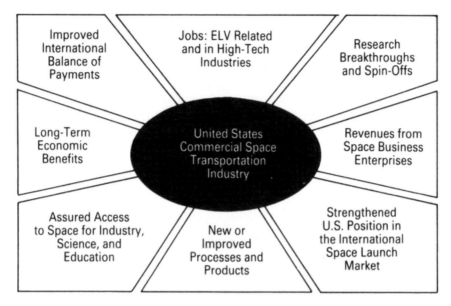

Exhibit 114. Benefits of a space transportation industry. Source: Office of Commercial Space Transportation, U.S. Department of Transportation (400 7th St. NW, S50, Rm. 5415, Washington, D.C. 20590).

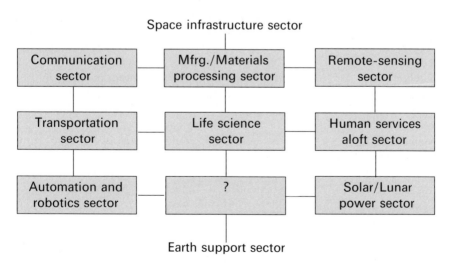

Exhibit 115. Space commerce model. Created by Dr. N. C. Goldman and adapted by Dr. P. R. Harris. See N. C. Goldman's "Space Law and Space Resources" in NASA SP-509, *Space Resources*. Washington, D.C.: U.S. Government Printing Office, 1992, Vol. 4, pp. 143–153 ... See also Goldman's book, *Space Commerce*. Cambridge, MA: Ballinger Publishing/Harper & Row, 1995.

platforms to mechanical servicing of same, including orbiting spacecraft; and extending to GEO (geosynchronous orbit) for space-based nuclear or solar power stations, storage depots, assembly/fueling facilities leading to Moon and Mars bases and settlements beyond; movement toward a space-based economy that mines and processes indigenous resources. [The author would extend orbital power and services into separate classifications.]

- **Communications sector**. This $3 billion satellite business faces problems of launch failures and rising insurance costs, plus strong competition from ground-based use of fiber optic cables and networks; its competitive advantage may be in lowered STS costs, and mobile, multi-point communication linkages that will expand satellite tracking of people, vehicles, ships, facilities, and resources. [To this, the author would add new developments in small satellites, plus the computer hardware and software businesses for satellite and spacecraft operations, along with the space management information systems. This category might also include space publications of all types, whether on the ground or aloft.]
- **Transportation sector**. Space transportation system (STS) that extends from rockets to the Shuttle Orbiter, to single-stage-to-orbit reusable vehicles, to an aerospace plane that climbs from Earth to orbit like an aircraft, becoming a spaceship in orbit. The next generation of spacecraft must lower the cost of space access, including advanced and cargo versions of the Orbiter or its replacement, plus a variety of expendable launch vehicles, and orbiting spacecraft necessary for lunar or Mars operations.
- **Remote-sensing sector**. The U.S. National Commission of Space predicted that advancing technology in observing systems and computers will revolutionize terrestrial science; enormous commercial applications range from weather forecasting and climate control, locating planetary resources/problems, to charting oceans and aiding archaeology. Leadership in this emerging field is shared by many entities and countries, such as NOAA (National Oceanic and Atmospheric Administration); private companies such as EOSAT (Earth Observation Satellite); and the French SPOT, which has high photographic resolution, as well as by ESA and Russian capabilities. [The author speculates enormous growth prospects generated by programs like the Mission to the Planet Earth, and soon a lunar outpost.]
- **Manufacturing sector**. Studies suggest this could become in the 21st century a multi-billion dollar industry, including the manufacturing of non-terrestrial materials. The new science of microgravity research, begun in industry and research institutes, has now extended to universities and NASA Technology Centers. [The author would expand this category to the whole realm of material processing from advanced and composite materials, to vacuum epitaxy technology.]
- **Earth support sector**. This will include privately, as well as publicly, owned launch facilities or spaceports; ground-based space vehicle maintenance centers; space services for propellants and energy sources; ground-based operations for satellite manufacturing, maintenance, and tracking; construction of habitat

modules and equipment for space deployment; space educational centers or academies to prepare spacefarers. [In this category, the author would include such enterprises as the International Space University, as well as a whole ground support system to emerge for deployment and service of spacefarers and their families.]

To the original Goldman model, the author proposes the following five additional classifications as emerging sectors:

- **Life science sector**. Commercial application of research on protein crystal growth, macromolecular crystallography, bio-services, space technologies, drug processing, cell research, closed ecological systems, and life support systems, etc. A NASA report suggests that the research will be in gravitational biology, biomedicine, biospherics, environmental factors, operational medicine, physiochemical/bioregenerative life systems, exobiology, and flight programs.
- **Human services aloft**. These will range from supplying food, cleaning, and similar hotel services to those in orbit; to travel and tourist agencies; to in-space support services that are psychological, sociological, educational, political, and social in scope; to entertainment and recreation, both in and from orbit. Any health care activities would be included here which ensure safety and survival in orbit, and which improve the quality of space life (see Appendix E).
- **Automation and robotics sector**. This will be more than supplying "tin collar workers"; many automated systems will provide information and assistance; others will be virtual reality systems. Some robots aloft may become friends and pets in isolated, confined environments. Primarily, A&R will be used for construction and maintenance, for transportation and monitoring, and a host of undreamed-of applications.
- **Solar/Lunar power**. This will probably become a 21st-century business producing billions of dollars in income by sending solar power to Earth from the Moon or vicinity (see Appendix B). Two methods are solar power satellites (SPSs) from lunar orbit advocated by Glaser [11] and a lunar solar power system (LSPS) from the Moon advocated by Criswell and Waldron.
- **Scientific and astronomical research sector**. Scientists, engineers, and astronomers in orbit doing basic or applied research will find commercial applications for their discoveries in a wide variety of fields and disciplines. For example, astronomers are already planning an international, multifunctional observatory, power station, and communication center on the Moon (*www.iloa.org*).

No wonder then that the trustees of the Robert A. Heinlein and Virginia Heinlein Prize Trust established a $500,000 annual award in 2003 for practical achievements in the field of *commercial space activities* (*www.heinleinprize.com*). Like so many others, the Heinleins shared a dream of human advancement in space through commercial enterprises. The science fiction books of the late Robert Heinlein still inspire many into aerospace careers, especially those who would be offworld leaders of tomorrow!

> **Note:** A blank box (?) was left in Exhibit 115 for the reader to insert his or her own candidates for other sectors of astrobusiness. Apart from space tourism and entertainment, what are your forecasts regarding future space commerce prospects?
>
> These might range from orbital education and training centers/programs/ software, to space law and astrolaw practices, to astrofarming and astroranching, astrofashions, to astrotherapy and astromedicine. Use imagination to conceive the human services likely to be needed on the Moon when outposts become settlements, and colonies become cities. Consider also what businesses the lunar colonists will undertake for their own benefit, and those back on the home planet.

Outer space offers unlimited economic potential for numerous products and services. The pace of commercial growth has led to the formation of CASE (Competitive Alliance for Space Enterprise) by Washington-based business people to influence government policy in creation of a single space office within the Federal agencies to help industry and entrepreneurs navigate the bureaucracy, and to streamline the regulatory process.

Furthermore, growth in space commercial activities on the ground or aloft will lead to new vocational fields and space careers, ranging from human factors and health services to architects and environmental engineers. S. Norman Feingold, a psychologist and author of *Careers Today, Tomorrow and the 21st Century and Beyond*, is resolutely optimistic about future job openings in space. Within 50 years, he predicted the development of space colonies and orbital tourism; the need for space specialists in law, biology, botany, medicine, police, and other non-terrestrial vocational activities [12].

9.2.2 Private/Public-sector cooperation in space commerce

Just as the early days of shipping, railroading, and flying were dependent on government subsidy, so too is space commerce. Today governments in spacefaring nations support astrobusiness through research and operating grants, deferred payments of launch costs, tax and investment incentives, free or cut-rate spacecraft rides, patent rights on government-sponsored research, and incubator programs to facilitate the transfer of space technology to private enterprise. The legislation from Congress that created the successful COMSAT entity through the Communication Act of 1962 may be a prototype for more space "public/private" corporations. In addition to reorganizing space agencies toward more commercial applications of their R&D, there may be need for another type of commercial space institution. For example, in France, there is CNES (Centre National d'Etudes Spatiales), a public establishment of industrial and commercial character that is responsible for formulating space programs, overseeing selected projects, and providing liaison with the country's foreign partners in joint space endeavors. CNES carries out its scientific activities through contracts with a dozen major research laboratories; it markets its products and services through companies it founded, such as Arianespace and SpotImage. To commercialize space fully will require new organizational creations, even world space corporations, possibly chartered by the United Nations or a global space agency.

The longer we operate in space and stay aloft, the more innovative and entrepreneurial will be the ways humans discover to capitalize on space as a place and as a resource. At the Russian space station, cosmonauts have already learned to make more than a hundred different alloys and metal mixtures thought impossible to manufacture, plus very pure crystal and glass lenses. Space-business devotees are constantly coming up with unique ways for extraterrestrial exploitation, from burials, banks. and Earth truck-tracking systems to lunar mass-drivers and large optical telescopes. Arthur C. Clarke, father of the communication satellite concept, proposes PEACESAT satellites to monitor military incursions, nuclear accidents, and other inappropriate activities which threaten world peace among the planet's 150 nations. He envisions two-dimensional communication networks from space contributing to the rise of the Global Family or Tribe in which members are linked electronically across national borders (which is happening with the Internet and mobile telephones). The need is for more *non-aerospace* corporations and associations to get involved in astrobusiness. One example is the American Society of Civil Engineers, which has established an aerospace division which sponsors both conferences and publications on space engineering, construction, and operations [13].

Aerospace consultants estimate worldwide space activities will produce over $20 billion in revenues by the year 2020! Roy Gibson, former Director General of both the European Space Agency and the British National Space Centre, offers an interesting observation on commercial challenges aloft [14]:

"The transition from developmental, government-funded activities to commercial exploitation is a gradual process. New methods of allowing for privatization, such as government solicitations based on performance capabilities rather than the purchase systems, have been suggested in ESA and NASA ... In the future, the private sector may be more strongly encouraged to take a larger role in space activities. With incentives, industry will take advantage of opportunities offered."

But it will be expanded human presence in space settlements and colonies that will truly open up the potential of offworld commercial enterprise. The lunar solar power system planners, for instance, project a community on the Moon of up to 4,000 "technauts" by the end of this century (Appendix B). Be assured that space explorers and traders will be among the leaders in humanity's trek through the Solar System. And then there is the unpredictable factor when private individuals get tired of waiting for governments to act, and move ahead in space businesses on their own, as Exhibit 116 intimates.

In 2006, another positive development occurred, as indicated above, when NASA announced it was seeking proposals from private enterprise to create and manage innovative space activities, events, products, services, and other educational methods to increase America's science and technological literacy. It wished to further the Space Act agreements with entrepreneurial organizations. In exchange

Exhibit 116. Private lunar enterprise: ISELA. Because of the slowness by the public sector in developing lunar resources, private initiatives are under way, such as the International Lunar Observatory project (*www.iloa.org*). The above artistic rendering was used in proceedings of the International Space/Lunar Exploration Conferences under the leadership of aerospace advanced planning engineer Michael C. Simon, chairman of ISE Research Corporation and author of *Keeping the Dream Alive*. The picture depicts a private ISELA lunar lander and commerce on the Moon. Source: International Space Enterprises Corporation (7345 Murphy Canyon Rd., Ste. K, San Diego, CA 92120; email: *msimon@isecorp*).

for collaborator investment to distribute NASA information more creatively, the Agency was willing to negotiate brand placement, limited exclusivity, and other opportunities in a strategic collaboration between the public and private sectors (*http://prod.nais. nasa.gov/cgi-bin/eps/synopsis.cgi?acqid=120084*).

9.2.3 Lowering space transportation access costs

Rockets have been called *dream machines* because they became the means for humankind to leave the home planet and go offworld [15]. They went from dream to reality in the 20th century, but are still too expensive to build and operate if they are *really* to open the new frontier in this 21st century! As long as getting payloads into orbit remains excessive costwise, space commerce and space settlement will be delimited. Presently, the Space Transportation System (STS) or Shuttle Orbiter costs taxpayers some $3.2 billion a year to operate, and absorbs almost half of NASA's budget. It is a very complex, technologically sophisticated system that is now costing about $4,500 per kilogram to get into orbit and requires 35,000 workers to maintain the fleet of four reusable orbiting vehicles and their expendable external fuel tanks. The Orbiter fleet, costing about $2 billion each to build, has been flying for more than two decades with only two launch failures, but these accidents were major in scope. Averaging 5-day to 16-day missions aloft, this orbiting laboratory can carry a crew of up to eight [16]. Over its 27-year history to date, this impressive spacecraft with its gallant crews has flown on more than 120 Shuttle flights, contributing to the building and servicing of the ISS and the Hubble Space Telescope.

Now that the Shuttle fleet is scheduled for retirement in 2010, NASA desperately seeks a less expensive substitute that will not only get us to the International Space Station—but also to the Moon. In its next configuration, designers are seeking to provide for a payload up to 11.5 tonnes and to a 51° inclination orbit. When Associate Administrator for Office of Space Access and Technology Dr. John Mansfield maintained that NASA is not only expecting to reduce costs by contracting STS operations to the private sector, it is actively searching for a Shuttle replacement system with reusable engines with a turnaround time between flights of two weeks which would require no more than 50 maintenance crew workers. With the aim of reducing spaceflight costs to $450 per kilogram to reach LEO, future space transportation will feature: (a) new composite materials in their construction, such as in engines and cryogenic tanks; (b) changed architecture by miniaturization of instruments within the small spacecraft initiative. Contrast current cost estimates of $50 million per vehicle vs. $400 million for larger spacecraft. In part, this is possible because of microcircuitry or application-specific integrated chips.

In August 1994, the U.S.A. issued a new *National Space Transportation Policy* to ensure reliable and affordable access to space. In addition to expendable launch vehicles (ELVs), NASA was directed to lead in the technology and development of next-generation reusable launch vehicles (RLVs), particularly *single-stage-to-orbit* or SSTO vehicles. The latter R&D program was called within NASA the X-33 and X-34, and provided the initial underwriting for industry to develop advanced launch systems for the Agency at reduced technical and business risk. In the contest for a Shuttle replacement by year 2012, corporate giant McDonnell Douglas teamed up with NASA to develop the DC-X and the DC-XA (known as Clipper Graham), a vertical take-off and landing vehicle, an artist's concept of which is shown in Exhibit 117. In the same exhibit is also a rendering of the Scorpius, an innovative low-cost launch concept of Microcosm, Inc., which uses 49 engines in 7 clusters or pods for 4 or more stages into orbit, carrying varying payloads from 100 kg to 27,000 kg. To reduce weight, the emphasis, even among entrepreneurial companies, is to utilize an all-composite cryogenic tank, one that is not only lightweight, but also reusable.

What the Ford "Tin Lizzie" did for automobile travel, is the spacecraft sought for space transportation. James Sloan of Information Universe envisions great synergism between the low-cost launch vehicle (LCLV) and the SSTO, offering mission models for a wide range of payloads to orbit. The LCLV would carry

Exhibit 117. Innovative designs of launch vehicles. To cut transportation costs into space, innovative designs for launch vehicles are required, such as those illustrated here. (Left) An artist's concept of the DC-XA (known as Clipper Graham), a reusable single stage to orbit launch vehicle built by McDonnell Douglas in conjunction with NASA, which took off and landed vertically. Unfortunately, at the end of its fourth test flight, the landing gear failed to deploy properly, and the vehicle toppled onto its side, exploded, and was completely destroyed in the ensuing fire. NASA decided not to rebuild the craft in light of budget constraints. Source: NASA Marshall Space Flight Center. (Right) Smaller entrepreneurial companies are developing a new generation of expendable launch vehicles intended to reduce the cost of launch to orbit by a factor of 5 to 10. One of these is Scorpius, now being developed by Microcosm Inc. The Scorpius launch vehicle depicted next (facing page) has 49 engines arranged in 7 clusters or pods, and takes 4 stages to reach upper orbit, plus an optional upper stage. Source: Microcosm Inc.

Sec. 9.2] **Macrochallenges on the commercial space frontier** 435

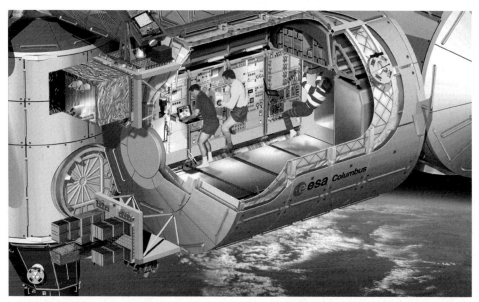

◄ **Exhibit 118.** Current ESA spacecraft. (Facing page) ESA uses the Ariane 5 launch vehicle. The basic version of the Ariane 5 has a lift capacity of 18 tonnes to low Earth orbit and 6.8 tonnes to geostationary transfer orbit. An imposing cryogenic stage 5.4 meters in diameter and 29 meters in length contains a 100-tonne thrust engine that carries 155 tonnes of liquid hydrogen/oxygen as fuel, flanked by two solid rocket boosters each 3 meters in diameter and more than 30 meters high. Ariane 5 can carry and then orbit three satellites at a time. Source: ESA/CNES/Arianespace-Service Optique CSG. (This page) A depiction of the ISS Columbus laboratory module. It is ESA's biggest single contribution to the International Space Station. The 4.5-meter diameter cylindrical module is equipped with flexible research facilities that offer extensive science capabilities. During its 10-year projected lifespan, Earth-based researchers, together with the ISS crew, will be able to conduct thousands of experiments in life sciences, materials science, fluid physics and a whole host of other disciplines, all in the weightlessness of orbit. Source: ESA—D. Ducros.

payloads in the 0.5-tonne to 4.5-tonne class. In Europe, planning and research have also been under way for a variety of advance space vehicles, such as Hotol (British), Hermes (French), Sanger (German). In the arena of expendable rockets, the European Space Agency launched a new Ariane 5 in 1996, pictured in Exhibit 118, and has research under way on a Crew Transport Vehicle, a capsule for crew of cargo transport to use on top of Ariane 5. ESA is working on *FESTIP* (Future European Space Transportation System Program), which includes both reusable and winged launchers using air-breathing engines (see Exhibit 118). The only heavy-lift vehicle in the world today is Russia's Energia, a rocket which may be too large for many missions ... In Japan work continues on spaceplane development, despite cancellation of the HOPE-X project in 2004; now called the High Speed Flight Demonstration (HSFD) project, the HSFD design bears little resemblance to the Shuttle-like HOPE-X ... China too is expanding its manned spacecraft capabilities (see Chapter 1). After the success of its Shenzhou 5 rocket, the *People's Daily* reported (October 24, 2003) that China's Jiuquan Satellite Launch Center is designed to launch large space stations in the future. The Chinese hope to re-establish human presence on the Moon within the next decade, and already have an agreement with Russia for a joint Mars mission in 2009 with a goal of sending people to the Red Planet in the near future ... Meanwhile, among the really far-sighted, the Interstellar Propulsion Society in La Jolla, California is considering plans for manned transportation to other *star systems* (www.digimark.net).

Meanwhile, to achieve its VSE goal of returning permanently to the Moon, NASA is planning the solid-fuel shuttle rocket Ares 1, with its capsule Orion to house the crew, according to the *National Geographic* magazine ("Space: The Next Generation," October 2007, pp. 104–125).

9.2.4 Entrepreneurial view on space access [17]

Currently, there is a difference of opinion on how to lower the cost of space access, and who should pay for it. The debate was evident among space advocates attending the National Space Society's International Space Development Conference in 2006. *NewSpace* entrepreneurs like Peter Diamandis and Burt Rutan are prepared to go it alone with just private investments, whether from billionaires, venture capital, or private equity funds. In fact, Rutan, designer and builder of SpaceShipOne, contends that taxpayer-funded space research makes no sense! He wants to get the aerospace industry off the "government dole". Yet, even advocates admit that smaller operations like his Scaled Composites, or Elon Musk's SpaceX, cannot alone fulfill the nation's *Vision of Space Exploration* goals. Space agencies have advantages of scale and funding that presently give them the leadership role in returning to the Moon permanently.

However, even big Agency contractors, like Lockheed-Martin Space Systems, are concerned that NASA's space efforts fail to inspire young people. Art Stephenson, vice president of Northrop-Grumman's Space Exploration Systems, laments that the Agency is risk-averse, not inspiring. But the conventional wisdom among the government's prime aerospace contractors is that the public sector has to first plow the path that private business will eventually take over. Thus, the strategy of the Agency's Commercial Orbital Transportation Services is perceived as a "breakthrough". COTS $500 million awards encourage private-sector participation in re-supplying the International Space Station. But this program is only the beginning of what is needed to attract small space firms with the purpose of lowering launch costs to orbit. Further, its challenge endeavor offers prizes for innovative technologies. Thus, the 2006 prizes for the *Lunar Lander Analog* will be administered through the non-profit X-Prize Foundation, as will the X-Prize Cup for the next round of hopefuls building space tourism vehicles.

NASA's Centennial Challenge included a $50,000 prize in 2005 for the space elevator game with a goal to develop wireless robot climbers and supercarbon nanotube tethers. In this first contest, none of the contestants won, but did demonstrate "breakthrough" technologies, such as an operational climber that power-beams. Thirty teams entered the 2006 competition because the prizes had risen to $200,000.

Even NASA's current research on a Crew Exploration Vehicle is being criticized because much of its hardware is Shuttle-derived. There is skepticism among taxpayers that the Agency does not have the resources to do all that the Administration has tasked it to do, and the political process is not likely to provide such. There is some consensus that, in the near future when CEV operations begin, the government will be responsible for the more difficult missions, such as spaceflights first to the Moon and Mars, while giant aerospace contractors and entrepreneurs will most likely build a commercial presence.

For a decade the entrepreneurial space transportation community has gathered in Phoenix, Arizona, every Spring for their informal *Space Access Conference*. According to Jeff Foust, founder of *The Space Review*, a free Internet service of essays and commentary about the final frontier, the agenda of these meetings is shifting. Until now their discussions and debates emphasized reusable launch vehicles, take-off and landing modes, launch modes, and engine designs. The transition is from entirely focusing on the technical aspects of spaceflight to the legal and financial issues faced by start-up aerospace enterprises. For example, there is concern about export control, specifically ITAR (International Traffic in Arms Regulations). Participants want the U.S. Congress to reform export control and security measures. Another illustration is the talk about needed liability insurance for RLV developers, especially for cargo and passengers. Federal law regarding insurance is a factor in the granting of an FAA permit or license for spaceflight. Space lawyers are already looking to influence state legislation regulation of personal spaceflight. But new space ventures require funding for their R&D. For those without wealth patrons or space prize winnings, this means more sophisticated business development with venture capital firms and equity capital funds. If a new company does not have an angel investor with long-term horizons, then space entrepreneurs have to prepare believable "cases for investment" that project ROI and satisfaction for such a consumer market as space tourism (see *www.thespacereview.com articles #840/1 of March 26, 2007*).

Perhaps strategic space planners in both the public and private sectors worldwide need to recognize the need for synergy among the various entities. After all, *the ultimate challenge is to drastically reduce the cost of accessing and developing space, so that space industrialization and settlement may progress by mass migration aloft*!

9.3 TECHNICAL SPACE A&R CHALLENGES

Were it not for our creating computers and automated systems, humanity may have never entered the Space Age. Automation and robotics play critical roles in spacecraft and space platforms. Astronauts and cosmonauts never could have built and maintained the International Space Station without A&R, as the robotic arms there confirm. Then AI expert systems and neuro-computers enhanced the use of simulated graphics, artificial intelligence, and distance learning. Robots in orbit can be programmed for quick decision-making by means of expert systems, as demonstrated with the Mars rovers. In 2007, the rover Opportunity was programmed to go down into the 800-meter wide Victoria Crater in search of the ancient Mars climate and its wet past. In this same year, NASA's JPL launched a Delta II rocket with the spacecraft Dawn on top. In 2011, this automated craft is scheduled to visit Vesta, the brightest and third largest asteroid, which has a dry surface that has been reshaped by basaltic lava flows and giant impacts, overlaying a rocky mantle and a denser core. Then in 2015, Dawn moves on to the largest asteroid, Ceres, with its thin dusty crust and icy mantle 60 km–120 km thick, overlying a rocky core. Thus, astronomers hope to learn more about protoplanet pieces that are the building blocks of planets like ours. Dawn contains a camera and two spectrometers that can detect the presence of elements and different minerals; scientists hope to map Vesta, and drill underground on its craters in search of evidence about ancient volcanic eruptions there. And so A&R continues to expand planetary exploration: the latest Mars probe, Phoenix, successfully landed in the Martian Arctic in May 2008.

Furthermore, expect A&R to play an even more vital role in offworld development by erecting space infrastructure. For example, in my science-based novel on lunar industrialization and settlement, I forecast a scenario in which an unmanned robotic mission would precede the permanent return of people to the Moon.[5] Thousands of mini-robots there could construct the basic lunar infrastructure before humans actually returned. Such automated extensions of human intelligence would learn from that experience, becoming in turn self-maintaining, self-repairing, and even self-replicating.

9.3.1 Space robotic possibilities

A&R may be used in lunar communications, transportation, construction, mining, and health services. Telerobotics could be used to create a circumferential electric grid around the Moon. Just imagine tiny robots building a lunar geodesic dome to house both persons and facilities, and then with the help of Kevlar and expert systems, covering the facility with regolith to protect the inhabitants from radiation! Anticipate these "tin collar workers" being employed in making lunar cement, processing lunar oxygen, and engaging in numerous other occupational activities that are too boring or dangerous for human beings. The human–machine interface will be especially valuable in lunar health care and medical practice. Imagine a robotic hand with finger tip sensors able to detect body cancers, providing the exact location and size of diseased cells. Mechanical helpers will augment the human health practitioners, especially for precise surgery. Consider a potential amputee on the Moon with microcomputers on that person's belt which enables the disabled individual to move an artificial ankle and foot like a perfect prosthesis.

---

[5] *Launch Out: A Science-based Novel about Space Enterprise, Lunar Industrialization and Settlement* by Philip Robert Harris. West Conshohocken, PA: Infinity Publishing, 2003, chs. 11–13/16 (*www.buybooksontheweb.com* or *www.drphilipharris.com*) ... Also see Appendix A, "Robots on Planet Moon" in Schrunk, D. G.; Sharpe, B. L.; Cooper, B. L.; and Thangavelu, M. *The Moon: Resources, Future Development, and Settlement*, 2008, Second Edition, pp. 201–233 (*www.praxis-publishing.co.uk*).

Sec. 9.3]   **Technical space A&R challenges**   439

Telemedicine will permit earthkind to provide patient care to spacekind, and *vice versa*; it will also permit first aid and medical services anywhere on the lunar surface! Offworld applications will advance the creation of intelligent machines that manifest artificial intelligence, robotic mobility, and computational speed. Terrestrial research is moving toward a convergence of computing and robotic skills with biosciences. Inventors are already applying nanotechnology to sensors and robots, leading someday to approximating life-like activities to mechanical "beings". Some scientists are researching artifical limbs capable of communicating with the human brain. Before the end of this century, expect evolution toward *Robo sapiens* as a result of genetic engineering blending humans, animals, and machines. In this millennium, robots will be used to terraform asteroids and other planets. No wonder the late Japanese genius Ichiro Kato, creator of the WABOT 2 robot with an IQ equivalent to a 5-year-old, remarked: "The robot is a system-engineering problem. But the human being is a microcosmic wonder!"

9.3.2 Spacebots' prospects

The spider-like robot pictured in Exhibit 119 underscores the trend of scientists who imitate life by now designing robots in the image of animals, as well as humans [18]. So, underwater robots are designed to look like turtles, while computer-controlled snake-like robots are similar to the spinal cord. Designers and technologists are inspired by nature in their robotic creatures, some of which lunar dwellers may adopt as their "pets". Years ago when your author mentioned this possibility in a national radio interview, listeners all over the country got excited by the very idea! Biologically influenced robots exploit fundamental physical principles. Such robotic designs on this planet or beyond are based on natural organisms as diverse as the animal world itself, allowing the device to mimic everything from hawks and dogs, to caterpillars and elephants!

Researchers at Carnegie Mellon University are working on a robotic medical device that moves inside the body like an inchworm; at DARPA, they have "Big Dog", a four-legged, semiautonomous robot that may someday work as a soldier or astronaut helper; at Clemson University, there is "Octor", a robot modeled on an elephant's trunk that can pick up a peanut or a tree; at the Neuroscience Institute they have an AI robot that can play soccer in a team; at Vassar College, New York State, they have "Madeleine", an autonomous underwater vehicle that looks like a marine reptile with four flippers; at Tufts University, engineers are developing a soft, crawling robot that someday will remove land mines or hard-to-reach machinery; at the University of Southern California, they are investigating SuperBot, a system of modular, Lego-like cooperative ants that plug into each other to create robots that can stand, crawl, roll, wiggle, climb, and even fly in a microgravity environment. And just to mindstretch readers, Rhex and RiSE function as six-legged robots capable of running, leaping over obstacles, climbing stairs, and, with modifications, walking up walls and trees.

What is the point here: well, robots will be big business in outer space ... The Japanese/European Spiderbot, for example, described above could revolutionize the

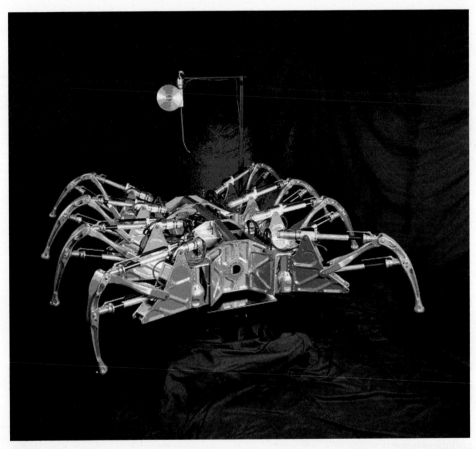

Exhibit 119. Space business through automation and robotics. Robots such as the above *Spiderbot* may do much of the offworld work on other celestial bodies. This one was launched in 2007 by a Japanese rocket on a suborbital flight that released a satellite. Spiderbot then deployed three palm-size, daughter satellites who pulled out a 360 square-yard triangular net with the mother-sat at the center; wheels gripped the net so the bots would not drift off in zero gravity! Robot designs no longer look just like their human creators (e.g., androids), but more frequently these technologies are inspired by nature, such as this spider-like robot. Source: NASA Headquarters Media Service.

satellite industry, permitting midget sats to be launched from small, less expensive rockets and then be assembled in space by robots.

9.4 LEGAL SPACE FRONTIER CHALLENGES

To ensure the industrialization and habitation of space, legal institutions, laws, and experts will become part of offworld governance. However, our planetary legal system and enactments need to be transformed for meaningful nonterrestrial applications. David Schrunk, an aerospace engineer and physician, maintains the traditional systems of lawmaking are archaic and ineffective, in need of reform [19]. In his book, *The End of Chaos*, this author argues for application of the scientific method to law making. He favors the use of competency and knowledge in designing and the enacting of new laws. Since laws are one manifestation of human culture (see Exhibit 20), perhaps only in outer space will people be free eventually to enact astrolaws and regulations that are suitable to their exceptional microgravity environment.[6]

For a functioning democracy, the rule of law must prevail, and a spacefaring civilization requires law-abiding citizens. But, offworld, which laws are to prevail? Just as in past centuries, the New World built upon the legal heritage of Europe, but created something unique to the North American continent, so too ground-based *space law* is now being created, to deal with legal issues related to orbiting satellites, launch vehicles, and space debris. In time, a new type of *astrolaw* will someday emerge in orbit, to meet the needs of people within a space culture (see Chapters 2–4). Those seeking to engage in space commerce currently face, both domestically and globally, a complex, convoluted, and often contradictory regulatory scheme. In the U.S.A., the centralization of laws, regulations, and agencies affecting space business practice is only at a primitive stage. Attorney Nathan Goldman has provided a review of both national and international practice, as well as future needs regarding space law [20]. Relative to international space treaties under the auspices of the United Nations, Declan J. O'Donnell, president of the World Bar Association, is among those who believe that some of these, such as the Moon Agreement, have proven to be inadequate, if not anti–space development in wording, and need to be replaced [21].

Nearly 30 years ago, a colonel in the U.S. Air Force legal division published an essay on "astronautical law" which is now recognized as the first comprehensive examination of the legal issues related to human movement beyond this planet. Its author, Brigadier General Martin Menter, was honorary director of the International Institute of Space Law, an ongoing forum for lawyers which sponsors colloquiums and publications on the subject of space laws and regulations. The organization is part of the International Aeronautical Federation headquartered in Paris [22]. Many IISL members have published and continue to write scholarly books on the legal dimensions of human actions in outer space [23].

9.4.1 Space laws and regulations

"Space law is actually a complex mixture of international and domestic laws that govern a wide spectrum of activities. They can range from the exotic, like creating the institutional framework for an international mining consortium, to the more

---

[6] See appendices in Schrunk, D. G.; Sharpe, B. L.; Cooper, B. L.; and Thangavelu, M. *The Moon, Resources, Future Development, and Settlement*. New York: Springer/Praxis, 2008 (especially Appendix A, "Robots on Planet Moon"; Appendix G "Quality standards for lunar governance"; and Appendix T, "Beyond our first moonbase."

routine, like drafting telecommunications agreements. The fields of law these activities can involve include administrative law, intellectual property, arms control, insurance law, environmental, criminal, and commercial laws, as well as the treaties and legislation written specifically for space" [24].

So explains attorney Joanne Irene Gabrynowicz, who teaches graduate courses in space law and policy at the University of North Dakota. She believes that space law provisions provide insights into the forces that come to bear on space programs and missions ... Presently, a private company or a nation needs to be concerned with both the international regimes and the regulations regarding space activities. These include, for instance,

- **UNCOPUOS**, the United Nations Committee on the Peaceful Use of Outer Space and its key treaties: "The Outer Space Treaty on Principles Governing Activities of States in the Exploration and Use of Outer Space, Including the Moon and Other Celestial Bodies" (October 10, 1967); the "Liability Convention on International Liability for Damage Caused by Space Objects" (October 9, 1973); "Convention on Registration of Objects Launched into Outer Space" (October 15, 1976); and the Moon Treaty or "Agreement Governing the Activities of States on the Moon and Other Celestial Bodies" (December 5, 1979).

  Of the five U.N. Treaties put forth by UNCOPUOS, the U.S.A. and Russia have signed four, excluding the last-mentioned Moon Treaty, in addition to bilateral agreements and the International Telecommunications Union. The agreements reached underscore the themes (a) of international cooperation and that space is to be used for the benefit of mankind, not for national annexation; (b) reserved for peaceful purposes; (c) that nations are responsible for the space activities of their nationals, but liability on default pertains not only to damages in space. In the case of joint space ventures among nations, the Liability Convention holds countries jointly and individually liable for damages caused by such collaborative efforts. Even the environmental protection of celestial bodies, as well as back-to-Earth contamination, is considered within the U.N. treaties. When the Sierra Club did sponsor a conference and proceedings on environmental ethics and the Solar System, space law and its applications were examined, but not in depth [25].

  Unfortunately, national interpretations of these international space agreements often differ. The American position is that nations are permitted to use—not own—space, and can mine and claim resources; nations do not have the right to own a site—but do own the ore extracted. An open question is whether "peaceful purposes" in space means non-aggressive and permits defensive uses, like the Strategic Defense Initiative of President Ronald Reagan's administration. There are other problems with these U.N. agreements. First, how representative of the global space and legal communities is the make-up of UNCOPUOS? Second, the times and situations have radically changed since these treaties were written. Do they represent today the best international thinking and consensus on their subject matter? Some international space legislation was a product of the "Cold War" between East/West ideologies. Third, may individuals claim rights to private property?

A case in point is the so-called "Moon Treaty", which was not agreed to by any of the spacefaring nations; in the United Nations only eight, mostly developing, countries actually signed it. Its content has met hardy opposition and controversy, often along North/South economic lines; scholars in many countries found its language deficient because of inadequate wording, vague concepts, and anti-developmental approaches to space resources. In the U.S.A., the National Space Society actively argues against its provisions. Yet, when the "treaty" came up for revision in 1994, the 37th session of UNCOPUOS met for ten days in June, and then decided to keep the *status quo*. The Committee's chairman, Ambassador Peter Hohenfellner, wrote to your author that UNCOPUOS recommended to the 49th U.N. General Assembly that no further action should be taken. In fact, his letter states (May 18, 1995), "This was endorsed by the General Assembly and, therefore, no developments are anticipated on this matter, in the near future." How informed were the Committee's members of the alternative viewpoints and recommendations for change in that 1979 document?

- **International agreements**, especially at a regional level, can occur among parties apart from U.N. involvement. For example, these are examples of the multilateral accords that have evolved: the European Space Agency Convention, the Intelsat Agreement, the Eumetsat Convention, and the International Space Station Intergovernmental Agreement. The ESA and its multinational agreements have played an important role in the evolution of space law, and may even provide a prototype for future government entities in orbit. Such understandings set the legal foundations among the signatories for a variety of space activities from administration to satellites, from design to maintenance. Sometimes the pact is bilateral in scope, as with the Treaty Banning Nuclear Weapons Tests in the Atmosphere and Underwater or the Liability Convention between the U.S.A. and the former U.S.S.R. The last-mentioned holds a spacefaring nation liable for injuries caused by its space-related activities, whether on the ground or in orbit.
- **National policies and rulings** regarding outer space can also differ. At the national level, space business and entrepreneurs face additional legal guidelines and constraints. Again using the U.S.A. as a case in point, generally a lack of coordination is inherent to government space policy and practice, complicated by traditional bureaucracy. Finch and Moore summarized the policy and regulatory situation regarding astrobusiness [26]:
  ○ **Administration policy**. Since Eisenhower, U.S. Presidents have issued national space policy directives. On July 4, 1982, for example, NCS-42 of President Ronald Reagan authorized private-sector participation in civil space activities with appropriate government supervision and regulation. On May 16, 1983, that same President issued another directive for the commercialization of expendable launch vehicles. Again on November 18, 1983, President Reagan designated the Department of Transportation as the lead regulatory agency for this purpose and promoted a Strategic Defense

Initiative for military space. But the first President George Bush failed to get the support of Congress behind his Strategic Exploration Initiative for civilian space, while President Bill Clinton never articulated a real national space policy. But the present administration of George W. Bush has set forth the *Vision for Space Exploration* on permanently returning to the Moon before year 2020 (see Chapter 1).

- **Regulatory agencies** also set their own organizational policies and rules within national policy and law which impact space commerce and industrialization. Currently, these are the major agencies influencing space business: NASA, Federal Communication Commission, and the U.S. Departments of State, Defense, Transportation, Commerce, and Energy. Within DOD, business may have to deal with the varied positions of the military services, and within a department, such as Commerce, there are sub-entities that can legally affect astrobusiness (e.g., National Oceanic and Atmospheric Administration, National Telecommunications and Information Agency). The FAA has also set guidelines for spaceplanes to fly and land. Instead of coordinating space regulations for business, the opening of space offices in various Federal agencies has only complicated bureaucratic actions. Perhaps the time has come for national governments to establish *a department of outer space activities* that coordinates all public and private-sector offworld activities!

- **Congressional actions** mainly influence astrobusiness in two ways. First, by monitoring and influencing governmental agencies that relate to space business activities (e.g., the various statutes requiring Congressional approval before Landsat was commercialized or the study of the space commercialization process by the Commerce Department). Second, by legislation directly regulating space activities, such as the streamlining of the approval process for new space ventures, or bills to encourage joint R&D activities to increase industrial innovations, such as in the use by NASA of automation and robotics. Finally, the President of the United States signs such legislation into law, as with the case of the National Aeronautics and Space Act of 1958 that separates military and civilian space activities; it was amended in 1988. Congress also passed the Comsat Launch Act of 1962 which authorized U.S. participation in Intelsat, an international communication satellite system; the Commercial Space Launch Act of 1984 to promote economic growth, peaceful entrepreneurial activity in space, and to authorize the Department of Transportation to regulate commercial launches within the country (amended in 1988 to require launch providers to obtain insurance to cover damage due to accidents and failures, such as the *Challenger* and *Columbia* Shuttle systems); the Remote Sensing Policy Act of 1992 which provides for the issuances of licenses for private operation and analysis of sensing data, requiring that only unenhanced data be made available to the governments of countries which have been imaged by this technology. Sometimes, Congress also establishes special Commissions to study and report on space-related problems and opportunities.

> Various legislation has been introduced into the U.S. Congress to further space developments, but many fail to receive the approval of both houses, and the legislation is not enacted. A few do get passed and have impact. Take the Commercial Space Amendment Act of 2004 which establishes regulatory oversight for the fledgling human spaceflight industry. Passed in the House of Representatives, the chairman of its Science Committee, Sherwood Boehlert (R-NY) commented: "As in most areas of American enterprise, the greatest innovations in aerospace are likely to come from small entrepreneurs. This is true whether we are talking about launching humans or cargo. And the Congressman's goal in this bill is to promote robust performance, to make sure that entrepreneurs and inventors have the incentive and capabilities they need to pursue their ideas."
>
> H.R. 3752 stipulates that the Federal Aviation Administration's Office of Commercial Space Transportation is responsible for regulating all commercial spaceflight vehicles, supposedly making it easier for spacecraft developers to take passengers aboard. It also extends government indemnification for the entire commercial space transportation industry, including licensed, non-experimental commercial human space launches ... The House of Representatives and the Senate passed this bill which was signed into law by the President (source: *Ad Astra*, the magazine of the National Space Society, April/May/June 2004, pp. 9–10, www.nss.org.)

   o **Court precedent** has always been a factor in legality. As people or institutions worldwide sue in court concerning space-related activities on the ground or in orbit, the judgments rendered contribute to the body of space law. For example, the case of *Martin Marietta vs. Intelsat* in a U.S. Federal court resulted in a finding that broadened interpretation of the Launch Act to the effect that it was the intent of Congress to provide protection for U.S. launch providers; therefore, waivers based on that Act and used by Martin, the provider, with Intelsat, the owner, were enforceable.

Two examples from the past illustrate how space business can be hamstrung by obsolete national policy and regulations:

- Art Dula, the Houston space lawyer who headed up the Space Commerce Corporation, had been trying since before 1992 to get the State Department to lift "Cold War" constraints on U.S. companies launching payloads on the then Soviet rockets. Similarly, if previous negotiations with Brazil had been completed to utilize Russian Proton launch vehicles at Alcantara near the equator, Dula's law firm would then have needed policy changes for American corporations to take advantage of these lower launch costs.
- The Cape York Space Agency in Australia not only had a contract with Glavkosmos to use Soviet rockets, but also wanted an American firm, United Technologies' USBI Division, to manage its facilities. Again, U.S. policy and regulations in the 1980s mitigated against American companies wishing to take

advantage of such opportunities. It had been hoped that a new, Presidential commercial launch policy would ameliorate both situations, while protecting to a degree the infant commercial launch industry in the U.S.A. by assuring that government payloads would only be launched on American vehicles.

In this post-Cold War era the above are moot points, for both the U.S. government and industry are seeking to cooperate with the Russian Federation and CIS countries, so as to foster democracy and market economies in these former totalitarian societies. Furthermore, entrepreneurs have been frustrated in their attempts to promote spaceports both in Australia and in Hawaii for lack of both government and public support. But these examples also illustrate that conditions, policies, and legislation must change to be relevant to contemporary human needs aloft. In 2007, for instance, some Western countries expressed concern when China used a guided missile to shoot down one of its own obsolete satellites. Supposedly, the outcry was against creating more space debris, but no national legal action was taken.

Presently, there is a growing body of space law and expertise worldwide which sometimes facilitates space enterprise, but more often constrains it. Professor Gabrynowicz [24] observed that much of what space law requires is unknown, and will change as technology matures, human experience grows, and new situations are encountered. She envisions space law as dynamic, with the next generation of such law agreeing on specific norms and addressing such vital questions as:

- Is sovereignty necessary to establish property rights?
- Are space resources, as well as space itself, the province of all humankind?
- If so, how are they to be allocated?
- How can non-spacefaring nations be assured of the use of space and its resources?
- How will the investments of spacefaring nations be respected?
- How will private space activities be allowed to operate?

Those concerned about space law issues should read the *Journal of Space Law* and the *Annals of Aviation and Space Law*, or newsletters such as those published by the space and communications law firm, Mosteshar MacKenzie (*www.Mosteshar.com* or email: *Mosteshar@ucsd.edu*).

9.4.2 Space settlements and governance

At the moment, there is no planetary policy for offworld colonization and administration. Yet, an International Space Station with a multinational crew is operational: a prototype of orbiting colonies envisioned by some. Further, by year 2020, several government space agencies are tentatively planning a lunar outpost, and private enterprise has grandiose schemes for a lunar base and industrial park before 2025. Many space planners worldwide expect an outpost on Mars or its moons by 2050. But even at the United Nations, there has been no serious discussion about how international partnerships could be formed and managed to ensure permanent settlements on the high frontier [21].

In the United States, one of the most significant legislative proposals in this regard was presented to the U.S. Congress as the Space Settlements Act of 1988. Introduced by Rep. George E. Brown, Jr. (D-CA), of the House Science, Space, and Technology Committee, this failed attempt was discussed in our opening chapters as a forerunner of what yet may come to be. The proposed H.R. 4218 would have amended the National Aeronautics and Space Act of 1958 to set the establishment of space settlements as a long-term mission for the space agency. Further, NASA would have to conduct a steady, continuing effort to explore all the technical and sociological issues relating to the achievement of settlements in space (*Space World*, June 1988, p. 3). The proposal came in response to various national recommendations (National Space Commission, 1986; Ride Report, 1987) and the first President Bush's then policy that a long-range goal for the United States was "to expand human presence and activity beyond Earth orbit into the Solar System." Although not enacted as originally proposed, Congressman Brown did manage to get the substance of his bill passed by means of an amendment to a NASA funding authorization during the 100th Congress session. S.2209 is now law and provides national legislative commitment to space exploration and settlement, requiring NASA to report every two years on its progress in that regard. Unfortunately, NASA has been rather lax in adhering to that mandate because of the defeat of the Space Exploration Initiative (see Chapter 2). Many space experts forecast that a whole new type of legislation will have to be enacted regarding space settlement and industrialization, administered outside the space agency itself. Conceivably, this undertaking may require an international agreement resulting in a global administration for this purpose (e.g., a Global Space Trust).

Apart from the hundreds of human-made satellites, as well as unmanned spacecraft, orbiting the Earth, space habitation raises the most intriguing non-terrestrial legal issues. In conjunction with the celebration of the 200th anniversary of the U.S. Constitution, the Smithsonian and its National Air and Space Museum organized two conferences of 30 law specialists regarding basic rights and freedoms which apply to space. The outcome was a "Declaration of First Principles for the Governance of Outer Space Societies" (see Appendix A). William J. Brennan, former chief justice of the U.S. Supreme Court, called for U.N. approval and insistence of these initial 11 principles to "ensure that the fundamental needs for life, individual freedom, liberty, justice, dignity and responsibilities inherent in self-determination are integral parts of humanity's exploration and settlement of space" (*Los Angeles Times*, May 22, 1987, I:27). In that address to the American Law Institute, Brennan noted that space law principles have to be developed in anticipation of future space settlements.

Former NASA and Smithsonian legal counsel Dr. George S. Robinson coauthored *Envoys of Mankind*, which offers a conceptual foundation for a new body of space law and constitution, as well as a review of the evolving body of national and international law concerning space. In the opening address to the Legal Forum on Space Law of Air/Space America (May 17, 1988), organized by your author, Robinson warned that establishing space habitats for lengthy human duration demands that planners and designers of space stations and future space settlements focus seriously on the social organization of space societies, including the values underlying whatever organizational principles are ultimately adopted. He contends that space law should clearly recognize the unique physical and psychological environment aloft where human biology and technology are totally integrated for survival. For the benefit of future spacekind progeny, Robinson advocates unparalleled principles of law to safeguard their lives as the truly bio-cultural envoys of earthkind. He also quoted Harrison Schmitt, former lunar astronaut and senator, "I have been governed in space by earthbound authorities, and within the constraints of survival, even rebelled against that governance." In his 2006 book, *Return to the Moon*, Dr. Schmitt also cites these U.S. legislations which would affect near-term lunar development [28]:

- the Communications Act of 1934 regulating radio frequency allocation and use;
- the Environmental Protection Act of 1970 that governs corporate and national activities, such as manufacturing and rocket launches;
- the Commercial Space Launch Act of 1984 which regulates space launches;
- the Commercial Space Acts of 1997 and 1998 with their regulations on launch licenses, return payload licenses, and protection of space assets as private property;
- the Commercial Space Launch Amendments of 2004 that focuses on human spaceflight, licensing for private space enterprises, and indemnification of space launches.

Although this significant forum of space lawyers was held some 20 years ago, these observations are largely relevant, so a summary is provided below of the other speakers. Co-sponsored with San Diego's National University Law School and under the chairmanship of Chuck Stovitz, a Beverly Hills attorney, the event's proceedings were not published, but the author was present to take these notes which in retrospect provide unique insights:

- **S. Neil Hosenball**, former NASA general counsel. Among the foreign competition to be faced in commercial space, consider that the People's Republic of China has a new Ministry of Aerospace, which now offers launch services at less cost than U.S. suppliers. The Russians not only offer similar low-cost launches, but provide quality remote-sensing photography services. The Japanese are planning to invest in an Australian international spaceport ... Currently the U.S. Congress is "hamstringing" private investment in space activities ... We need to get NASA out of operations which could be taken over by the private sector, thus freeing up the Agency to concentrate on science, technology, and exploration.
- **Art Dula**, partner in Dula, Shields, Ecknert, Houston attorneys. We had set up the Space Commerce Corporation to market Soviet rocket services (see Exhibit 108). Our foreign technical team was the first to be permitted to check out the Baikonour Cosmodrome near Tyuratam in Central Asia. We were not doing this for the Russians, but for American business who have a need to buy less

expensive transportation services. Although this may only be a temporary service, we have to overcome opposition from U.S. governmental officials and existing policies on export licenses.
- **H. Cushman Dow**, vice president and general counsel of General Dynamics, San Diego. In attempting to market the services of our Atlas–Centaur, we face competition not only from other American launch craft, the Shuttle itself, but also from foreign competitors, principally the French Ariane and possibly the Japanese. The latter are backed by their governments, which means they need not make a profit and may be state-subsidized. It is only in the United States that commercial really means "commercial" or private sector. The expendable launch business in the U.S.A. is dependent on cooperation with NASA and the military, who control the facilities to be used, and upon Congressional action regarding liability insurance.
- **Theodore Harper**, Graham & James, attorneys, Los Angeles. In building their aerospace plane, the Japanese seek international partners, but not the United States. The trend in space business is toward many new players and more international partnership, but not with Americans, who insist on applying U.S. contract law and export control regulations. [Fortunately, this practice is changing in favor of international partner agreements, but absurd Federal regulations on transfer of technology is an impediment to space developments.]
- **Dan Byrnes**, adjunct professor, Pepperdine University Law School. Presently, only $500 million is available in liability insurance for space accidents. We need the establishment of a Space Industrial Advisory Council to counsel the U.S. President on space commercial policy.

These excerpts underscore accelerating changes within planetary societies that alter perspectives which impact emerging space law. Imagine the legal implications of an international venture to build and operate a lunar base! Conditions now are more favorable to international cooperation in space, and global joint technological ventures that may really open the high frontier (see Exhibit 114). Finally, there is a body of legal precedents in the international agreements among various national space agencies, as indicated in Exhibit 120.

The emerging fields of space and astro laws are the subject of continuing scholarly research and discussion, such as at the Institute of Air & Space Law in McGill University, Canada, which offers graduate degrees in such law, or at regional institutes of space law, or among professional societies of attorneys and their publications. For five years, your author tried to further this dialogue as founding editor of the journal *Space Governance* [29].

9.4.3 Future space administration and governance

The global space community is gradually recognizing that some type of international jurisdiction or entity has to be created to manage and coordinate international enterprises offworld. Again, there has been no serious discussions, even within the United Nations and aerospace associations, as to what form this new legal body

Exhibit 120. Interagency space agreements: SOHO. This is an artist's depiction of this solar observatory that resulted from a legal agreement between ESA and NASA. It was launched from Cape Canaveral on December 2, 1995, on an Atlas IIAS rocket. Located 1.5 million kilometers from Earth, SOHO enables investigators from the U.S.A. and Europe to study the structure of the solar interior, as well as the physical processes that form and heat the solar corona. Source: ESA Solar–Terrestrial Science Program (STEP).

should take. The first challenge would seem to be on the Moon where a demonstration model might be established for the common benefit of the many partners involved, whether from the private or public sectors. What initial governing body is created for that purpose will have implications for exploration and development on Mars, as well as other celestial places. However, some individuals and organizations have given the matter serious consideration in their discussions and publications.

One such proposal in the matter of *space governance* has been promoted by both the World Bar Association and the United Societies in Space Inc. of Castlerock, Colorado. This strategy, discussed in Chapter 7, would create a *space metanation* to protect the *common heritage of humankind* (CHOM). Such a concept was not foreseen in the 1967 Outer Space Treaty signed by 100 nations, including the U.S.A., but was later incorporated into the Moon Agreement of 1979. CHOM as a "term of art" was defined in the U.N. agreement on the Law of the Seas in 1982, signed by the U.S.A. in 1995. Under the rules of treaty law, it would apparently be applicable also to astrolaw, or the laws and regulations developed in the future by offworld inhabitants. The difficulty comes in the application of this concept which requires an active sharing of space resources and properties for the "benefit of humankind". Some interpret such provisions as curbing space development, for there would be no incentives for investors and entrepreneurs to develop the high frontier with a hope of getting a return on their efforts aloft.

In contrast, the USIS proposal would place the new space institution under U.N. Trusteeship Council provisions, with trustees appointed from the spacefaring nations for a hundred years. Both the principles of international law and the concept of CHOM would be protected under this new entity through a constitution acceptable to all nations, including those in developing economies. This innovative plan suggests a new Treaty on Jurisdiction in Outer Space to be passed under the auspices of the United Nations [30]. It also envisions the establishment of *space authorities* that may or may not be placed under metanation or U.N. control for the purpose of issuing bonds, processing leases on land and facilities, as well as promoting new enterprises on various planets and asteroids. The first such would be a *Lunar Economic Development Authority* (LEDA), its legislation and scope being comparable with that of the Tennessee Valley Authority which came into being by an Act of the U.S. Congress. Such strategies would provide answers to some of the critical questions raised above by Professor Gabrynowicz, and be more pro-development.

> "The challenge for members of the legal profession worldwide is whether the law as currently applied will be used to impede or facilitate space commerce, whether its practitioners will build upon terrestrial legal foundations to fashion systems of laws relevant to space travelers and dwellers!"

9.5 POLITICAL CHALLENGES ON THE SPACE FRONTIER

The Space Age literally came into being and turned into a space race because of political decisions connected with the Cold War. It was the Soviet launching of

452 Challenges in offworld private enterprise

Exhibit 121. Multinational space synergy. Historic link-up: crews of STS-71 and Mir 18/19. In June 1995, the American and Russian crews of these missions gathered aboard the Shuttle in the space science lab module for this inflight portrait. To identify those in the photograph, hold the picture vertically, with the socked feet of Anatoly Solovyev at the bottom. Then go clockwise to view Gregory Harbaugh, Robert Gibson, Charles Precourt, Nikolai Budarin, Ellen Baker, Bonnie Dunbar, Norman Thagard, Gennady Strekalov, and Vladimir Dezhurov. Try to distinguish the astronauts from the cosmonauts in this original example of space synergy. Source: NASA.

Sputnik 1, the world's first artificial satellite on October 4, 1957, followed by the first unmanned probes of the Moon with the Russian Lunas, and the first man in space, cosmonaut Yuri Gagarin, that prompted political responses within the American administration. John Logsdon, the premier space policy analyst, has documented the political influences leading to national decisions to send men to the Moon through project Apollo to the current Space Shuttle program [31]. As director of the Space Policy Institute at George Washington University, he led the political debate for internationalization of the U.S. space program: namely, to bring global partners into the costly Shuttle and Space Station programs. In an ABC network television discussion of the matter, Logsdon is quoted as saying that the country does not have to be "first" in every aspect of space exploration:

> "Why shouldn't we go with the Japanese and the Germans and so forth? This is a human problem, not a United States problem. I think the days of racing to a place

are over ... We can't cooperate [on a space station project] with others until we have a strong program ourselves. And that—a strong articulation of purpose and priorities—can only come from the President" (*Los Angeles Times*, September 12, 1988, p. 4).

As the two superpowers struggled for preeminence as spacefaring nations, the political debate over the issue of international participation increased, but is seen today as quaint history [32]. Americans are less afraid that joint space ventures with foreigners, especially the former Soviets, may weaken the nation, militarily or technologically, by giving away secrets. On the other hand, the sheer complexity and cost of new space technological ventures are forcing many countries to form consortia, as we discussed in Chapters 1 and 2. The competition became less between the space programs of various countries than with other parts of an economy seeking their share of shrinking national resources. Thus, in the same media program cited above, Senator William Proxmire (D-Wis.) talked about the exorbitant costs to build and run the Shuttle/Space Station programs, suggesting back in the 1980s that NASA seek alternatives by sharing the projects with other nations, as well as the private sector.

Two decades ago, political issues exacerbated the situation between the superpowers and their endeavors on the high frontier. For example, in the U.S.A. leaders disputed whether to invest public monies primarily in "Star Wars" or commercial space activities, or both. A 1988 Congressional Budget Office report rightly observed that the U.S. space program was at a crossroads, noting that "civilian space is becoming an all or nothing proposition" because of the large price tags associated with proposed space projects. It was a combination of such factors that made it impossible for the first President George Bush to garner initial support for his policy on the Space Exploration Initiative. Geopolitical changes in the past decade altered the myopic national focus of space planners. When Mir was in orbit, the Shuttle docked at that Russian station and astronauts worked there with cosmonauts. Further, Russian cosmonauts, along with ESA/Japanese astronauts, now fly on the Shuttle, and representatives of many nations serve cooperatively on the International Space Station. The new era of collaboration is symbolized in Exhibit 119.

The political reality of the 21st century is that governments and space agencies seek *space synergy*, primarily to share expertise and risk, as well as to save costs. Thus, many countries enter into international agreements for joint space exploration, Even the current U.S. space policy, *Vision for Space Exploration*, is based on the premise that the Americans will not return *alone* to the Moon permanently, but this undertaking will include formation of partnerships with other nations and even the private sector. Henceforth space missions will be multinational in scope, and the crews, consisting of both genders, will be multicultural and multidisciplinary in backgrounds! Furthermore, there are long-duration spaceflight policy issues to be resolved worldwide that range from cross-cultural training of crews to matters of privacy.

9.5.1 Influencing space policy

Policy has been defined as a definite course of action taken to achieve a goal, to facilitate, to build consensus, or for expediency. In adopting policy that gains public support, the art of compromise is common practice. Politics, on the other hand, is viewed as the art or science of governing, of managing political affairs, power, and decisions. Whether policy or politics, the key is developing and implementing strategies for moving human systems ahead, for accomplishing tasks. Political scientists and politicians are supposed to be the experts in such matters, but increasingly those in leadership positions are turning to competent professionals in strategic issue management [35].

Dr. Nathan Goldman believes there is a political dimension to all human enterprise whether on this planet or in space [33]. In the U.S.A., for instance, this University of Houston and Rice University professor maintains that policy decisions are involved in everything from getting Presidential or Congressional support for a particular mission scenario, to setting a national priority, such as to build a lunar base or explore Mars. Goldman identifies three models of policy science which apply to space activities:

(1) **Incremental decision-making** or business-as-usual. Present arrangements are the starting point for future alterations, changes come in increments through debate and compromise, as well as forming coalitions of support to realign matters, as happened with the long-standing efforts to build the International Space Station.
(2) **Non-incremental decision-making** or Schulman model. This is evident in big projects that need a "critical mass" of support before receiving full public funding, such as the major leap forward with the decision to invest in the building of a Space Shuttle where the very size and scope of the program makes it non-divisible once started.
(3) **Strategic decision-making** or Huntington model. This is when a crisis causes a major political choice, such as when President Kennedy, reacting to Soviet space advances, made the decision to go to the Moon after stating, "We are in a strategic space race with the Russians, and we have been losing."

The third model tends to place the emphasis on external factors, while the other two focus on internal factors. For this latter approach to succeed, John Logsdon, North American editor of *Space Policy*, the prestigious journal published in Oxford, believes that well-developed preliminary plans and consensus on feasibility have to exist before the decision is made.

In Europe, ESA used to set policy through its Council of Ministers. However, in 1995, the European Union notified the Agency that it would take charge of Europe's space policy. The story in *Space News* stated [34]:

> "Frustrated that billions of dollars of government investment has not brought a secure commercial position for Europe in space-based telecommunications—either in satellites or the lucrative ground equipment market—the EU Commission and the European Parliament propose to assume control of Europe's

Apparently, the EU is pressuring ESA, primarily a research agency, to pay more attention to space commerce and less to space science and technology. The proposal enacted in 1996, and resulted in the EU Commission licensing global satellite telephone systems."

Perhaps, this is the case today as budget reductions and escalating costs stimulate all spacefaring government officials to rethink the "go-it-alone" policy in space [35]. The current position forms alliances with other nations to share in space undertakings, encouraging the private sector to join with the public sector in space development. However, analysis of a recent American space policy initiative demonstrates the difficulties: "Project Pathfinder" was announced by President Ronald Reagan in 1988 as a policy to develop "pathfinder" space technologies which would enable the country to return to the Moon and eventually fly to Mars successfully. The President proposed to spend $1 billion on this program; Congress in FY 1989 appropriated only $40 million, a 60% cut in the original request. This initiative was not a pathfinder. Obviously, the Federal government will no longer monopolize U.S. space business because it will not fund such programs alone.

A Resources of the Future symposium held in June 1986 concluded that space policy makers need to understand better the interrelationship of economics and technology. By adopting a policy of space collaboration over competition, there are more political and economic benefits to be gained (e.g., in Earth observation and mapping, space transportation and tourism, in addition to lunar science, industrialization, and settlement).

9.5.2 Influencing lunar policy

Returning to the Moon for a permanent settlement and development of its resources is a policy issue to which one or a combination of the above theoretical models may be applied. It is now the national policy in the United States, and other countries are moving to join the venture. Many organizations have taken a policy stance on this, such as the National Space Society. Back in 1995, NSS got on the information highway by establishing a *Return to the Moon Homepage* on the World Wide Web (*http://www.ari.net/back2moon.html*). As part of their mission to advance people living and working in space, the Society's officers and members regularly testify on behalf of space legislation in the Congress. For example, on May 25, 1995, the NSS Executive Director testified that the "ultimate purpose of NASA should be to empower individuals and private organizations to go into space for their own reasons." At the Second Annual International Exploration Conference sponsored by NSS in that same year, participants agreed to promote a *Lunar Consortium*, made up of organizations represented at this meeting. They pledged to promote a variety of lunar return strategies, including the *Lunar Economic Development Authority*. Worldwide major meetings, such as the International Lunar Exploration Conferences, bring together a cross-section of supporters for this strategy (*www.spaceage/pub.com/ILEC*). Private initiatives may also promote lunar development, such as the International Lunar Observatory Association (*www.iloa.org*). One of most influential efforts on space policy and political decisions is the International Lunar Exploration Working Group established in 1995 at a Meeting in Germany. ILEWG includes representatives from the major space agencies in the world, and aims to develop international strategy and consensus by space professionals for exploring and developing the Moon. In 2003, the Group joined in the International Lunar Conference held in Hawaii which has a more global representation from the private sector. In ILC's 2004 convocation in India both organizations issued a declaration that the U.N.'s Moon Treaty be "revisited, refined, and revised" in the context of contemporary impetus for lunar expeditions, both robotic and human. In 2006, after the COSPAR assembly in Beijing, both ILEWG and ILC representatives conferred there with Chinese space scientists and engineers.

Like many others in the global space community your author believes a key rationale for lunar industrialization will come from the need for clean, non-polluting solar energy to replace organic fuels, such as provided by a lunar solar power system (Appendix B) or solar power satellites. Building a lunar base is the kind of 21st-century macroproject that requires a broader and more integrated space constituency than now exists. Since this massive undertaking probably involves spending a total of $100 billion over 20 years or more, a *case for investment* has to be communicated that will attract worldwide political and public backing, especially on the part of the media and the young generation who will pay for and will implement the policy. The only governmental group with a viable plan in this regard is the European Space Agency, but its initial scientific focus is too narrow to gain mass enthusiasm, notwithstanding the necessary expenditures by financially pressed member states. However, ESA's useful lunar reports do go beyond previous American and Russian studies, and recommend participation by other international partners, as well as the need for eventual industrial development on the lunar surface [36]. The problem is that ESA has been slow in carrying out these strategies, and seemingly has not convinced the European Union to adopt such policies.

Establishing a lunar base by a spacefaring consortium would necessitate development of a *global space ethos* among the public in the nations involved, so as to establish an international entity that will achieve the following [37]:

(a) a convincing *rationale* for this endeavor that includes a combination of "paybacks" to investors and taxpayers, whether in terms of science, commerce, resource utilization, habitation, or whatever arguments make sense;
(b) a legal, financial, and administrative mechanism for nations and their institutions to work effectively together in such a venture, such as a *Lunar Economic Development Authority*;
(c) a fully operational *Space Transportation System* would have to be in place, able to move people and cargo at more reasonable cost than today's shuttles and rockets;
(d) a *site plan* that takes advantage of the best location in terms of geology, has concerns for biological and sociocultural problems, and makes infrastructure provisions to ensure human survival and quality of life for an expanding population in a gravity-free environment;

(e) the *technology* for innovative robotic applications, habitat construction, mining, oxygen production, and mass-driver activities, but particularly for lunar telecommunication on the Moon and with Earth;
(f) a *lunar personnel deployment system* that provides for recruitment/assessment, orientation, on-site support, and re-entry guidance (see Chapter 6).

The montage of NASA photographs in Exhibit 122 illustrates some of the "infrastructure" needs if both humans and robots are to function effectively on the Moon. Building first a lunar outpost, then a base with an industrial park, infers a commitment by *earthkind* toward space colonization: a lunar foundation should become the "launch pad" for human expansion into the Solar System.

For that type of commitment to long-range developments on the Moon by an international consortium comprising both public and private sectors would require strategic policy decisions within the countries and organizations participating. To accomplish the kinds of lunar enterprise described above, as well as those in Chapter 10 and Appendix B, policy decisions for financing and governing the operations must come first. Conceivably, the lunar ambitions of China, Japan, and India may fast-forward the process. The predictions of my colleague Dr. Goldman are now being confirmed that new players from private enterprise will arise and positively influence such decisions: they are already to be found in the communication satellite industry, commercial launching companies, space service firms and entrepreneurs, solar and nuclear energy scientists, and even in the hotel, tourism, and entertainment businesses. They will be found in downsized aerospace and defense complexes seeking a new outlet for their technologies and well-trained technicians. Hopefully, such private enterprise will form an economic network that will translate into a political network with clout. The process is under way by forming coalitions with space entrepreneurs, advocates, and scientists, and possibly defense supporters.

Using again the case of the U.S.A., for a critical mass to coalesce in support of such decisions regarding far-sighted mega-space programs, more is needed than Presidential or Congressional statements and studies with visionary goals and initiatives. The basis for national consensus may already be found in the findings and recommendations of those numerous studies paid for by American taxpayers (referenced in Chapters 1 and 2). But to ensure "national will" toward the creation of a *spacefaring civilization*, supporters must expand recruitment beyond members of space agencies and defense departments, beyond aerospace contractors and space activists. Globally, those favoring development of the space frontier now number only several million, when one counts only patrons in government and industry employees, pertinent trade and professional associations or citizen societies, plus fans of space movies or television programs (such as *Star Trek* and *Star Wars*). When such groups band together, minor political space battles have on occasion been won. One researcher discovered that the level of public support for NASA programs varies according to economic conditions, perceived benefits, and the success of ongoing space events. In North America, Canada, the United States, and Mexico need to reassess public spending priorities, so that more is invested in

Exhibit 122. Providing lunar infrastructure. In the upper rendition SAIC artist Pat Rawlings shows a future telescope on the lunar surface which has an automated walking mobility platform. In the lower rendition, Rawlings shows a possible reusable Oberth Lunar Lander firing its advanced cryogenic fuel engines to resupply the needs of a lunar outpost (source: NASA Johnson Space Center, Houston, Texas).

Exhibit 122 (*cont.*). Commercial activities in a 21st-century lunar industrial park (source: NASA Johnson Space Center, Houston, Texas).

domestic and space programs, and less on the military and wars. Space is a place for humanity to cooperate together in peaceful pursuits—not in competition and imperialism.

To get national and worldwide public investment in the enormous expenditures and risk behind present space plans demands mobilization of civic champions on a scale yet to be experienced. It requires assertive, synergistic leadership by the pro-space movement to reach beyond their own membership and enlist public backing at a level beyond that given by the peoples of different countries in World War II. Mass communication and education is essential to get cultural commitment by the citizenry as a whole to the peaceful exploration and development of the next frontier! An example of this was when CNN's Lou Dobbs left his network feature (CNNfn.network) to join Space.com, an informative, daily space news report on the Internet. Furthermore, utilizing space resources will demand political decisions and new institutional arrangements by which more people *actually* invest money in space enterprise, be it through the purchase of stocks, bonds, tax incentives, or whatever it takes—even lottery tickets! Entrepreneurs are trying to use new communication technologies to reach out and enlist a broader space constituency as Exhibit 123 confirms.

> **NEW LUNAR PLANS, 2008**
>
> - NASA has just awarded two planning grants to design systems and antennas that would lead to placing telescopes on the far side of the Moon, away from Earth interferences. Astronomers envision radio telescopes there looking out into the galaxies. The site was chosen because it is protected from distortions caused by the Earth's turbulent atmosphere and radio signals. Thus, scientists hope to discover how galaxies formed from gas clouds. The ultimate aim is to understand better how the universe originated.
> - JAXA has put its first satellite in orbit around the Moon which will gradually move into orbit closer to the lunar surface. The year-long observational probe called SELENE is the largest lunar mission since the Apollo days. The object is to place the main satellite Kaguya in a circular orbit at an altitude of 60 miles above the Moon itself, and then deploy two smaller satellites into elliptical orbit. The aim is to gather data about the Moon's origin and evolution. Soon India hopes to follow with its unmanned Moon mission, followed by a manned one there in 2015!

9.5.3 Influencing commercial space development

In 1990, the Denver firm of Stratton, Reiter, Dupree & Durante made a proposal to the American Aerospace Industry on "Building a Broader Space Constituency." These public relations experts recommended that the aerospace industry should back a national marketing endeavor to

- re-establish strong Congressional support for the space program;
- undertake a review of all previous public opinion surveys on attitudes toward space projects, and promote an additional study of the public, media, and Congress to ascertain possible linkages, particularly with reference to space exploration and settlement;
- chart, based on a full evaluation of the current attitudes, a clear course to enlist greater public space support, possibly through targeted direct mail, phone banks, road shows, paid and free media advertising.

Unfortunately, that counsel was ignored and the American aerospace industry continues to lose market share and jobs within the space field, while there is a shortage of competent aerospace engineers.

In a lecture at the California Space Institute (May 11, 1988), Dr. Brenda Forman, then of the Lockheed Corporation, suggested that civil space programs needed a well-organized lobbying effort to move the President, Congress, and voters. Such a strategy to enlist the silent majority is under way with *Spacecause*, a non-profit organization founded in 1987 to build a lobbying network among public, academic, and commercial space industries so that government will give policy priority to civilian space activities; it complements SPACEPAC, which was formed in 1982

Exhibit 123. *THE SPACE SHOW.*

Innovating with *podcasting*, Dr. David Livingston weekly produces *The Space Show*. The program focuses on timely and important issues influencing the development of space commerce and tourism, as well as other relevant space subjects. The method has Dr. Livingston interviewing invited space experts, and later these guests answer questions from telephone callers or listeners via the computer. The show and its host serve as a platform for knowledgeable, educational exchanges on humanity's space future and opportunities.

Streaming audio requires a computer with Microsoft Windows Media Player or compatible equipment (technical help available through *webmaster@The SpaceShow.com*). Each week a bulletin is issued with a briefing on the upcoming topic and guest, as well as listener participation instructions and details of future editions (*www.thespaceshow.com/newsletterfinal.htm*). Podcasting subscription information is also available (*http://wwwgigadial.net/public/station.11253/rss.xml*). Listeners may talk to the host and guests by calling a toll-free number (1/866 687-7223), or by sending an email during the program (email: *drmlivings@yahoo.com*). CD-ROMs of programs may also be purchased.

In addition to your author, other guests have included Dr. Albert Harrison, space psychologist; John Spencer, founder of The Space Tourism Society; Manny Pimenta, DVD creator of the *Lunar Explorer Virtual Moon Simulations*; Dr. David Schrunk, senior author of *The Moon: Resources, Development, and Settlement* (2008); Professor Haym Benaroya of Rutgers University on lunar settlement engineering; Dr. Alan Hale, astronomer; Dr. Molly McCauley on the economics of space as a natural resource; Dr. Elgar Sadeh of University of North Dakota on space management; Dr. David Criswell of the University of Houston on space-based energy; Rick Tumlinson, founder of the Space Frontier Foundation; and many more space professionals!

Congratulations should be given to the enterprising Professor Livingston for producing such a program. He also teaches at the University of North Dakota's Space Studies Department; yet, every Sunday he creates a global space forum for the space community worldwide!

Source: The Space Show, PO Box 95, Tibuton, CA 94920 (tel.: 415/435-6018; email: *drspace@thespaceshow.com*; *www.thespaceshow.com*).

to support political candidates with outstanding civilian space-voting records or to endorse public servants who merit recognition. In 1992, Spacecause joined with the National Space Society in facilitating passage in the U.S. Congress of the *Commercial Space Competitiveness Act* with provisions to

- extend the limitations on launch liability;
- launch a voucher demonstration program to improve competition among companies;

- set up a space transportation infrastructure matching grants to develop private spaceports, payload integration facilities, etc.;
- anchor tenancy and termination liability for private companies when government defaults on legal contracts, allowing for more predictability in space ventures;
- open up surplus or otherwise available federal launch and payload support facilities for use by private companies on a cost-reimbursable basis;
- protect the intellectual property of companies doing business with NASA;
- create a *Commercial Space Achievement Award* to be administered by the U.S. Secretary of Commerce.

Such endeavors are attuned to Forman's contention that politics and technology should be mutually supportive. For an adequate and coherent national space policy, this international marketing expert advocates building bridges of information and understanding between legislators and space scientists. For her, political action on behalf of space development must come from the bottom up in society, focusing on space-related government committees of agencies at all levels. Forman believes that the American public loves space, but has yet to perceive the need for political activism if space resources are going to be effectively utilized to resolve Earth-based problems, related to poverty, hunger, and health. In a democracy, the political system is a heritage that must be skilfully worked if space industrialization and settlement is to become a mainstream reality.

From his research, Alan Marshall of Massey University in New Zealand has concluded that once extraterrestrial materials are perceived as economically valuable, Solar System development will expand rapidly. But he cautions that politico-legal regimes should not proceed with it in an imperialistic manner, and advocates creation of *space environmental ethics* to protect offterrestrial bodies [38]:

> "The challenge is to awaken the world's political leaders to the potential of space resources and human destiny on the high frontier, so they will enable legislation and mechanisms for creating a spacefaring civilization!"

SPACE ACTION PLANS

Political institutions and forces can undermine the best efforts of space entrepreneurs and innovators. Thus, public lobbying is necessary to influence politicians to support space enterprise. That is best accomplished by checking the websites and then joining non-profit space organizations, such as the American Astronautical Society (AAS), the American Institute of Aeronautics and Astronautics (AIAA), the Association for the Advancement of Science and Technology (ASCONT), the British Interplanetary Society (BIS), the *Challenger* Centers for Space Science Education (CCSSE), the International Astronautical Association (IAA), the International Lunar Observatory Association (ILOA), the International Space

University (ISU), the National Space Society (NSS), the Personal Spaceflight Federation (PSF), The Planetary Society (TPS), the Space Frontier Foundation (SFF), the Space Nursing Society (SNS), the Space Tourism Society (STS), and Students for the Exploration and Development of Space (SEDS). Through such non-governmental organizations, collective actions can be pursued that should have an impact on mindsets, decision-making, and funding, which should promote developments on the high frontier. To foster a spacefaring civilization, action planning should be focused on

(1) Influencing the global media to inform and educate the world population on the necessity for living and working offworld.
(2) Encouraging government legislators, agencies, and departments to streamline and coordinate their bureaucracies, so as to facilitate space entrepreneurship and provide the necessary space infrastructure, such as establishing a single Department of Space Transport, Resources, and Settlement to include all levels of government.
(3) Informing venture capitalists, private equity funds, stock brokers and investors of the opportunities in space commerce.
(4) Persuading leaders of spacefaring nations of the need for cooperative, global institutions to promote joint ventures beyond Earth and for the benefit of humanity, such as orbital, lunar, and Mars development authorities, administrations, or foundations.

9.6 CONCLUSIONS ON OFFWORLD PRIVATE ENTERPRISE

Safely establishing the human presence on any frontier involves coping with challenges and risks. Moving humanity beyond its planetary cradle may represent the boldest venture in the evolution of our species. Ingenuity and innovation will be stretched in the transformation of Earth-bound mindsets and culture, policies and procedures, laws and regulations, and education and business. The task requires multidimensional actions, demanding multidisciplinary thinking and methodologies. This chapter reviewed only three dimensions of the challenges aloft, and sought to show their interrelatedness. Space commerce, law, and politics cannot be viewed separately, but require a more integrated approach by business, legal, and political leaders, in conjunction with scientists, engineers, and academicians. This analysis, though, demonstrates some of the complexities and opportunities inherent in space industrialization and settlement, particularly for large-scale technology venturing and entrepreneurialism.

9.6.1 Space education

The next generation must be prepared to face offworld challenges, for it is their future! In this book, I have referred many times to *space education* as it is being pursued in colleges and universities, particularly through the Universities Space Research Association and the International Space University. There are now many youth space camps and schools and prototypes of programs, like the *Challenger* Centers for Space Science Education, all of which need to be expanded globally. However, there is a critical need for *mass education* at primary and secondary levels worldwide to make the younger generation aware of the opportunities of *living and working in space*! After all, these children may be the future Martians, so education departments within space agencies must expand their outreach programs to school districts.

As part of its EuroMir 95 mission, the ESA arranged for young people in its member states to come to Disneyland Paris for a space education/entertainment event that culminated with a live satellite link-up with the then Russian space station Mir. Dr. Sally Ride, first American woman astronaut in space, had a NASA project involving cameras installed on the Shuttle Orbiter, which was exclusively operated by students; these cameras instantaneously relayed high-resolution data to classrooms in high and middle schools throughout the country. Ride hoped by this method to "give students their own piece of the space program." The Young Astronauts Council currently operates "Space School", a distance-learning method that uses Space Age technology to teach math and science, as well as to beam quality educational adventures. This educational programming comes from Seattle, Washington, and through satellite television is transmitted into classrooms nationally. Entrepreneurial companies are already designing games for kids to learn about space living through simulations and virtual reality. Canada's Apogee Books has been producing DVDs and popular publications to interest youngsters in space exploration (*www.apogee-books.com*). Aerospace and science museums have special space exhibits and often Imax films to help the public appreciate the necessity of going off the Earth to utilize space resources. Expansion of such creative efforts, including movie and television programs, are essential to introduce young people to tomorrow's opportunities in space, so as to ensure humanity's expansion aloft for Millennium III (see Exhibit 124)!

High-frontier prospects in the 21st century are only dimly perceived, as humankind struggles like infants to leave our cradle, Earth. For human enterprise in space to succeed and flourish, synergy or cooperation become the key ingredients between generations, between the public and private sector, between planners and policy makers, between professionals and technicians, as well as among organizations and nations.

> "There are countless planets, like many island Earths ... Man occupies one of them. But why could he not avail himself of others, and of the numberless suns? .... When the sun has exhausted its energy, it would be logical to leave it and look for another, newly kindled star in its prime" (Konstantin Tsiolkovski).

Exhibit 124. Educating tomorrow's spacefarers. Space education of youngsters on the ground or in orbit, such as is being done through the International Space University, creates the next generation of *spacefarers*. Original artwork by Howard Cook/Air Works (tel.: 719/472-076). Source: International Space University, Strasbourg, France (*www.isu.edu*). This illustration was created by the Canadian Space Agency.

> **WE CLIMB TREES STILL***
>
> They climbed trees once
> To get closer
> To the stars
> —Jumped and fell short.
>
> So, they dreamed of
> Winged chariots and horses and men,
> But wax wings burn
> —Again they fell short.
>
> So, they returned to the trees
> To gaze at the stars and to think.
> We built balloons then, and
> Planes and rockets.
>
> Although this time we did not fall,
> The stars remain beyond our reach;
> —And still we climb trees
> To get closer
>
> Source: Nathan C. Goldman, Ph.D.,
> Attorney/Author, Houston, TX.
> * In honor of Robert Goddard's
> Anniversary Day.

9.7 REFERENCES

[1] Shuttleworth, M. "Bringing Software Down to Earth," *The Economist*, June 9th, 2007, pp. 33–34 (*www.economist.com/*).
[2] Robinson, G. S. "Bicentennial of the Bill of Rights and Rethinking Space," unpublished address presented at the Air/Space America Legal Forum, 1998, National University, San Diego, CA ... Robinson, G. S.; and White, W. M. *Envoys of Mankind*. Washington, D.C.: Smithsonian Press, 1986.
[3] Kinsley, A. P. (ed.) *Space: A New Era* ... Hayes, W. C. (ed.) *Space: New Opportunities for International Ventures*. San Diego, CA: UNIVELT/AAS, 1989; 1980.
[4] McKay, M. K.; McKay, D. S.; and Duke, M. B. (eds.) *Space Resources*. Washington, D.C.: U.S. Government Printing Office, 1992, 5 vols. See Vol. 4 on *Social Concerns* (boxed set available via *www.univelt.com*).
[5] U.S. Congressional Budget Office, *The NASA Program in the 1990s and Beyond*. Washington, D.C.: U.S. Government Printing Office, May 1998.
[6] When a consultant to NASA headquarters, your author (Harris) was instrumental in encouraging its administration to publish for the public an annual report on the "spinoff"

benefits of its space technology. As a result, the magazine *Spinoffs* was published yearly by NASA (Center for Aerospace Information, National Technical Information Service, 5285 Port Royal, Springfield, VI 22161; tel.: 401/859-5300, ext. 241) ... See also *The Civil Space Program: An Investment in America*. Washington, D.C.: American Institute of Aeronautics and Astronautics (*www.aiaa.org*).

[7] Boorstin, D. *The Discoverers*. New York: Random House, 1984.

[8] Harris, P. R. *The New Work Culture*. Amherst, MA: Human Resource Development Press, 1998, pp. 254–255 ... Stevenson, H. H.; Roberts, M.; and Grousebeck, I. *New Business Ventures and Entrepreneurialism*. Homewood, IL: Richard D. Irwin Publishers, 1985.

[9] Foust, J. "Current Issues in NewSpace," *The Space Review*, March 5, 2007 (*www.spacereview.com/article/823/1*) ... Oberg, A. "Who Needs NASA? Private Sector Can Do It Cheaper," *USA Today*, Editorial, April 5, 1998 (*www.reston.com/NASA/editorial/04.05.98/oberg.html*) ... NASA News, "NASA Extends Bioastronautics Contract with Wyle Labs," *America Online*, July 5, 2007 ... David, L. "New Rocketbelts: High Time for Technology," Space.com (*www.space.com/businesstechnology/060802_rocket_belt.html*) ... Schwab, D. "Once-in-a-Lifetime Flight Pushes the Envelope," *La Jolla Light*, March 8, 2007, p. 14.

[10] Haskell, G.; and Rycroft, M. (eds.) *New Space Markets*. London: Kluwer Academic Publishers, 1998 ... Gump, D. P. *Space Enterprise beyond NASA*. New York: Praeger, 1990 ... Harr, M.; and Kohil, R. *Commercial Utilization of Space*. San Diego, CA: UNIVELT/Battelle Press, 1990 ... Logsdon, J. M. *Space Inc.: Your Guide to Investing in Space Exploration*. New York: Crown Books, 1988 ... Goldman, N. C. *Space Commerce: Free Enterprise on the High Frontier*. Cambridge, MA: Ballinger, 1985 ... O'Leary, B. *Space Industrialization*. Boca Raton, FL: CRC Press, 1982, 2 vols.

[11] Glaser, P. E.; Davidson, F. P.; and Csigi, K. *Solar Power Satellites: A Space Energy System for Earth*. Chichester, U.K.: Wiley/Praxis, 1998.

[12] Sheffield, C.; and Rosen, C. *Space Careers*. New York: Quill Publishing, 1984.

[13] For information on the Aerospace Division of the American Society of Civil Engineers (345 East 47th St., New York, NY 10017), contact Dr. Stewart W. Johnson (email: *STWJohnson@aol.com*).

[14] Gibson, R. "Commercial Space Activities: An Overview," *Space Commerce*, 1(1), 1990, 3–6.

[15] Miller, R. *The Dream Machines: An Illustrated History of Spaceships in Art, Science, and Literature*. Malbar, FL: Krieger Publishing, 1993 ... For current SpaceShipOne information, visit *www.eBay.com*; *www.upsiral.com*; *www.pycos.com*; *www.redzip.com*; and *www.thefreedictionary.com*

[16] Reichhardt, T. (ed.) *Space Shuttle: The First 20 Years*. Washington, D.C.: Smithsonian Institution/DK Publishing, 2001 (*www.dk.com*) ... Harland, D. M. *The Space Shuttle: Roles, Missions, Accomplishments*. Chichester, U.K.: Wiley/Praxis, 1998.

[17] Leahy, B. "Space Access: The Private Investment vs. Public Funding Debate," *Ad Astra*, May 12, 2006 (published by National Space Society, *www.space.com/adastra*).

[18] LaFee, S. "A Bot's Life: Biology Proves a Natural for Robotic Design," *San Diego Union-Tribune*, July 12, 2007, pp. E1 and E8.

[19] Schrunk, D. G. *The End of Chaos: Quality Laws and the Ascendancy of Democracy*. Poway, CA: Quality of Laws Press, 2005 (*www.glp.com*; email: *docscilaw@aol.com*).

[20] Goldman, N. C. *American Space Law: International and Domestic*. San Diego, CA: UNIVELT, 1995 (*www.univelt.com*) ... Doyle, S. E. *Origins of International Space Law and the International Institute of Space Law*. San Diego, CA: UNIVELT, 2002 (*www.*

univelt.com) ... Jasentuliyana, N. *Space Law: Development and Scope*. Westport, CT: Praeger Publishing, 1992.

[21] O'Donnell, D. J. "Astrolaw: The First Thousand Years," *Space Governance*, **9**, 2003, 11–19 ... O'Donnell, D. J.; and Harris, P. R. "Is It Time to Replace the Moon Agreement?" *The Air & Space Lawyer*, **9**(2), Fall 1994, 3–11 (published by the American Bar Association) ... "Space-Based Energy Needs a Consortium and Revision of the Moon Treaty," *Space Power*, **13**(1/2), 121–134 (email: *djopc@quest.net*; *philharris@aol.com*).

[22] International Institute of Space Law (IISL) in conjunction with the International Astronautical Association (IAF), 3–5 Rue Marios Nikis, Paris Cédex 16, France (*www.iisl.com*; *www.iaf.org*), which publishes the prestigious international journal, *Acta Astronautica*.

[23] Benko, K. and Schrogel, W. L. (eds.) *International Space Law in the Making: Current Issues on the Peaceful Use of Outer Space*. New York: U.N. Publications, 1994 ... Meredith, P. L.; and Robinson, G. S. *Space Law: A Case Study for the Practitioner*. Dordrecht, the Netherlands: Martinus Nijhoff, 1992.

[24] Gabrynowicz, J. I. "A Beginner's Guide to Space Law," *Ad Astra*, **6**(2), March/April 1994, 43–45 (published by the National Space Society) ... Christol, C. Q. *Space Law: Past, Present, and Future*. Boston, MA: Martinus Nijhoff, 1991.

[25] Hargrove, E. C. *Beyond Spaceship Earth: Environmental Ethics and the Solar System*. San Francisco, CA: Sierra Club Books, 1986.

[26] Finch, E. R.; and Moore, A. L. *Astrobusiness: A Guide to Commerce and Law*. New York: Praeger, 1884.

[27] Heppenheimer, T. A. *Colonies in Space* and *Toward Distant Suns*. Harrisburg, PA: Stackpole Books, 1977; 1979.

[28] Robinson, G. S. "Natural Law and a Declaration of Humankind Interdependence," *Space Governance*, **2**, Part I June 1995, Part II December 1995. See also Schmitt, H. W. *Return to the Moon: Exploration, Enterprise, and Energy in the Human Settlement of Space*. New York: Springer/Copernicus, 2006, chs. 8/9/11–13.

[29] The journal *Space Governance* is still being published, but only annually, by United Societies in Space, Inc. and back volumes are available (contact: Declan O'Donnell, 499 S. Larkspur Road, Castlerock, CO. 80104; *www.space-law.org* or *usis.org*; email: *djopc@quest.net*) ... The best existing journal on this subject is *Annals of Air and Space Law* published by McGill University (3661 rue Peel, Montreal, Quebec, H3A 1X1, Canada; *www.aasl.mcgill.ca*).

[30] For those interested in reading the proposed "Model Treaty for Jurisdiction in Outer Space" go to Appendix D in *Living and Working in Space* by Philip R. Harris. Chichester, U.K.: Wiley/Praxis. 1996, Second Edition, pp. 360–368.

[31] Logsdon, J. M. *The Decision to Go to the Moon: Project Apollo and the National Interest*. Cambridge, MA: MIT Press, 1970 ... "The Space Shuttle Program: A Policy Failure?" *Science*, May 30, 1986 ... Bormanis, A.; and Logsdon, J. M. (eds.) *Emerging Issues for Long Duration for Human Space Exploration*. Washington, D.C.: The Space Policy Institute, George Washington University, December 1992 Report.

[32] McDougal, W. A. *A Political History of the Space Age*. New York: Bantam Books, 1986 ... For current commentary on this subject, we recommend the Elsevier journal *Space Policy* (*www.elsevier.com/locate/spacepol*).

[33] Goldman, N. C. *Space Policy: An Introduction*. Ames, IA: Iowa State University Press, 1992.

[34] deSelding, P. B. "European Union Takes Charge of Space Policy," *Space News*, **5**(44), November 1995, 1–30.

[35] Byerly, R. (ed.) *Space Policy Revisited*. Boulder, CO: Westview Press, 1989.

[36] ESA Lunar Study Steering Group. *International Lunar Workshop: Towards a World Strategy for Using Our Natural Satellite*, November 1994 (ESA SP-1170); *Mission to the Moon: Europe's Priorities for the Scientific Exploration and Utilization of the Moon*, June 1992 (ESA SP-1150),

[37] Schrunk, D.; Sharpe, B.; Cooper, B.; and Thangavelu, M. *The Moon: Resources, Future Development and Settlement*. Chichester, U.K.: Springer/Praxis, 2008, Second Edition.

[38] Marshall, A. "Development and Imperialism in Space," *Space Policy*, **11**(1), 1995 ... "Ethics and the Extraterrestrial Environment," *Journal of Applied Philosophy*, **10**(2), 1993.

SPACE ENTERPRISE UPDATES

- Galaxy forums were launched in 2008 by Space Publishing Company, a private enterprise. The aim is for participants to explore "Galaxy education in the 21st century". Our own Milky Way Galaxy has 200 billion plus stars, a 225 million year period of revolution, 100,000 light-year diameter, with a massive black hole, weighing as much as 4 million Suns, at its center. The global sessions will promote galaxy awareness, education, physics, and astrophysics.
- The Space Access Society holds annual conferences on technology, politics, and businesses that open the high frontier for private ventures. The 16th convocation in Phoenix, Arizona, March 2008, featured leaders in NewSpace enterprises and aerospace regulation industries. Entrepreneurs and investors also exchange with officials from NASA, the Commercial Sector, FAA Office of Space Transportation, and the USAF Research Laboratory.

Exhibit 125. Why the Moon? The essence of the current U.S. National Space Policy, *The Vision of Space Exploration* is summarized in this NASA poster.

10

Lunar enterprises and development

Of all the sites in the Solar System, the Moon is the most logical first place to establish a permanent offworld human settlement. Why? Because it is nearby—less than three days with existing propulsion technology. It has a wealth of resources that can be used to support a broad range of human activities from science and engineering to commerce and tourism. It has an abundant continuous, and virtually unlimited supply of solar energy. It offers the opportunity for a greatly expanded program of space exploration. It is within our technological reach!
   David Schrunk, Burton Sharpe, Bonnie Cooper, and Madhu Thangavelu [1]

After a hiatus of some 40 years since humanity's first lunar landing, we are now planning to return to the Moon permanently, hopefully by year 2020. No *united* vision, plan, or strategy for exploring, settling, and industrializing the lunar surface has come forth from the world's space agencies, despite numerous conferences, studies, and reports on the subject. However, NASA has undertaken to implement a national space policy to do just that, called the *Vision for Space Exploration*. The Agency is beginning to seek collaboration from public and private entities within the global space community in this latest orbital venture. But to build a lunar transportation system and outpost within a dozen years would involve facing up to harsh economic, technological, and political realities, leading to the realization that the only way *earthkind* can afford this macroproject is by means of international cooperation with major participation by private enterprise. If the limited resources of China, Europe, India, Japan, Russia, and the U.S.A. were combined into a joint technological enterprise on the Moon, then synergistic activities there would encourage competing space constituencies to work together on lunar development. Exhibit 125 (opposite) gives the principal reasons for doing this.

The Moon can serve as the best space station and launch pad to Mars and beyond. Space science proponents would then be able to use this lunar platform for their experiments, both in astronomy and life sciences, as well as to launch from there additional planetary missions. Simultaneously, those favoring human exploration and "manned" missions might use the Moon as a laboratory for study of extreme environments and habitats, comparable with what is happening today among cooperating countries in Antarctica. In addition, champions of lunar industrialization could undertake research programs relating to manufacturing, mining helium-3, or energy transmission by power beaming, and other uses of *in situ* resources, such as lunar oxygen produced from ilmenite [2].

The uniqueness of the Earth–Moon system provides an advantage for humanity to use this lunar laboratory to create a spacefaring civilizaton for the new millennium! The Moon's characteristics as a resource to humanity are illustrated next in Exhibit 126. The Moon is an enormous piece of accessible real estate, comparable with the largest continent on Earth. As meteorites struck the lunar surface, they shattered solid rock which mixed with the soil creating *regolith*, a fine powder or cosmic dust which can be from three to sixty feet in depth. Apparently, it also contains a lot of oxygen, so the prize will go to whomever can extract the oxygen most quickly. Regolith has other bounty to be obtained—hydrogen, carbon, nitrogen, potassium and other trace elements. When asteroids crashed into the Moon, there is the likelihood of accompanying deposits of platinum and other valuable metals [3]. A key question, according to Dr. Paul Spudis, a planetary scientist at Johns Hopkins University, is whether the Moon's resources can be used profitably [4].[1]

In 1969, humanity's first lunar envoys installed a commemorative plaque with these words: "Here men from Planet Earth first set foot on the Moon. We came in peace for all mankind." Then in December 1972, astronauts from the last Apollo mission unveiled another commemorative plaque on the front landing gear of the lunar module: "Here man completed his first exploration of the Moon. May the spirit of peace in which we came be reflected in the lives of all mankind." And within a dozen years, our species will return permanently to "Sister Moon", so called by a 13th-century visionary, Francis of Assisi! A great deal of the Earth's character has been influenced by its relatively close position to this Moon: it is responsible for two-thirds of our ocean tides; its gravitational pull on Earth's equatorial bulge causes the Earth's axis to precess, or wobble like a child's spinning top. This precession causes the star that is closest to the north pole of the sky to change over time, and it moves the position of the equinoxes on the sky. But it exerts a stabilizing influence on the Earth, preventing the other planets from making Earth's axis go through extreme changes of obliquity (axial tilt) like those of Mars (see Exhibit 126).

The Moon is our "beachhead" for exploring and settling our Solar System! Your author presents here a near-term strategy for utilization of its resources for the benefit of our planet's inhabitants. By turning outward in peaceful lunar development, the human family can develop both its potential and that of its offworld bounty!

---

[1] For NASA video on *regolith challenges*, go online to this website: *http://one.revver.com/watch/292585* and *206790* ... Also see Appendix B "Lunar regolith properties" in Schrunk, D. G.; Sharpe, B. L.; Cooper, B. L.; and Thangavelu, M. *The Moon: Resources, Future Development and Settlement*, 2008, Second Edition, pp. 235–255 (*www.praxis-publishing.co.uk*).

Sec. 10.1] **Reclaiming the Moon: rationale** 473

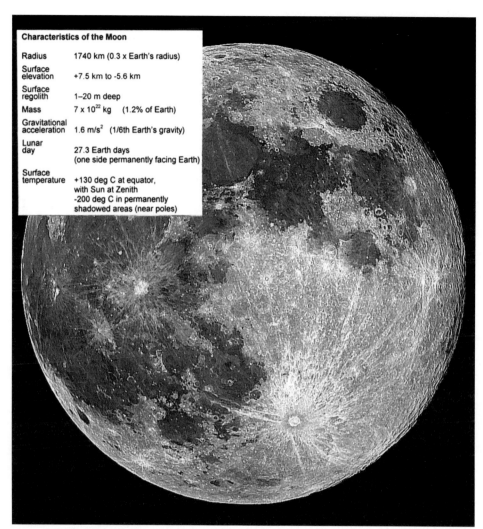

Exhibit 126. Characteristics of the Moon. Utilizing lunar resources is feasible, but the problem has been well summarized by Wernher von Braun: "We can lick gravity, but sometimes the paperwork is overwhelming!" Source: NASA. For further information, contact the International Lunar Exploration Working Group through Dr. David McKay at NASA/JSC (*dmckay@snmail.jsc.nasa.gov*).

10.1 RECLAIMING THE MOON: RATIONALE

As this is written, we are celebrating the 50th anniversary of Sputnik, the satellite that launched our Space Age. It is fitting that the global space community is again preparing to move beyond Earth and back to its sister Moon, possibly establishing in the process a twin-planet economy. Recently, the magazine of the National Space Society, *Ad Astra*, devoted a whole issue to "Reclaiming the Moon: The First Steps"

[5]. In these pages, space writers made a case as to why humanity should do this. Jeff Foust gave a recent review of U.S. policy on returning to the Moon permanently, *Vision for Space Exploration* (see Chapter 1). That writer explained that NASA's implementation plan entitled *Global Exploration Strategy*, was developed with the participation of some 14 space agencies which identified 180 potential objectives for lunar exploration, divided into 23 categories ranging from astronomy and lunar geology, to commercialization and technology testing. Six exploration themes emerged for such endeavors: human civilization, scientific knowledge, exploration expertise, global partnerships, economic expansion, and public engagement. Within that context, NASA then engaged in a *Lunar Architecture* study that favors the building of an international Moon base. Before 2020, a decision is to be made as to site for the first outpost: the current consensus is a location in the lunar north or south pole. Some favor the Shackleton crater near the south pole where the Sun hangs low on the horizon for most of the lunar day, thus enabling use of solar power both on the Moon, and possibly power beaming of space-based energy to the home planet in the future. Others favor a base near Malapert Mountain also near the south pole because of the similar sunlight advantage and direct view of the Earth [5].

Lunar scientists estimate that some 20 billion tons of ice may be present in the permanently shadowed regions of craters near both poles. In 2008, a mission is planned for a Lunar Reconnaissance Orbiter to gather data confirming which is the best location to start a research outpost. Both international and private partners will help to determine the eventual size of this base which I have called *Lunar World* in Exhibit 125. The current thinking is to start with crews of four to be rotated every six months until permanent occupants, dubbed Selenians, determine to stay longer. A versatile lunar lander, yet to be built, is critical for bringing humans, robots, and equipment to the lunar surface. By 2013, the Agency hopes to have made decisions on this spacecraft, plus a heavy-lift launcher, possibly the Ares V, with the Orion Crew Exploration Vehicle which is presently being built by Lockheed-Martin. Significantly, this is all viewed as a long-term, evolutionary process.

In this same 2007 Spring issue of *Ad Astra*, Jeanna Bryner discussed the split in astronomer thinking on orbital or lunar-based telescopes. That year a series of meetings were inaugurated called "Astrophysics Enabled by Return to the Moon" in which both alternatives were examined, noting that lunar telescopes erected on a permanent platform would benefit by the Moon's lack of clouds or any blurring atmosphere (see Exhibit 129). Astronomers are attracted to the far side of the Moon which is free of radio pollution, but for which no base is yet planned. Thus. it is more likely that a deep-field infrared observatory will be erected on the lunar south pole where the necessary living, power, and communications infrastructure will exist in the near term. An upright telescope could be built that has a 330-foot scope of the Moon that enables astronomers to view the galaxies in a way impossible on Earth. Semi-autonomous robots are expected to service the anticipated facility.

Again in this same issue of the NSS publication, Andrew Chaikin wondered what it would be like soon for Earth's inhabitants to look up on the Moon, knowing that people are actually living and working there. That writer sees the Moon as a spectacular world, a cosmic library full of secrets about our universe to be decoded. While

Sec. 10.1]   **Reclaiming the Moon: rationale   475**

NASA has the ability to turn our dreams into machines, Chaikin wants care to be taken in choosing dreams wisely that will take us down the path of becoming a *multiplanet* species, living out our DNA destiny! The Moon, then, is the stepping stone, a place to teach and prepare people for living in other worlds beyond their home planet!

There is so much to be learned in this big lunar laboratory: human–robotic relations, closed-loop recycling, telemedicine, teleducation, and scientific and technological innovations. Literally, a permanent, sustainable lunar settlement means changing both human culture and species! Almost four decades ago, the *New York Times* did a special supplement on "The Moon as a New Frontier" [6]. The feature opened with:

> "A new age began when Man first stepped onto the Moon at 10:26 PM (Eastern daylight time) on July 20, 1969 ... The Moon is no longer merely a disembodied orb, a subject of myth, an abstraction in the sky. It has been touched and now is a place."

Yes, a place for human exploration, some 356,400 km from Earth at its closest, just 2% the volume of Earth and only 1.2% of its mass. The Moon's daytime temperature can reach over 120°C, while at night it plunges to −230°C. The Moon spins at a constant speed (in the same time it takes to orbit Earth), but orbits at variable speed (due to its eccentric orbit); its slight wobbles back and forth are called librations. This orbiting land mass, 3,476 km in diameter, is presently three days away from Earth. As the Moon revolves around the Earth, our orbiting planet revolves around the Sun. Regarding the Moon's origin, there are three basic theories: it broke away from the Earth; it was formed at the same time as our planet from the same whirling cloud of dust, and other material, but evolved with less density; it was formed elsewhere, passed near the Earth, was slowed down by debris, and captured by its gravity. The Moon's multiple value to humanity extends from an object of study and a base for examining our universe, to a place for building laboratories, factories, and habitats (see Exhibit 127). Others propose to utilize its raw materials, as well as to establish a refueling stop for launching spacecraft out into the Solar System. There are those who envision the lunar surface as a listening post for us to better decipher cosmic messages, from particles to subatomic objects.

The second man on the Moon Buzz Aldrin has long argued that the Moon could provide both rocket propellant and shielding for human structures, as well as supplying our space needs because of its weaker gravitational field (about one-sixth that of Earth). He would like to see a fleet of "space tugs" built to create "Trans Lunar Rendezvous", using the Moon as a refueling depot. Little wonder that Buzz Aldrin wrote a book for children *Reaching for the Moon*, for it will be the generations of youth who will make the most of lunar opportunities [7].

As we prepare to return permanently to the lunar surface, author James Hansen reminds us that two-thirds of today's Earth population were not alive when the first humans visited the Moon [8]. And only four of the six astronaut commanders who piloted a spacecraft down to a lunar landing are among the living. Of the 12 moon-walkers who had a great time there, 9 are still with us, all in their 70s. The implication

Exhibit 127. Lunar world. An artist's concept of a future lunar base, beginnings of a major industrial park, and settlement on the Moon. Oxygen propellant derived from lunar raw materials could play a key role in reducing the amount of mass launched into low Earth orbit to support a lunar base program and thus cut costs. The illustration shows a pilot plant designed to manufacture liquid oxygen on the Moon. Uses for the pilot plant oxygen include LOX reactant for fuel cells and to supplement oxygen requirements for life support systems. The painting was done for NASA by Eagle Engineering artist Mark Dowman. The concept's principal investigator was Eric Christianson. Source: NASA.

is that earthkind needs a massive re-education as to the Moon and its potential. Further, this second time, we depend on established technologies and the Apollo missions legacy. Perhaps Harrison Schmitt, one of the last men on the Moon, put it best when he said:

> "that return will be comparable to the movement of our species out of Africa some 150,000 years ago ... a return to the Moon today will be comparable to the permanent settlement of North America by European immigrants ... Apollo was our evolutionary path to the future" [9].

Schmitt, a former U.S. Senator from New Mexico, recently gave these reasons as to why we must go back to the Moon. They include (1) clean abundant energy; (2) stepping stone to Mars; (3) species survival; (4) expanding understanding of the universe; (5) save Earth from threats from space; (6) education in space exploration stimulates the mind in unique ways; (7) lunar tourism and settlement.

When Drs. Buzz Aldrin and Harrison Schmitt met at a Lunar Base Symposium in Houston, Texas (June 1999), they concurred with the event's organizer, Rick

Tumlinson, founder of the Space Frontier Foundation, that going back to the Moon is an opportunity for a "new synergy between government and the private sector." This time, Tumlinson believes the Moon is close enough to Earth for private enterprise to play a big role there, and to make it profitable. (The space between the two celestial bodies is called *cislunar*.) Just imagine if the peoples of this planet were to work together on this venture for the benefit of all!

"Ten Reasons to Put Humans back on the Moon" was the title of an online essay by science writer Robert Roy Britt (*www.space.com/moon* 12/8/03). In addition to the above motivations, he added these:

- satisfy the soul by the exploration quest for new knowledge;
- bring nations together in an offworld technological enterprise;
- gather rocks from the "attic of the Earth" for further scientific analysis;
- study the catastrophic effects of asteroid impact to answer survival questions about our cosmic shooting gallery;
- spur technology advances, including for health and economic benefits.

Finally, a worldwide meeting of lunar experts in 2003 gathered for the fifth time, calling now for a sequence of technological, exploratory, and commercial joint missions that would culminate in establishment of a permanent human presence on the Moon. Further, these ILC/ILEWG conferees issued *The Hawaii Moon Declaration* which underscores the importance of lunar development for humanity in the 21st century. This is a statement put together by many representatives from spacefaring nations [10]:

> "The Moon is currently the focus of an international program of scientific investigation. Current missions underway or planned will lead to the future use of the Moon for science and commercial development, thereby multiplying opportunities for humanity in space and on Earth. We need the Moon for many reasons: to use its resources of materials and energy to provide for our future needs in space and on Earth, to establish a second reservoir of human culture in the event of a terrestrial catastrophe, and to study and understand the universe. The next step in human exploration beyond low Earth orbit is to the Moon, our closest celestial neighbour in the Solar System."

REALiTY CHECK: THE U.N.'s OUTER SPACE TREATY

The exploration and use of outer space, including the Moon and other celestial bodies, shall be carried out for the benefit and interest of all countries, irrespective of their degree of economic or scientific development, and shall be the province of all mankind.

Outer space, including the Moon and other celestial bodies, shall be free for exploration and use by all States without discrimination of any kind, on the basis of equality and in accordance with international law, and there shall be free access to all areas of celestial bodies. There shall be freedom of scientific investigation in outer space, including the Moon and other celestial bodies, and the States shall facilitate and encourage international cooperation in such investigation. (Source: Article I, the United Nations 1967 Outer Space Treaty.)

10.1.1 International lunar agreements and initiatives

The legal context for any nation or consortium to carry out activities on the lunar surface is evident in the above U.N. Outer Space Treaty of 1967 (see Section 9.4). In December 1979, a second document was drafted entitled *Agreement Governing the Activities of States on the Moon and Other Celestial Bodies*. Because of its antidevelopmental provisions, many of the leading spacefaring nations, including the U.S.A. and the former U.S.S.R., did not sign this document prepared by the Committee on the Peaceful Uses of Outer Space (COPUOS). Only nine nations, largely representing developing economies, signed that resolution, while five others have yet to ratify it. In 1984, this so-called treaty went into force after being ratified by only five nations. But, in the summer of 1994, the Moon Treaty or Agreement came up for review in the U.N. General Assembly, and within the U.S. State Department. In a private letter to the author (May 18, 1995), its chairman, Peter Hohenfellner, informed me that at the 37th session of COPUOS (June 6–16, 1994) this Moon Agreement revision was considered and it was recommended that no further action be taken, which the 49th General Assembly of the United Nations endorsed. While he says no new developments are expected on the issue in the near future, Hohenfellner did find interesting material on the subject which the United Societies in Space, Inc. (USIS) submitted, commending their proposal for a Lunar Economic Development Authority (see Section 10.2.2). Unfortunately, this Committee failed to sufficiently review the position of critics of this unpopular lunar document, such as the opposition from the National Space Society. The latter's position was that if lunar resource development were to be controlled by a monopolistic international organization, it would slow the process, discourage entrepreneurs, and possibly delay lunar settlement. This lobbying, especially over property rights, led to the U.S. Senate voting not to ratify the agreement.

Realistically, any international plans for lunar development will have to consider the implications of these two U.N. space agreements (see Appendix A). Eilene Galloway, director of the Paris-based International Institute of Space Law, notes that, while only nine nations originally approved this Agreement, its provisions are a problem for commercial entities and entrepreneurs wishing to exploit the Moon's natural resources. While the Outer Space Treaty accepted by the international community forbids national appropriation by means of sovereignty and is accepted in U.S. Constitutional Law, the claims of institutions and individuals on the Moon and its resources are still unclear. It would appear that exploiters or developers may remove such resources, but have no private property rights over them. Thus, S. Neil Hosenball, the U.S. representative to COPUOUS made the case for the establishment of an "international regime" to deal with resources above or below the Moon. While the Outer Space Treaty makes the point that the Moon is the "province of all mankind", the Moon Treaty uses the terms "common heritage of mankind", requiring an international regime for the "equitable sharing" of lunar resources [11].

Since then a variety of nations have undertaken a series of lunar missions. Several outcomes from annual international conferences point to a growing global consensus emerging on lunar development, particularly relative to exobiology on the Moon. Concurrently, energy scientists worldwide, including those within the Russian Academy of Science and Japan, are showing exceptional interest in lunar solar energy, whether beamed from the Moon or its orbiting satellites. Leadership is coming from a strategic and international partnership being formed by proponents of wireless power transmission. Speaking at the International Astronautical Federation's 45th Congress, Dr. David R. Criswell, co-inventor of the lunar solar power system (see Appendix B), observed [12]:

> "By mid century 2050, lunar power industries can be sufficiently experienced and profitable to diversify into a wide range of other products and locations, other than solar power beaming. Specialized industries on asteroids and other moons will arise. Mankind can begin the transition to living independently off Earth. People can afford to move to space and return, allowing the womb of biosphere Earth to the evolution of other life."

10.1.2 Asian lunar initiatives

- **Japan** has the oldest mission efforts relative to the Moon in all Asian countries. Furthermore, Japan's Lunar and Planetary Society proposed to institute evolutionary lunar programs, involving orbiters and landers, roving robots and telepresence, astronomical projects, habitat studies, and even tourism. As far back as 1994, Japanese business leaders and scientists urged that their government's Space Activities Commission invest in a 30-year, 3-trillion-yen undertaking to build a Moon station by 2024, entirely constructed by robots! Japan's Science and Technology Agency welcomed the proposal which included solar power generation. In July 1994, a task force report of that Commission accepted those recommendations for Japan's space policy to include the building of a manned station and observatory on the Moon, preferably with its international spacefaring partners. In March 1996, the Mitsubishi Corporation made a major investment in the ambitious plan of LunaCorp (Arlington, Virginia) to provide interactive robotic exploration of the Moon, as well as high-definition lunar video for space theme parks, television networks, and scientific research. In 1997, its space agency planned Lunar A, a lunar orbiter and penetrator probe. Exhibit 128 indicates the scope of imaginative Japanese macroplanning for the Moon.

    But the Japanese found that there was a big gap between their ambitions and planning vs. actually doing and performing projects in outer space. The Japanese Aerospace Exploration Agency (JAXA) has had a series of launch postponements for two lunar missions, Lunar A and Selene, costing some $400 million. Lunar A was designed to hurl missile-like seismic penetrators into the Moon's surface near its far side, to study the Moon's internal make-up. The Selene mission (now named Kaguya) is designed to gather more scientific data on the Moon's origin and composition. It consists of a main orbiting satellite and two smaller ones. Selene was launched on September 14, 2007 from

Exhibit 128. Japanese lunar macroplanning. For over 15 years, Japan has been calling for an international initiative to develop a program for sustainable exploration of the Moon. This artist's concept shows one idea for an unmanned lunar mission to explore a previously unexplored area, such as the vicinity of a central peak in a crater. Such a mission would incorporate a lander and a roving vehicle with technologies for obstacle avoidance and safe landing, precise pin-point guidance and navigation, and capable of wide area exploration of the rough lunar terrain. Such an unmanned lunar surface landing is a key element of Japan's overall space strategy, which was once the most ambitious in Asia but has recently fallen behind China. Source: Japan Aerospace Exploration Agency (JAXA).

the Tanegashima spaceport, to provide serious remote-sensing data. Project Scientist Hitoshi Mizutani told *Space News* (7/20/04) that lunar exploration goes beyond its scientific value: "it provides inspiration, human extension to other worlds." Thus, Japan is actively seeking international partners beyond NASA and looking to form cross-Pacific ties, so as to carry out its aspirations for the high frontier. Even its activities within the annual Japan–United States Science, Technology, and Applications Programs involve the private sector in North America. JAXA's focus now seems to be on perfecting lunar instrumentation; with this objective in mind Selene will make use of 14 such science instruments.

Irrespective of whether there will be a "space race" between Japan, China, or India, their engagement in lunar enterprise should also include alliances among progressive Asian nations within that region, such as Indonesia, South Korea, Thailand, and even Vietnam.

- **China** has a three-phase plan under way through its China Lunar Exploration Program (CLEP). China's National Space Administration initiated the first step in a one-year Moon probe in 2007 using their Chang'e 1 orbiter. This mission, named after a Chinese goddess who lives on the Moon, originated from the Xichang Satellite Launch Center in the southwestern province of Sichuan. Laun Enjie, chief commander of CLEP, reported that Chang'e 1 carries eight primary instruments to photograph and map the lunar surface, probe its depth, study regolith chemical composition, and analyse the environment around the Moon. A payload data management system even includes 30 songs popular in the country, while the whole mission is estimated to cost about $180 million. This is a serious investment in lunar exploration; Laun stated that if the first mission is successful, the Chinese hope by 2012 to land a lunar rover, and by 2017 return a lunar sample. Ouyang Ziyuan, a leading lunar scientist, elaborated that their automated lander would land on the Moon, allowing robots to snag and return soil samples ... For manned missions to the Moon, the Chinese are depending on improvements in its heavy-lift Long March 3-A rocket which is to be used in this mission. They are planning to build such a spacecraft (with 3,000–4,000 tons of thrust) at a new launch site on Hainan Island in the China Sea. The new booster may be ready in eight years, and be able to lift up to 26 tons into orbit. Meanwhile, CNSA's third manned spaceflight, Shenzhou 7, is scheduled for 2008.

  Perhaps to gain more Congressional support, NASA administrator Michael Griffin said, in September 2007, that he thinks China will beat the United States in getting back to the Moon.[2] Like the rest of the world, China has massive needs for clean energy, so it might agree to a global effort to tap space-based energy. China wants to become a major player in space development, and may be open to becoming a partner in an international venture to utilize lunar resources, as well as to create communication protocols for that purpose.

- **India** is an active player in the multi-nation movement toward lunar missions. By 2008, its Satish Dhawan Space Center expects to launch Chandrayaan 1 atop their Polar Satellite Launch Vehicle XL. The satellite will be placed in orbit around the Moon for possibly up to two years, and contain a deep-space antenna system. It's devoted to high-resolution remote sensing of the lunar surface, including gathering infrared, X-ray, and low-energy gamma ray imagery of the Moon. Its payloads contains a number of Indian experiments, in addition to some from ESA, Bulgaria, and the U.S.A. Also incorporated is a Moon impact probe to test a future lunar landing; the device is designed to hit a predetermined location on the lunar surface. This will be a demonstration of state-of-the-art technology consisting of a highly sensitive mass spectrometer, a video camera, and a radar altimeter. Together these instruments hopefully will detect possible gases in the exosphere and provide video images of a prospective landing site. This unmanned Indian mission will carry two NASA scientific devices to find minerals and ice on the Moon. The Indian Space Research Organization has an agreement with the Agency to carry this payload on its 1,600-pound spacecraft, along with three instruments from European research

---

[2] Hedman, E. R. "China and India Want to Play," *The Space Review*, December 3, 2007 (*www.thespacereview.com/article/1014/1* or email: *ehedman@ldcglobal.com*).

centers. The international partnerships evident in this mission likely will lead to more scientifically and technically challenging cooperation on future lunar undertakings! At least that is the aim of ISRO chairman, K. Kasturirangan, after Chandrayaan 1 ends its two-year mission. Now India, like China, is indicating a desire to become part of the International Space Station program. Will the next step for both countries be as partners in developing lunar infrastructure?

All of the above Asian lunar missions are unique, and while there is some overlaps in coverage, they all advance lunar science and exploration. The combined data obtained will contribute to advancing human knowledge (see Section 1.5 for space case studies on China, Japan, and India)!

10.1.3 European lunar initiatives

As previously discussed, European space organizations have been pursuing lunar studies for about 20 years, especially in conjunction with colleagues at international space conferences [13]. For example, the European Space Agency's International Lunar Workshop concluded on June 6, 1994 that both logic and timing make it apparent that the Moon is "a natural, long-term space station" and the "testbed for any plans of human expansion into the solar system", as long as such endeavors protect the lunar environment. Along with international delegates to those discussions in Switzerland, the consensus was that the world's space agencies must coordinate their lunar plans in a phased evolutionary approach. That year's ESA declaration optimistically observed that "current international space treaties provide a constructive legal regime within which to conduct scientific exploration and economic utilization of the Moon, including establishment of scientific bases and observations." The "Beatenberg Declaration" that emerged endorsed the four-phase lunar initiative which we explained in Section 7.2.2 (see also the case study in Section 1.4). Exhibit 129 illustrates two such science prospects.

The Europeans announced that their approach to the Moon (ESA BR-101):

- is founded on long-term objectives and a phased approach;
- would substantially increase scientific knowledge and would exploit advanced technologies;
- would afford wide public participation through advanced communications;
- would preserve the lunar environment.

Their four-stage program includes (1) lunar explorers; (2) permanent robotic presence; (3) use of lunar resources; (4) lunar human outpost. ESA expects to use its upgraded Ariane 5 transport system to deliver both lunar orbiters and landers, as well as to support possible geostationary transfer orbit.

The most recent ESA lunar accomplishment was SMART 1, a spacecraft that punched into the Moon's barren soil in September 2006. On a volcanic plain near the Lake of Excellence, thus ended a three-year mission of testing space technology while examining our celestial neighbor. At CFHT atop Mount Mauna Kea, Hawaii, their brand new WIRCAM recorded the impact site with its infrared mosaic. The ESA

Sec. 10.1] **Reclaiming the Moon: rationale** 483

Exhibit 129. Lunar science opportunities. ESA studies identify specific astronomy/astrophysics prospects on the Moon's surface, such as illustrated above in the lunar telescope and the solar interferometer. Source: ESA reports, *Mission to the Moon* (SP-1150, 1992) and *International Lunar Workshop* (SP-1170, 1994). Also see N.A. Budden's *Tools of Tomorrow: Catalog of Lunar and Mars Science Payloads* (RP-1345) available from NASA/JSC (Mail Code SN2, Houston, TX 77058).

coordinator on this small mission for advanced research technology (SMART), Detlef Koschny, operated from the project's ESOC in Darmstadt, Germany. The purpose of the mission was to take a close look at the permanently shadowed lunar craters of the Moon's south pole, scanning them for possible ice. That resource would be needed not only for lunar dwellers, but also when broken into hydrogen and oxygen for providing breathable air and drinkable water, as well as being useful for making rocket fuel. When it was launched on an Ariane 5 rocket in September 2003 ion propulsion was used for this ESA lunar probe. Project scientist Bernard Foing emphasized that SMART 1 data were being shared with other countries as a first step in collaboration for future missions to the Moon. The information gained included surveys of lunar resources, polar illumination data, and characterizations of future landing sites. One aim is to develop common technical standards for future landers and orbiters. The next ESA reconnaissance missions are the Mars Sample Return in 2011, and the ExoMars in 2013 which would involve rover exploration. Are such missions simply replications of what other space agencies have under way, and should limited resources be combined and concentrated initially on lunar development (for further ESA information, email *Bernard.Foing@esa.int*)?

- *Russia* has been engaged in lunar research for some 50 years [14]. After Soviet scientists shocked the world in 1957 with their first orbiting satellite, Sputnik, they began to devise plans the very next year for sending a small spacecraft to the Moon. Under the leadership of their great rocketman, Sergei Korolev, plans were even drawn up to fly to Mars and Venus! When their first automated moonshot missed in 1959, they followed up with the unmanned Luna 2 which hit the middle of the lunar surface. A decade ahead of the Americans, this research led to a soft-landing on the Moon for Luna 9 in the Ocean of Storms on February 3, 1966. For four days that spacecraft sent back pictures of the rocky and cratered lunar surface!. That same year, Venera 3 hit the planet Venus, while Luna 10 continued the automated survey of the Moon, becoming the first lunar satellite. With the death of the "great designer" Korolev, the driving force went out of this stunning venture into outer space.

  For Russia, the Moon race had two sequels (according to author Brian Harvey): the unmanned program which concluded in 1976 was a great success! In 1970, just over a year after Apollo landed the first men on the Moon, the Russians soft-landed Luna 16 in the Sea of Fertility, extending a robotic arm that drilled into the lunar surface to collect a rock sample, and three days later returned it to the "motherland". In 1972, Luna 20 brought back more samples from the Apollonius highlands, and in 1976 Luna 24 brought back a core rock sample weighing 170 grams from the Sea of Crises. But the robotic missions that were the most impressive were Lunas 17 and 21; their 760 kg Lunokhod rovers were solar-powered with cameras, lasers, and other special equipment. These craft were steered by a ground crew back in Moscow's mission control! They accepted commands to drive, swivel, cross craters, and explore the terrain. Lunokhod 2 spent five months roving some 37 km before using up its power ... Unfortunately, the Russian manned lunar endeavors did not land cosmonauts on the Moon. Since 1964, their government set a goal of putting one of their citizens on the lunar surface. A giant Moon rocket was built, the N-1, which had 30 engines to fire the immense rocket. A second powerful rocket, Proton, was also built to send Russians to the Moon. Their preliminary flights lost any significance when the American Apollo program put two of their astronauts on the lunar surface in 1969. The Russians never again gained the lead in the space race between the two superpowers, and so turned their attention away from the Moon to Mars missions. With the collapse of the Soviet Union in 1991, their space scientists shifted from research on space mirrors, to the big Energia heavy-lift rocket, and the Mir space station.

In the 21st century, Russian space scientists are again turning their efforts back toward the Moon, especially in terms of international partnerships for lunar development. The Russian space agency's budget has gradually increased, as has consultation with China and India on their lunar exploration plans. Both the head of Roscosmos, Anatoly Perminov, and chief of RSC Energia, Nikolai Sevastianov, made bold lunar statements in 2005, especially about putting humans back on the Moon. For instance, in 2006 the latter talked about a Russian lunar base intended to mine helium-3 on the Moon as a rich source of energy. Energiya's director spoke of a new spacecraft design, Kliper (in partnership with ESA), which could possibly serve to transport helium, and proposed building an orbital tug, Parom, that could be useful in assembling elements for a Moon mission. But the Russian government has yet to fund their plans for further lunar exploration, but then neither has any other foreign government expressed interest in financing such ventures. Thus, the focus of Russian space missions is presently on cooperation in International Space Station missions which produce income via NASA contracts ($1 billion annually). NASA's Robotic Lunar Exploration Program (RLEP) comprises a series of robotic missions to the Moon, starting in 2008, paving the way for later human exploration; RLEP begins with the launch of Lunar Reconnaissance Orbiter which carries the Lunar Crater Observation and Sensing Satellite (LCROSS) to hopefully confirm the presence or absence of water ice in a permanently shadowed crater at the Moon's South Pole. It remains to be seen how much of the $104 billion that NASA plans to spend for a permanent return of humans to the Moon by 2020 will be shared with Russian space scientists and engineers, or how much the Russian Federation would be willing to invest in a world lunar enterprise. But whatever the global space community agrees to do together on the Moon, there is much to learn from Russian space and lunar technologies (for further insights, see Section 1.3).

In an essay on "Space Travel: Sealing Wax and String," *The Economist* (January 13th, 2007, pp. 71–72) refers to the "junior partners" in achieving American space goals. Led by the European Space Agency, these include Australia, Britain, Canada, China, France, Germany, Italy, India, Japan, Russia, South Korea, and the Ukraine. The ESA strategy is to launch more robotic and science missions to the Moon and Mars. This London-based magazine favors the construction and array of astronomical telescopes on the Moon, and gives support for U.K. small satellites and miniaturized space instruments. The latter capability might then be used to build and launch MoonRaker, a low-cost lander intended for the lunar surface to date geological samples found there, as well as MoonLite to listen to the noise generated by missiles fired into the Moon!

10.1.4 U.S. lunar initiatives

The U.S.A. has had two great lunar initiatives. The first was the Apollo missions to the Moon, begun in 1961 when President John Kennedy committed the country to lunar conquest within a decade, and prematurely terminated by President Richard Nixon in 1972 (see Section 8.2) [15]. Unfortunately, there was no follow-up on this magnificent milestone in human accomplishment, in terms of lunar industrialization and settlement. Many, like your author, argued that the American taxpayer deserved a return on their enormous Apollo investment, nearly $20 billion! As one columnist, Charles Krauthammer, lamented:

> "Ours was the generation that first escaped gravity, walked on the Moon, visited Saturn—and then, overtaken by an inexplicable lassitude and narrowness of vision, turned cathedrals of flight into wind tunnels."

No wonder so many space advocates have grown cynical over the failure of the United States to vigorously carry out the recommendations of numerous national commissions and studies regarding space enterprises. Apart from media feats like Tom Hanks' *Apollo 11* and his IMAX film, *Magnificent Desolation*, there were some hopeful signs during the past 36 years of lunar hiatus that the U.S. government would seek ways to provide a return on the public sector's original expenditures for the Apollo missions to the Moon. These positive actions toward ROI were

(a) Since going back to the Moon permanently requires a robust, less expensive *space transportation system* with capability in both LEO and GEO, R&D funding was spent on new rocket technology which utilizes composite materials that will lower the cost to orbital access (see Exhibit 114).
(b) Further, the low-cost Clementine 1 mission sponsored by the U.S. Department of Defense's Ballistic Missile Agency, in conjunction with the Navy and NASA, took 1,500 pictures in February/March 1994. This produced the first global digital map of the Moon, multispectral imaging data of 34 million square kilometers. This automated exploration also found a mountain top that might sometime be valuable for the first human settlement aloft; on this plateau the Sun never sets and a future colony would have full-time solar power by building a high collecting tower there [16].
(c) After more than 25 years since the Apollo missions, NASA launched Lunar Prospector into polar lunar orbit in June 1997 where it crash-landed in the Moon's southern region in September 1999 (see Exhibit 130). It was part of the Discovery Program, a new way to do smaller, faster, cheaper innovative missions via competition among private contractors; in this case the award went to Lockheed Corporation, whose Dr. Alan Binder conceived, designed, and developed the proposal for using a new LLV2 spacecraft. Supplementing previous automated lunar data collection missions, this one provided a global,

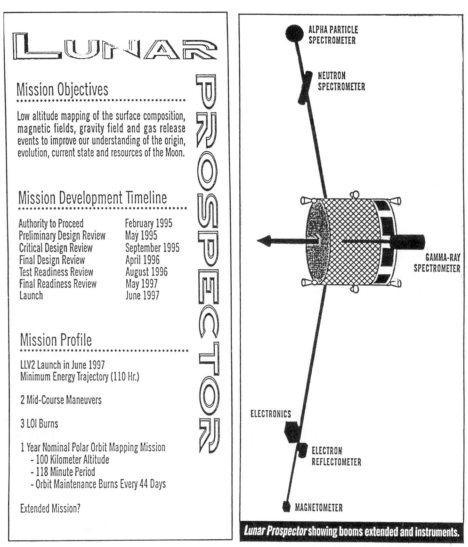

Exhibit 130. Lunar Prospector Mission Profile. The last unmanned lunar mission of the 20th century by NASA, summarized above, produced critical information on the lunar poles and confirmed the presence of ice there [17]. Source: P. Dasch, "Lunar Prospector," *Ad Astra*, May/June 1995, 32–33; *www.nss.org*). For current information on lunar missions, consult *The Space Report: The Guide to Global Space Activities*, published by the Space Foundation (*www.TheSpaceReport.org*).

low-altitude mapping of the lunar surface composition, gravity fields, and gas release events. The Prospector payload collected information that significantly improved our understanding of the evolution of the lunar highland crust, basaltic volcanism there, and lunar resource mapping. Mission adviser Dr. James Arnold,

University of California-San Diego, noted that the new, simpler approach contrasted with previous NASA Solar System exploratory missions, which often took too long and were too costly [17].

(d) National space legislation: in 1995, a "Back to the Moon Bill", part of an Omnibus Commercial Space Act, was put to Congress; while limited in scope, it would create a legal regime for NASA to purchase lunar data from private enterprises, allowing commercial companies to conduct innovative lunar probes on their own designs. Sadly, it never became national legislation or policy.

The second great American lunar initiative is ongoing; it is part of the national policy called the Vision for Space Exploration (see Section 1.2 and Exhibit 123). Inaugurated by the second President George W. Bush in 2006 and endorsed by an act of the U.S. Congress, this endeavor tasks NASA to move toward unmanned lunar probes by 2008, and manned missions by 2015, with a goal of returning permanently to the Moon by 2020. The plan is both incremental and cumulative to develop the necessary "lunar architecture", beginning by again scouting the Moon with automated lunar trailblazers until astronauts can begin to establish a lunar outpost. As previously explained (Section 7.5), the "Global Exploration Strategy" emerged in 2006 with the participation of 14 space agencies and assistance from over 1,000 experts. This produced 200 objectives that might be pursued on the lunar surface. There was agreement that its principal objective is to create a sustainable human and robotic presence on the Moon that opens significant opportunities for science, research, and technological development. Both the VSE and GES make it clear that other national partners in this lunar enterprise are welcome to share the risk, the costs, the glory, and the eventual benefits! With a $16 billion budget until 2010, the Agency's Exploration Systems Mission Directorate has been transforming "the vision" into a *bona fide* space program that includes signing contracts and building hardware. All this has to be done while NASA and its partners complete the International Space Station, retire the Shuttle fleet, and replace it with Ares and Orion spacecraft, now in the testing and building stage. A series of sorties to the Moon's poles and equator, reaching lunar sites that were never visited by Apollo expeditions, are planned for both robotic and human missions in the next dozen years.

The outpost, which will be developed into a larger base, apparently is to be located at one of the lunar poles, probably in the south to start. The Lunar Reconnaissance Orbiter to be launched in 2008 will determine the exact site. The Lunar Architecture Team is working on a series of expeditions that would culminate in spacefarers living on the Moon for up to six months. A lunar transportation system is being planned so that each lunar landing, unmanned or manned, will include critical pieces of the new lunar infrastructure. By the end of the next decade, the second group of moonwalkers are expected to include seven astronauts who will bring their own habitats and will likely stay at the start for at least a week. A second-year mission would deliver power generation equipment, an unpressurized rover, and other infrastructure. The gradual build-up on the Moon will include extending human stays for up to 14 days initially, then 30 days, and so on until eventually mission durations are likely to be six months or more. Initially, the emphasis will be on lunar resource utilization or ISRU aimed at self-sustaining operations. Much will depend on available U.S. funding, and agreements at to the contributions of other national space agencies to this macroproject.

The imaginative *Vision for Space Exploration* is about more than merely returning to the Moon for the benefit of earthkind [18]. As Dr. Paul Spudis of Applied Physics Laboratory reminds us, it is to increase human knowledge about the objects trapped in gravity by our star, the Sun, and everything else. In the future, the material trapped in the Sun's vicinity could be incorporated into our way of life. VSE's goal is to extend humanity's reach beyond low-Earth orbit and destinations within cislunar space, and then farther out into our Solar System. In the process, we will learn new skills and technologies that will eventually be commercialized. It could occupy human ingenuity throughout the 21st century!

Meanwhile, Mars advocates are encouraged because VSE implies that the Red Planet is the next logical target for manned missions beyond the Moon. While continuing to Mars will be most dangerous and costly, the knowledge from establishing a lunar base should facilitate that process, especially in terms of a launch pad. Members of the Mars Society estimate that a Mars manned mission would be possible for an investment of $100 billion, and that one-third of the NASA budget now should be devoted to that objective. The search for microbial life there is seemingly worth the effort. In the meantime, robotic exploration of Mars proceeds: in August 2007, a Delta II rocket launched a lander Phoenix on a nine-month journey to study its icy soil. On May 25, 2008 the spacecraft successfully landed on the frozen plains of the Martian Arctic, NASA's most northerly landing on the planet, and began three months analyzing the water ice–rich deposits there. The $420 million mission will be worth it if it helps us to understand the history of water on the planet, a key to unlocking the story of past climate change. Such efforts confirm that the purpose of VSE is more than establishing a human presence on the Moon—it is to use what we learn there to go beyond, starting with Mars [19].

At NASA's JPL, Dr. Martin Lo designed the flight path for the Genesis mission in 2002 to collect solar wind particles. Then he recommended formulation of an *interplanetary superhighway* to make space travel easier. Lo envisions a place to construct and service science platforms around one of the Moon's Lagrange points. The latter are landmarks of this interplanetary superhighway to planets, asteroids, and comets. Dr. Lo thinks spacecraft could easily be shunted to and from these lunar stations for maintenance and repairs. By using this "freeway" through the Solar System, the amount of fuel needed for future space mission could be slashed (*www.genesismission.org*).

10.1.5 Private lunar initiatives

Far-sighted business leaders and entrepreneurs recognize that we are in the process of creating a twin-planet economy, and that new wealth can be gained on the Moon which will eventually contribute to a thriving and prosperous society on both bodies. But to take advantage of the lunar opportunity is a very costly and risky investment. Without the aid of the public sector, private lunar enterprise in the beginning will rely on government contracts, grants, tax incentives, and other forms of subsidy. In 2004, the Presidential Commission on the *Vision for Space Exploration* issued a report highlighting the importance of private enterprise in this national endeavor. But will the U.S. Congress and Administration provide any incentives for that participation? Dr. Wendell Mendell, Lunar & Planetary Exploration chief for NASA, made it clear in an email to the author (12/12/06) that the Agency does not believe that it should be or can be in charge of settling the Moon. Obviously, this will be the business of private enterprise, starting first with lunar contractors, tourists, and then recruiting lunar dwellers. Again, in the exploration of the New World in the 17th century, it was the trading companies who signed up, transported, and initially supported the colonists. The research of Jonathan Karpoff at the University of Washington confirms that historically in the exploration of previous frontiers, private expeditions were better organized, more adaptable, suffered fewer fatalities, and made the better use of the latest technologies! And so it is likely that will be the case in the settlement of the Moon and Mars.

In the 21st century, there are currently few lunar entrepreneurs because of the enormous costs involved in financing a venture, and the inability to raise sufficient funds for lunar projects. This stopped a joint venture exploration mission by LunaCorp with Radio Shack from sending the Superstar telecommunications satellite to map the Moon. That also was the experience of Dennis Laurie, CEO of TransOrbital, Inc. (*www.transorbital.net*). That was the first commercial company to receive license approval from both the U.S. State Department and NOAA to launch a mission to the Moon, beyond geosynchronous orbit. The strategy was to send into orbit, and then land a private small spacecraft, Trailblazer, on the lunar surface. With photovoltaic cells on each side, power could be independent of orientation and polar orbit. The craft was reconfigurable in design, and capable of carrying a host of payloads [20]. The primary purpose was high-definition lunar imaging of the Moon and nearby objects of interest that might prove of value in planning future lunar missions and landings. Apart from the income produced from the sale of its lunar images, TransOrbital's market was to be the selling of offworld advertising and public relations by Internet packages, television documentaries, brand-naming rights, as well as transportation to the Moon of business cards, messages, and personal or commercial products, plus sale of post-launch extension products. In conjunction with International Space Company Kosmotras, TransOrbital managed in 2003 a successful test launch of its Trailblazer satellite from the Baikonour Cosmodrome in Kazakhstan on a Dnepr LVSS-18 intercontinental ballistic missile. Although TransOrbital had the support of both Space Age Publishing and the Lunar Enterprise Corporation, it could never obtain sufficient sponsors or investors to make a final launch to the Moon, or achieve its $150,000 a month projected income stream. TransOrbital was a magnificent commercial space effort, ahead of its time and pocket book.

Entrepreneurs have proposed many business projects for the Moon, most of which are premature. For example. Dave Dietzler wants to use the lunar surface for farming (email: *Dietz37@msn.com*), while others want to dig down into that surface and build storage units for archives from Earth, as a precaution against catastrophe on the home planet. Another imaginative proposal comes from the Alliance to Rescue Civilization to develop a DNA repository for all life on Earth to be stored at a lunar base (*www.arc-space.org/*). ARC would also deposit below the lunar regolith a compendium of all human knowledge. Its founder, NYU professor Robert Shapiro argues that civilization must protect the things it values [21]. More imminent is the Centennial Challenge Prize offered in 2006 by NASA and the X-Prize Foundation for private-sector innovators. Within context rules, the $2.5m Lunar Lander and Vertical Takeoff awards go to the competitor providing the best rocket demonstration of a trip from low Earth orbit to the Moon. The Agency hopes to contract with the winner/s, and intends to continue the program with other lunar challenges as a spur to space entrepreneurs.

Universities are examining prospects for their own lunar missions, seeking alumni support (e.g., *Stanford on the Moon* and *Iowa on the Moon*). The former was started by alumnus Steve Durst, founder of Space Age Publishing, to introduce the university to lunar enterprise. The latter refers to Iowa State University whose alumnus, Dr. David Schrunk, has urged professors and students at his Alma Mater to consider research relative to *in situ* resource utilization (ISRU), directed specifcally toward a lander and robot in the equatorial region of the Moon. This aeronautical engineer/physician recommends that a consortium of universities might be formed to design experiments, including lunar manufacturing (email: *docscilaw@aol.com*).

Non-profit organizations have also dabbled with projects for the Moon. The oldest venture, which coordinates efforts of some ten other groups, is the Artemis Project (*www.asi.org*). This 1997 ambitious plan of the Lunar Resources Company proposes a four-step mission: (1) the crew assemble spacecraft at a LEO space station, mating it with a Lunar Transfer and Descent Stack; (2) the Lunar Transfer Vehicle then ferries the Descent Stack to the lunar orbit, and (3) the DS lands on the Moon, providing a habitat for spacefarers as the LTV remains in lunar orbit during the exploration period; (4) the crew level the stack as a solar power station and antenna, then spend ten days exploring before returning to Earth. It makes for a wonderful graphic, but sufficient funds were never raised for the venture, while VSE is actually moving ahead with NASA's lunar return.

A more realistic plan was put forth in 2007 by Steve Durst on behalf of the International Lunar Observatory Association (see Section 7.4 and Appendix E). ILOA is in the process of fundraising from sponsors, affiliates, and investors to erect a multifunctional lunar observatory at the Moon's south pole near Malapert Mountain, possibly before 2010 (see Exhibit 131). SpaceDev Inc. has been contracted to develop a lunar lander demonstrator model. The ILOA lunar commercial communications center will include a power station for transmission to Earth of real-time astrophysical data and videos.

In 2005, a Lunar Commerce Roundtable was held to discuss the prospects for business endeavors on the Moon (*www.lunarcommerceroundtable.com/*). The participants examined how such enterprises could contribute to the global economic growth foreseen in the Vision for Space Exploration (see Section 1.2). A key conclusion was that publicly funded exploration should facilitate an open approach to lunar development that provides flexibility for private and public-sector stakeholders. In lunar

Exhibit 131. Planned International Lunar Observatory. Private enterprise is now fundraising for this non-governmental telescope on the Moon (see Appendix E). Source: International Lunar Observatory Association (480 California Ave. #203, Palo Alto, CA 94306; *www.iloa.org*).

entry and exit, these entities need to practice both collaboration and innovation in creating a commercial market on the Moon where government is only one of the many customers. This "eighth continent" is both accessible and offers commercial opportunities in energy, transportation, mining, construction, manufacturing, entertainment, advertising, branding, and sponsorship. With the assistance of robots, markets will emerge that are profitable and beneficial to humans there and on the home planet!

But such cislunar undertakings are fraught with risk, both on the ground and

Exhibit 132. LUNAR ENTERPRISE FORECASTS.

If a lunar macroproject is undertaken by a global government consortium or trust, then provisions should allow for participation by private enterprise. On the other hand, some lunar commercial possibilities should be left entirely to the private sector to underwrite, manage, and develop. Non-profit organizations, such as universities, professional and scientific societies, humanitarian associations, should be encouraged to become lunar sponsors in their fields of expertise. To optimize a lunar economy for the benefit of humanity, here are some areas for R&D:

- space transportation systems to and from the Moon, including spacecraft and fuel depots;
- lunar transportation systems across the Moon's expanse and beyond, including lunar fuel production from oxygen;
- lunar supply systems in terms of food, water, materials, inflatables, and equipment until such can be produced on the Moon itself;
- lunar concrete, water, and *in situ* resources that will contribute to the construction of infrastructure there;
- lunar dweller system habitats, regenerative life support systems, radiation shielding, food management, waste management, health services, thermal controls, and so on;
- lunar science and engineering provisions to further research by scientists, astronomers, biologists, geologists, resource mapping and analysis, behavioral science experimentation, etc.;
- lunar communication systems for use on the Moon and with the home planet;
- lunar energy production and distribution, especially scale-up of a solar electric power system, and eventually a lunar solar power system to beam energy to Earth;
- lunar construction systems utilizing bi-planet materials and technologies;
- lunar mining and manufacturing systems, such as with regolith and helium-3;
- lunar governance and administration that encourages peaceful, commercial, and collaborative development of the Moon's resources;
- lunar financial, real estate, insurance, liability, and legal services, etc.;
- lunar educational and cultural systems for both Selenians and programs to be transferred from the Moon to Earth, or *vice versa*;
- lunar entertainment and performing/creative art provisions for inhabitants and visitors;
- lunar security and peacekeeper provisions, including conflict resolution and a justice system to deal with delinquent, deviant, and criminal behavior;
- lunar tourism industry and provisions for visitors from Earth;
- lunar design and manufacturing of spacecraft to go to Mars and the asteroids;
- lunar manufacturing of telescopes, tools, and instruments made of Moon materials;
- lunar resources, products, and services for the benefit of the Earth's inhabitants;

> - lunar production and deployment of free-floating structures and vehicles in outer space;
> - lunar enterprises to help create a spacefaring civilization that can be hardly imagined now, as human knowledge and potential expand offworld!
>
> Expert consensus on human development of the Moon in the 21st century.

Source: Dr. Lawrence L. Kavanau, DOD deputy administrator who helped forge the vision and policy that led to the Apollo program that landed humans on the Moon (this aerospace engineer of rare business acumen died in 2005); Dr. David G. Schrunk, aerospace engineer and radiologist, senior author of two editions of *The Moon: Resources, Future Development and Settlement* (2008), and founder of the Quality of Laws Institute; and Dr. Philip R. Harris, behavioral scientist and management/space psychologist, author of three editions of this book, *Space Enterprise: Living and Working Offworld* (2008) and the classic *Managing Cultural Differences* (Seventh Edition, 2007).

aloft, requiring more user-friendly legal and regulatory practices, as well as helpful financial incentives and provisions if lunar partnerships are to flourish. To succeed in a bi-planetary economy, Exhibit 132 predicts the activities that should be pursued. Each will require macrothinking and macromanagement.

10.2 LUNAR ADMINISTRATION AND GOVERNANCE

Moving humanity beyond Earth raises serious issues relative to leadership and organization in outer space. These go beyond mere matters of science, technology, legality, and finances. This "giant leap for humankind" is altering the species, especially in terms of biological and cognitive adaptability. Within that context, we may then consider space policies, settlement, and commerce. Governance, especially of space settlements, will shape human futures into an extraterrestrial civilization. Thus thinks Professor Yehezkel Dror, a policy planner and governance expert. As an astute social scientist, he observed [22]:

> "The repercussions of moving into space are largely inconceivable, and efforts to predict them on the basis of a very different past are extremely doubtful. Therefore, applying NASA's experience beyond some technologies to the problem of building a future for women and men beyond Earth offers a much too conservative perspective. Similarly cost–benefit analysis in terms of contemporary economic realities misses the implications of settling human beings outside of Earth on all aspects of thinking and living, individually and collectively. Hence, new social structures are needed, with novel core capacities meeting the requirements of moving humanity into space. This is all the more crucial because space settlement is only one of the extreme changes that add up to a radically novel epoch into which humanity as a whole is inexorably moving, with tremendous potential for better or worse!"

Among the characteristics of space governance, Dror suggests that it be (1) *global* in scope requiring cooperation among spacefaring nations; (2) *inspirational*, with the

Sec. 10.2] **Lunar administration and governance** 495

Exhibit 133. Lunar industrialization. Entepreneurs will discover a variety of commercial endeavors to pursue on the Moon, from building lunar structures and facilities, to mining and manufacturing, to providing goods and services almost beyond our present imagination. In this artist's concept, a lunar mining facility harvests oxygen from the resource-rich volcanic soil of the eastern Mare Serenitatis. The high iron, aluminum, magnesium, and titanium content in the processed tailings could be used as raw material for a lunar metals production plant. This image produced for NASA by Pat Rawlings, (SAIC). Source: NASA Johnson Space Center.

seeking of knowledge, the search for the new and unknown, the education of the masses as to the necessity of going offworld; (3) *long-term perspective and persistent*, that requires radically different governance systems which are as democratic as possible; (4) *planned in terms of mega-project resources and management*, that may involve generations and innovative methods of financing, as well as learning; (5) *positive re-enforcement tools* which allow for human cultural differences, frailties, and abuses, but maintains civilized standards; (6) *cognitive abilities in leaders* who are competent in managing large-scale enterprises; (7) *new rationale that espouses positive, democratic values* for the long-term benefit of humanity and the common good. This will necessitate creating new lunar institutions, leaders, and professionals who act within the context of existing space law (see Appendix A)!

10.2.1 Background observations

Previous multinational negotiations relative to U.N. space treaties have recognized the need for some type of international regime to deal with the exploration and development on the Moon. No nation has yet offered a specific proposal as to what this entity might be and do. Some scholars have referred to the possibility of duplicating the Antarctica model in which various countries entered into an international agreement as to their operations in that remote region. Thus, various national outposts or bases were established there, primarily for the purpose of exploration and science. However, under that arrangement, some significant science has been achieved, but little development of the areas resources or population. Dissatisfied with provisions in the Moon Treaty, which most spacefaring nations have not signed, the space law community would prefer to have the nations actually engaged in lunar enterprises confer and agree on some legal and regulatory institution that would facilitate cooperative lunar development within the global space community (see Section 9.4).

Various specialists familiar with the challenge have offered thoughtful solutions to this need, which becomes more pressing as space agencies from Canada, China, Europe, India, Japan, Russia, and the United States move ahead on a series of lunar missions, culminating in the placement of humans back on the Moon to stay. Until lunar governance and legal matters are resolved, public–private partnerships, venture capital, and private equity financing, will be restrained in promoting lunar commercial enterprises. Legally, we need transnational answers to basic questions, such as proprietary rights relative to lunar resources, including who may sell or profit from such; who may issue licenses or permits for lunar mining and other business activities. One possible model suggested by Sadeh, Benaroya, Livingston, and Matula is the U.S. Deep Seabed Hard Minerals Resource Act of 1980 [23]. This American Federal legislation establishes an interim regime that provides legal protection for U.S firms, pending an international agreement on deep-sea mining and resource activities. The professors argue that comparable legislation for lunar commerce is needed until international agreement on lunar resources is achieved. For the latter, these four authors offer an interesting analysis of three possible solutions (government, business, and technology models). Then these scholars suggest that dual-use technology deals with lunar projects and technological developments that are so expensive, long term and large scale, they require substructuring of the macro venture into smaller— yet profitable—independent units. To that end, they recommend formation of a Lunar Development Corporation, seemingly modeled somewhat after LEDA which will be discussed in Section 10.2.2. This public corporation would include teams of management scientists and engineers, and financial, legal, and other experts. LDC would ensure that the right technologies, materials, and workers are available when needed to successfully complete lunar projects.

In a previously cited book, *Return to the Moon*, former U.S. Senator and Apollo astronaut Harrison Schmitt devoted a chapter to "Law: Space Resources". Here the last man on the Moon examines relevant laws and precedents as applied to lunar development. He concludes that the present situation is fraught with serious impact on the private entity's ability to maintain space operations, especially as related to safety concerns with its personnel, access to capital, and customer decisions on which business is so dependent. Unfortunately, this knowledgeable scientist and statesman offers no recommendation as to what kind of international regimen might solve the problems of lunar enterprise.

Sometimes, professional article and science fiction precede space realities, giving us insights for future decisions [24] For example, in his writings Krafft Ehricke laid out a detailed plan for creating a polyglobal civilization, including a lunar city called *Selenopolis* ... In his book, *Welcome to Moonbase*, Ben Bova made a persuasive case on how to establish lunar bases and how to manage a dual economy there with Earth. His scenario envisioned two bases, one founded by Russia on Mare Nubium and another by the United States on Mare Vaporum. The latter's development is illustrated for the period 2015–2018, remarkably similar to what is now under way under the VSE plan. It would be operated by Moonbase Inc., made up of stockholders from 15 nations, plus large multicultural corporations ... Phil Harris' *Launch Out* offered a science-based scenario for space enterprise, lunar industrialization, and settlement. His 2010 prognosis for *Selenians* living and working in a future *lunar world* would be made up of several areas at the start: a Krafft Ehricke Lunar Industrial Park, a Konstantin Tsiolkovski Educational and Research Center, underground multicultural living communities of Eastasia and Euroamer, as well as a Gerard O'Neill Health and Wellness Center and a Carl Sagan Astronomical Institute. This macroproject on the Moon would be under the administration of a Global Space Trust, guided by a civilian, democratic management council of multidisciplinary experts who adhere to the *Declaration of Interdependence for Governance of Space Societies*. The GST Fund to underwrite these costs for an eventual lunar population of 2,000 would include contributions from both the world's public and private sectors. Further financing would come from the sale of GST/LEDA bonds and an international lottery. In addition to its own spaceport, the Trust would operate four ground-based facilities and services for those in orbit: transportation, communication, and supply bases, plus the Unispace Academy located in Hawaii near the East–West Center there, for the training and preparation of all GST spacefarers. All stages of this undertaking on the Moon would be coordinated through a Lunar Economic Development Authority, as described next.

10.2.2 Lunar Economic Development Authority (LEDA) proposal [25]

To meet the need for an interim regime to develop a twin Earth–Moon economy, the co-founders of United Societies in Space conceived a strategy based on an existing model: the Tennessee Valley Authority and the many port authorities around the world (see Section 7.2.2). The entity was called the Lunar Economic Development Authority, and could be a quasi-public or private or combined corporation [25]. For those objecting to the term "authority", other words could be substituted, such as corporation, council, or foundation. The point is that the global community would create a centralized institution, whether in conjunction with the United Nations or not, to coordinate humanity's activities on the Moon, thus avoiding overlapping efforts and expenditures. Here are some of the services that LEDA might provide:

(a) issue bonds to underwrite lunar enterprises, including a transportation system, a base, and an industrial park;
(b) lease surface and mining rights for private or public-sector lunar macroprojects and collect fees therefrom;

(c) coordinate and facilitate international endeavors by space agencies, scientific organizations, and private corporations on the Moon or its vicinity;
(d) provide a lunar administrative structure for oversight supervision of terrestrially sponsored undertakings and communities on the Moon, so as to protect the environment and interests of its owners, humanity;
(e) contract and supervise necessary infrastructure provisions, such as a lunar transportation system, a lunar power system, and a lunar personnel deployment system;
(f) operate a lunar spaceport for all users, possibly with a landing fee to support infrastructure development;
(g) act as a clearinghouse of data about lunar information, such as conditions, resources, sites, and programs for future investors, project sponsors, and settlers;
(h) conduct public information, outreach, and development programs on Earth to encourage investments in lunar resource utilization and lunar resettlement.

Rather than depending on the home planet's taxpayers to finance construction of lunar infrastructure, LEDA would underwrite such development through income-producing activities such as outlined above, or by contracting with the world's financial systems for loans, etc. On Earth, ships and airplanes pay for the privilege of sailing or flying into urban ports; it would seem reasonable, then, that spacecraft from both public and private sectors worldwide might also be charged a fee someday by LEDA for the privilege of landing on the Moon. When tourism reaches the lunar surface, such fees could contribute significantly to the lunar economy.

Right now for development purposes, there is no legal or financial mechanism, no less a technological infrastructure, for such interplanetary undertakings that provide transnational, global participation. Even those who drafted the 1979 Moon "Treaty" envisioned some type of outer space regime or *authority* to oversee and regulate the "orderly development and exploitation" of extraterrestrial resources. Writing in 1994 on "Lunar Industrialization", Prof. Haym Benaroya of Rutgers University forecast that such commercialism could employ 3%–12% of our population in new jobs, both on the ground and aloft [26]! But he too foresaw the need for some type of Space or Lunar Industrialization Board (in this case within the U.S. government) to set policy, as well as coordinate and oversee economic, legal, and technical aspects of resource development on the Moon, and later Mars. In that same year, Lubos Perek, a former chief of the U.N. Outer Space Affairs Division, made a significant case for improving the management of outer space activities. He argued that to manage extraterrestrial resources, a U.N. International Space Center (UNISC) needed to be formulated.

Dr. Nathan Goldman, a Houston attorney and author of *American Space Law*, commented [27]:

> "The Lunar Economic Development Authority, similarly, will be structured to create an international regime that would encourage, as well as regulate, (rationalize) the habitation and commerce on the Moon. The LEDA fills in the blanks of

incomplete space law with details that can make space available for human development in a very short time."

More recently, Dr. George S. Robinson, former associate counsel of the Smithsonian Institution, called for a *Declaration of Interdependence* between earthkind and spacekind, perhaps establishing an International Organization for Spacekind Cultures (IOSC, see Appendix A). This space law scholar sees as its purposes to [28]:

(1) provide an interdisciplinary, international, and transnational body of recognized experts to continuously review interactive relationships between Earth dwellers and spacefarers;
(2) grant international agreements of recognitions and capacity (IARCs) to those space communities that satisfy the requisites for home rule as set by IOSC or something comparable (such as, the proposal of United Societies in Space to found a space *metanation*, possibly under U.N. auspices)
(3) refer case situations of conflict to the International Court of Justice, or a transnational court yet to be founded for this purpose.

The above citations underscore the growing consensus that utilization and development of space resources require creation of a new entity to coordinate global space enterprise and governance, whether within or outside the existing United Nations [29].

Certainly, some institution has to be devised to foster lunar development, preferably one which in scope goes beyond government sponsorship or even a combination of national space agencies. A Lunar Economic Development Authority should be intersectoral, representing the interests of public and private sectors on a planetary scale. It should encourage participation of transnational consortia, whether from universities, corporations, space associations, or agencies. The macrothinking here should go beyond the European Space Agency proposal for an International Lunar Quinquennium meeting every five years to discuss lunar projects (see Section 7.2.2).

In conclusion, the U.N.'s Outer Space Treaty was ratified by some 90 nations. Today, few states or their commercial entities would seriously consider participating in a venture which was not perceived in accordance with the international agreement. Before returning to the Moon by 2020 under the Vision for Space Exploration plan, *it is essential for some accord among space agencies and private enterprise as to how we are going to proceed there for the benefit of humanity as a whole.* At least, LEDA is a proposal to spur discussion until some formal concurrence is achieved within the next ten years [30].

10.2.3 The strategy of space authorities

Although your author believes that to further human enterprise in space, a viable solution to this challenge is to establish space authorities, now for the Moon, and eventually for Mars, and other planets, as well as for stations and platforms in orbit, even asteroids. It is a way to use our "interplanetary common" for the benefit of earthkind in the 21st century [31]. Then, we would put institutions in place to empower scientists, engineers, entrepreneurs, or settlers to go aloft and utilize space resources. To facilitate living and working in isolated, confined, sensitive environments, like the Moon, the legal and governance prototypes already exist (e.g., in the Antarctic Treaty (1958–1961), with its protocols and organizations, as well as in the Tennessee Valley Authority). The Antarctic Treaty provides the legal framework for the area south of 60°S latitude on this planet, reserving the region for peaceful purposes and encouraging international cooperation in scientific research there. The other model is the TVA, authorized by the U.S. Congress in 1947 with a Board of Governors appointed by the President and confirmed by the U.S. Senate. When it was founded, the United States Government not only donated land and facilities for the new entity, but vested enough sovereignty in the TVA so that it might obtain more land, including controversial seizures by eminent domain. The Authority's objectives were to conserve assets for the benefit of the American public in general, and specifically to provide electric power for the benefit of the people in the region served. To achieve its objectives, in 1948, TVA issued $50,000,000 (U.S.) worth of bonds @ 3.2% interest rate, secured by a blanket debenture of its assets. Thirty years later these debentures were retired with no defaults, rollovers, or commissions having been paid. Today, TVA is one of the largest, most successful power producers in the world, a strategy worth emulating if resources on the space frontier are to be transformed for the betterment of the people of "Spaceship Earth".

This type of quasi-governmental *service authority* is a proven, respected, and traditional venue for underwriting and managing both public and private undertakings across jurisdictions and borders. It has been gainfully used to construct terrestrial infrastructure from air and sea ports, to building bridges, toll roads and convention centers. Port authorities have been successfully constituted across the U.S.A., from New York to San Diego. The New York Port Authority, for instance, crosses state lines to serve the metropolitan area's transportation needs. NYPA has its internal police force, which can arrest those who fail to comply with state, local, and Authority regulations. The approach is justified because of the size, value, and complexity of port facilities relating to transportation, safety, docking, food spoilage, longshore personnel traditions, and union contracts ... The new Denver International Airport Authority was also financed by Municipal Airport Revenue Bonds totaling $275,000,000, but these were government-guaranteed.

Spaceport Authorities are another example of the same strategy in use internationally from Florida to Australia. Why not adopt a comparable mechanism to finance and promote a *space infrastructure* which might supplement or replace direct taxation for space exploration and commerce? ... An interesting historical point is that the U.S. Federal Statutes, authorizing the inauguration of the U.S. space program, began with a section on Police Authority (42 U.S.C. 2456) with the power for personnel to arrest citizens and bear arms for that purpose; that statute also created the agency that eventually became the National Aeronautics and Space Administration.

Thus, legal precedent exists and might be tested for application in space by the immediate incorporation of a global *Lunar Economic Development Authority*, whether this be accomplished by private enterprise, government, or a combination thereof; whether it be within or without the United Nations, whether it be under national or international law. Should LEDA prove to be a successful prototype, then it might be replicated next by the establishment of a *Mars Economic Development Authority (MEDA)*, and the model eventually repeated for the development of orbital stations or cities, other planets, or asteroids in our Solar System.

There are various scenarios as to how such *space authorities* might come into being within a decade:

(1) Assuming a Lunar Economic Development Authority is the prototype, incorporate it in one or more states or nations. Thus, profit and/or non-profit organizations might combine their strengths to undertake macroprojects on the Moon. There is ample precedent for this among world corporations and foundations seeking to protect global commons. One scenario is for the U.S. Congress to provide legislation constituting a Lunar Economic Development Authority, essentially to conserve national interests and promote development of the Moon and its resources for the benefit of its citizenry and to cooperate with other nations in this goal. The charter might be similar to that of TVA, and possibly some existing space assets might be transferred from NASA or DOD to the new Authority to provide security for the lunar bonds sold for investments on the Moon. LEDA, in turn, might legally contract for services from NASA or other federal and state agencies, or from universities and corporation in the private sector at home or abroad. The Communication Satellite Act of 1962 is another precedent for such action, for it established COMSAT to cooperate with other countries to develop an operational satellite system, as well as to provide services on a global scale to others. Given the trend toward "privatization" of public property, imagine if the assets turned over by Congress to the new lunar Authority were to be two spaceports now functioning at Cape Canaveral in Florida and Vandenburg AFB in California; both built and paid for by taxpayers could then produce bond revenue and other income flows if operated by LEDA!

(2) Another scenario would form a global consortium by spacefaring nations committed to lunar enterprise who sign an international agreement to establish a Lunar Economic Development Authority. In this approach, LEDA acts on behalf of the participating countries in financing and macromanaging resources on the Moon. The precedent for this already also exists in such agreements as INTELSAT, which established a global satellite communication system signed by governments or their designated public or private telecommunications entities.

(3) Although any of the above solutions might precipitate desired action toward near-term lunar development, many prefer a strategy whereby spacefaring nations work through the United Nations to found LEDA. At the very least, the U.N. would be the logical organization to call a summit conference of

spacefaring nations in an attempt to achieve some international consensus on this important matter of humanity's move offworld!

Admittedly, the Outer Space Treaty implies that nations which place their citizens into space, such as on the Moon, have a responsibility to exercise some form of control over them. That is relatively easy with a few government-sponsored astronauts living in facilities provided to them by that entity. But what happens with settlers who are many decades in orbit? Further, how can this control be exercised when private citizens gain access to low-cost spacecraft and begin to migrate on their own in ever larger numbers to the lunar surface? It would appear the relevance of this 40-year-old Treaty will diminish. Better to have a global consensus and legal provisions in place before the masses move to the Moon and beyond.

LUNAR ECONOMIC DEVELOPMENT AUTHORITY

The purpose of this Corporation (LEDA) shall be to promote the Moon as a place to live and work as a society of peoples and to help create and maintain a consensus governance authority at the venue of the Moon, including its useable orbits, and to educate people on the benefits, burdens, and responsibilities of living and working in space. The Corporation will serve as the agent of humankind in space and at the Moon, as well as the agent for all of the sponsor nations, to develop the Moon for humankind. It is the intended business of the Authority to administer each nation's rights under the *1967 Outer Space Treaty*. (Source: Article III, Articles of Incorporation for the Lunar Economic Development Authority, Inc.).

Interestingly, a space policy analyst for the *Washington Dispatch*, Mark Whittington, has written that a *Lunar **Exploration** and Development Authority* would be helpful in opening up the high frontier [32]. This proposed LEDA has the emphasis on exploration, not economic development as does the above LEDA. He considers such a worthwhile mechanism for carrying out White House Space Transportation Policy (STP). The Administration's committee which examines ways to implement its Vision for Space Exploration is chaired by Admiral Craig Steidle. That group is concerned about how to "open space enterprise, markets, and ultimately self-supporting activities," which is also the purpose of this LEDA proposal. This strategy, like the other LEDA, is to encourage commercial development on the Moon, and through that improve the Earth's economy. STP states the government will refrain from activities that have commercial applications, so as to involve the private sector in the design and development of space transportation systems. Whittington maintains that his LEDA offers an innovative way to explore space, one using the entrepreneurial and commercial strength of private enterprise.

In that same issue of *Lunar Enterprise Daily*, Anatoly Perminov, head of Russia's Roskosmos, has written that to explore Mars with humans, there must be an International Space Station and a Moon base. According to *RIA Novosti*, he believes that ISS provides the laboratory for long-duration, microgravity training, and the Moon for Martian environmental simulation studies. He envisions ISS as a spaceport for lunar-bound spacecraft (*http://rian.ru*).

10.2.4 Creating lunar social systems

Dr. Ben Finney, when a sociologist at the University of Hawaii, contributed to the NASA publication on *Space Resources*, previously cited in this book (SP.509, 1992). His theme was "Planning for Lunar Base Living", but his perspective was that of the *social* sciences and systems (i.e., creation of a *human* community on the Moon, see Chapter 3). Perhaps this quotation best summarizes his thesis:

> "But going back to the Moon presents a social, as well as a technical challenge. As Krafft Ehricke well recognized, in addition to developing low cost spaceflight, methods for processing lunar and other space materials to manufacture, safe and ecologically sound habitats, we also must develop systems of social organization for living in space. The experience of small, isolated groups in highly stressed environments points to the need for developing social systems that will enable people to live and work productively in space. The composition, organization and governance of the first lunar communities will be vital to their success, and ultimately to realizing the goal of living permanently in space. We need to start now on a research program directed to developing social systems designed so that people can live safely and productively on the Moon."

That is exactly what Dr. James Grier Miller proposed in applying living systems theory to humanity offworld (see Section 3.6). Finney, co-editor of *Interstellar Migration and the Human Experience* (cited in Chapter 2), offered five recommendations relative to establishing a lunar community of diverse spacefarers:

- use an integrative approach to living systems, one that is multidisciplinary and combines both biological and social science research;
- make this planning of an appropriate lunar social system part of a larger iterative program of learning how to live beyond Earth, whether in orbit, on the Moon or Mars, and other celestial bodies;
- conduct realistic simulations and experiments of space social systems before they are put into operation aloft;
- include self-designed plans by those who actually will have to live on the lunar surface and encourage them to be active participants in this R&D;
- facilitate planning for lunar community autonomy while, at the start, the lunar dwellers will be very dependent on earthkind for materials, supplies, and equipment, encourage local initiative, especially in the innovative utilization of lunar resources and creation of a culture that is appropriate to the environment and situation up on the Moon. (For instance, NASA has already designed software tools called *SpaceNet* for supplying and tracking needed inventory on the lunar

surface, so that the astronauts there will know when to request necessities for re-supply.)

Less and less control and monitoring of the explorers should be exercised, so they become more independent and responsible for their own wellbeing offworld. As their communities mature, grow in size and competence, spacekind should be encouraged to develop their own solutions and enterprises. One example is the matter of rules, regulations, and laws: the less such is imposed on these pioneers, the better they can formulate a governance system appropriate to their experience (e.g., that is how *astrolaw* will emerge). Actually, the goal is to cultivate interdependence between earthkind and spacekind.

10.3 LUNAR EXPLORATION AND SCIENCE

Human nature is to explore: our mammal ancestors began to do so between 100 and 85 million years back in time, long before the asteroid arrived some 65 million years ago. Explore means traveling to an unknown or an unfamiliar place or region for the purposes of discovery. Certainly, that definition fits the Apollo Moon missions, and what is being now undertaken to implement the VSE policy to return to the Moon permanently. In the past exploration was undertaken by adventurers, scientists, navigators, and even the military to establish jurisdiction over a territory. Lunar exploration, however, will be led first by engineers, scientists, and technologists until such time as all those "others" follow.

Given the circumstances of the Moon, lunar explorers will have to be knowledgeable, experts in several fields. Lunar exploration is planetary in scale, seeking information not only about our Solar System, but also the composition and history of our own Earth. Exploration on the Moon is a high-technology investment that should produce multiple benefits for humanity. It will affect the aspirations, education, and motivations of future generations of today's youth, ultimately it will involve everyone on the home planet. But the considerable expense and risks in lunar activities demand international cooperation on a level never achieved before in the human family.

The Moon is a natural laboratory of some 38 million square kilometers. Study of its geological processes will help us better understand both our Sun and Solar System, their evolution and that of our twin sister planet. As our closest and most reachable neighbor in the universe, the Moon is a test bed for learning new skills and developing new technologies, as indicated previously in Exhibits 127 and 129. It is an orbital platform for Earth observation, offering scientists an ideal location for projects in exobiology and radiation biology, lunar ecology and environment, and eventually human physiology and psychology in an isolated, confining environment. It is also a place for electromagnetic and ionizing radiation, an ideal location to investigate their biological importance in cosmic and solar radiation. The protection of lunar dwellers means inventing radiation monitoring, shielding, and solar-flare shelters, as well as health-monitoring systems. It also means dealing with moondust, composed of half silicon dioxide (rich glass bombarded by meteorites), nd the other half mainly iron, calcium, magnesium, olivine, and pyroxene). Apollo astronauts reported this fine dust covered their spacesuits and seemed to smell like "gun powder". Then there are moonquakes, possibly tidal in origin and some 700 km below the lunar surface; vibrations from meteorite impacts; thermal quakes from the expansion of the frigid lunar crust; shallow moonquakes up to 30 km below the lunar surface. Ah yes, the challenge of living and working on the Moon demands careful synchronization of scientific, technical, robotic, and human capabilities. The facilitating of synergy among lunar explorers may prove to be our hardest task to accomplish there! But it also an opportunity to live in a world of our own peaceful creation!

10.3.1 The Antarctica Model vs. terraforming

Presently, there is an international agreement for multinational bases on that continent for the purpose of scientific research [33]. Forty-five Antarctica signatories agreed to suspend territorial claims and disputes there, to forego all military and mining activity, to protect the environment, and to preserve the continent as a "natural reserve, devoted to peace and science". That experience does provide an analog for humans who will be living and working on the Moon in somewhat comparable circumstances [34]. Similarly, scientists from one country may wish to establish research outposts on the Moon as they do now in Antarctica. Currently at the latter's South Pole Station, the United States has built a new habitat with many amenities for 200 occupants. Only about 1,000 people live in this remote land year round, adding another 3,000 during the summer when the weather improves. But while some worthwhile scientific knowledge has been gained under that model, there has been no economic development on that remote, icy continent for the past 50 years! With climate change concerns, some 60 countries are finally planning to spend some $1.5 billion there, plus 10,000 researchers will visit during this International Polar Year of 2007. Space technology benefits Antarctica in some ways—besides mobile phones—such as GRACE, or the Gravity Recovery and Climate Experiment which measures minute changes in the Earth's gravity produced by the thickening and thinning of the ice sheets.. Yet, the Moon, like Antarctica, might hold the future to life on Earth and possibly beyond.

Humanity wants to do more than science and astronomy on the Moon: we want to industrialize and settle that planet, and use it as a launch pad into the universe. We hope to use the Moon as the first planet for terraforming [35]. As Martyn Fogg explained, our goal should be to engineer planetary environments for the better. This British author argues that alien worlds can be transformed by humans into human-habitable planets, like Earth. That is why Carl Sagan proposed to terraform the planet Venus. Thus, the Moon can become a terraforming laboratory for Mars and beyond. There is terraforming expertise to be acquired on and under the lunar surface, as we alter the Moon to suit living creatures, but with due regard to preservation of its environmental integrity. New technology applications will have to be created in terms of closed environments and life support, and to find ways of using local resources to provide sufficient water, power, and communications.

On the 50th anniversary of the first landing on the Moon by our species, 2019 might be an appropriate time to announce an *International Lunar Year!*

10.3.2 Science role in exploration

John Connolly of NASA's Johnson Space Center has said, "We are going back to the Moon to relearn the art of exploration." In that same article, NASA's Ames Research Center scientist Chris McKay gave his reasons for science to be involved in that process [36]:

- science is needed to provide data to make human exploration safer;
- the Vision for Space Exploration needs to be science driven and the science community should shape the lunar choices;
- before humans arrive back on the Moon, robotic missions should precede them, especially to collect scientific data and measurements are needed, especially about topography;
- robots can provide vital information about lunar hazards, ice, vacuum.

This distinguished planetary scientist seeks a well-connected, well-developed lunar science community that achieves some consensus now in setting forth the science agenda for the Moon. Others, such as engineers, might counter that their knowledge and skills should be given priority. In any event, to succeed within that huge lunar-learning laboratory, a multidisciplinary approach will be needed. It will be a combination of many fields of knowledge and skill that will enable lunar settlement by people. The Moon is more than a place for astrobiology, astrophysics, and astronomy! Assuredly, behavioral scientists and health care experts, as well as life science and habitation researchers will play a critical role in human survival. So will those innovators who develop instruments, like microdosimeters whose sensors measure radiation energy in individual blood cells, thus detecting harmful levels of radiation that may endanger human health. Scientists at the ASRC Aerospace Corporation in the Kennedy Space Center are working on an electromagnetic shield that would form a protective force field around a habitat. They propose 5-meter inflatable spheres made of strong fabric and coated with a conductor to repel positively and negatively charged ions contained in cosmic and solar radiation, especially dangerous during swift solar storms.

In August 2007, NASA announced the following science projects as part of its VSE plans ("Astrophysics: Hitching to the Moon, *The Economist*, August 11th, 2007, p. 73):

(1) Placing a *radio telescope* on the far side of the Moon to examine the early universe, including the study of large-scale structures, such as galaxies and stars. Such a lunar-based telescope could detect long wavelengths that cannot be observed on Earth because of atmospheric absorption. This device might prove

useful to detect extraterrtiral life, to map the stars and exoplanets that circle the stars, and to engage in studies that are almost unimaginable today. Joseph Lazio of the U.S. Office of Naval Research proposes an array of three telescopes, each 500 meters long, whose Y-shaped arms would be covered in plastic film and could be rolled out on the surface of the Moon

(2) Examination of *solar winds*, a stream of charged particles ejected from the Sun which interacts with the tenuous lunar atmosphere close to the Moon's surface. Headed by Michael Collier of the NASA Goddard Space Flights Center, the study will analyse the resulting bombardment, the low-energy X-rays it produces on the lunar surface.

(3) Wider dispersion of new, more sophisticated *lunar reflectors* beyond those dropped on the Moon previously by Apollo and the Russian Luna missions which are clustered around the lunar equator. Such reflectors are used in geophysics and geodesy (e.g., to research how the Moon's gravitational fields shift in time). Now Stephen Merkowitz of Goddard and Douglas Currie of the University of Maryland want to put improved reflectors spread over the Moon.

No wonder ILEWG has begun to award the "Young Explorers Prize" to innovative young people engaged in space science research (*www.esa.int* or *www.katysat.org/*)! And then there are those, like SETI and the National Institute for Discovery Science, who hope lunar exploration will advance the search for extraterrestrial life (*www.access.nv.com/nids* or email: *nidstaff@anv.net*).

10.4 LUNAR SETTLEMENT AND INDUSTRIALIZATION

Lunar enterprise will reach the Moon through contractors who will build the initial infrastructure there for science, settlement, and industrialization. Some of these "technauts" and their robots will be sent by big aerospace corporations, and others by start-up companies, like Bigelow Aerospace with their inflatable buildings. As a case in point, consider the matter of constructing a lunar base, such as the one depicted in Exhibit 125. This was a proposal of Lockheed Missiles & Space Company, Inc., in cooperation with Bechtel and Science Applications International Corporation. The artist's conception is based on assumptions that it will serve as a permanent center for scientific, industrial, and mining operations. Capable of expansion, the first stage begins as a lunar outpost with the following features:

- Living, working, and recreational facilities to support a crew ranging from 20 to 30 people.
- Greenhouses designed to recycle life support air and water supplies while supplementing food requirements.
- Shielded plant growth facilities to assure an adequate seedling population for the greenhouses if a crop fails.
- Utility workshops that enable technicians to repair and maintain equipment in a shirt-sleeve environment.

- Burial of the outpost or base in lunar soil for added protection against hazardous solar flare radiation.

This is an example of innovative, macroengineering planning for early 21st-century lunar development by three private corporations that are ahead of the curve. Now consider answers to these practical questions:

(a) Who would operate and pay for the lunar transportation system to get to and from the Moon?
(b) Who is the customer(s) and how is this huge enterprise going to be paid for?
(c) Under whose authority is this base to be built? That is, assuming the proponents could raise the money for this endeavor, how do private companies get international permission to use the Moon for this purpose?
(d) If and when such an outpost and/or base became functional, whether it is by private or public or combined initiative, who or what supervises or manages this operation?

In trying to answer these critical questions, remember that under existing space treaties, the U.S.A. and NASA on its own would not seem to have the power to authorize, or even to contract for such development. Right now, Article II of the U.N.'s Outer Space Treaty (1967) would seem to preclude it:

> "Outer space, including the Moon and other Celestial Bodies, is not subject to national appropriation by claim of sovereignty, by means of use or occupation, or by any other means."

Former astronaut and U.S. Senator Harrison H. Schmitt has written:

> "The mandate of an international regime would complicate private commercial development of lunar efforts. The Moon Treaty is not needed to further the development and use of lunar resources for the benefit of humankind—including the extraction of lunar helium-3 for terrestrial fusion power."

Now chairing NASA's Advisory Council, there would be many space lawyers who would challenge that scientist's interpretation, particularly with reference to the original Outer Space Treaty (see Section 9.4).

For decades, NASA has invested in numerous conferences on lunar activities and facilities, and published many documents on the subject with detailed plans. For example, a classic written in 1979 under the direction of Dr. Gerard K. O'Neill was *Space Resources and Space Settlements* (SP-428). Then in 1985, its Lunar and Planetary Institute released *Lunar Bases and Space Activities in the 21st Century* edited by Dr. W. W. Mendell. In 1988, the Office of Exploration sent another annual report to the NASA Administrator, *Beyond Earth's Boundaries: Human Exploration of the Solar System in the 21st Century*. Again in 1992, the Agency issued five volumes entitled *Space Resources* (SP-509) centered on strategic planning for a lunar base,

Exhibit 134. Lunar shelters. After humans return to the Moon permanently, the strategy is not only to build more than one base, but to have the lunar surface dotted with scientific instruments and facilities to carry out science and commerce. On its 38 million square kilometers, there will be small life support, communication, and equipment storage stations or refuges, possibly like the one depicted above. Source: NASA/John Frassanito Associates.

edited by Drs. Mary Fae and David S. McKay with Michael B. Duke. Then in 1996, NASA issued Technical Memorandum 4747 on *Lunar Limb Observatory: An Incremental Plan for Utilization, Exploration, and Settlement of the Moon* authored by Paul D. Lowman of its Goddard Space Flight Center. These are all remarkable and valuable documents representing extensive scholarly research by aeronautical experts. The last-mentioned provides unique insights on five stages of lunar exploration: (1) site selection and certification; (2) emplacement of a robotic lunar observatory; (3) opposite limb missions or seven automated launch and observation missions to various locations on the lunar surface; (4) lunar base establishment; (5) permanent human settlement on the Moon. Such lunar historical publications should be the basis of all current strategic planning about returning to the Moon in the next decade by both the public and private sectors. But how many of today's NASA engineers and scientists engaged in planning to implement the Vision of Space Exploration are even familiar with such research? Are we ignoring history as we reinvent the wheel?

Exhibit 135 displays one key ingredient of the transportation system: a lander with life support systems. Eventually, it is hoped that its propellant will be produced from lunar minerals. For every pound of propellant thus offloaded during descent, an equal amount of scientific and other equipment can be brought to the Moon's surface.

Exhibit 135. Possible lunar lander. At this writing, space agencies have not finalized the spacecraft that will be used to land humans permanently on the lunar surface by 2020. The indications are that this autonomous vehicle will have a habitation module, a cargo container, as well as a piloting area. One version might be like the one depicted above. Other issues to be resolved are whether the lander is to be able to return to lunar orbit for reuse, or whether expendable landers will be used for habitation. Source: NASA.

In January 2007, NASA held a press conference on its preliminary lunar base plans, so as to insure crew survival and reusability. The main strategy seemingly will be to pick a primary base site, and then plan for sending to the Moon both cargo and manned missions in the next decade. Lunar experts favoring the south pole location, Paul Lowman, David Schrunk, and others urge Malapert Mountain as this site over the alternative, Shackleton Crater. The exploration emphasis will be on crew safety, both in vehicles and backup spacecraft for rescue missions, if necessary. For astronauts returning to the lunar surface to stay in an environment without atmosphere, these are some of the detailed matters being worked on today by lunar strategists and designers:

- designing a less costly space transportation system of reusable spacecraft, capable of launching large payloads to the Moon and back on a regular schedule;
- building a lunar spaceport near the outpost site for landing and liftoff;

- sequence, type, and storage of cargo from an autonomous lander before humans actually arrive;
- unloading procedures and equipment (e.g., a crane) to remove cargo from the landers;
- excavation equipment to dig holes and cover the hab units with protective shielding;
- developing cryogenic propellant depots for the lunar surface, orbit, and LEO;
- designing lunar-based power and energy systems;
- designing various infrastructure for the Moon, possibly inflatables, such as storage warehouses;
- creating a lunar transportation system to move personnel, equipment, propellant, etc.

And this is just a sampling of the type of planning that has to go into a macroproject of this scope! No wonder one newspaper editorial called it "the costly frontier": yes, it will be expensive until the return on this huge investment begins to be realized (for more information, check out *www.lunarbase.rutgers.edu/index.php*).

Finally, there are many emerging technologies that could be employed in lunar development. For example, lunar planning would be advanced by use of virtual reality and augumented reality software, as well as by sharing computer resources and even unmanned aerial vehicles on the Moon.[3]

10.4.1 Lunar start-up enterprises

Back in Exhibits 131 and 132, we indicated some possibilities for lunar start-ups. First, it will be in building housing and facilities, as well as providing survival systems for the lunar pioneers. These will relate to radiation protection, lunar dust, safety and wellness, water and food supplies, waste disposal, and especially communication systems, both on the Moon and with Earth. Next the real industrialization processes will get under way: turning lunar soil with its oxides into oxygen and fuel; constructing transportation systems on the lunar surface; installing detection sensors against hazardous conditions, materials and objects, etc. In subsequent stages of lunar development, there will be a variety of undertakings, from scientific experiments and telescopes, to arranging for solar energy supplies, mining helium-3, and a host of necessary activities. The latter will include broadcasting from the Moon, lunar education and training, and eventually tourism and settlement services. Should the initial lunar outpost be founded at the lunar south pole to take advantage of its shadowed craters, almost continuous Sun, and view of the Sun, a second site might be developed at the lunar north pole for a continuous communications center and observatory. Someday these two locations may be connected by a lunar "railroad".

Beginning with the individual responsibilities of lunar project managers and contract supervisors, regulators and administrators will likely grow in number. So

---

[3] See *The Economist Technology Quarterly*, December 8, 2007, 31 pp. (*www.economic/reprints* or *course@fosterreprints.com*).

will a security system of peacekeepers until a full governance system can be instituted. Whether in ground support services or orbital endeavors, terrestrial systems will have to be adapted for budgeting, funding, accounting, reporting, information management, and technology transfer. This is what we discussed as *macroplanning and macromanagement* in Chapters 7 and 8.

10.4.2 Earth support enterprises

Obviously, spaceports will play a big role in lunar exploration, as Derek Webber's case study reminded us in Section 7.2. Whether on the ground or in orbit, they will be critical for more than maintaining and launching spacecraft. Spaceports will be where spacefarers are initially housed, fed, and processed before going offworld. An emerging market will be expertise in constructing lunar spaceports, beginning with the initial port of entry. Contractors will be required to build each aspect of the lunar transportation system, from the ground to the lunar surface and around the Moon itself.

In his book *From Footprints to Blueprints*, Michael Ross examines many dimensions of lunar development and private enterprise [37]. This Canadian engineer reminds us that all lunar activities will require support facilities and services on Earth, to a greater or lesser extent. Sometimes the stakeholders will have the necessary technical expertise within the sponsoring organization. But with the increased scope and complexity of lunar enterprise, outside support will become necessary. He forecasts these emerging Earth markets providing (1) operational performance on the terrestrial surface; (2) monitoring of in-flight performance; (3) research and testing laboratories; (4) technology transfer consultation; (4) lunar base simulations and mock-ups for testing layouts, ergonomics, and technical functioning; (5) facilities and services for lunar spacefarers, whether crew, workers, or settlers and eventually tourists (e.g., medical and psychological evaluation, etc.). To this end, the American Society of Civil Engineers has already recommended establishment of a Lunar Center for Extraterrestrial Engineering and Construction [38].

As always in any orbital undertaking, lunar space suits will have to be improved beyond those used in Apollo missions some 40 years ago. That complete garment with helmet, gloves, and boots weighed over 200 Earth pounds; today manufacturers can produce a suit of new, lightweight materials with improved life support and communication functions, as well as a radiation dosimeter, such as presently used for EVAs on the International Space Station. In addition, lunar surface vehicles and shelters will have to be designed, built, and shipped to the Moon with provisions for complete life support systems therein. Vendors will need to devise mechanisms for inventory control, and monitoring, both on the ground and aloft, for all such critical equipment.

Telepresence. One of the greatest opportunities for private enterprise is to create the necessary communication systems for people to use on the lunar surface, and between the Moon and Earth. The present mission control approach of space agencies will be obsolete by the time lunar operations are under way. Therefore, one promising possibility is the new technology available for telepresence.[4] To cut down on costly and sometime dangerous business travel, the old technology of videoconferencing has been updated and gone high-tech, with increased speed and quality of transmission, as well as improved, high-definition screens and reception. It creates the impression among participants that they are in the same room, though individuals or groups may be very far apart. It would be the perfect communication technology to reduce the distance in cislunar space! If the R&D were undertaken now between earthkind and spacefarers on the International Space Station, then it might be feasible to introduce a working system for the benefit of lunar workers and settlers by the end of the next decade. Such telepresence could not only improve business and management conferencing between Earth and the Moon, but benefit families and friends in both locations to maintain relations. It would be a counterpoint to the isolation and alienation that humans might experience on the Moon, so cut off from their fellows on the home planet. Presently, the leading manufacturers are Hewlett-Packard and Cisco who have yet to appreciate the market prospects for their technology and services offworld. Hollywood's Dream Works is in the emerging business, using telepresence systems to make movies cheaper and faster by allowing creative types worldwide to confer without actually traveling to meet one another. Now imagine if this technology could also be used to transmit to the Moon the latest movies, sports, or entertainment events! The market will be there to capitalize on in 20 years for the entrepreneurs who start now! Consulting firm Frost & Sullivan forecast that this innovative technology will develop into a $1.24 billion terrestrial market by 2013. Do some "mindstretching" about the market income prospects beyond Earth!

Schrunk *et al.* foresee modular assembly in low-Earth orbit.[4] This strategy would build initial operational capabilities for a lunar habitation base. Modular components would be brought to LEO for assembly in conjunction with EVAs from ISS.

10.4.3 Lunar base location and expansion

NASA, ESA, and other partners are conducting site searches for the best location of the initial outpost and/or base on the lunar surface. Some consensus is building for the south pole near Malapert Mountain, but site confirmation will come after automated mapping missions are completed. Some planners favor a lunar lava tube site which offers radiation protection and a constant temperature environment. There is much continuing, unresolved discussion among the experts on the size, design, and composition of the first lunar outpost. Obviously, much of this material would be prefabricated and have to be initially transported to the lunar surface for assemblage

---

[4] "Behold Telepresence: Far Away but Strangely Personal," *The Economist*, August 24th, 2007, pp. 57–58. Also see Appendix R, "MALEO," and Appendix T "Beyond our first moonbase," in Schrunk, D. G.; Sharpe, B. L.; Cooper, B. L.; and Thangavelu, M. *The Moon: Resources, Future Development, and Settlement*, 2008, Second Edition, pp. 457–475; 497–504 (*www.praxis-publishing.co.uk*).

by both robots and humans. Whatever the composition of these first structures, they will be buried largely underground and covered with lunar regolith for radiation protection. Once we move beyond space agency needs for a small crew of astronauts, the base will unfold as more facilities are provided for scientific research, health care, and industrial operations. When the lunar population expands and commerce is under way, undoubtedly there will be satellite structures erected away from that base at other locations on the Moon.

It was encouraging in 2005 when several Italian companies prominent in the space business sponsored an international conference in Venice on "Moon Base: A Challenge for Humanity" (*www.moonbase-italia.org*). But until NASA and its partners issue a lunar architecture report possibly before 2010, we can only speculate as to how this generation will proceed with its return and settlement of the Moon. However, the scientific and engineering communities have provided the public with what to expect, as our many chapter references confirm.

We can anticipate a *lunar scenario* somewhat like this [39]:

- initially, there will be a series of unmanned missions for the purpose of transporting robots, cargo, and shelter payloads to the Moon, plus telerobotic lunar rovers capable of being driven from Earth or programmed for preliminary lunar construction tasks, such as erection of telescopes and an observatory;
- the first manned lunar landing since Apollo 17 will initially focus on astronaut deployment and completion of own outpost shelter; establishment of life support, communications, and transportation systems; inspection of previously placed telescopes; returning robotic collections of further lunar soil and rocks to Earth for additional analysis (petrographic, chemical, and radiometric); on-site human inspections and evaluations; further emplacement and maintenance of geophysical and astronomical instruments, plus similar precursory activities toward base expansion;
- technology research and development by technauts trained to begin the utilization of lunar resources, from oxygen and water (H-E extraction), to solar energy and mining prospects (helium-3), to broadcasting special programs to Earth's inhabitants, to lunar farming and experiments on a "critter colony", etc.;
- expansion of the lunar population beyond contractors and technauts, to tourists and recruiting settlers live on the Moon for a year or more will involve pre-departure and onsite training of families, and providing sufficient support services while aloft and on return to the home planet (see Chapter 6.)

H. H. Koelle, a respected German researcher at the Technical University of Berlin, published "Steps toward a Lunar Settlement" [40] on the subject. As chairman of a lunar development subcommittee for the International Academy of Astronautics, he had access to some of the best global thinking on the topic as he built a lunar data bank, so his insights are significant. After presenting a compelling rationale and objectives, he examined three options for lunar transportation, recommending a single-stage-to-orbit (SSTO) tanker and a ferry from LEO to a spaceport on the Moon. Dr. Koelle favors beginning with a lunar laboratory which over 15 years would be developed into an outpost and refueling center. As this was enlarged into a permanent base and settlement, he envisions leasing space to commercial interests who would pursue development of lunar resources, from liquid oxygen and hydrogen to solar energy and astronomy. In the next 50 years, he forecasts a lunar population of some 2,000 and a settlement that is 75% self-sufficient by year 2100.

As lunar dwellers and their infrastructures increase, our dispersal beyond Earth into our Solar System will gradually lead to the creation of a spacefaring civilization!

10.5 CONCLUSIONS ON LUNAR DEVELOPMENTS

Flush with excitement over initial Space Age successes, Patrick Moore, British broadcaster and author, wrote a book in 1975 about *The Next Fifty Years in Space*. One chapter was devoted to "The Lunar Base", which he then forecast would become operational between the years 1995 and 2000. With hindsight, one might ask why the U.S.A. did not follow up with plans for such an undertaking after the Apollo mission series ended? There were many political, economic, and social factors that prevented the nation from achieving such a vision by the end of the 20th century. Now space experts from NASA and the aerospace industry estimate that a lunar outpost might be functional by year 2020. From my viewpoint both as a management consultant and as a space psychologist, it would seem that *mindset* may prove to be the stumbling block, delaying and hindering the exploitation of space resources in general, and lunar resources in particular. To cultivate positive mindsets in both the world's political and space communities, as well as mass public support, here are some suggested activities to pursue:

- **Place individual space missions in a larger context:** to think in terms of not just individual missions, but within the context of a broader strategic plan for lunar science and industrialization; emphasizing to the public near-term returns on investment.
- **Act synergistically in global cooperation:** today no one country or space agency can effectively undertake space macroprojects without forming international partnerships; the sharing of talent and resources ensures long-term "paybacks" from space enterprises.
- **Facilitate intersectoral and interdisciplinary planning:** space macroprojects require going beyond traditional interfaces, such as public and private, science and business; investment will be forthcoming, for example, if goals include both science and industrialization, involving industry, universities, and entrepreneurs with government agencies in planning for a permanent lunar return.

The above review of current trends underscores why the next major space investment and undertaking by the U.S.A. and its spacefaring partners should be the Moon, underscroing reimbursement to be realized for all of humanity, especially the home planet itself. Many "Mission to Planet Earth" goals can be achieved best by using the Moon as a platform for scientific, environmental, and economic advantage. The Moon can become the laboratory for international cooperation, the launching pad of humanity into the universe. Expert consensus for this is emerging as confirmed in this statement from an International Academy of Astronautics report [41]:

> "We believe the time has come that these global trends should induce responsible governments to take action deciding to continue the development of lunar resources and consequently to assign an existing multinational space organization (or establish a new one) the responsibility of returning people to the Moon permanently and developing its resources for the benefit of mankind."

The author concurs in these recommendations, but is convinced that a "new multinational space organization" or regime must be formed to sustain joint lunar exploration, one that goes beyond—yet collaborates—with existing national space agencies. To this end, we have proposed here the creation of a Lunar Economic Development Authority, and have explored several alternatives for accomplishing this goal. LEDA is considered a prototype for future space authorities that could be constituted to develop eventually other planets and asteroids in our Solar System, as well as for constructing orbiting colonies as proposed by visionaries, such as the late Wernher von Braun and Gerard K. O'Neill [42]. This strategy offers a bridge over troubled waters in contemporary space policy and law, and a mechanism for constructing and really financing lunar infrastructure through public participation. This legal entity with its own Board of Directors would enable national sovereignties to act synergistically in the exploration and development of the high frontier. Not only could it issue revenue bonds for this purpose, but such a lunar development entity could literally build a "bridge between the two worlds of the Earth/Moon system" by:

- leasing land, facility, and equipment rights;
- fundraising and fee collection from investors and developers;
- site planning and permits for habitats and industrial parks;
- zoning and inspection to protect lunar environmental and ecological concerns;
- long-term management and peace-keeping within lunar settlements;
- administering a lunar personnel deployment system, while regulating tourism.

Such practical matters have already been researched by Charles Lauer when at the University of Michigan College of Architecture with reference to real estate aloft. He has written extensively on the financial, legal, regulatory, and design aspects of business parks in orbit that have implications for a lunar industrial park [43]. Now he is a founder of Rocketplane, Inc. (see Exhibit 151).

Yet another strategy for moving toward industrialization of the Moon would be for the U.S. Administration and Congress to convoke a *White House Conference on Lunar Enterprise* that would build appropriate consensus, as well as lunar policy and strategic planning (see the Epilogue). Another action would be for the United Nations to sponsor a global summit for the world community to consider a lunar

Sec. 10.5] **Conclusions on lunar developments** 517

development agenda, possibly under the aegis of the U.N. Office for Outer Space Affairs. Like previous convention sessions, this might be called *World Space Congress 2010*. If such official leadership is not forthcoming, then transnational enterprise will act to fill the vacuum and promote space resource development and commercialization [44]. Since the beginning of the Space Age, the military of several countries have lusted for bases on the high ground, especially the Moon. Better to delimit their role to helping with space transportation, construction, and peacekeeping!

The prophetic Krafft Ehricke envisioned lunar industrializaton as our *extraterrestrial imperative*, warning of the consequences of growth or no-growth policies (described in Exhibit 136).

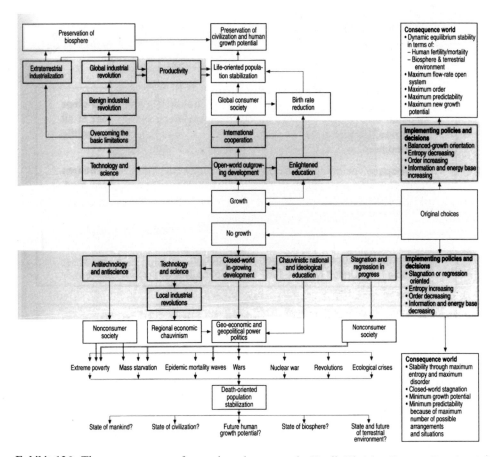

Exhibit 136. The consequences of growth and no growth: Krafft Ehricke. Source: Reprinted with permission of Marsha Freeman from her "A Memoir of Krafft Ehricke's Extraterrestrial Imperative," *21st Century Science and Technology*, **7**(4), Winter 1994/1995, 34 (PO Box 16285, Washington, D.C. 20041).

Exhibit 137. Future possibilities. Humans offworld will engage in a variety of activities, some traditional and some never done before. This artist's concept depicts a possible scene as humans begin exploration of the surface of Mars. The scene is just after sunrise, above the Valles Marineris system of enormous canyons. On the canyon floor six kilometers below, early morning clouds can be seen. The astronaut depicted on the left might be a planetary geologist seeking to get a closer look at the stratigraphic details of the canyon walls. On the right, the geologist's companion is setting up a weather station to monitor Martian climatology. In the far right is a six wheeled articulated rover, which transported the pair of astronauts here from their landing site. Source: NASA.

Obviously, pursuing lunar development is a growth policy for our world. Right now within the human family, there is much chaos and international destruction. How much better it would be for humanity to turn people outward and upward toward the stars in space exploration and development! It might pull the international community together in common cause, and give youngsters hope for offworld participation! The high frontier opens up all kinds of future possibilities, as Exhibit 137 illustrates.

10.6 REFERENCES

[1] Schrunk, D.; Sharpe, B.; Cooper, B.; and Thangavelu, M. *The Moon: Resources, Future Development, and Settlement*. Chichester, U.K.: Springer/Praxis, 2008, Second Edition (*www.praxis-publishing.co.uk*).

[2] In addition to the above book, we recommend: Harrison M. Schmitt's *Return to the Moon*. New York: Springer/Copernicus/Praxis, 2008 ... In 1984, a NASA study at the California Space Institute took place on the theme, "Strategic Planning for a Lunar Base"; edited by Drs. Mary Fay and David McKay with Michael Duke, the proceedings were published in 1992 under the title, *Space Resources*, by the U.S. Government Printing Office, Washington, D.C. These four volumes of NASA SP-509 are still relevant and available from *www.univelt.com* (see especially Vol. 4 on *Social Concerns* in which the contribution of your author appears) ... For updates on lunar development, subscribe to *Lunar Enterprise Daily*, Space Age Publishing (*www.spaceagepub.com* or email: *news@spaceagepub.com*). Also see *www.lunarlibrary.com* developed by Ken Murphy, who graduated from the International Space University, plus the blog site, *Out of the Cradle* which includes Murphy's lunar bibliography.

[3] Science & Technology, "Spaceflight: Objective Moon," *The Economist*, May 28th, 2005, pp. 79–80 (*www.economist.com*) ... "The Fans of Mars," *The Economist*, March 11th, 2000, pp. 85–86.

[4] Spudis, P. D. *The Once and Future Moon*. Washington, D.C.: Smithsonian Institution Press, 1996 ... See "The New Race to the Moon," *Science*, **300**, May 2, 2003 (*www.sciencemag.org*) and the "blog" discussion on "To the Moon, Mars, and Beyond," March 4, 2007 (*www.kelloggserialreports.blogspot.com/*).

[5] *Ad Astra*, **10**(1), Spring 2007 (the magazine of the National Space Society, 1620 "I" St. NW, Ste. 615, Washington, D.C. 20006; *www.nss.org*). See articles by Jeff Foust, "Moon Base: The Next Step in the Exploration of the Solar System," pp. 28–31; Jeanna Bryner, "Lunar Observatories: Grand Plans vs. Clear Problems," pp. 32–35; Andrew Chatiklin, "Building the Future," pp. 36–37 ... In *Ad Astra*, November/December 2001, Philip Harris, "Time for a Return on our Apollo Investment," pp. 31–32 (*www.drphilipharris.com*).

[6] "The Moon: A New Frontier," *New York Times* Special Supplement, August 3, 1969, 20 pp. (*www.thenewyorktimes.com*) ... For an update, see Johnson-Freese, J.; and Handberg, R. *Space, the Dormant Frontier: Changing the Paradigm for the 21st Century*. Westport, CT: Greenwood Publishing Group, 1997.

[7] Aldrin, E. (Buzz) *Reaching for the Moon*. New York: Harper Collins, 2005 ... with John Barnes, *The Return*. New York: Forge Books, 2000 ... Also see Tom Hanks' IMAX/DVD, *From Earth to the Moon*, or Walt Disney Television's two-hour movie, *Plymouth*, about the first lunar colony (tel.: 818/5608844 or fax.: 818/566-6566) ... Websites: check out online *www.moonsociety.org* or *www.TheMoonPeople.org* or *www.moonbasesanalog.net* ... Then download "pdf" files on your computer from *www.adobe.com/products/acrobat/readstep2.html* ... Inquire about attending the annual International Lunar Conference at *www.spaceagepub.com* or the yearly International Space Development

Conference at *www.isdc2007.org* (subsequently change the year in this website for future sessions)! ... The following open online wiki-type encyclopedia is worth a visit: *www.Lunapedia.org* ... If you want to listen to a Real Audio radio show like *Space Beat*, contact *http://expert. ccpurdue.edu/-Jstudy/*

[8] Hansen, J. *First Man: The Life of Neil A. Armstrong.* New York: Simon & Schuster, 2006 ... See also Pyle, R. *Destination Moon: Apollo Missions in the Astronaut's Own Words.* Washington, D.C.: Smithsonian Books/Collins, 2005.

[9] Schmitt, H. H. *Return to the Moon: Exploration, Enterprise and Energy in Human Settlement of Space.* New York: Springer/Copernicus/Praxis, 2005 ... "My Seven: Why We Must Go Back to the Moon," *National Geographic*, July 2004 (*www.nationalgeographic.com/magazine/0407*) ... Cernan. D.; and Davis. T. L. *The Last Man on the Moon: Eugene Cernan and America's Race into Space.* New York: St. Martin's Press, 1999.

[10] Durst, S. M.; Bohannan, C. T.; Thomason, C. G.; Cerney, M. R., and Yuen, L. (eds.) *Proceedings of the International Lunar Conference 2003/International Lunar Exploration Group 5.* San Diego, CA: UNIVELT/AAS, 2004, Vol. 108 in "Science and Technology Series" with CD-ROM supplement.

[11] Galloway, E. "Status of the Moon Treaty," *Space News*, August 3–9, 1998, pp. 21–22. See also Glen Reynold's "Return of the Moon Treaty," *Ad Astra*, May/June 1994, pp. 27–28 (*www.nss.org*).

[12] Criswell, D. R. "No Growth in the Two Planet Economy," International Astronautical Federation Proceedings (8.1.704) at the 45th Congress, Jerusalem, Israel (October 9, 1994).

[13] Ockels, W. J. *A Moon Programme: The European View.* Noordwijk, the Netherlands: European Space Agency, May 1994 (ESA BR-101), 14 pp ... ESA Lunar Study Steering Group Report, *Mission to the Moon* (ESA SP-1150, 1992) ... ESA Press Release on "International Lunar Workshop Declaration" (NR-12-94, June 6, 1994).

[14] Harvey, B. *Russia in Space: The Failed Frontier?* ... Hall, R.; and Shayler, D. J. *The Rocket Men.* Chichester, U.K.: Springer-Praxis, 1997; 2001.

[15] Reynolds, D. W. *Apollo, The Epic Journey to the Moon.* New York: Harcourt, Inc./Tehabi Books, 2002 ... Light, M. *Full Moon.* New York: Alfred A. Knopf, 2002 ... Krauthammer, C. "Clinton is Cutting NASA to Pieces," Washington Post Writers Group in the *San Diego Union-Tribune*, April 9, 1995, p. G-5.

[16] Spudis, P.; and Bussey, B. *The Clementine Atlas of the Moon.* New York: Cambridge University Press, 2005 (*www.spudislunarresources.com*).

[17] Godwin, R. (ed.) *Surveyor Lunar Exploration Program.* Burlington, Ontario: Apogee Books, 2006 ... Binder, A. B. *Lunar Prospector: Against All Odds.* New York: Ken Press, 2005.

[18] Spudis, P. A. "A Moon Full of Opportunity," *www.thespacereview.com* January 27, 2007, 23 pp. (article #791/1) ... Berger, B., "The Vision at Three Years and Counting," *Space News*, March 12, 2007, 4 pp. ... Britt, R. B. "Commentary: The Reasonable Cost of Putting Humans on the Moon and Mars," *www.space.com/news* January 14, 2004 ... Malik, T. "For the President's Moon-to-Mars Commission" *www.space.com.news*, May 11, 1994, 3 pp.

[19] "Spaceflight: Objective Moon," May 28th, 2005 , pp. 79–80; "The Fans of Mars," March 11, 2000, pp. 85–86, *The Economist* (*www.economist,com*) ... Cowing, K.; and Sietzen, F., *New Moon Rising.* Burlington, Ontario: CGPublishing/Apogee Books, 2004 ... Belbrano, E. *Fly Me to the Moon.* Princeton, NJ: Princeton University Press, 2007.

[20] Blaise, W. P.; and Stroke, P. "The First Commercial Lunar Mission," *Ad Astra*, January–February 2003, 38–42 ... Blaise, W. P.; and Radley, C. F. "The Trailblazer Class of Low

Cost Space Vehicles," 1st Responsive Space Conference, Redondo Beach, CA (AIAA LA Section/SSTC), April 1–3, 2003, 16 pp. (*www.aiaa.org* or *www.transorbital.net*).

[21] Shapiro, R. *Planetary Dreams: The Quest to Discover Life beyond Earth*. New York: John Wiley, 1999 ... Burrows, W. *The Survival Imperative: Using Space to Protect Earth*. New York: Forte/Tom Doherty, LLC, 2006.

[22] Dror, Y. "Governance for a Human Future in Space," in Krone, B.; Morris, L.; and Cox, K. (eds.) *Beyond Earth: The Future of Humans in Space*. Burlington, Ontario: Apogee Space Books, 2006, pp. 41–45. In the same volume, also see S. E. Bell and D. L. Strongin's "Evolutionary Psychology and Its Implications for Humans in Space," pp. 78–83 ... P. J. Werbos, "Strategic Thinking for Space Settlements," p. 23 ... T. L. Matula and K. A. Loveland "Lunar Commercial Development for Space Exploration," p. 29 ... F. Hsu and R. Duffey, "Managing Risks on the Space Frontier," p. 30.

[23] Sadeh, E.; Benaroya, H.; Livingston, D.; and Matula, T. "Private–Public Models for Lunar Development," in Durst, S. *et al.* (eds.) *Proceedings of the International Lunar Conference, 2003/International Lunar Exploration Working Group 5*. San Diego, CA: UNIVELT/AAS, Vol. 108, 2004, pp. 349–357. In this same volume, see Magelsse, E. C.; and Sadeh, E. "Political Feasibility of Lunar Base Mission Scenarios," pp. 359–369 ... H. L. Starlife's "Property Rights in Space," pp. 407–411.

[24] Ehricke, E. A. "The Anthology of Astronautics," *Astronautica*, **2**(4), November 1957 ... "Lunar Industrialization and Settlement: Birth of a Polyglobal Civilization" in Mendell, W. W. (ed.) *Lunar Bases and Space Activities of the 21st Century*. Houston, TX: Lunar and Planetary Institute, pp. 731–756 ... Bova, B. *Welcome to Moonbase*. New York: Ballantine Books, 1987 ... Harris, P. R. *Launch Out: A Science-Based Novel about Space Enterprise, Lunar Industrialization and Settlement*. West Conshohocken, PA: Infinity Publishing, 2003 (*www.buybooksontheweb.com*).

[25] O'Donnell, D. J.; and Harris, P. R. "Legal Strategies for a Lunar Economic Development Authority," *Annals of Air and Space Law*, 1996, 121–139 (published by the Institute and Center of Air & Space Law, McGill University; *www.islaw.ca*) ... Harris, P. R.; and O'Donnell, D. J. "Creating New Social Institutions to Develop Business from Space," *48th International Astronautical Congress Proceedings, Turin, Italy, October 5–10, 1997*, #IAA-97-IAA.3.1.01 (published by the International Aeronautical Federation, 3–5 rue Mario-Nikis, 75015, Paris) ... O'Donnell, D. J. "International Space Development Corporation," *Space Governance*, **9**, 2003, 16–18 (published by United Societies in Space, Inc., 499 South Larkspur Dr., Castlerock, CO 80104).

[26] Benaroya, H. "Lunar Industrialization," *Journal of Practical Applications in Space*, **VI**(1), Fall 1994, 85–94 ... Petek, L. "Management in Outer Space," *Space Policy*, **10**(3), August 1994, 189–198.

[27] Goldman, N. C. "A Lawyer's Perspective on the USIS Strategies for Metanation and the Lunar Economic Development Authority," *Space Governance*, **3**(1), July 1996, 16–17, 34 ... Smith, M. L. "The Compliance with International Space Law of the LEDA Proposal," *Space Governance*, **4**(1), January 1997, 16–19.

[28] Robinson, G. S. "Natural Law and a Declaration of Humankind Interdependence," *Space Governance*, **2**(1/2), 1995, June and December issues.

[29] O'Donnell, D. J. "Overcoming Barriers to Space Travel," *Space Policy*, **10**(4), November 1994, 252 ... Harris, P. R. "A Case for Lunar Development and Investment," *Space Policy*, **10**(4), August 1994, 187–188.

[30] The Lunar Economic Development Authority strategy was inaugurated by Attorney Declan O'Donnell and Management/Space Psychologist Dr. Philip Harris, and in 1997 was incorporated in the State of Colorado. In 1998, Dr. Michael Duke was elected LEDA

president with Dr. Brad Blair as Chairman of the Board; both are now at the Colorado School of Mines ... In 2000, LEDA issued a private memorandum relative to "LEDA Development Bonds". In this document, the reason for issuing the bonds was explained as a movement to maintain a private enterprise paradigm in both space governance and lunar development. For further information, contact Attorney O'Donnell (email *djops@quest.net*).

[31] Cleveland, H. *Birth of a New World Order: An Open Moment for International Leadership.* San Francisco, CA: Jossey-Bass Publishing, 1993.

[32] Whittington, M. "Lunar Exploration and Development Authority (LEDA) Would Help Open Space Frontier," *Washington Dispatch* article summarized in *Lunar Enterprise Daily*, **5**(59), March 30, 2005, p. 1.

[33] "Briefing Antarctic Science: To Coldly Go," *The Economist*, March 31st, 2007, pp. 85–87.

[34] Harrison, A. A.; Clearwater, Y. A.; and McKay, C. P. (eds.) *From Antarctica to Outer Space.* New York: Springer-Verlag, 1991.

[35] Fogg, M. J. *Terraforming: Engineering Planetary Environments.* New York: SAE Press. 1995.

[36] David, L. "Back to the Moon: Uniting Science and Exploration," *www.space.com/scienceastronomy/060516*, May 16, 2007, 5 pp ... "Moon-Based Instruments Could Change How We See Earth," *www.space.com/businesstechnology/070116*

[37] Ross, M. *From Footprints to Blueprints: Development of the Moon, and Private Enterprise.* Bloomington, IN: AuthorHouse, 2005 (*www.AuthorHouse.com*).

[38] Hart, P. A. et al. (eds.) *A Center for Extraterrestrial Engineering and Construction (CETEC)* (proceedings of Space '90 Conference). Reston, VA: American Society of Civil Engineers, 1990, p. 1198 ... Also see from same proceedings and source, W. R. Sharp's article on "Mining and Excavating Systems for a Lunar Environment," p. 294 ... plus Daga, A. W.; and Daga, M. A. "A Preliminary Assessment of the Potential Lava Tube-Situated Lunar Base Architecture," p. 568.

[39] Lowdon, P. D. *Lunar Limb Observatory: An Incremental Plan for Utilization, Exploration, and Settlement of the Moon.* Greenbelt, MD: NASA Goddard Space Flight Center, 1996, Technical Memorandum 4757 ... Alfred, J.; Bufkin, A. et al., *Lunar Outpost*. Houston, TX: Johnson Space Center, 1989, 60 pp. ... Bialla, P. H. (ed.) *Low-cost Lunar Access.* Arlington, VA: American Institute of Aeronautics and Astronautics, 1993 ... Robotics Insitute, *Entertainment Based Lunar Rover Concept*. Pittsburgh, PA: Carnegie-Mellon University, 1994, 148 pp. ... Heiken, G. D.; Vaniman, D. T.; and French, B. M. *Lunar Sourcebook*. Cambridge, U.K.: Cambridge University Press, 1991, 736 pp. ... Wendell, W. W. (ed.) *The Second Conference on Lunar Bases and Space Activities in the 21st Century.* Washington, D.C.: NASA Conference Publication 3166, 1992.

[40] Koelle, H. H. "Steps toward a Lunar Settlement," *Space Governance*, **4**(1), January 1997, 20–25 ... Koelle, H. H. (ed.), *Handbook of Aeronautical Engineering*. New York: McGraw-Hill, 1961.

[41] Koelle, H. H. (ed.) *Recommended Lunar Development Scenario.* Paris: International Academy of Astronautics Subcommittee on Lunar Development, January 1995 ... See also Schmitt, H. H. "The Case for Establishing a Human Presence on the Moon," *Space Governance*, **4**(2), July 1997, 118–131 ... "Return to the Moon and Stay," *Ad Astra*, **7**(3), May/June 1995 (whole issue of the National Space Society devoted to this theme), and again "Reclaiming the Moon: The First Steps," *Ad Astra*, **19**(1), Spring 2007 (*www.nss.org/adastra*).

[42] Neufeld, M. J. *Von Braun: Dreamer of Space, Engineer of War.* New York: Alfred A. Knopf, 2007 ... Stuhlinger, E.; and Ordway, F. L. *Wernher von Braun: Crusader for Space.*

Malabar, FL: Krieger Publishing, 1994, Vols. 1/2 ... O'Neill, G. K. *The High Frontier: Human Colonies in Space*. Princeton, NJ: Space Studies Institute, 1989.

[43] Lauer, C. et al. "A Reference Design for a Near Term Lower Earth Orbit Commercial Business Park," *Legal & Regulatory Aspects of Lower Earth Orbit Business Park Development*. Reston, VA: American Institute of Aeronautics and Astronautics, 1995 ... *Briefing Book: Lower Earth Orbit Business Park* (available from Peregrine Properties, 540 Avis Dr., Ste. E, Ann Arbor, MI 48108). [Author note: such a strategy can also be applied to a lunar business park.]

[44] Howerton, B. A. *Free Space: Real Alternatives for Reaching Outer Space*. Port Townsend, WA: Loompancis Unlimited (or Box 1197, zip code 98368, USA, or contact author at Public Information Office: *www.nascar.org*).

SPACE ENTERPRISE UPDATES

The Google Lunar X-Prize (GLX-Prize): in both aviation and space technologies, money prizes have been offered to encourage innovative enterprises and achievements. The original Ansari X-Prize Foundation challenge developed by Dr. Peter Diamandis to promote suborbital flights was won by Burt Rutan of Scaled Composites in 2004. His SpaceShipOne carried by WhiteKnightOne was the first privately funded manned trip into space. Then Robert Bigelow offered the O-Prize, or American Space Prize, for the first private firm to do two orbits by January 10, 2010. Our contributor, Dr. Tom Matula refers us to the website: *www.bigelow aerospace.com/multiverse/space_prize.php* ... The GLX-Prize is a total $30 million, international competition to safely land a robot on the Moon, then travel over 500 meters over the lunar surface, while transmitting a gigabyte of data back to Earth and carrying a payload of 500 grams. But entrants must succeed by December 31, 2012, or the first prize drops to $15 million and second place is $5 million. If these deadlines are not met and the landing is attained by December 31, 2014, then a bonus prize of an additional $5 million will be awarded. By July 7, 2008, nine teams had entered the competition (*www.XprizeFoundation.org*).

Dennis Laurie, CEO of Transorbital, the forerunner of a comparable lunar mission, observes that the original X-Prize did not cover the cost of the project, but implied prospects for new suborbital business, possibly as much as a billion dollars of downline income within a few years. However, with the GLX-Prize competition, Laurie estimates the whole enterprise, if successful, would cost the entrant $60 million plus without any discernible lunar business opportunity. So he questions whether this endeavor will attract enough entrepreneurs when initially most business activity on the Moon will be government-dominated, and any private ventures would have to be coordinated with national space agencies. However, this experienced executive does see a marketing possibility in interactive telvision via the lunar landers and rovers. He expects that Google is hoping in this way to generate future revenues for itself and any production company involved. Their objective, as well as the X-Prize Foundation, is to inspire space entrepreneurs, and educate the public by their team efforts. Eventually, the venture might build an international audience and attract potential customers.

Exhibit 138. Space light: the Moon as seen from Earth beckons us!. *Spatium Lux* is an original painting by the astronomical artist, Dennis M. Davidson. His art is available from Novaspace Galleries (*www.novaspace.com*).

Epilogue

The opening of a new, high frontier will challenge the best in us. The new lands waiting to be built in space will give us new freedom to search for better governments, social systems, and ways of life, so that our efforts during the decades ahead enable our children to find a world richer in opportunity.

Dr. Gerard K. O'Neill, author of
The High Frontier: Human Colonies in Space, Princeton, NJ

The motive for sending humans into space comes down to rediscovering the importance of realizing our potential as a people ... It has been said that three main motives for going offworld are national security, economics, and glory. But human potential in space means developing capabilities not yet in existence. Everything we have today stems from the above three benefits in the form of knowledge, discoveries, inventions, and human creations of every kind which were once not in existence ... Exploring space enables us to achieve our potential, to advance society, attain goals that improve our chance for survival and growth.

Frank Stratford, Founder of MarsDrive, Melbourne, Australia[1]

TRANSFORMING SPACE VISIONS INTO REALITIES

Human dreams and ideas span time, often taking centuries before being transformed into worthwhile activities. Some of our forebears dimly perceived the spectacular

---

[1] *The High Frontier* is available from *www.amazon.com/books* or the Space Studies Institute, PO Box 82, Princeton, NJ 08542 ... The Frank Stratford quotation was adapted from "Our Potential in Space," *The Space Review*, October 9, 2007 (*www.thespacereview.com/article/977/1*).

achievements which this generation has witnessed since the dawn of the Space Age. Hopefully this book, *Space Enterprise*, has provided readers with insights into the challenges which lie ahead in exploring and settling offworld. That process will herald a higher state of consciousness for our species.

In the future, our descendants may remember the 21th century primarily for proving that humanity is not Earth-bound and is able to live and work in a microgravity environment. The last five decades may be viewed as a watershed period for commercial space and living aloft. It was a period when nations shifted from space competition to cooperation, from a space race to forming joint ventures for international macroprojects. The satellite industry not only turned our world into a *global village* by its communication capabilities, but demonstrated that it could be a profitable enterprise. Furthermore, orbital imaging and sensing has shown myriad practical applications on Earth, even in protecting our planet's environment. The Russian space station Mir became a platform for true international cooperation by agreements which brought aboard Europeans, Japanese, and even Americans. Today, the International Space Station expands the opportunities for some 16 national partners to practice *synergy*. Now spacefaring nations have much to gain in forming partnerships in lunar missions, particularly toward the goal of returning humans to the Moon permanently by year 2020. Perhaps this Vision for Space Exploration was best expressed over 20 years ago in these prophetic words:

> "To lead the exploration and development of the space frontier, advancing science, technology, and enterprise, by building institutions and systems that make accessible vast new resources and support human settlements beyond Earth orbit, from the highlands of the Moon to the plains of Mars" (*Pioneering the Space Frontier*, 1986, p. 2).

To actually implement such lofty goals requires global *transformational leadership* in both the public and the private sector now and in the centuries ahead [1]. The business community at large, not just aerospace and communication satellite companies, must lead in the creation of a space ethos that supports an enlarged and well-funded space program, both in the public and private sector. Yes, space is a place for fulfilling dreams, as well as for acquiring knowledge and promoting free enterprise!

But how? Specifically, as a case in point, how can America further capitalize on its $20 billion investment in the Apollo lunar landings? How can all nations get payback on their total space expenditures, especially through the *utilization of space-based resources*? Some innovative answers may be gleaned from the reports and recommendations of various space studies previously cited in the ten chapters of our book. Apart from technical and economic insights, especially for the establishment of a lunar base, these studies include proposals for

- building public consensus and financial support for the space program;
- initiatives within the private sector to foster the peaceful use of space by its exploration and industrialization;

- legislation that would transform nations' space agencies, as well as their policies and procedures so as to facilitate private space enterprise;
- promotion of educational and research endeavors that prepare the next generation of spacefarers for offworld challenges!

At this juncture, the justification for peaceful and commercial development of space resources is more human and scientific, than economic or political. The rationale for moving forward on the space frontier has to do with discoveries which maintain technological excellence, security, and leadership in a knowledge culture [2]. Space undertakings should aim at benefiting the Earth's peoples, especially in the Third World, by technology transfer within the twin planet economies of Earth–space. Our aspirations should be to actualize our potential by extending human presence permanently into our universe. One proposal from Kim Peart in Tasmania is worthy of implementation: namely, the formation of a *Solar Peace Corps* to take a proactive role to ensure peace and security within our Solar System, especially through utilization of the Sun's energy and Solar System resources. The aim is to connect Earth's children to the wealth of a solar economy.

For those readers who internalize the principal message of this book, here are three dimensions of a personal action plan to participate toward creation of a *spacefaring civilization*:

1. National, regional, and global convocations on space enterprises

Individuals and organizations can raise the public's awareness by sponsoring space enterprise conferences at both the local and world levels. Although this can be accomplished in actual group meetings, the best prospects for raising public consciousness on the necessity of space exploration and development may be the Internet and international television. Think back to the global media encounters sponsored by rock stars, environmentalists, and others with a humanitarian cause. Suppose supporters were to promote a *Global Space Day* that included international television and computer exchanges about humanity's future beyond Earth. The primary objective would be to further understanding and consensus on improving the quality of life for this planet's inhabitants by peaceful, commercial exploration and use of space-based resources. The second purpose would be to help earthkind appreciate the importance of human migration to the Moon. The impact on world citizens would be greater than present space gatherings among only the professional elite. It is the masses of our planetary inhabitants who need education about the necessity for our moving beyond Earth.

2. Alternative funding of space enterprises

New options must be pursued for financing space ventures, other than through the taxes and annual governmental budget allocations. That traditional public sector approach will not obtain the $700 billion which the National Commission on Space estimated is required over the next five decades to open up the space frontier. Nor will the $200 million needed to build a lunar outpost be secured by the usual financial methods. Where are funds of that magnitude to come from, especially with huge national deficits and legislative spending restrictions? The history of both the Shuttle and the Space Station to date has been that of government cutbacks which undermined NASA designs and safety in mission planning.

Creating a space ethos implies getting the masses of citizens involved in some way or other in space ventures, as indicated in Chapter 1. In a democratic, free enterprise society, what better way to accomplish this than as a "financial investor". Innovative ways for space financing must be sought that provide citizens and entrepreneurs with financial incentives, like tax rebates, the sale of bonds, or opportunities for private equity funds. To capitalize on the enormous public interest and goodwill generated by the space program in the past 50 years, alternative or supplementary funding possibilities should be explored, including the authorization of stock sales in limited R&D technological space partnerships or trading companies. Recall that back in the Sputnik days, the COMSAT offering on the stock exchange was oversubscribed by the public!

Public lotteries to support scientific exploration and civilizing ventures in newly opened frontiers are part of national experiences. Since the 15th century, European countries have used the lottery device to raise capital for public works. In 1612, the English used this means to support the Jamestown settlement. In the New World, the colonists and first citizens of the American republic employed this mechanism to fund the establishment of higher education, including Harvard, Kings College (Columbia), Dartmouth, Yale, and other universities. In the 19th century, Americans again used lotteries to open up the Western frontier. During the present decade in the U.S.A., for instance, lotteries have become popular again within states to fund public services, particularly education. Today, many foreign countries, such as Australia and Mexico, successfully utilize lotteries or games of chance as a means of raising money to accomplish social goals.

If income produced from new funding sources is to alleviate the tax burden of central governments relative to space expenditures, the investment scope must be vastly broadened. That is what underlies the proposal to establish space authorities, such as a *Lunar Economic Development Authority*, discussed in Chapter 10. More creative methods of external financing of space enterprise will occur with the formation of innovative institutions for that purpose. With the proper *space ethos* in a country, extraterrestrial endeavors would be perceived as a primary national interest and asset. The public generally does not fully appreciate the handsome paybacks that resulted from previous space investments. To ensure citizen involvement in underwriting civilian space ventures, more research is needed both by government and universities on this subject.

Were more private space capitalization encouraged, then public policy makers and world leaders would be challenged to cooperate in setting disbursement objectives for the money so raised. The public is more likely to contribute enthusiastically by purchasing space bonds, stocks, or lottery tickets if the initial funds raised were devoted exclusively or primarily to offworld economic, international, and scientific use, in preference to "star wars"–type activities. For example, the initial target might be in the area of *space transportation systems*. That is, to build the space "highway" for the first few hundred kilometers up into low Earth orbit, the most difficult part of interplanetary travel. Global participation in financing joint space ventures could provide advanced aerospace planes and reusable launch vehicles capable of operating in geosynchronous orbit or beyond ... Just as the Conestoga wagons and railroad opened up Western resources to the nation, so will these less expensive space vehicles bring resources from orbit back to benefit the home planet.

There already exist basic constituencies to enhance success for alternative forms of space promotion and financing, such as among

- the 3,000,000 members of 50 space advocacy groups worldwide who have an estimated aggregate budget today of more than $30 million;
- the millions of space media fans from *Star Trek* television viewers and many motion pictures like *2001* and *Apollo 13*, to the worldwide audience who witness the satellite televising of space feats or watching television productions, such as Disney's *Plymouth* series about the first lunar community;
- the millions of people who make up the global space community: *aerospace* workers and contractors, astronomers and engineers, professors and students, etc.

Before his death, Gerard O'Neill, the visionary scientist, predicted that it will be private capital that will eventually finance space industrialization and colonization. The continued internationalization of space activities will attract global investment

3. Reorganization of National Aeronautics and Space Agencies

The emergence of a new work culture based on knowledge calls for the organizational renewal of varied space administrations within the spacefaring nations. Not only do they need to cooperate more effectively on planning joint ventures, but there may be a need now for creation of a Global Space Administration, Authority, or Agency. Such an entity could coordinate the combined efforts of both the public and private sectors in space development worldwide. Such an international institution might prevent overlapping missions, facilitate cost savings, and concentrate efforts on space macroprojects with the best prospects for ROI. With a modernized charter, this space clearinghouse and research center might obtain more creative financing and planning of space activities, particularly with reference to space technology transfer, as well as attracting more venture capital and licensing *space trading corporations*. In past centuries, great trading corporations were formed by rulers and/or private investors to facilitate exploration and commerce in unknown or foreign lands. The 21st century may replicate this approach by international space-trading entities, comparable with existing multinational communication satellite corporations.

Citizen involvement in any of the above three strategies would contribute to humanity's offworld progress. Michael Simon, when president of International Space Enterprises, maintained that government and industry should do more real joint space venturing together. This engineer and entrepreneur made a case for space commercialization and lunar development. Within a free enterprise, government would encourage the private sector to greater responsibility and risk by

- incentives for taxpayers who invest in space enterprise;
- policies promoting innovative space entrepreneurialism;
- mechanisms for improving space market responsiveness;
- opportunities for achieving large-scale commercial benefits;
- initiatives which encourage synergy among companies, universities, and government entities engaged in working together to apply space research and transfer technology.

Perhaps Simon best stated the case for investment in space development in his volume, *Keeping the Dream Alive* [3]:

"The era in which we live presents humanity with three great challenges: to live in peace, to bring economic prosperity to all people, and to offer tomorrow's generations an exciting future of physical and spiritual growth. During its relatively brief existence, the Space Program has emerged as a central force in our quest to meet all of these challenges. By breaching the bonds of our home planet, we have taken the tentative early steps to become an advanced interplanetary civilization. The impact of the embryonic space age on our lives, already great, will expand and intensify in the years to come, as our horizons become as limitless as the Universe itself."

The U.N. has already designated those who go aloft as humanity's *envoys*, as illustrated in Exhibit 139. In creating a spacefaring civilization, these words of George S. Robinson and J. M. White highlight the global paradigm shift under way [4]:

"Our embryonic envoys have been essential intelligence agents for greater understanding of this survival vision—a total view. Through our efforts to propagate our envoys into the cosmos, through their own personal preparation and adjustments, and also through our remote biotechnological reception of their new transglobal outlook, our envoys have helped us begin to understand the systematic, dynamic, multidimensional, and continuous nature of the cosmos."

EXERCISING TRANSFORMATIONAL LEADERSHIP

Since our species is in transition to space-based living, this necessitates profound changes in sociology, biology, philosophy, government, and law (see Appendix A). Space technological advances are the drivers of a wholly new offworld environment and creation of a space culture. Thoughtful citizens concerned about humanity's destiny want to participate in the process, beginning with the formulation of a space ethos. But it also requires a new type of leadership which has been characterized as *transformational*.

Exhibit 139. Orbital envoys of humankind. Every spacefarer represents the human family offworld, whether worker, tourist, or settler. The hopes of our species in the future depends on their performance aloft, and they are expanded by our robotic creations in space. In this artist's concept, two astronauts have prepared an autonomous robot to collect samples from the Martian surface. Once the robot is released, crewmembers located back at base will either remotely control the robot's activities or allow it to use its own logic systems to perform sample collection. Traveling at their side is another robotic assistant, ready to perform any tasks required. Behind them stands their unpressurized rover, relaying their communication back to the base and waiting to take them to their next destination. Source: NASA.

Transformational leaders, according to N. M. Tichy and M. A. Devanna, recognize the need for changes, such as has been examined here in *Space Enterprise: Living and Working Offworld* [5]. Furthermore, such leaders create and communicate the vision of these desired changes so that a critical mass of people find them acceptable. Then, this leadership personally mobilizes commitment into foresighted strategies which are converted into actualities. So, too, can transformational leadership renew the space program worldwide, restructure space agencies, refinance space undertakings. Transformational leadership can promote synergy among spacefaring nations, as in joint transnational human missions to the Moon, Mars, and Venus before the end of this century.

When humans are engaged in such missions of long duration, André Bormanis and John Logsdon and their colleagues remind us that a whole range of space policy issues need to be addressed, such as [6]

(1) the uniqueness of the space environment;
(2) the selection, composition, and interactions of space crews;

(3) space inhabitants as microsocieties with standards, laws, ethics, and values;
(4) the medical and scientific experimentation under way in orbit;
(5) the spacefarer's survival and quality of life, including communication and privacy rights, health care, pregnancy, deviant acts, death, and risk management;
(6) the space explorer's environmental responsibilities relative to contamination, management of waste and debris, and other such ecology issues.

The exercise of authentic global leadership within all segments of both the public and private sector could transform citizen goodwill into a space ethos that permeates our lives toward opening up the high frontier. When the majority of the world's population perceive the economic and human advantage to be gained there, then energies will be directed at its development and settlement. As astrophysicist and author David Brin reminds us, *science* and its child, *technology*, are cooperative endeavors requiring knowledge to be shared, especially when applied beyond Earth [7]. The message of this book is simply that space is the place where *human emergence* can truly occur, as implied in Exhibit 140 [8]!

Recent research has provided insight into swarm or collective intelligence, the self-organizing swarm behavior so evident in all living creatures. When it occurs, there is a movement in concert, as demonstrated among flocks of swarming birds. This will happen to human beings when our collective intelligence perceives the movement of ever larger groups into outer space. But for now there is the gradual gathering of information and experience regarding living and prospering offworld.

Exhibit 140. Human emergence in space. It is your author's conclusion in his latest book, *Toward Human Emergence*, that humans can only actualize their potential as a species offworld (*www.hrdpress.com*). Source: The above illustrations are from the Foundation for the Future, with artwork by Robert McCall. (*www.futurefoundation.org*).

For travelers, it is not enough to see the horizon alone. We must make sure of what is beyond the horizon, and go there together.

Kemal Ataturk

REFERENCES

[1] Wibbeke, E. S. *Global Business Leadership: GEOLeadership for the World Marketpace.* Burlington, MA: Elsevier/Butterworth-Heinemann, 2008 (*www.books@elsevier.com/business*).

[2] Harris, P. R. *Managing the Knowledge Culture* (2005); *Toward Human Emergence* (2008). Amherst, MA: Human Resource Development Press (*www.hrdpress.com*).

[3] Simon, M. C. *Keeping the Dream Alive.* San Diego, CA: Earth–Space Operations Press, 1987.

[4] Robinson, G. S.; and White, J. M. *Envoys of Mankind.* Washington, D.C.: Smithsonian Institution Press, 1986.

[5] Tichy, N. M.; and Devanna, M. A. *The Transformational Leader.* New York: Wiley, 1986.

[6] Bormanis, A.; and Logsdon, J. M. (eds.) *Emerging Policy Issues for Long-Duration Human Exploration.* Washington, D.C.: Space Policy Institute/The George Washington University, December 1992.

[7] Hargrove, E. C. (ed.) *Beyond Spaceship Earth: Environmental Ethics in the Solar System.* San Francisco, CA: Sierra Club, 1986.

[8] Harris, P. R. *Toward Human Emergence.* Amherst, MA: Human Resource Development Press, 2008 (*www.drphilipharris.com*).

A SPACEFARER'S CREDO

"When we fly around the Earth at eight kilometers a second, 400 kilometers up, we see our Earth as a whole planet. We observe the oceans, forests, mountains, cities, and roads—we absolutely do not see the borders between nations. The time has come for all people of the Earth to work together to build a bright future. Let's start! (Yuri Romaneko, October 10, 1989, cosmonaut with the then record aloft of 420 days, including 326 consecutive days in an orbital environment!

SPACE ENTERPRISE UPDATE

Who will be the first family to live offworld? On the Moon, it is likely to be spacefarers such as an astronaut or cosmonaut, a contractor or technaut, and occur sometime after 2025. For example, there are twin brothers, Mark and Scott Kelly, both NASA astronauts and shuttle commanders, who are likely candidates for this honor. Right now they depend on their spouses and relatives on the ground for their psychological support (Mark is married to a U.S. Congress Representative, Gabrielle Giffords, D-AZ). Imagine if the Kelly twins were living on the Moon, and wanted to bring their wives and children to live with them on the lunar surface for a year or more!

Exhibit 141. Living and working offworld. Though no concrete plans are underway for a human mission to Mars, several studies have looked at the possibilities for such visits. This artist's concept, by Pat Rawlings, depicts Pavonis Mons, a large shield volcano on Mars' equator overlooking the ancient water eroded canyon in which the base is located. Hardware seen here include the Mars explorer, a traverse vehicle, a habitation module, a power module, greenhouses, central base, lightweight crane and trailer, launch and landing facility, water well pumping station, a maintenance garage, tunneling device, water well drilling rig, large dish antennae, mast antenna, even a Mars airplane. Source: NASA.

Appendix A

Governance issues in space societies
George S. Robinson, D.C.L. (Doctor of Civil Laws)

For much of the lay public, and even for too many in the legal profession, when the term "space law" is mentioned, thoughts of special treaties and international conventions, as well as acts of diplomacy involving a multitutde of related and unrelated sensitive issues, come to mind. Adventurous "peaceful" activities in space conducted for the "benefit of mankind" are formulated and assessed by a multitude of professions and disciplines, including politicians, diplomats, scientists, engineers, budget analysts, and the like. They are also viewed by lawyers attempting to determine where these activities, extant or simply in the proposed stage, fit into many of the amorphous and frequently preactory and imprecise provisions, phrases, terms, and single words appearing in international space law, public and private.

Dr. George S. Robinson[1]

A.1 BEYOND IMPERIALISM: BEYOND SPACE COLONIES

In the 21st century, human exploration of space, as well as human migration into and settlement of near and deep space, will offer an extraordinarily rare opportunity to sever the endless cycle in human history of economic and religious imperialism, colonialism, denial of basic human and human*kind* rights, and the subsequent violent confrontations that inevitably follow those practices. While global civilizations are acutely aware of the negative and even horrific aspects of traditional imperialistic practices leading to the establishment of colonies, there also are positive lessons to be learned and applied in pursuing the exploration and settlement of space. Perhaps we

---

[1] Unpublished paper, "Public Space Law, the Practitioner, and the Private Enterprise," October 2006 (available from the author by emailing *astrolaw@aol.com*). [For an author biography see "About the contributors" page in the prelims.]

Appendix A: Governance issues in space societies

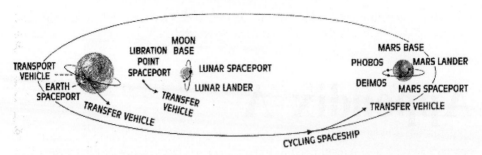

Exhibit 142. Bridge between two worlds: Earth/Moon system. Interplanetary exploration for the New Millennium depends upon first developing a space infrastructure in our own Solar System. As *spacekind* develops a new interdependence, our species offworld will emerge differently from *earthkind*. Source: *National Commission on Space, Pioneering the Space Frontier*. New York: Bantam Books, 1986.

should begin by formulating an acceptable and workable *Migratory Manifesto* that creates an infrastructure between terrestrial humans and human*kind* in space, including their supporting national governments and international governmental organizations. Such an infrastructure must also address the governing relationships of space societies/communities, as well as their private entrepreneurial underwriters. We are literally building a bridge between worlds, as Exhibit 142 illustrates.

Governing space societies will involve an infinitely greater complexity of rights and responsibilities for a multitude of interested and involved parties. The application of traditional line-of-command structures, some direct and some diffuse, as in the manned space programs of the United States, Russia (even former Soviet Union), and China, will prove increasingly inappropriate. Even in the multilateral agreement under which the International Space Station operates, there is an overall line-of-command structure governing relations between Earth governments and Space Station participants of various national origins. Nevertheless, in certain discreet ISS component habitats and laboratories, governance of societal/commercial/military activities can be unique to that habitat or laboratory.

Governing requirements are multifaceted, of course, and those requirements are reflected in the laws that are formulated and brought to bear on space society relationships, now and in the future. Prior history of exploration involved the conventional practices of economic imperialism and embraced both private commercial activities (e.g., charters and governmental acquiescence for a percentage of the profit: the Virginia Company, Hudson Bay Company, and the East India Company are examples) and the need for military involvement primarily to provide protection of the private commercial interests and an administrative infrastructure for routine matters in remote colonies. In addition to the positive and negative aspects of governance in the colonies that were established and evolved from imperialistic practices, we also see the precedent lessons of governing practices unique to international and domestic free trade zones, as well as certain trust territories established and administered under the aegis of the United Nations. Presently, in the United States' strategy of space exploration, which involves government, military, and commercial development partners, the matter was made crystal clear with the issuance of the August 2006, U.S. National Space Policy.[2]

A.2 GOVERNANCE IN THE ABSENCE OF HUMAN RIGHTS AND DUTIES

Learning from the negative and positive lessons of our previous history of imperialism and colonialism on this home planet, we can look back for constructive guidance. Yet today, we still note an absence of carefully articulated and crafted, basic human and human*kind* rights relative to responsibilities for those inhabiting near and deep space. Contrary to the beliefs of certain military pragmatists, space is not "just another place" ... another military high ground. It is a critically important and alien arena for ongoing the evolution of our species, along with the evolution of space cultures and civilizations.

Presently, the emerging space societies are significantly devoid of governance principles that assure our envoys of human*kind* that living and working in space will carry with them such individual and societal values as freedom of speech, peaceable assembly, and the right to petition governments for redress of grievances. These types of individual and societal rights, particularly those not yet identified for humans and human*kind* who are living and working aloft, must not rest on inferences alone. Otherwise, spacefarers will never be protected and safeguarded during the unfolding of space migration and settlement.

In the 1980s, an effort was pursued through the Smithsonian Institution's National Air and Space Museum in Washington, D.C., to attempt to formulate a "Declaration of First Principles for the Governance of Outer Space Societies."[3] The participants in this effort represented an extraordinary array of disciplines, hard and soft. The objective of the participants was not to draft some form of a constitution for the governance of space societies, a responsibility reserved for sovereign governments. Rather, the purpose was to formulate a type of pronouncement, a *Migratory Manifesto*, that would assist national and international governmental authorities who were responsible for establishing and implementing space policy, so as to secure the fundamental rights, responsibilities, and unique survival requirements of humans and human*kind* residing in orbital societies, whether temporarily or permanently. The founding of spacefaring civilizations requires some form of shared governance responsibility, transitioning ultimately to some form of independent, self-governing space society.

Ultimately, a "Declaration" emerged from these deliberations that reaffirmed faith in the fundamental freedoms and inalienable human rights of individuals living

---

[2] See G .Robinson, "The U.S. National Space Policy: Pushing the Limits of Space Treaties?" *German Journal of Air and Space Law*, 2007, 45–57 (ZLW 56, Jg. 1).
[3] See G. Robinson, "Re-Examination of Our Constitutional Heritage: A Declaration of First Principles for the Governance of Outer Space Societies," *High-Tech Law Journal*, 3, 1989, 81–97 (published by the School of Law, University of California, Boalt Hall, Berkeley, CA. 90024).

in a space environment. As may be seen in the boxed material at the end of this appendix, that statement significantly asserted that governance of and by space societies and civilizations should reflect the *will and special requirements* of those living in space—not just those values subtending governance structures that evolved in the communities of Earth. These provisions anticipated unique individual and societal survival requirements for those who live and work offworld, whether for a short time or permanently. Some guidelines will be similar to those shared with humans and societies on Earth; others will be unique to the space environment, whether in our Solar System, or ultimately, beyond. Obviously, survival requirements offworld are significantly different, if not advanced *in extremis*, from those of its progenitors who remain primarily on Earth. Their governance structures and underlying values are very likely, at least unique to that location with its Earth–alien, life support requirements. The first major application will likely be a lunar base to be built within a decade.

Today, we live enshrouded with fears of apocalypse, fears of Armageddon. The future of our species is in large part in our own hands, but that future rests on our ability to expand our ecotone into space: to send our successive and evolved generations in varying forms of human biotechnological integration off Earth. And we must recognize that these current and future generations of human*kind*, of transhumans, and ultimately post-humans,[4] will have their own unique individual and societal needs.

A.3 TRANSITIONING PRINCIPLES FOR GOVERNANCE OF SPACE SOCIETIES

The world's spacefaring civilizations clearly are in the process of establishing what they only sense to be unique societies and, ultimately, diverse civilizations in space. In fact, we are formulating the basic technological and pragmatic requirements for creating a "cradle" in which new civilizations can ultimately be born and nurtured. Nevertheless, values, principles, and motivating factors wrapped in historical precedence indigenous to Earth are largely being relied on to shape those prospective cultures with their different values and principles. Frequently, what we propose is totally irrelevant to the needs and the requirements of offworld societies and civilizations, trying to survive in completely synthetic and alien life support environments. In many respects, it might be said that cultural recidivism is being relied on to develop the principles underlying space social constructs, including the legal foundations for human and human*kind* evolution off Earth. This deficiency is classically represented in the multilateral agreement governing the cultural/social/commercial/military

---

[4] For our discussion, "transhuman" is defined as a human permanently altered for adaptation to a specific environment, but still remaining a human. A "post-human" has several definitions depending on thematic contexts. For the present discussion, a "post-human" can be defined as a non-human entity of unprecedented physical, intellectual, and spiritual capacities.

aspects, as well as operational objectives and control, currently evidenced at the International Space Station.[5]

In many, if not most, respects we seem to focus only on the fact that our technology is allowing us access to near and deep space. Only comparatively slight thought seems to be directed toward the process of laying the foundations for a new and distinct culture embracing our envoys of "human nature and essence" beyond the home planet. In most respects, we seem to be content dragging old and frequently irrelevant interpersonal values and interactive survival principles into space communities, while applying them to the inhabitants, expecting them to satisfy the same needs as those of us remaining on Earth.

In order to evolve some basic understanding of the unique needs and values of humans in orbit, constructive interactive relationships and dialogue between and among actual spacefarers with those individuals and organizations remaining on Earth. Thus, an international conference should be considered for this purpose, co-sponsored primarily by non-governmental research and educational organizations. The aim of this convocation would be to identify and formulate the biocultural constructs necessary for effective interactions, which facilitate human evolution and survival offworld. Such a global exchange might also be promoted electronically as representatives construct an interdependent infrastructure governing *earthkind* and *spacekind* relationships. Participants in these deliberations should include interdisciplinary and cross-disciplinary delegates, such as evolutionary biologists, astrobiologists, philosophers, theologians, economists, cultural and physical anthropologists, historians, psychologists, human factor experts, astrophysicists, engineers, and other representatives of the pragmatic/empirical disciplines, as legislators, and jurisprudents. This assemblage should also include experts in artificial intelligence, telepresence, teleportation, biorobotics, genetics, and the like.[6]

A.4 SPACE GOVERNANCE: PRELIMINARY CONCLUSIONS

Space governance in the 21st century will reflect legal regimes incorporating values essential for the success of private commercial endeavors in space, as well as governmental public responsibilities offworld in both domestic and international contexts. Governance of space societies will reflect both old principles of economic imperialism from our past, as well as economic and fiscal principles of a steadily evolving and unifying form of transglobalism. In the future, perhaps some form of a private transglobal entity with quasi-sovereign authorities may provide certain task aspects of space society and governance. On the other hand, if there is recidivism of ethnic,

---

[5] *Agreement among the Government of Canada, Governments of the Member States of the European Space Agency, the Government of Japan, the Government of the Russian Federation, and the Government of the United States of America concerning Cooperation on the Civil International Space Station* (available at *ftp://ftp.hq.nasa.gov/pub/pao/reports/1998/IGA.html*).
[6] See G. Robinson, "No Space Colonies: Creating a Space Civilization and the Need for a Defining Constitution," *Journal of Space Law*, **30**(1), 2004, 169–179.

religious, economic, political/military, cultural parochialisms, and the like, governance characteristics, familiar to some historians, will certainly be destined for serious confrontational, cyclical changes during this century. Expanding global and interplanetary communication capabilities will impact decision-making in all aspects of space governance. But such exchanges and choice should subject to direct input and effective influence over those decisions by global citizenry, as well as those who are actually living aloft.

Finally, space governance will have to accommodate unique survival requirements of space communities populated by not only knowledge workers, but also highly advanced artificial intelligence. The latter will be embodied in human biotechnologically integrated entities that may not meet or satisfy Linnaean taxonomic principles and binomial nomenclature (i.e., they will have evolved from *Homo sapiens sapiens* to *Homo alterios spatialis*). They will be true post-humans: *spacekind* in every sense requiring self-governance with values and principles alien to humans, human*kind*, and even transhumans regardless of their locations in the universe.

DECLARATION OF FIRST PRINCIPLES FOR THE GOVERNANCE OF OUTER SPACE SOCIETIES*

Preamble

On the occasion of the Bicentennial of the Constitution of the United States of America and in commemoration and furtherance of its values we, the undersigned petitioners,

- Bearing witness to the exploration and inevitable settlement of outer space;
- Recognizing the universal longing for life, liberty, equality, peace and security;
- Expressing our unshakable belief in the dignity of the individual;
- Placing our trust in societies that guarantee their members full protection of law, due process, and equal protection under the law;
- Reaffirming our faith in fundamental freedoms;
- Mindful, as were our nation's founders, of the self-evident truth that we are endowed by our Creator with certain inalienable rights;
- Recognizing the responsibility of a government to protect the rights of the governed to exist and to evolve;

Do assert and declare in this petition the intrinsic value of a set of First Principles for the Governance of Outer Space Societies and, at the beginning of this third century of nationhood under our Constitution, resolutely urge all people of the United States of America to acknowledge, accept, and apply such First Principles as hereinafter set forth.

Article 1

A. The rule of law and fundamental values embodied in the United States Constitution shall apply to all individuals living in outer space societies under United States jurisdiction.
B. Appropriate constraints upon and limitations of authority shall be defined so as to protect the personal freedom of each individual, such as the right to

reasonable privacy, freedom from self-incrimination, freedom from unreasonable intrusion, search, and seizure, and freedom from cruel and unusual punishment.
C. Toward this end, the imperatives of community safety and individual survival within the unique environment of outer space shall be guaranteed in harmony with the exercise of such fundamental individual rights as freedom of speech, religion, association, assembly, contract, travel to, in, and from outer space, media and communication, as well as the rights and petition, informed consent and private ownership of property.
D. The principles set forth here should not be construed to exclude any other such rights possessed by individuals.

Article II
A. Authority in outer space societies, exercised under principles of representative government appropriate to the circumstances and degree of community development, shall reflect the will of the people of those societies.
B. All petitions to the United States Government from outer space societies under its jurisdiction shall be accepted and receive prompt consideration.
C. The United States shall provide for an orderly and peaceful transition to self-governance by outer space societies under its jurisdiction at such time as their inhabitants shall manifest clearly a belief that such a transition is both necessary and appropriate.
D. In response to aggression, threats of aggression, or hostile actions, outer space societies may provide for their common defense and for the maintenance of essential public order.
E. Outer space societies shall assume all rights and obligations set forth in treaties and international agreements, relevant to the activities of such societies, to which the United States is a party and which further freedom, peace, and security.
F. The advancement of science and technology shall be encouraged in outer space societies for the benefit of all humanity.
G. Outer space societies shall protect from abuse the environment and natural resources of Earth and space.

* **Editor commentary:** Reprinted with permission by the Smithsonian Institution Press from "Declaration of First Principles for the Governance of Outer Space Societies," in *Envoys of Mankind* by G. S. Robinson and H. M. White, © Smithsonian Institution, Washington, D.C., 1986, pp. 266–270. The above Declaration was based on the United States Constitution, and was drafted by scholars assembled for two conferences at the National Air and Space Museum, Washington, D.C., in December 1986 and November 1987. It is reprinted here as a basis for discussion on governance of space communities, especially by adaptations by other nationals. The above book by attorneys Robinson and White also contains a proposed "Treaty Governing Social Order of Long-Duration or Permanent Inhabitants of Near and Deep space", as well as a proposed "Spacekind Declaration of Independence". For further information on this space philosopher and attorney, Dr. George Robinson, see "About the contributors" section (p. xxxiii).

With reference to Robinson's speculations on the possible need for quasi-public space authorities, the proposal for a Lunar Economic Development Authority has been discussed in Chapter 10. Interestingly, in 2005, NASA suggested consideration of forming a Lunar Exploration and Development Authority. Both concepts would be viable if international in scope, with partners from the world space agencies and private enterprise. P.R.H.

A.5 REFERENCES

Robinson, G. S. "Space Law in the 21st Century and Beyond," in Krone, B.; Morris, L.; and Cox, K. (eds.) *Beyond Earth: The Future of Humans in Space*. Burlington, Ontario: Apogee Books, 2006, pp. 46–52.

Robinson, G. S. "Transcending to a Space Civilization: The Next Three Steps toward a Defining Constitution," *Journal of Space Law*, **32**(1), 2006, 147–175.

Robinson, G. S. "Forward Contamination of Interstitial Space and Celestial Bodies," *German Journal of Air and Space Law*, 2006, 382–398 (*ZIW*, **55**, 3).

Robinson, G. S. "Future Private Commercialization of Space Resources: Foibles of Applicable Law," *Annals of Air and Space Law*, **XXVII**, 2002, 2–34 (published by McGill University, Montreal, Canada).

Robinson, G. S. "Space Treaties: Documents of Transcending Principles?" *Cosmos*, **12**, 2002, Editorial, iii–iv (journal of the Cosmos Club of Washington, D.C.).

Robinson, G. S. "Natural Law and a Declaration of Humankind Interdependence," *Space Governance*, Part I June, and Part II December, 1995 (journal of United Societies in Space, Inc.; www.usis.org).

SPACE ENTERPRISE UPDATE

Clearly, space jurisprudents and practicing lawyers with a broad spectrum of interdisciplinary expertise must keep up with and stay ahead of, anticipated technological and biotechnological users of interstitial space. All the more because the "users" are actually part of interstitial space and involve highly advanced artificial intelligence, sophisticated forms of telepresence (and perhaps even teleportation) cybrspace among the "celestial bodies ...

Are we creating autonomous entities who or which can be held individually accountable under a regime of positive law based upon existing natural law theory, or will be faced with interacting with entities having principles and values deriving from some unique legal theory of interspecies interaction as yet unknown to us?"

Source: George S. Robinson, "Space law for humankind, transhumans, and post humans: Is there need for a unique theory of natural law principles, *Annals of Air & Space Law*, **XXXIII**, 2008, 287–323 (journal published by Graduate Law Faculty, Institute of Air and Space Law, Montreal, Canada, where the author received his Doctor of Civil Laws degree in space law).

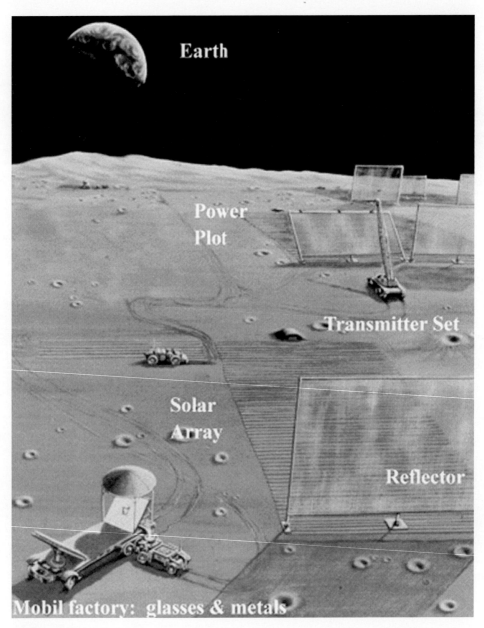

Exhibit 143. Demonstration lunar power base. In this artist's concept, sunlight hits the solar converter, which transmits power via underground cables to a microwave generator, which in turn illuminates a microwave reflector. All such reflectors, when viewed from Earth, overlap to form a "lens" that can direct a narrow power beam toward Earth. Source: Dr. David R. Criswell and Dr. Robert D. Waldron © 1991. From "Results of Analysis of a Lunar-Based Power System to Supply Earth with 20,000 GW of Electric Power," paper presented at the Solar Power Satellite Congress, Paris (August 26–30, 1991). See Appendix B.

Appendix B

Space-based energy: Lunar solar power
David R. Criswell, Ph.D. and Phillip R. Harris, Ph.D.[1]

> *The Space-based Solar Power Study Group universally acknowledges that a necessary pre-requisite for the technical and economic viability of space-based energy is inexpensive and reliable access to orbit ... The need for a large-scale, clean new source of electricity is evident.*
>
> Space-based Solar Power Report, 2007[2]

B.1 INTRODUCTION

Since the 1960s, extensive lunar studies have been under way by both Russian and American scientists, first in conjunction with the Lunik and Apollo missions, and now with the *Vision for Space Exploration* national policy in the 21st century. Adjoining NASA's Johnson Space Center, the Lunar and Planetary Institute has continued further research, conferences, and publications on the Moon and its resources [1]. NASA and ESA have had numerous conferences and publications on this matter, providing both universities and contractors with funding to further such investigations [2]. For example, over four decades ago, General Dynamics Space Systems Division completed a NASA report in 1965, "Operations and Logistics Study for Lunar Exploration Systems for Apollo (LESA)," and in 1979, "Lunar Resources Utilization for Space Construction" [19].

In the 1990s, the Stafford Synthesis Group report focused on life support and propulsion technologies needed for Moon and Mars manned missions. In that same

---

[1] For author biographies see "About the contributors" page in the prelims.
[2] Adapted from "The Chicken and the Egg: RLVs and Space-based Solar Power" by Taylor Dinerman in *The Space Review*, November 19, 2007 (*www.thespacereview.com/article/10041*).

decade, unmanned missions, such as Clementine and Prospector, furthered the mapping of lunar resources and prospects. The Europeans have also carried out their near-term program for the Moon as a means to stimulate global economic development [3]. Between 1994 and 2003, a series of international lunar exploration or development conferences have brought together researchers who emphasize scientific and commercial enterprises on or about the Moon, featuring both human and robotic activities [4]. On these occasions, the assembled delegates support technologies to reduce access costs to the Moon, so new lunar transportation systems will enable us to benefit from its resources. More recently, authors have written useful books detailing lunar resources, technologies, and base possibilities [5].

Now in the first decade of this century, the U.S.A. has undertaken a national policy, The Vision for Space Exploration, that mandates a permanent, human return to the Moon by year 2020 , eventually proceeding to Mars and beyond (see Chapters 1 and 10). NASA is busy planning and implementing new programs and technologies to achieve this goal, while seeking international partners for this macroproject. Simultaneously, the space agencies of Russia, China, Japan, and India are planning unmanned missions to or around the Moon to be executed within the next dozen years. The global space community is beginning to appreciate that there are sufficient scientific, technological, commercial, and exploration justifications to warrant a huge investment in offworld lunar enterprises for the benefit of humanity in the near future. The 2007 NSSO report of the Space-Based Power Study places special emphasis on immediate development of reusable launch vehicles, if space solar power is to be feasible.

Indeed, a principal rationalization for going back to the Moon seems to be economic by means of lunar industrialization for the benefit of Earth's peoples. There is growing realization that it is possible to develop a twin-planet economy through these two celestial bodies [6]. One such prospect has been the renewed interest in space-based energy, such as solar power. As far back as July 1989, this awareness was evident in the *Report of NASA Lunar Energy Enterprise Task Force*. The rekindling of such attention resulted in a book, *Solar Power Satellites: A Space Energy System for Earth* [7]. For lunar economic development, research confirms that the Moon can be used as a platform for energy production of several types, whether on or around the Moon. This appendix will make the case for one such undertaking from the Moon itself, rather than surrounding satellites: a lunar solar power system right on the lunar surface!

The development of a lunar solar power system (LSPS) presents manifold challenges as a large-scale, space-based enterprise (Exhibit 143). First, there is the need for global communications about the rationale and case for investment to supply solar power to the Earth from the Moon. Second, the technical community must cooperate in *integrated planning* of an operational lunar power base which provides a quick growth rate, lower power costs, and higher energy capacity [8]. To collect solar energy on the Moon and beam it to Earth is a macroengineering project calling for research demonstrations, as well as formation of joint ventures for major funding and planning of innovative space infrastructures (see Chapters 7 and 8).

The concept of lunar solar energy transmission is a 21st-century global power system which will not introduce new pollution into the biosphere, nor deplete existing organic resources. A window of opportunity now exists to provide this clean energy solution to improve the human condition and add new net worth to Earth. The LSPS could offer a viable electric power supply to enhance the standard of living on the home planet, especially in economically developing countries. Engineers and technicians worldwide should be encouraged to initiate R&D into space energy resources. The challenge goes beyond engineering and extends into many other realms, such as finance, communication, sociology, law, and politics [9]. For instance, any beaming of solar energy to the home planet will have considerable legal, environmental, and safety ramifications. By implication, the undertaking means application of a new type of *macrothinking in both strategic planning and macromanagement*.

B.2 RATIONALE FOR A LUNAR SOLAR POWER SYSTEM (LSPS)

The advantages of this innovative solar energy system over conventional power are summarized in Exhibit 144. Looking along its last row, the case for lunar power is highlighted as follows [10]:

- offers maximum useful power level of >100,000 GW;
- developing this lunar resource may require 10–20 years for R&D to produce an almost endless supply of energy; its only limiting factor appears to be stray microwaves, while producing no pollution;
- high upfront initial investment, but long-term cost trends are downward with robust ROI forecast;
- risks in achieving project goals are seemingly relatively low.

Almost three decades ago, 4.43 billion inhabitants of this planet used 10,300 GW of power (average 2.33 kW in 1980). The developed nations used most of that energy (6.3 kW/person), while other countries used less (1.0 kW). Today only 10% of the world's population consumes most of the energy, but dramatic population and economic growth in the Third World is producing escalating demands for energy, and so an increase in polluting fossil fuels. Not only is increasing global affluence causing more energy to be consumed worldwide, but catastrophic climate change is exaggerating the energy situation [11]. For a projected planetary population of 10 billion by the year 2050, the global electric power capacity must be increased tenfold at a time when coal and oil as a source of power may be eliminated. Exhibit 145 estimates the power needs of the next century in terms of population (billions), kilowatts of electrical power per person (at 1,000 watts per kW), and total power necessary (GW).

Earthlings now operate within a thermal economy in which 15% of the gross world product is invested in energy sources that are continually in crisis because of

Exhibit 144. Comparison of 21st-century energy systems. This analysis shows the advantage of setting a goal by 2040 to obtain 20,000 GW of electricity from lunar solar power.

Power system	Maximum output (GW/yr)	Maximum useful power level (GW)	Time to exhaust (yr)	Limiting factors	Polluting products	Long-term cost	Risks to reach goal
Bioresources	100	1,000	<10	• Mass handling • Nutrients • Water • Land use	• CO_2 • Biohazards —methane —disease	Up	Not possible
Coal	<1,500,000	20,000	100	• Supply • Pollution	• CO_2 • Ash acids • Waste heat	Up	Not possible
Oil	<100,000	5,000	<30	• Supply	• CO_2, acids • Waste heat	Up	Not possible
Nuclear fusion	>500	500 Base load	100s	• Accidents and terrorists • Social acceptance	• Radioactives —spent fuels —components • Waste heat	Up	High
Nuclear fission	>20,000	>200,000	1,000s (D-T) <100s (D-He$_3$)	• Engineering demo • First wall life (D-T) • Power balancing*	• Radioactive components	Unknown	High

Appendix B: Space-based energy: Lunar solar power

Hydroelectric	2,000	2,000	<1,000s	• Sites and fill-up • Rainfall • Dam failure	• Sediment • Flue water	Up	Not possible
Ocean thermal	<1,000,000	<20,000	<100	Deep cold waters	• Power balancing	Up	Not possible
Terrestrial solar power	>20,000	>20,000 Not tied to peak or base load	>10^8	• Clouds • Power storage • Power distribution • Power balancing*	• Waste heat • Production waste	Down	Moderate
Solar power satellites	2,000	2,000 Base load	>10^9	• Orbital debris • Shadowing	• Sky lightning • Orbital debris	Down	Not
Lunar power system	≫20,000	>100,000 Base and load following	>10^9	Stray microwaves	None	Down	Low

Source: David R. Criswell, Institute for Space Systems Operations, University of Houston [10].
*Balancing local and interhemispheric heat loads.

550 Appendix B: Space-based energy: Lunar solar power

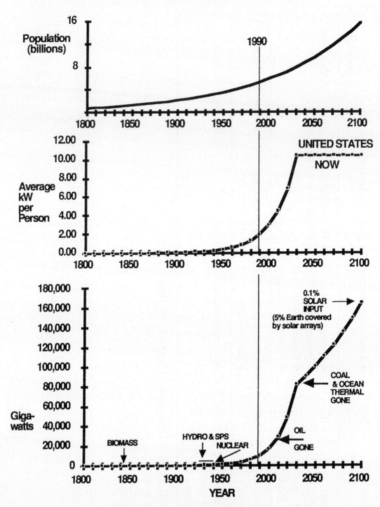

Exhibit 145. 21st-century power crisis. Further comparison of growing world population and the energy needs of individuals and how these could best be met by lunar solar power. Source: David R. Criswell and Robert D. Waldron [12].

declining resources, increasing prices and pollution, terrorism and war. Sometimes, international trade and economic development are hostage to Middle East, Nigerian, or Russian oil supplies, as well as other regional problems. In a July 1990 presentation to the first Lunar Power System Coalition workshop, Dr. M. N. A. Peterson, professor emeritus of the Scripps Institution of Oceanography, emphasized that most concerns underlying Earth observation and modeled predictions are *directly related to energy needs*, such as [13]:

- atmospheric CO_2 and other greenhouse gases;
- air quality, acid precipitation, and oxidants;

- fossil fuel reserves and associated tensions;
- petrochemical needs and other energy options;
- waste and disposal;
- stratospheric ozone depletion;
- water resources and river use;
- transportation and food production;
- levels of economic wellbeing and incentives for environmental degradation and population growth.

Peterson, a former chief scientist for the National Oceanic and Atmospheric Administration, maintained that all these factors can be related to economics and the availability of energy. Then in a paper presented to a national conference on Earth Observation and Global Decision Making, this distinguished geologist and oceanographer expanded on this theme that every environmental concern addressed there had a connection to energy, its conversion to useful forms, its distribution, or the efficiency and costs of its use. The late Dr. Peterson realized that the proposed LSPS program can solve the problems posed in the above situations by providing a major new source of energy that is dependable, sustainable, economical, environmentally enhancing, available worldwide, and capable of growth. When director of Pacific Ocean Policy Institute in Honolulu in 1993, Peterson began examining the connection between the oceans and alternative energy sources, including possible offshore rectennas that might receive lunar power beams.

Preliminary analysis by the system inventors, Criswell and Waldron, indicate a projected annual net revenue from lunar power could be a billion dollars a year, with an internal rate of return on investment exceeding 30% if the power is sold at 0.1 $/kWh. Exhibit 146 shows the power build-up to 20,000 GW (GWe) and spread out beyond the year 2050. Only a lunar solar power system can meet those requirements at reasonable cost!

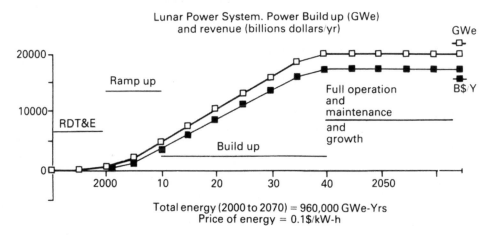

Exhibit 146. Lunar power system. Power build-up (GWe) for the next 50 years with forecasts of anticipated revenues for an LSPS. Source: David R. Criswell and Robert D. Waldron [14].

The LSPS strategy is timely and synergistic because:

(1) it fits within the current NASA Vision for Space Exploration program, while offering the space agency additional reasons for returning to the Moon that will gain public support, and provide some economic justification for further space exploration;
(2) it coincides with national needs to become less dependent on foreign petroleum and gas, especially from OPEC, as well as to halt the environmental damage of fossil fuels, in addition to depletion of the world's forest and ecological reserves;
(3) it provides for a transfer of defense technology (Strategic Defense Initiative), which would involve scientists, engineers, resources, and investment to a "defense of the planet", thus fostering peaceful, leading-edge technology, research, and development;
(4) similarly, it offers the hard-pressed aerospace industry and high-technology companies a new market for investment, income, and jobs by transfer of technology to lunar power and related technology;
(5) it presents a new leadership opportunity to the U.S. Department of Energy and the Environmental Protection Agency for solving energy shortage problems with an environmentally safe alternative, possibly in conjunction with NASA, as well as other international and private partners
(6) it builds on existing research and technology, such as in microwaves and beaming, so as to encourage new applications, even of DOD's Strategic Defense Initiative studies;
(7) it will not only foster new career opportunities for astrophysicists, as well as aerospace and energy engineers, but renew existing industies, such as corporations in aerospace and satellite communications, while creating wholly new *astrobusiness*, from lunar construction to law;
(8) it contributes to the development of a *two-planet economy*.

> Lunar industries lay the foundation for migration of humanity to space habitats. The needs, time, resources, knowledge, educated people, and growth path now exist to move mankind to more interesting engineered venues off Earth. The next fifty years can see the continued stifling poverty of billions of people, or we can create a wealthy, healthy future aloft. (Source: David R. Criswell, "Net Growth in a Two-Planet Economy," *45th Congress of International Astronautical Federation, Jerusalem, Israel, October 9–14, 1994* (IAA-94-IAA.8.1.704).)

Space-based energy, such as lunar solar power, challenges many existing institutions to consider its implications for furthering their concerns and objectives, such as:

- the Electric Power Research Institute by investigating how LSPS could benefit the utility/energy industry;

- the Off-Shore Industry Association by considering how ocean platforms could be utilized as rectennas to receive microwave power beams from the Moon;
- the Sandia National Laboratory applying its expertise in terrestrial solar energy to studies of solar power from the Moon or its vicinity.

No wonder that in a letter to one of the authors (Harris, August 10, 1990), Dr. Michael B. Duke, then Deputy for Science, NASA Lunar and Mars Exploration Program Office, stated:

> "I have a feeling that it is a concept like the Lunar Power System that will get us back to the Moon and into space in a big way."

After retirement from the Agency, this planetary scientist was president for a time of the Lunar Economic Development Authority, before becoming a professor at the Colorado School of Mines ... In a seminal letter to the author (Harris, October 27, 1993) when Vice President of the United States, the future Nobel Prize winner, Al Gore wrote:

> "Thank you for sending a copy of your articles published in the *Earth–Space Review* in which you propose an alternative space-based energy program. As you requested, I have forwarded your letter and articles to the Department of Energy for evaluation ... Your views will be very helpful as I work with the President to shape policies of concern to us as individuals and as a nation."

Of course, that was 15 years before the ex-Presidential candidate and Nobel Prize winner's internationally acclaimed film on climate change called *An Inconvenient Truth*. Unfortunately, in it he never mentioned the possible solution to a global deteriorating environment: space-based energy!

Much has been published about the international energy crisis. In an article on "Tomorrow's Energy" in *Parade* Magazine, Dr. Carl Sagan, the late noted astronomer, observed that solar energy is a safe, promising solution to the world energy dilemma [15]. Then the eminent space journalist, Leonard David, wrote an astute analysis of space power beaming, excerpts of which are reproduced in Exhibit 147.

In the long term, there are severe limitations on the current means of producing electrical power on the home planet. Offterrestrial energy options are promising and inexhaustible, whether solar power satellites (SPS) as advocated by Peter Glaser, or the lunar solar power system (LSPS) described by Criswell and Waldron. Former astronaut and U.S. Senator Harrison Schmitt is a proponent of helium-3 mined from lunar soil in fusion reactors as a potential power source for use on Earth or in distant parts of the Solar System [16]. With major adverse impacts on the Earth's environment at a minimum, the technological feasibility for exploiting solar energy at lunar facilities is worth pursuing, particularly for its economic potential. Yes, the initial investment in any space-based energy system would be very costly, but over time there would be a great return on such: the payback would be endless, clean, and less expensive energy for the benefit of all humanity!

Exhibit 147. SPACE ENERGY COMMENTARY.

Some call it wireless transmission. Others tag it power beaming. Years ago it was referred to as the solar power satellite program. A growing international choir of experts foresees these terms as the 21st century solutions to what looks like an eco-catastrophe in the making ... In the late 1800s, Russian space pioneer, Konstantin Tsiolkovsky ... envisaged utilization of space solar power, not only to sustain humanity's entry into space, but for "locomotion throughout the Solar system" ... At the close of the 19th century, Nicolas Telsa, pioneer of alternating current, presented a lecture before the British Royal Society on what he called "wireless power transmission"—the sending of energy via waves. These two concepts came together in 1998, with a proposal floated by Peter Glaser when vice president of Arthur D. Little Inc. ... which sketched out how large arrays of solar cells could be positioned in geostationery Earth orbit ... Solar energy drawn from satellites' photovoltaic cells would be converted into microwaves, and then relayed to power receiver antennas on Earth for reconversion into electricity.

A 1975 test carried out by Raytheon, a Massachusetts-based electronic company, in association with NASA's Jet Propulsion Laboratory, transmitted microwave energy from a ground station to a target over a mile away. At the target, the beam was converted into direct current electricity to light up a ten-kilowatt array of light bulbs ... The success of the test kicked off $20 million worth of research conducted over three years, carried out by NASA and the Department of Energy ... Infighting at the time was evident between the White House and the Congress, and the various agencies evaluating Solar Power Satellites. That bickering eventually crippled the project ... The prestigious National Research Council issued its critique of SPS. They called the concept too *costly and too uncertain for a research program in the next decade* ... The past few years have seen a renaissance in power beaming ... The First Annual Wireless Conference was held in San Antonio, Texas in 1993 ... organized by the Center for Space Power at Texas A&M University. Established in 1987, the Center has as its goal to demonstrate that power in space is indeed viable and can be a commercial activity. One of their studies reported at the Conference provides a step-by-step learning process to evaluate the technology, environmental impact, and feasibility of power beaming. One category involves point-to-point power beaming on the ground. In this area, the government of Alaska is exploring controlled experimentation of wireless energy transmission in remote sites throughout the region. Research centers are located in other far-flung areas—Siberia and Antarctica might benefit as well ... Related to the Earth-to-satellite power beaming work is the Fission Activated Laser Concept (FALCON) being developed at the Department of Energy's New Mexico-based Sandia National Laboratory. Its team leader, Ronald Lipinski suggests that "After we have a larger presence in space, it makes sense to beam power back to Earth."

While some researchers contemplate the design of power-transmitting satellites, others contend that the most efficient space-based power system is sitting right in front of us. The Lunar Power System Coalition believes that the Moon can be transformed into a natural solar power satellite, sporting fields of solar cells made from lunar resources. Under cloud-free lunar skies, tens of terawatts of pollution-free power could be gathered and broadcast to the Earth. The coalition draws its get-up-and-go from David Criswell, director of the Institute of Space Systems Operations at the University of Houston, and aerospace engineer, Robert Waldron of Hacienda Heights, California. Criswell says, "my argument is that I can design components—the microcircuits, the glass, the photovoltaics—right there on the Moon out of fairly simple industrial processes; then we are a lot further down the road than making these same components on Earth. Why transport materials from the Earth or Moon to build large platforms in Space? Why not build the solar power collectors and transmitters on the Moon from lunar materials?" Criswell argues for deploying a set of automated lunar landers at select areas across the lunar terrain. These craft would collectively focus their radio beams at Earth receivers, thus demonstrating that the Moon could provide a stable platform for transmitting even greater power levels ...

Speaking at the Earth Summit held in Rio de Janeiro, 1992, Michael Willingham, a United Nations technical advisor, remarked that some consider the environmental crisis to have been brought about by humanity's insatiable need for energy. He added, "Perhaps the ultimate answer lies in returning to the heavens for a new, safe, and plentiful energy source from space."

Source: Adapted from Leonard David, "More Power to You," *Final Frontier*, March/April 1993, pp. 20–23, 48.

B.3 LUNAR SOLAR POWER SYSTEM PROPOSAL

The Criswell/Waldron concept as reviewed at numerous scientific gatherings worldwide utilizes solar power for the benefit of earthlings, as illustrated in Exhibit 148. A mature system would channel tens of terawatts of free and pure energy, using proven basic technology with passive and low-mass equipment. Further details are presented below in Exhibit 149 on lunar solar power in terms of the major elements involved, their function and advantages, as well as the specific challenges related to each. The transmissions would be at the speed of light over great distances for use where and when needed without physical connections. Essentially, the system involves the use of solar reflectors in lunar orbit to collect sunlight by thin-film photovoltaics which are then converted into thousands of low-intensity microwave beams at power stations on the Moon. From large-diameter, shared synthetic apertures on the lunar surface, these microwaves are beamed to receivers located anywhere on the Earth, including in economically developing nations. Finally, ground electric power distribution systems complete the process begun aloft in the Solar System.

The co-inventors projected LSPS development stages in terms of costs over a 20-year period. Originally, this was based on a timeframe from 1990 to 2010. Realistically, however, significant research funding for such a macroplanning is not likely to be available before 2050. A functioning lunar base with the necessary

556 Appendix B: Space-based energy: Lunar solar power

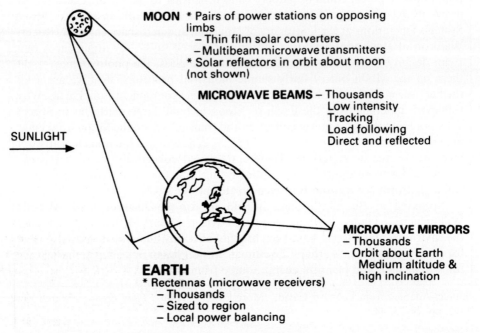

Exhibit 148. Concept: lunar power system. The LSPS is projected in this illustration as having a capacity of producing 100,000 GW. Source: David R. Criswell and Robert D. Waldron from their United States Patent granted to the co-inventors in 1990.

infrastructure has to be in place before LSPS operations would get under way. It should be noted that most of the investment is ground based for rectenna construction. An international effort would be necessary which integrates private and public-sector contributions to this macroproject, beginning with an adequate government-funded space transportation system to the Moon and back.

Criswell and Waldron envisioned a five-stage enterprise to develop a lunar solar power system:

Stage 1 Engineering research, analysis, design, and development relative to LSPS production processes/options; transportation elements in cislunar space; lunar base and space facilities required aloft; Earth-based LSPS facilities and rectennas.

Stage 2 Assuming governments and private enterprises make provisions for space transportation to and from the Moon, as well as scientific and production facilities on the Moon, then LSPS researchers would be in a position to build, operate, and demonstrate rectenna production capacities, possibly under private-sector funding.

Stage 3 Start-up and build-up in emplacement rate of lunar power capacity and expansion of rectennas; such production expansion would probably occur under a private consortium of various terrestrial organizations; negotiations for sale of power to multiple customers on Earth.
Stage 4 Continued build-up of rate for lunar power capacity.
Stage 5 Steady solar power production increase to an annual capacity of >500 GWe.

In developing a case for investment in this macroproject, the LSPS originators also provided additional information on the challenges facing this enterprise, as summarized in Exhibit 149.

Obviously, a large-scale engineering program like lunar solar power is dependent on international cooperation both on the ground and in space. For a technological venture of this scope to succeed requires the formation of global partnerships in this *space-based energy macroproject*, not only among nations and public-sector agencies, but with transnational organizations in the private sector. An energy global consortium might be developed among spacefaring nations: namely, at present, besides the U.S.A., potential partners might be Australia, Canada, China, Japan, European Space Agency, India, and Russia. The Russians have unique capabilities in lunar studies, heavy-lift launch systems, and long-duration space-living experience. There has been some previous Russian interest in lunar solar power, particularly among the fusion engineering community [18]. However, recent interviews by one of the authors (Harris) with key members of their Academy of Sciences confirmed that although Russian space scientists are fixated on Mars missions, their energy scientists are interested in pursuing studies on lunar solar power. Given the current socioeconomic crises throughout the Commonwealth of Independent States, the Russians would have to be convinced (a) that their research would be underwritten partially by external sources; (b) of lunar profits in the future. Before they would undertake technological venturing to the Moon with other nations on this scale, the Russians would have to be convinced of future space-based energy as a means for that nation to prosper after their present petroleum and gas resources run out!

To transform this innovative concept into a realistic plan for implementation, readers are encouraged to consider what they might do to further development of a lunar solar power system. Perhaps such future activities may involve undertaking a research study, or writing a professional paper, or networking with others similarly interested in this macroproject. Co-inventors Criswell and Waldron have outlined some proposed near-term actions:

(1) convening, through an appropriate sponsor, a national and/or international action workshop on lunar solar energy;
(2) additional technology and economic investigations by other investigators of LSPS macroeconomic models, and demonstrations of technological possibilities (e.g., microwave power beaming, orbital reflectors, rectenna options, lunar construction of solar cells and reflectors);

Appendix B: Space-based energy: Lunar solar power

Exhibit 149. LSPS: details and challenges. The major elements for a lunar solar power system are compared as to function, advantages, and challenges.

Major element	Function	Advantages	Challenges
Sun	Power source	Functioning and no ash Immense energy supply	Day/Night cycle
Moon	Platform resources	Natural resource —place —components —logistics —good electronic environment Minimal transportation Non-intrusive	Construction and logistics —remoteness —exotic environment Utilization of resources
*Thin-film photovoltaics	Solar to electric power ($\geq 5\%$)	Little material for much power Fast production	High production rate
*Power plot	Collect electric power	Local small area (100 m^2)	Iron gathering and refining Wire making
*Microwave transmitters	Electric to microwave power	Highly redundant Many beams Beam pointing and bore sight control Limit maximum beam intensity	Precision beam forming
*Segmented lunar reflectors	Redirect sunlight	Focused beams Minimal stray energy Low cost per beam Immense beam range Limit maximum beam intensity	Surface shaping Efficient production

Appendix B: Space-based energy: Lunar solar power

*Industrial base (option)	Make emplacement machinery	Option to minimize transport and start-up cost	Design and low-volume diverse production
Earth rectennas	Microwave to electric power	Output high-grade electricity Base and load following Minimal waste heat Multi-use of land	Minimize production costs (LPS cost driver) Power balance
Microwave reflectors	Illuminate rectennas hidden from Moon	Very low mass Simple Wide range of orbits	Pointing and control
Lunar orbital mirrors (option)	Illuminate lunar power bases —New Moon —eclipse	Eliminate power storage —on Moon —on Earth Increase output	Materials transport to orbit Manufacturing in lunar orbit Pointing and control
Space rectennas (option)	Power facilities Power fast ships	Service inner Solar System Very low mass, simple, and rugged	Minor
Organization	Analyses Commitments Funding Implementation	New net wealth to Earth Eliminate pollution Expansion beyond Earth Advance science	International cooperation Upfront investment

Source: David R. Criswell and Robert D. Waldron [17].
* *Note*: Climate change crises in the early 21st century confirm the need for a long-term, clean fuel source from outer space (see Emma Duncan's "Clean the planet," *The World in 2008*, *The Economist*, p. 24).

(3) action planning for large boosters for transportation into orbit, for lunar construction materials, for large-scale habitats and facilities on the Moon.

The artistic renderings in Exhibit 150 illustrate some other lunar possibilities for industrialization.

To promote the LSPS concept, the first step was taken in July 1990 when a peer-review workshop held in La Jolla, California, confirmed that the proposal was feasible. Some 25 distinguished scientists, engineers, and other professionals established a Lunar Power Systems Coalition (LPSC) for interested individuals and institutions under the leadership of the late Dan Greenwood, CEO of Netrologic Inc. in San Diego. Research papers, proposals, and a newsletter were prepared and

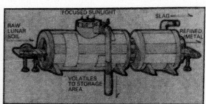

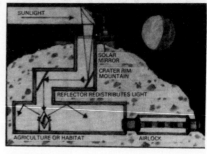

Exhibit 150. Lunar industrialization possibilities. In addition to the forecasts described in Chapter 10 and this Appendix, other technical prospects are illustrated here while operating a lunar base. The basic infrastructure would require advanced lunar transportation and habitat systems, then construction might proceed with liquid propellant tank farms, storage areas for volatiles and processing plants for lunar materials, while solar energy is concentrated by a mirror system used to heat, fuse, and partially vaporize lunar material (such as the above visual on the left). Source: *Space Resources* edited by M. F. McKay, D. S. McKay, and Michael Duke. Washington, D.C.: U.S. Government Printing Office, 1992 (NASA SP-509, five volumes available now from *www.univelt.com*).

distributed, while sponsors and investors were sought. However, LSPS is an idea ahead of its time; so far, insufficient vision and financial resources have curtailed the furtherance of this enterprise and the basic research to be pursued. For space-based energy to come into being in this century, support must be obtained beyond government space agencies and energy departments, such as from global energy corporations, foundations, and others. A *Lunar Economic Development Authority* would facilitate this process were such an entity to come into being.

However, prototype lunar power research would seem appropriate for integrated doctoral studies, as well as by national, corporate, and university laboratories. The need is for multilevel technical analysis, interdisciplinary research that produces SPS and LSPS pilot or demonstration projects. For example, the Aerospace Division of the American Society of Civil Engineers conducts annual conferences and publishes proceedings relative to engineering, constructions, and operations in space. These multidisciplinary gatherings are ideal forums for considering lunar challenges in construction, industrialization, and settlement, as well as for the building of a solar power system from the Moon to Earth. There are also the Wireless Power and Transmission Conferences which have been held periodically to develop a strategic international partnership in that regard. In 1995, these sessions were held at Kobe University and the papers were published (email: *kaya@kobe-u.ac.jp*). The SUNSAT Energy Council and its journal *Space Power* organized the Space Power Network (SPN) to promote Internet discussion of future power generation and supply systems (email: *canough@bingvaxa.cc.binghamton.edu*). However, much more has to be done by the global media to enlighten the general public, the energy industry, and political leaders about the possibilities for development of space-based energy!

B.4 LUNAR POWER BASE TECHNOLOGIES

When Peter Glaser (1977) originated the concept of solar power from space, it was to come from many satellites in geosynchronous orbit (SPS). The NASA Lunar Energy Task Force (1989) confirmed that this is technically feasible if satellites are placed in lunar orbit. While SPS does not require a return to the Moon, economic feasibility may necessitate the use of lunar materials and support services. LSPS, on the other hand, envisions the Moon as a natural platform that is unique, suitable, and available as a power station. Lunar power will not only site large arrays of solar collectors on its surface, but will utilize its resources as well. It is predicated on establishing a lunar base through a combination of public and private funding. The visionary 1979 General Dynamics study assumed the production of 10 GW of solar power per year requiring 400 humans on the Moon [19]. The Criswell and Waldron LSP model is based on 560 GW annually or 56 times more lunar power than that earlier analysis [20]. Assuming the same level of automation to human workers, that would project up to 4,000 people living and working on the Moon for Millennia III! That implies space colonization on a grand scale, far beyond the outpost and base now conceived by NASA. To build facilities sufficient for that number to live and work safely on the Moon will tax the ingenuity of more than building contractors. Biological and behavioral scientists will have to be involved in providing life support systems and human factor planning much beyond traditional industrial engineering (such as was discussed previously in Chapters 2, 7, 8, 9, and 10). Obviously, experts in space communications and high-frontier defense technology can contribute much insight on antennas, rectennas, satellites, linking systems, transponders, and other applications for the peaceful and commercial applications of microwave and beaming technologies of lunar power. Continuing progress in wireless technology research also will contribute to this project.

In his 1979 previous study for a preliminary supply base on the Moon, the senior author (Criswell) projected a communications and control center within the lunar habitat area. Initially, the needs and deployment of the lunar crews would not be unlike those at scientific outposts in Antarctica or other polar regions [21].

B.5 CONCLUSIONS

We have sampled the Moon and possess sufficient understanding to make use of many lunar resources. At the end of the inaugural LPS workshop in 1990, our colleague from Rockwell International, Dr. Robert Waldron, provided a statement, *A Lunar Power System Coalition Charter*, worth repeating here [13]:

> "Global population trends coupled with legitimate aspirations for improved standard of living for current and future inhabitants of Earth will require substantial increase in energy consumption well into the 21st Century."

Waldron, now deceased, concluded that only a lunar power system could provide that amount of energy in an affordable manner, while avoiding the environmental and ecological damage of fossil fuels. For this conversion to extraterrestrial energy sources to occur will take generations of commitment and investment. The time has come for the technical community *to lead* in this change toward greater utilization of lunar resources. No other global investment on the horizon could provide such enormous returns over many decades as lunar solar power. If *space is the place for synergy*, then why not prove it by joint, interdisciplinary studies of space-based energy utilization. If humans are to return to the Moon permanently before year 2020, the 1961 words of President John F. Kennedy bear repetition now:

> "We choose to go to the Moon in this decade to organize and measure the best of our energies and skills."

What is puzzling is why so many within the scientific community and government leadership do not recognize the potential of offworld energy. This was strikingly underscored when UNESCO and the Foundation for the Future held a seminar in Paris, on *Humanity and the Biosphere: The Next Thousand Years* (September 2006) [22]. Here scholars from five continents gathered, including energy experts, policy and management specialists, plus professional futurists, to consider critical factors that will impact our biosphere for this millennium. In the context of our planet's long-term future, the assembled brain trust were examining global environmental issues, as well as how to conserve biological diversity and cope with drastic climate change. Strangely for such a convocation, the 246-page report showed little discussion on the energy crises, especially that caused by use of fossil fuels. And there seemingly was no mention of the prospects for humanity to utilize space-based energy by year 3000. Perhaps these omissions will be addressed in their next session planned for 2008?

Within the global space community, there is recognition of the need and potential for offworld energy. For example, in 2006, the Aerospace Technology Working Group published a volume, *Beyond Earth: The Future of Humans in Space* [23]. One contributor, Dr. Paul Werbo, formerly of the National Science Foundation and the U.S. Department of Energy, did address "Strategic Thinking for Space Settlements: Energy from Space." In his second section on "Energy from Space: New Options and a Critical Need on Earth," Werbos makes these salient points (*www.werbos.com/energy.htm*):

- Among the requirements for sustainability in space that could provide a huge revenue stream is space solar power.
- In the 1990s, NASA's John Mankin revived scholarly and governmental interest in space-based energy with a program called SERT which reviewed, verified, and made recommendations from the National Academy of Science on the proposed technologies. By 2002 his cost estimates for the best designs came in at 20 cents per kW-h, good enough for niche markets—but not to compete initially with coal or nuclear fission for the big baseload power markets on Earth.
- The 2000 Millennium Project of the American Council for the United Nations recognized that the number one way to improve the human condition was to find an alternative energy source that was a non-fossil, non-fission, affordable baseload electricity on a scale large enough to meet needs on Earth (*www.stateofthefuture.org*). This is just what proponents of space-based energy are proposing.
- The not-too-new concepts proposed by Gerard O'Neill, David Criswell, and others for using non-terrestrial materials to produce energy deserve further exploration and fine-tuning.
- The future of space-based energy is dependent on an efficient infrastructure, especially lowering the cost of access to space, possibly to $200/pound or even lower, such as would be feasible by using the new reusable lunar rockets being worked on presently by Boeing and Kistler Aerospace, or ion rockets already proven for deep space by Japan.

The Global Exploration Strategy [24] issued by 14 space agencies in May 2007 was a milestone in its implications for orbital synergy (*www.bnsc.gov.uk/home.aspx*). Their joint report, *The Framework for Coordination*, cited the Moon as our nearest and first offworld goal:

> "As a repository of four billion years of solar system history, it has enormous scientific significance. It is also the base from which to study Earth and the universe, and to prepare humans and machines for venturing farther into space ... Space exploration also offers significant entrepreneurial opportunities for creating new technologies and services. These advances will encourage economic advancement and the creation of new business.

That is exactly what the advocates of space-based energy are trying to do! Among the challenging technologies identified in this Framework for the next era of exploration is *efficient power generation and energy storage! ... The Moon has a strong place in the cultures of many peoples, and it instinctively appeals to human imagination!*

Humanity cannot develop its potential except by going off this planet, and Sister Moon is where we take our first emergent steps [25, 26].

CASE STUDY DISCUSSION QUESTIONS

(1) Why does lunar industrialization, such as a lunar solar power system, provide a strong rationalization for return-to-the-Moon missions, while justifying further space exploration?

(2) Why do present terrestrial environmental concerns make study of an LSPS feasible as an alternative form of 21st-century energy?

(3) Why do current research developments in automation/robotics, micro-beaming, thin-film solar arrays, solid-state microwave electronics, and microcomputers have implications for the successful installation of an LSPS and facilities on the Moon?

(4) Why is so much of the upfront funding for LSPS required on the ground?

(5) Why would such LSPS terrestrial infrastructure investment on Earth be of benefit to the developing economies and their peoples?

(6) Why are national and international consortia essential to develop, procure, and implement elements for LSPS production and rectenna research?

(7) Why are graduate students and industrial investigators well advised to undertake research in some aspect of developing space-based energy?

(8) What can be done to convince the executive and legislative branches of the government, the public utilities and energy associations, the media and general public, that support now for an LSPS makes sense and is good public policy?

(9) Why does the Moon offer a reliable platform for (a) scientific observation; (b) energy production: (c) stepping stone into the universe?

(10) Would you like to live and work on the Moon? If "yes," why?

Appendix B: Space-based energy: Lunar solar power

> **CHANGING CLIMATE: GREENHOUSE EARTH**
>
> Earth is hospitable to life because its atmosphere works like a greenhouse, retaining enough of the Sun's heat to allow plants and animals to exist. The natural climate control system depends on the trace presence of certain atmospheric gases, most important being carbon dioxide, to trap the Sun's radiation. But the system has been jolted. The burning of carbon-laden fossil fuels has hiked atmospheric CO_2 to levels unprecedented during human history. Greenhouse Earth is growing warmer. How much warmer depends on the human response.
>
> Climate is changing more rapidly than ever before. Human activity is the main cause. Burning of fossil fuels (oil, gas, and coal) has flooded the atmosphere with heat-trapping carbon dioxide, triggering a 1°F spike in the average global temperature in the past century, largely in the past 30 years. Already impacts include melting glaciers, intensifying storms, and a rise in sea level. Unless CO_2 emissions are slashed, the planet will likely heat up even faster, fundamentally changing the world in which we live. (Source: *Climate Change*, a map insert in the *National Geographic* magazine, October 2007 (*www.npr.org/climateconnections*).)
>
> *So, why not eliminate the negative effects of fossil fuels by obtaining space-based energy?*

B.6 REFERENCES

[1] Mendell, W. W. (ed.) *Lunar Bases and Space Activities of the 21st Century*. Houston, TX: Lunar Planetary Institute, 1985.

[2] McKay, M. F.; McKay. D. S.; and Duke, M. B. (eds.) *Space Resource*. Washington, D.C.: U.S. Government Printing Office, 1992, 5 vols. (available now from *www.univelt.com*).

[3] ESA Lunar Steering Committee Group, *International Lunar Workshop: Towards a World Strategy for the Exploration and Utilization of Our Natural Satellite*. Paris: European Space Agency, 1994 (ESA SP-1170); *Mission to the Moon: Europe's Priorities for the Scientific Exploration and Utilization of the Moon*, 1992 (ESA SP-1150).

[4] Simons, M. S. (ed.) *Proceedings of the International Lunar Exploration Conference*. San Diego, CA: International Space Enterprises, 1994 Vol. 1, 1995 Vol. 2 ... Durst, S. M.; Bohannan, C. T.; Thomason, C. G.; Cerney, M. R.; and Yuen, L. (eds.) *Proceedings of the International Lunar Conference 2003/International Lunar Exploration Working Group 5*. San Diego, CA: UNIVELT/AAS, Vol. 108, 2004 ("Science and Technology Series"; *www.univelt.com*).

[5] Angelo, J.; and Easterwood, G. W. *Lunar Base Concepts*. Melbourne, FL: Orbit Books/Krieger Publishing, 1989 ... Ross, M. *From Footprints to Blueprints: Development of the Moon and Private Enterprise*. Bloomington, IN: Author House, 2005 ... Schrunk, D.; Sharpe, B.; Cooper, B.; and Thangavelu, M. *The Moon: Resource, Future Development, and Settlement*. Chichester, U.K.: Springer/Praxis, 2008, pp. 174–175 (*www.praxis-publishing.co.uk*).

[6] Criswell, D. R. "Net Growth in a Two-planet Economy," *Proceedings of the International Astronautical Federation Congress, Jerusalem, Israel, October 9–14, 1994* (IAF 94-IAA8.1.704).

[7] Glaser, P. E.; Davidson, F. P.; and Csigi, K. *Solar Power Satellites: A Space Energy System for Earth*. Chichester, U.K.: Wiley/Praxis, 1998 (*www.praxis-publishing.co.uk*). Note contributions by Dr. David Criswell: "Lunar Based Solar Power System," p. 227; "Solar Power System Based on the Moon," pp. 559–621.

[8] Criswell, D. R. "Lunar Solar Power System: Scale and Cost Versus Technology Level, Boot-strapping, and Cost of Earth-to-Orbit Transport," *Proceedings of 45th International Astronautical Federation Congress, Oslo, Norway, October 2–6, 1995* (IAF 95-R-2.02).

[9] O'Donnell, D. J.; and Harris, P. R. "Space-Based Energy Needs a Consortium and Revision of the Moon Treaty," *Space Power*, **13**(1/2), 1994, 121–134.

[10] Criswell, D. R. "Lunar Solar Power and World Economic Development," *Solar Energy and Space Report, World Solar Summit, July 3–9, 1993*. Paris: UNESCO Publications.

[11] Keller, E. A. *Introduction to Environmental Geology*. Upper Saddle River, NJ: Pearson/Prentice Hall, 2008 ... Nielsen, R. *Little Green Handbook: Seven Trends Shaping the Future*. New York: Picador Publishing, 2006.

[12] Criswell, D. R.; and Waldron, R. D. "International Lunar Base and Lunar-Based Power System to Supply Earth with Electric Power," *Acta Astronautica*, **29**(6), August 1993, 469–480 (published by the International Astronautical Federation, Paris).

[13] Bodie, S. (ed.) *Lunar Solar Power Workshop Proceedings, July 9–11*. San Diego, CA: Lunar Solar Power Coalition/Netrologic Inc., 1990 ... Pederson, M. N. A.; Criswell, D. R.; and Greenwood, D. "Clean Sustainable Power from the Moon," paper presented at the National Press Club, Washington, D.C., October 24, 1990 ... See also co-inventors Drs. D. R. Criswell and R. D. Waldron, *A Power Transmission System*, United States Patent, granted in 1990.

[14] Criswell, D. R.; and Waldron, R. D. "Lunar Solar Power System: Options and Beaming Characteristics," *Proceedings of the 44th International Astronautical Federation Congress, Graz, Austria, October 16–22*. Paris: International Astronautical Federation, 1993 (IAF 93-R.2.430).

[15] Sagan, C. "Tomorrow's Energy," *Parade Magazine*, November 25, 2990, pp. 10–15 ... David, L., "More Power to You," *Final Frontier*, March/April 1993, pp. 20–23, 48.

[16] Schmitt, H. H. *Return to the Moon: Exploration, Enterprise, and Human Settlement of Space*. New York: Springer/Copernicus/Praxis, 2006.

[17] Criswell, D. R.; and Waldron, R. D. "Lunar Power Systems to Supply Solar Electric Power to the Earth," *25th Intersociety Energy Conversion Engineering Conference, Reno, Nevada, August 11–17, 1990* (paper #900279/243).

[18] Sarkisyan, S. A. et al. "Socio-Economic Benefits Connected with the Use of Space Power and Energy Systems," *Proceedings of the International Astronautical Federation*. New York: Pergamon Press, 1986 (#IAF-85-188).

[19] General Dynamics, *Operations and Logistics Study of Lunar Exploration Systems for Apollo* (1969, NASA Contract #NAS9-1555-60); *Lunar Resources for Space Construction* (NASA Contract #GDC-ASP79-001). San Diego, CA: General Dynamics Convair Division (now Lockheed-Martin).

[20] Criswell, D. R.; and Waldron, R. D. "Results of Analysis of Lunar-based Power System to Supply Earth with 20,000 GW of Electric Power," paper presented at SPS '91 Conference, Paris, August 1991.

[21] Harris, P. R. "Personnel Deployment Systems: Managing People in Polar and Outer Space Settings," in Harrison, A. A.; Clearwater, Y. A.; and McKay, C. P. (eds.) *From Antarctica to Outer Space: Life in Isolation and Confinement*. New York: Springer-Verlag, 1991, pp. 158–188 ... "Promoting the Coalition of Lunar Solar Power," *Space Policy*, May 1991.

[22] Foundation for the Future, *Humanity and the Biosphere: The Next Thousand Years*. Seminar Proceedings, September 20–22, 2006. Bellevue, WA: Foundation for the Future in conjunction with UNESCO/Division of Ecological and Earth Sciences (*www.future foundation.org*; *www.unesco.org/science/earth* or *www.unesco.org/mab*).

[23] Werbos, P. J. "Strategic Thinking for Space Settlements: Energy from Space," in Krone, B.; Morris, L.; and Cox, K. (eds.) *Beyond Earth: The Future of Humans in Space*. Burlington, Ontario: Apogee Space Books, 2006, pp. 162–168 (*www.apogeebooks.com*) ... Also see Krone, B.; and Krone, S. *Ideas Unlimited: Capturing Global Brainpower* and Harris, P. R. *Launch Out*. West Conshohocken, PA: Infinity Publishing, 2003; 2007.

[24] The Global Exploration Space Strategy, *The Framework for Coordination*. Press Release by BNSC Media Centre, June 9, 2007, 31.05.07; 20.0407; 15.05.06 (*http://bnsc.gov.uk/home/aspx?nid=3191*).

[25] Harris, P. R. *Toward Human Emergence*. Amherst, MA: Human Resource Development Press, 2008 (see Epilogue).

[26] Obtain the book and CD entitled *2007 State of the Future* by J. Glenn and T. J. Gordon, published by The Millennium Project, World Federation of U.N. Associations (4421 Garrison St. NW, Washington, D.C. 20016; *www.stateofthefuture.org* or email: *jglenn@igc.org*) ... *ENERGYVILLE*, an online, interactive, futuristic energy game developed by The Economist Group in conjunction with Chevron Corporation (*www.willyoujoinus.com* or *www.economist.com*).

SPACE ENTERPRISE UPDATE

Two fundamental realities will drive space exploration forward. First, wealth is accumulating in the hands of ambitious and visionary individuals, many of whom view space simultaneously as an adventure and a place to make money. What was once affordable only to nations can now be funded by individuals.

Second, corporations and investors are realizig the resources of Earth are limited, and are running out. But everything we hold of value on Earth—metals, minerals, energy and real estate—is in near-infinite supply in space. As space operations become more affordable, companies will set their sights on extra-terrestrial resources, and what was once thought of as a vast wasteland, will become the next "gold rush" ... The next 50 years will be when we establish ourselves as a space-faring civilization.

Source: Peter Diamandis, M.D., chairman and CEO of the X-Prize Foundation, and co-founder of the International Space University, "The Next Space Race," The World in 2008, *The Economist*, 2007, p. 153.

Appendix C

Learning from space entrepreneurs
Thomas L. Matula, Ph.D.[1]

> *Entrepreneurs are discoverers and promoters who perceive and pursue opportunity, thus creating start-up, high-risk ventures with great potential to meet human needs. Such personalities are needed to organize, operate, and assume risks in offworld enterprises.*
>
> Philip R. Harris[2]

C.1 DAWN OF THE SPAGE AGE

Entrepreneurs have a long association with the development of space. The first efforts to develop practical reusable rockets date to rocket mail ventures in the 1920s and 1930s. However, neither the technology nor markets were right for space commerce, so space quickly became a government activity, first in Germany, then in the United States and Russia. But the dream of space enterprise never faded and one of President's Kennedy's goals in his famous speech on national space goals (May 1961) was the development of a network of communication satellites [1]. Although not as well remembered as the goal of landing a man on the Moon, it has turned out to be a legacy that perhaps surpasses it.

On July 10, 1962 Telstar, the first commercial satellite was launched, followed on August 31 by President Kennedy signing the Communications Satellite Act of 1962 into law [2]. The modern era of space commerce had begun. Today space commerce is a $110 billion industry, with communications satellites accounting for the majority of space commerce revenues. Communication satellites have become so woven into the

---
[1] For author biography see "About the contributors" page in the prelims.
[2] For more on the opening quotation, refer to P. R. Harris' *New Work Culture*. Amherst, MA: Human Resource Development Press, 1998, ch. 11 (*www.hrdpress.com*).

fabric of the world economy, they have become all but invisible. Live pictures from the most remote regions of the world, even from battle zones, are taken for granted by television viewers. International phone calls, once rare and expensive, are now routine while the Internet allows anyone with a computer to surf websites and communication instantly with anyone, anywhere in the world. All these activities are not taken for granted, yet all are dependent on communication satellites and space commerce.

Although the modern space commerce era began in 1962, activity in the first couple of decades was limited to large corporations. The high levels of investment, high levels of technology risk and long payback periods made it an environment that was not suited to small entrepreneurs. It wasn't until the late 1970s that the first space entrepreneurs emerged. In this appendix, we will discuss some lessons learned from those first 25 years of space entrepreneurship.

In the early 1980s the climate for space entrepreneurs was one of optimism. NASA's Space Shuttle was just beginning its service life and with it a new optimism for space ventures was being created. Books like G. Harry Stine's *The Third Industrial Revolution* predicted a revolution in material science and manufacturing resulting from access to zero-G research laboratories and manufacturing facilities. It appeared that a new age of space technology was dawning with the creation of new orbital industries and numerous opportunities for entrepreneurs.

The first and biggest challenge space entrepreneurs faced was one they are still familiar with today, the high cost of launching payloads. In the early 1980s, options were limited to reimbursing NASA for payloads launched on expendable launch vehicles, the NASA Space Shuttle, or going to Europe and purchasing a launch on an Ariane rocket. All had very high price tags that greatly limited space access for small entrepreneurs interested in space. But entrepreneurs also recognize challenges as opportunities, so the high cost of space access was recognized as a business expense by several early space firms. In this analysis,, we will compare the story of two firms, Space Services Incorporated and Orbital Sciences Corporation, that responded to the challenge, so that lessons may be learned for later generations of space entrepreneurs.

C.2 SPACE SERVICES OF AMERICA INCORPORATED

Space Services Incorporated (SSI) was organized in Houston, Texas, by David Hannah, Jr. in 1979 with the goal of developing a new low-cost launch vehicle [3]. Hannah believed that expendable launch systems were expensive because they were based on launch vehicles originally developed for the military as intercontinental ballistic missiles (ICBMs). He believed that by developing a family of launchers designed specifically for commercial needs, the cost of launch could be greatly reduced. SSI's new vehicle was named Conestoga after the horse-drawn wagon made famous in opening up the American frontier. It was assumed it would have a similar role in opening up the space frontier.

SSI had success initially by locating a number of investors from the oil industry used to taking gambles in high-risk ventures who saw space as a new frontier for investment. Their first attempt at Conestoga was based on a design by Gary Hudson called the Percheron [4]. It failed in its only attempted launch on May 8, 1981. However, the launch was valuable in other aspects. As the first attempt to launch into space, SSI had to pioneer the process of obtaining the permits and licenses from the government which were required to launch its vehicle. Because there was no centralized authority for such undertakings at this time, the corporation had to deal with numerous state and federal agencies before they were allowed to go ahead with the launch. Their experience eventually led to the creation of the Associate for Space Transportation at the Federal Aviation Administration, as the Federal government agency responsible for licensing commercial launches as a means of simplifying the licensing process [5].

Following the failure of the Percheron, SSI regrouped and hired former Astronaut Donald "Deke" Slayton, Jr. as CEO. They also replaced the liquid-fuel Percheron with a solid-fuel designed based on the Castor 4A motor used as strap-on boosters for the Atlas and Delta launch vehicles [6]. The switch in rockets was successful and resulted in the first launch into space of a privately funded launch vehicle on September 9, 1982, from Matagorda Island in Texas [7]. However, this first launch was also marked by tragedy as one of the major investors, Toddie Lee Wynne, who owned the launch site, died of a heart attack that morning before the launch. His will left the ranch to the State of Texas as a park, forcing SSI to locate a new launch site.

Space Services Incorporated regrouped again and signed an agreement with NASA to operate from NASA's Wallops Island Flight Center. Also, with development of the Conestoga delayed due to funding and the lack of firm customers, the firm switched its focus to marketing suborbital flights using NASA's Black Brant sounding rocket which was renamed Starfire. Its first launch was on March 29, 1989 at White Sands Missile Range, marking another milestone in space commerce as the first commercial launch licensed by the FAA Associate for Space Transportation [6].

However, funding issues continued to plague SSI due to a lack of viable markets for its services. In 1990, its principle investors stopped funding it and the firm was sold to EER Systems in December 1990, a major defense contractor. The company continued to be active under the new owner winning a contract from NASA for the Commercial Experiment Transporter (COMET) program in 1991. On October 23, 1995, a new version of the rocket, now known as Conestoga 1620 was launched from NASA Wallops Island Flight Facility, but broke up in flight only 46 seconds into the launch [8]. NASA stopped further funding of the program and EER Systems closed out the Conestoga. In 2004 the former employees of Space Services Incorporated of America combined with the founders of Celestis Inc. to form a new company Space Services Incorporated to continue the pioneering tradition of Space Services Incorporated of America. Today Space Services Inc. contracts to provide payloads, as on the recent flight of a UP Aerospace sounding rocket from Spaceport America in New Mexico [9].

C.3 ORBITAL SCIENCES CORPORATION

Orbital Sciences (OSC) was founded on April 2, 1982 by three graduates of Harvard Business School, David Thompson, Bruce Ferguson and Scott Webster [10]. Their analysis of the environment for space commerce identified a different opportunity. Instead of focusing on development of a complete new launch system for commercial payloads, they identified a need created by the new Space Shuttle. The Space Shuttle was designed to replace the existing Expendable Launch Vehicles (ELVs) in the launch of all satellites including communication satellites. However, the Shuttle was limited to low-Earth orbit while communication satellites, military satellites, and space exploration craft required much higher orbits.

The initial business model for OSC was to fill this need by securing from NASA a contract to develop a liquid-fuel upper stage for the Space Shuttle designed to place these satellites in their orbit. On October 20, 1983 they signed an agreement with NASA to develop the Transfer Orbit Stage (TOS) and secured $2 million in venture capital. They also signed an agreement with Martin Marietta to build the TOS [11]. With this NASA contract, the firm was able to secure $50 million in venture capital to finance the TOS. They also acquired a second contract with the USAF to produce guidance modules for the Minuteman Missile [11]. OSC continued to bid on additional contracts with NASA and the USAF.

Following the *Challenger* accident in 1986, NASA canceled the contract for the TOS. But by this time OSC had acquired a number of DOD-related contracts. In 1988, the Defense Advanced Research Project Agency (DARPA) signed a key contract with OSC to develop the Pegasus launch vehicle, a revolutionary air-launched system designed to allow quick liftoff into orbit of small payloads from anywhere on Earth [11]. The first Pegasus was successfully launched on April 5, 1990 from Edwards Air Force Base. In April 1990, OSC also conducted an Initial Public Offering (IPO) for $32.5 million becoming the first space commerce firm to be publicly traded [11]. Since then OSC revenues and market revenue have continued to expand. Today OSC is listed on the New York Stock Exchange and has a market capitalization of $1.48 billion with revenues of $501.5 million for the first 6 months of 2007 [11].

C..4 LESSONS LEARNED FROM THESE SPACE ENTREPRENEURS

For those entrepreneurs planning to enter space commerce, there are several lessons to learn from these cases of SSI and OSC. Lessons that reinforce the basics of any good business strategy. Orbital Sciences was the clear winner. Today they are a publicly traded corporation on the New York Stock Exchange with a market value of $1.48 billion. By contrast the reformed SSI only exists today as a privately held firm with only a small handful of employees. What are the insights to be learned from these two pioneering space firms?

First, space entrepreneurs spend a great deal of time discussing business models without really understanding their foundation. In essence they treat them as something magical, like finding the right formula for rocket fuel. Actually a business model is simply the value exchange process in which value is generated for the customer. In his book *Good to Great: Why Some Companies Make the Leap and Others Don't* [13], Jim Collins point out that the key to sustainable success for any firm is not its business plan but its management team [12]. A weak business plan may be easily salvaged by a strong management team, while a great business plan will fail if the wrong team is in place to implement it.

How did OSC and SSI differ on their business team? OSC's core management team consisted of three individuals well trained in business. They view space commerce as a business opportunity, but did not get lost in any specific technology solutions, or let their passion for space overrule their business training. Instead, they identified a clear opportunity created by the Space Shuttle, the TOS, and moved forward to take advantage of the prospect. Unlike other space commerce firms they didn't let the technology drive their products, but the market. Once they secured the contract for the TOS, they outsourced the technical development to Martin Marietta, a firm well qualified to perform OSC objectives. It should also be noted this was the winning formula with the X-Prize: Paul Allen simply contracted with Scaled Composites, an established aerospace contractor, to build SpaceShipOne.

By contrast SSI spent a large amount of time and money on doing hardware development within the firm. Their management team focused on acquiring managers with technology skills, not the critical business skills needed. Their team was heavy in the area of engineering and technical expertise, weak in finance and marketing. They did achieve a number of historical milestones as a result, but none really provided the revenue needed to move forward. The company spent years in fundraising and in finding customers for its products. After a number of years of poor performance the investors grew tired of the lack of revenue and sold out to another firm.

The learning from this is that space commerce firms normally must be run by business professionals, not technology experts. A minimum level of technical knowledge is needed, but the key technology task should be outsourced. This wisdom is being proved today with Bigelow Aerospace and Blue Origins, business professionals with limited knowledge of space technology who have simply hired engineers and subcontractors to perform the necessary technical work. Basically these engineering skill should be viewed as another resource to acquire as needed.

The second key lesson and difference between the two firms was their approach to revenue. SSI followed a "Field of Dreams" approach in the belief that if they built their system, customers would appear to buy it. In the reality of the business environment, this is often a path to failure. At the start, OSC didn't build anything without a firm contract in hand to cover the research and development costs. This guaranteed revenue made it much easier to raise the capital necessary to accomplish the work as it eliminated a large measure of the uncertainty for the revenue streams. This also is a lesson to today's space entrepreneurs that a signed contract, preferably from a major corporation or federal government, for services is worth far more than any projected revenue streams for future customers. Investors are risk-averse by nature.

The third lesson is sustainability and flexibility. Both start-up companies faced major challenges that impacted their original products. The *Challenger* accident resulted in NASA banning liquid-fuel rocket stages like the TOS from future Space Shuttle missions. The loss of a major investor and their launch site forced SSI to focus on sounding rockets. The difference is the OSC never built their business around a single product. Once they acquired the TOS contract, they didn't stop seeking new contracts for other products and services. Instead, they redoubled their efforts. Thus, when the TOS project failed due to *Challenger*, they just left it behind and moved on to other enterprises, including their highly successful Pegasus Launch Vehicle.

By contrast SSI kept focused on its one core idea of the Conestoga and repeatedly kept returning to it. Other projects, like its Starfire sounding rocket, seemed to be viewed as mere stopgaps until it could raise funding for another go at the Conestoga. This narrow focus prevented it from branching out into new endeavors and pursuing new markets. In short, it prevented SSI from adapting to the changes in the space commerce environment that were taking place in the 1980s and 1990s.

C.5 CONCLUSIONS

In summary, the principal lesson to learn is that space commerce is a business, and should be pursued as a business. Passion and vision are useful, but only if they are tempered by a willingness to recognize the realities of the business environment within which the organization operates. Firms managed by individuals with strong backgrounds in business are also more likely to succeed for this reason. Firms managed by engineers tend to get too focused on the technology of the projects they are working on, often becoming blind to other, better, opportunities. Technology should be viewed as merely a tool to service markets—not the other way around. Without documented proven markets, the best technology in the world will become simply a curiosity. Therefore, the key to successful space firms is no different than any other business firm: namely, identifying markets that both exist and are willing to purchase the products and services offered, and then the firm focuses on providing them with what they need.

Editor commentary: The above case does indeed provide valuable advice for those new to space business. It supplements what I wrote about in Chapter 9. But many of the *NewSpace* entrepreneurs not only reject the overdependence of the traditional aerospace industry on government contracts, but they seek to provide space services and products beyond the usual rockets, spacecraft, and launch systems, while obtaining new sources of investment or acquisitions. The innovative trend can be seen in the next exhibit which applies some of Dr. Matula's lessons. When I first met Chuck Lauer, a real estate tycoon, his concern was applying the procedure of that industry to orbital and lunar habitats. Now consider what this entrepreneur is presently doing.   P.R.H.

Exhibit 151. PRIVATIZING SPACE TRANSPORTATION.

When the Ansari X-Prize in 2004 went to Burt Rutan's Scaled Composites for making the first privately funded manned trips to space, the others that had been vying for the $10 million prize money, though no doubt disappointed, kept plugging at their design of commercial spacecraft, Some, Rocketplane Global among them, are now racing to offer the first suborbital tourist flights; I chronicle these efforts in my recent book, *Rocketeers*... Then in August 2006, the company won a contract through NASA's new Commercial Orbital Transportation Services program to provide an orbital ship for servicing the International Space Station. The firm began channeling its resources away from its suborbital Rocketplane XP, to orbital spaceships under development by Rocketplane's newly acquired Kistler Aerospace. The changed Rocketplane will have the same business jet look, but the modifications will enable the craft to fly in orbit... Chuck Lauer, their co-founder and business development manager, has an innovative business plan for space tourism to attract new investors. The firm believes in building on existing, tested hardware to achieve its goals.

Rocketplane now uses the fuselages of Learjet 26 airplanes with its GE CJ610 jet engine. But its redesign has added a new delta-shaped, strengthened wing assembly and tail needed for 4-G spaceflight. The Rocketplane XP would take off from a long runway and fly to a launch altitude of 25,000 feet. Then the plane calls for a pilot to shut down the jets and light the rocket engine for a 70-second boost into space at a speed three and a half times that of sound, thus giving passengers three minutes of weightlessness ... As the spaceplane left and re-entered the atmosphere, the XP's computers would blend their RCS controls with standard airplane controls on the ground. The XP's computers would fly the craft from boost to reentry, with a pilot taking over only in an emergency and for landings. On the return, the pilot would also restart the jet engines at 20,000 feet. Meanwhile, Lauer and partner, Burnside Clapp, got a fresh infusion of cash for their company from businessman and space enthusiast George French, who now serves the firm as its President. Together they followed a brilliant strategy with the State of Oklahoma which provided generous tax credits when the corporation moved its headquarters to that state, so as to offer new job opportunities there. Further, the $18 million Rocketplane received in tax credits are transferable. Having beaten out their competitors for this "O-Prize", French then sold some of the credits for $13 million which furthered the XP business plan. When Clapp became skeptical that flying tourists into space was not a viable business operation, he resigned. Then Lauer and French brought in David Urie to lead their design team. Urie had 30 years experience as an engineer and manager of Lockheed-Martin's famed "Skunk Works" of innovators. A former Shuttle astronaut, John Herrington, was hired to be chief pilot when the spaceplane becomes operational.

Fortunately, some 80 miles away from Will Rogers Airport at Burns Flats, the company found the perfect 13,500-foot runway at a base of the former Strategic Air Command. By June 2006, the Federal Aviation Administration certified the place as an Oklahoma commercial spaceport. So, all is in place for the Rocketplane XP to fly into space, and meet the challenges of its principal competitors: Virgin Galactic's SpaceShipTwo, Blue Origin's New Shepard, SpaceX, and EADS Astrium from the European Aeronautic Defense and Space Company ... With all this competition, it seems possible that within the next decade or two, a suborbital passenger service to space could drop to the cost of an ordinary expensive vacation: a Caribbean cruise, say. But entrepreneur Lauer sees a future market not only for space tourism, but use of suborbital spaceplanes for point-to-point transportation on Earth!

Source: Adapted from Michael Belfiore, "The O-Prize: Will Rocketplane Launch Spacecraft from Oklahoma," *Air & Space*, November 2007, pp. 36–41 (Smithsonian magazine). See also Belfiore's book, *Rocketeers: How a Visionary Band of Business Leaders, Engineers, and Pilots Is Boldly Privatizing Space*. Washington, D.C.: Smithsonian Books/HarperCollins, 2007.

Perhaps the biggest hope for 21st-century space development may come from innovative space entrepreneurs! Coming from the bottomless well of human ingenuity, innovative strategies tap into hitherto neglected intellectual capital and connect it better with financial capital. Space innovators are in the vanguard of private-sector *astrobusiness*. That is why the X-Prize Foundation and Google established a $30 million prize for the first private-sector team to land and operate a rover on the Moon! As the foundation's innovator, Dr. Peter Diamandis, observed:

> "Real breathroughs require risk and the ability to absorb failure, and large organizations are incapable of such risk taking."

When speaking on the culture of improvement, SpaceX's Elton Musk said it best:

> "There is a culture here that celebrates the achievements of individuals—and it is too often forgotten in history that it is individuals, not governments or economic systems, that are responsible for extraordinary breakthroughs."

Perhaps it will be innovative entrepreneurs who will really implement the vision of space exploration [14]!

C.6 REFERENCES

[1] Kennedy, John F. "Special Message to the Congress on Urgent National Needs," 1961 (retrieved July 30, 2007 from www.jfklibrary.org/Historical+Resources/Archives/References+Desk/Speeches/JFK/003POFO3NationalNeeds525196.htm).

[2] Satellite Chronology, "Telstar," 2007 (retrieved from http///roland.lerc.nasa.gov/dglover/sat/telstar.html).

[3] NASA History Division, "Communication Satellites: Making the Global Village Possible," 2007 (retrieved July 30, 2007 from http://www.hq.nasa.gov/office/paoHistory/satcomhistory.html).

[4] Space Services of America Records, 1979–1991 (retrieved October 1, 2007 from *http://siris-archives.si.edu/ipac20/ipac.jsp?uri=3100001~!265546!0*).
[5] *Encyclopedia Astronautica*, Percheron (retrieved August 10, 2007 from *www.astronautix.com/lvs/percheron.htm*).
[6] Cushman, J. H. "Early Risks of Business in Space," *New York Times*, August 26, 1990 (retrieved September 7, 1970 from *http://query.nytimes.com/gst/fullpage.html?*).
[7] *Encyclopedia Astronautica*, Conestoga, 1620 (retrieved August 10, 2007 from *www.astronautix.com/lvs/conestoga*).
[8] Butricia, A. J. "The Commercial Launch Industry, Reusable Space Vehicles, and Technological Change" (retrieved August 16, 2007 from *www.hnet.msu.edu/business/bchweb/publications/BEHprint*).
[9] Celestis Foundation, "About Us" (retrieved August 31, 2007 from *www.memorialspaceflights.com/asp*).
[10] *Ibid.*, "The Legacy Flight".
[11] Company Histories, "Orbital Science Corporation" (retrieved August 31, 2007 from *http://www.fundinguniverse.com/company.action?sym=ORB*).
[12] Company Information, Orbital Science Corporation (retrieved October 16, 2007 from *http://www.quote.com/stocks/company.action?sym=ORB*).
[13] Collins, J. *Good is Great: Why Some Companies Make Long Leaps and Others Don't*. Boulder, CO: Collins, 2001.
[14] See "Something New Under the Son: A Special Report on Innovation," *The Economist*, October 13th, 2007, 20-page insert (*www.economist.com/special* reports).

SPACE ENTERPRISE UPDATE

Dr. Peter Diamandis has forecast that the next race into space will be led by entrepreneurs, members of the private sector, who envision the high frontier as an opportunity for expansion and vast wealth creation. Some of these leaders have already emerged from among the new billionaires who benefited from high-tech innovations. Beside the NewSpace enterprises, many of the start-up companies will come out of the older aerospace technologies and corporations. Let us examine just one: XCOR Aerospace based at the Mojave Spaceport and Civilian Aerospace Test Center in California. It has improved on traditional rocket technology to create the EZ-Rocket for manned flying vehicles. Its engines are fueled with isopropyl alcohol and liquid oxygen. The alcohol is stored in an external composite fuel tank, and the LOX is contained in an insulated internal aluminum liquid oxygen tank. The EZ-Rocket is incorporated into XERUS, a multi-mission suborbital spacecraft with many safety features. Its pilots include Lt. Col. Dick Rutan (USAF Retd.), former NASA astronaut Richard Searfoss, and private astronaut Michael Melville. Founded in 1999, the company has become a leader in development of reliable, low-cost rocket engines and reusable launch vehicle technology. Its diverse team of aerospace builders innovate with rocket propulsion hardware systems. Its markets are servicing microgravity research, suborbital tourism, and microsatellite delivery to LEO. The business income comes from a mix of government and private contracts, as well as private investments (*www.xcir.com*).

Exhibit 152. Lunar medics. A health care team on the Moon as depicted by artist Pat Rawlings: space nurses in action aloft. Source: Center for Advanced Space Studies, Houston, TX.

Appendix D

Health services aloft: Space nurses
Linda M. Plush, M.S.N. and Eleanor A. O'Rangers, Pharm.D.[1]

Telemedicine—the practice of medicine from a distance using advanced information and communications ssytems—has been developed on an international scale by the USA and Russia to allow clinical consultation and medical education between academic and clinical sites, linked together through an Internet-based telemedicine testbed... While general paramedic-level knowledge among the crew will be normal on every mission, there will be times when sick or injured crew-members will require more extensive care... The best medical care will be available through telemedicine links from Earth to help the crews on board, which still have the responsibility for treatment.

Marsha Freeman[2]

PART 1 HEALTHCARE OFFWORLD: HISTORICAL PERSPECTIVES AND CURRENT STATE OF AFFAIRS RELATIVE TO MEDICAL CARE KNOWLEDGE AND CAPABILITIES IN THE MANNED SPACEFLIGHT PROGRAM AND BEYOND

Introduction: The foundation of the field of space medicine has been integral to acquiring an understanding of the physiological effects of spaceflight. As we look forward to the time when interplanetary travel becomes a reality, this emerging discipline will become critical to ensuring that human beings can survive in the microgravity environment, as well as live and thrive on other worlds. We begin now with a review of the evolution of space medicine and knowledge.

---

[1] Eleanor A. O'Rangers, Pharm.D. and Linda M Plush, R.N., MSN, CSN/FNP, FRSH, are co-founders of Space Medicine Associates, LLC, and founding members of the Space Nursing Society. For author biographies see "About the contributors" page in the prelims.
[2] Marsha Freeman's *Challenges of Human Space Exploration*. Chichester, U.K.: Springer/Praxis, 2000, §7.2, p. 205 (*www.praxis-publishing.co.uk*).

D.1 DAWN OF SPACE MEDICINE

Until the advent of German rocketry during World War II, little serious thought was given to the possibility that humans could achieve spaceflight, much less that space health care would evolve as a new area of medical specialization. "The earliest conference addressing possible medical concerns associated with spaceflight in the United States can be traced back to 1948, when Air Force Major General H. G. Armstrong organized a meeting of physicians and astrophysicists at the USAF School of Aviation Medicine" [1]. As interest in aerospace medicine began to grow, the Aeromedical Association admitted a Space Medicine Branch in 1951, providing broader legitimacy to this emerging specialty [2]. During the 1950s, the U.S. Air Force and Navy expanded their space medicine training programs and initiated research into the field in the areas of life support, tolerance to acceleration (g) forces, and reactions to confinement,topics that were initially extensions of aviation research [3].

With the launch of Sputnik in October 1957, a new sense of urgency infused the space medicine community, as the possibility of humans traveling into space suddenly drew nearer. In 1958, the National Academy of Sciences–National Research Council Committee on Bioastronautics proposed several potential physiologic problems that humans would face in the space environment (Exhibit 153) [4]. Much controversy existed in the scientific and medical communities at the time regarding the fate of humans in space. It is interesting to note that many of the NAS–NRC predictions were later borne out when more systematic physiology investigations were performed in space. As the pace of space activities accelerated post-Sputnik, little time was afforded to the deliberate investigation of space physiology and medicine in laboratories and ground simulations; indeed, only mission-critical issues pertaining to life support, health, and safety were simultaneously developed along with launch vehicles in the early days of the "space race".

D.2 SPACE MEDICINE ISSUES IN PROJECTS MERCURY, GEMINI, AND APOLLO

Shortly after the creation of the National Aeronautics and Space Administration in mid-1958, astronauts were selected for the first U.S. manned spaceflight program, Project Mercury. All astronaut candidates were military test pilots by presidential mandate; President Dwight Eisenhower's rationale was that a group such as this would have refined motor skills, prior demonstration of quick decision-making skills, and the ability to effectively confront threatening situations in the air [5]. Candidates were subjected to a battery of interviews and medical examinations intended to select out any hidden medical problems. Criteria were primarily adapted from those used for military aviator qualifications, with the addition of parameters physicians guessed would be important for astronaut adaptation and functioning in space. Seven astronauts were eventually identified from a pool of 100 test pilot candidates [6, 7].

Exhibit 153. PREDICTED EFFECTS OF WEIGHTLESSNESS.

Anorexia	Demineralization of bone
Nausea	Renal calculi
Disorientation	Motion sickness
Sleepiness	Pulmonary atelectasis
Sleeplessness	Tachycardia
Fatigue	Hypertension
Restlessness	Hypotension
Euphoria	Cardiac arrhythmias
Hallucinations	Postflight syncope
Decreased g tolerance	Decreased exercise capacity
Gastrointestinal disturbance	Reduced blood volume
Urinary retention	Reduced plasma volume
Diuresis	Dehydration
Muscular incoordination	Weight loss
Muscle atrophy	Infectious illnesses

Source: National Academy of Science–National Research Committee on Bioastronautics, 1958.

The overall objective of Project Mercury was to prove that an astronaut could be safely launched into low-Earth orbit and could also return to Earth [8]. Life support systems, including maintenance of cabin air pressure and temperature, removal of metabolic by-products such as carbon dioxide, and availability of food and water were pioneered during this program. Moreover, Mercury helped to diminish many theoretical concerns that some scientists harbored regarding spaceflight. For example, they learned that astronauts were capable of performing activities in space and were able to swallow food and water. Nevertheless, Project Mercury also helped verify that spaceflight is not without its medical hazards: weight loss due to dehydration and cardiac dysfunction, manifested as dizziness and orthostasis post-flight, were documented [8].

It was during the Project Mercury period that President John Kennedy announced plans to send astronauts to the Moon and back—all by the end of the 1960s [9]. With a substantial amount of enabling work yet to be done, NASA planned for the next incremental space program, Project Gemini. With a larger spacecraft, two astronauts would travel into low-Earth orbit. The major objectives of this program included medical issues for the first time, such as (1) the successful implementation of extravehicular activity (EVA), since the upcoming Apollo program would land astronauts on the Moon, and (2) the assessment of physical endurance and health during a prolonged spaceflight in order to gauge performance over the maximum duration of a lunar mission [10]. The cardiovascular changes observed in Project Mercury were also evaluated in more detail during Gemini; the effects were thought to be primarily due to dehydration that occurs as a consequence of exposure to microgravity. New medical issues—bone loss and anemia—were observed during this program, but these were not considered "mission-limiting" conditions for the relatively short duration of the lunar missions (up to two weeks). Project Gemini also confirmed that astronauts could productively function in the spaceflight environment [11].

With the successful completion of Project Gemini, all enabling technologies and abilities had been developed to successfully launch astronauts to the Moon during Project Apollo. Indeed, a successful lunar landing was accomplished during the Apollo 11 mission on July 20, 1969. In total, Project Apollo included 29 astronauts, 12 of whom walked on the lunar surface. A number of medical objectives were established during this program. First was ensuring the safety and health of crewmembers; with the inability to return quickly to Earth once out of low-Earth orbit, the possibility of inflight illness was of significant concern. Apollo implemented preflight health stabilization (isolation) and the ability to minimally treat illness onboard the spacecraft. Another objective was the protection of terrestrial inhabitants from lunar microbial contamination. Due to the possibility of microscopic life existing on the Moon, astronauts on early Apollo flights had to don isolation suits following splashdown. They were subsequently decontaminated and underwent strict isolation for several weeks postflight as well. These practices were discontinued on later Apollo flights once NASA had assurance that the returning astronauts had not contracted "moon germs". A third objective was to acquire additional data on the physiologic effects of spaceflight; building on the observations of Projects Mercury and Gemini, astronauts during Project Apollo collected data on cardiovascular function, metabolic balance, and radiation exposure [12]. In addition, vestibular disturbances and the associated syndrome known as "space motion sickness" was identified (technically, cosmonaut Titov first reported dizziness and a feeling of tumbling on the second Russian human spaceflight [13]).[3] In fact, several crew members on Apollo missions 8 and 9 experienced stomach awareness and vomiting; one crewmember was so debilitated that portions of his flight plan had to be altered [14].

D.3 THE SKYLAB PROGRAM

As the lunar missions came to a close, NASA embarked on the development of its first manned space station: Skylab. Launched in May 1973, Skylab hosted three separate crews over the following year; the final crew established an endurance record of 84 continuous days in space. This was the first U.S. space program in which long-duration spaceflight physiologic adaptation and habitability issues could be investigated, and a wealth of data on crew health was collected. Further confirmation of space motion sickness was made, with the additional finding that this syndrome, unlike its terrestrial counterpart, could not necessarily be predicted or prevented, even with prophylactic medication [15]. Cardiovascular deconditioning was also

---

[3] For information on the biomedical challenges of the other major spacefaring nations, consult David Harland's *The Mir Space Station* (1997) and Brian Harvey's *Russia in Space* (2001) published by Praxis of Chichester, U.K. (*www.praxis-publishingco.uk*).

examined in detail and was found to stabilize after about six weeks of long-duration spaceflight. The deconditioning, as suspected, was primarily due to adaptive fluid shifts and losses that occurred when astronauts entered the microgravity environment [4]. Bone loss, particularly in weight-bearing bones, was confirmed and muscle wasting was seen as well. In contrast to cardiovascular deconditioning, these effects were not found to be self-limiting, even with countermeasures, for up to three months in space [4, 16]. Increased urinary excretion of calcium (probably associated with bone loss) was also observed; this could predispose astronauts to urolithiasis [17].

Much of NASA's knowledge of the physiologic effects of long-duration spaceflight was derived from the Skylab program and would be confirmed on shorter duration Shuttle flights. Unfortunately, Skylab hosted only three crews before its program was discontinued in 1974 and NASA turned its attention to the development of the Space Shuttle.

D.4 SPACE SHUTTLE PROGRAM

It has been over 25 years since the first successful orbital flight of the Shuttle Transportation System (STS), or Space Shuttle, on April 12, 1981. The Shuttle was designed to be partially reusable, unlike earlier U.S. space vehicles. In addition to the U.S., the European, Japanese, and Canadian Space Agencies have flown astronauts on the Shuttle, while Russia has flown cosmonauts. The Shuttle also boasts of having flown the oldest astronaut, John Glenn, who was 77 years old at the time of his flight in 1998. Physiologically, he was reported to respond similarly to his younger crewmates and returned to Earth without incident [18].

Physiologic changes experienced by crews who have flown up to 18 days in low-Earth orbit have been similar to those observed in earlier U.S. manned programs. The Shuttle has had several dedicated space physiology and life science flights that have continued to refine our understanding of adaptation to microgravity. Countermeasure development, to mitigate the effects of microgravity, have also been studied on the Shuttle, though it is important to note that definitive treatment strategies to mitigate the health risk associated with sustained exposure to microgravity remain largely elusive.

The Shuttle program was devastated by the loss of the *Challenger* orbiter on January 28, 1986, when an O-ring seal in one of the solid rocket boosters (SRBs) failed at liftoff, causing a breach in the SRB joint, allowing a flare to penetrate the large external tanks and thus igniting the fuel inside. The resulting explosion cost the program the lives of seven astronauts, including the first teacher in space, Christa McAuliffe. The program went on hiatus for two years while the accident was thoroughly investigated and new procedures for emergency egress of crewmembers, among other *safety enhancements*, were implemented. The Shuttle Program has dominated the U.S. human spaceflight program and continues to fly today, though in the wake of the loss of a second orbiter, *Columbia*, in 2003, plans were put in place to phase out the Shuttle by approximately 2010 as part of a renewed long-term commitment to deep-space exploration and settlement that includes lunar settlement and human Mars exploration [19].

Current operations for the Shuttle focus on completion of the International Space Station (ISS). Prior to the *Columbia* tragedy, there had been discussion of prolonging Shuttle missions for up to 30 days (while docked to the ISS) in order to extend the research capabilities of the three-member ISS crew [20], but given NASA's new focus, this has not been pursued.[4]

D.5 THE SHUTTLE–MIR PROGRAM AND THE INTERNATIONAL SPACE STATION

Recognizing its limited experience with long-duration spaceflight, NASA forged an agreement with the Russian Space Agency in the 1990s to fly U.S. astronauts aboard the Mir space station [21]. Since the 1970s, Russia had amassed a significant record of long-duration accomplishments in their Salyut space stations and the Mir. Six astronauts spent time on that Russian orbital laboratory, gaining valuable—and sometimes harrowing—experience with long-duration flight. Of particular note, Shannon Lucid broke the U.S. record for long-duration spaceflight (188 days) previously set by Skylab astronauts (Shannon's record was subsequently broken by ISS astronauts Daniel Bursch and Carl Walz, who returned on June 19, 2002, after 196 days in space [22]; the current U.S. record holder is Michael Foale, with a total of 374 days in space [23]). John Blaha [24] experienced a bout of depression during his stay, and Norm Thagard [25] also reported feelings of cultural and social isolation. Jerry Linnenger [26] survived a fire onboard the station, and Michael Foale [27] endured a spacecraft collision with Mir and subsequent station depressurization (see Chapters 5 and 6).

The Shuttle–Mir program served as a testbed for U.S. astronauts as they prepared for long-duration stays on the ISS, which began construction in late 1998. The ISS was intended to carry seven crewmembers from the U.S., Russia, the European Space Agency, and Japan. One of the major objectives for the ISS has been to continue to develop countermeasures necessary for interplanetary travel. Unfortunately, due to significant cost overruns [28] and the subsequent redirection of NASA following the *Columbia* accident, the ISS currently subsists with a three-member crew, who spend the majority of their time maintaining the station, rather than conducting research in a variety of disciplines, including medical. In addition, budgetary constraints had forced the cancellation of a Crew Return Vehicle [29] which was proposed to return crewmembers to Earth in the event of an emergency (normal procedure calls for crew return via the Shuttle). In its place, a Russian Soyuz spacecraft remains docked to the ISS [30]. In the event of significant crew injury, however, a return to Earth on the Soyuz is challenging. Three crewmembers sit in a fetal position in the spacecraft. There is no way to supply oxygen to injured crew-

---

[4] For more information on ISS, consult D. M. Harland and J. E. Catchpole's *Creating the International Space Station* (2002), published by Springer/Praxis (*www.praxis-publishing.co.uk*).

members, much less artificially ventilate them. It also carries limited medical supplies and minimal food. Re-entry exposes the crew to high g-forces, possibly up to 10 g's, which could be agonizing to an injured crewmember. Finally, the Soyuz lands in the middle of the Kazakhstan steppe and may require prolonged search and recovery of the spacecraft. Because of the obvious limitations presented by a return to Earth in the Soyuz, some NASA flight surgeons debated maintaining a position of crew "stand and fight" when medical intervention was required, opting to return sick or injured crew only when absolutely necessary; others felt that a "stabilize and transport" option was preferred at all costs.

D.6 TO THE MOON (AGAIN) AND MARS

With NASA's focus returning to interplanetary travel, the need to "stand and fight" with robust medical capabilities and effective countermeasures to the physiological impact of microgravity are more crucial that ever for ensuring a sustained long-duration human presence in space. Researchers at NASA had outlined medical areas of concern in a project known as the Critical Path Roadmap: Human Exploration and Development of Space. This project was tasked with defining a strategy for assessing, understanding, mitigating, and managing the risks associated with long-duration spaceflight. Several areas of research were identified, including the highest priority areas of bone loss, psychosocial interactions/crew selection, immune function, and other chronic effects of high-energy radiation exposure, including cancer. The Critical Path Roadmap was re-envisioned in 2005 as the Bioastronautics Roadmap: A Risk Reduction Strategy for Human Space Exploration [31] to more closely align with NASA's revised focus on long-duration missions; medical risks were defined for missions of up to 1 year on the ISS, a 30-day lunar surface mission, and a 30-month journey to Mars. These areas of research complement programs being run through the National Space Biomedical Research Institute, which partners with NASA on the research and development of spaceflight countermeasures. Other programs in which NASA is supporting medical risk reduction include NEEMO (NASA Extreme Environment Mission Operations) project, which sends groups of NASA employees and contractors to live in the analog space environment of the Aquarius undersea habitat (located off Key Largo, Florida) for up to three weeks at a time [32] and the Exploration Medical Capability Element [33], which is managing the development of autonomous health care capability to prevent, monitor, diagnose, and treat an ill or injured crewmember who must be managed in a microgravity environment. This work involves the creation of closed-loop control algorithms for ventilation, drug delivery, fluid management, and hypothermia [34].

However, despite our knowledge of space physiology that has evolved over the 40-plus years of manned spaceflight, we have yet to develop the countermeasures to significantly mitigate many of the effects of microgravity. Yet the feasibility of ultimately developing countermeasures for artificial gravity is only speculative at this time. These countermeasures may significantly impact the success of planned lunar and Martian exploration. There are several critical research areas which must constitute the focus of NASA over the next decade in order to enable a sustained human presence in space, including

- defining the psychological and physical impacts on the ability to perform and handle stress, including better definition of crew selection criteria to optimize collective crew performance;
- establishing policies regarding crew death and body disposal;
- characterization of acute and chronic radiation exposure and the cumulative effects of both on cognition, overall performance, cancer risk, etc.;
- identifying and testing countermeasures to prevent or halt skeletomuscular loss;
- environmental factors associated with long-term missions, including space suit design for extended extravehicular activity on the lunar and Martian surfaces;
- characterization of the pharmacokinetics and pharmacodynamics of medications in order to develop microgravity and reduced gravity-appropriate dosing recommendations.

Curiously, in spite of a decade of research collaboration between NASA and the NSBRI, substantial progress towards effective countermeasure development has yet to be fully realized. Furthermore, insufficient research is being directed toward *preventing health problems in space*, as well as space fitness and wellness programs for those living and working in orbit. Such efforts are important if mental health aloft is to be normal.

It is also important to note that traditionally, astronauts are not medically trained; few receive marginal training during mission preparation for handling medical contingencies and ultimately must rely on ground flight surgeon support. Among the career backgrounds of both cosmonauts and astronauts, physicians have been most dominant. Space nurses, for example, have yet to achieve orbit in any spacefaring nation! Whether medical personnel should be included as part of an interplanetary exploration crew will need to be resolved. The need for true autonomous medical care and self-care, including telerobotic surgery and telemedicine, will become mandatory for interplanetary travel! Shear distance from Earth, offworld weather, and signal loss due to position can cause communication problems, making it difficult or impossible to communicate with ground personnel. For instance, if a signal is delayed by more than 0.7 seconds to a surgical robot, a ground-based surgeon will begin to have problems controlling the remote device. Time delays from Earth to Mars range from approximately 15 to 40 minutes, making urgent "real-time" medical consultation all but impractical [35] In order to provide adequate medical coverage for exploration missions, many experts believe it will be necessary to include a clinically trained physician or other health care provider, such as a nurse, psychologist, or pharmacist, as part of the crew, to augment automated onboard and ground medical support. Certainly, when it comes to the matter of planning a lunar settlement, inclusion of a space nurse would be essential.

A critical undeveloped area of research is the mental health of spacefarers, before, during and after spaceflight, along with the critical role the behavioral sciences might contribute in this regard (see Chapters 3 and 6). Space psychology will be increasingly important to the success of long-term spaceflight and settlement [36]. Exhibit 154 underscores such concerns relative to traveling to Mars and back.

Obviously, we are only beginning to appreciate the mental health needs of spacefarers, especially on long-duration space travel.

D.7 HEALTH CHALLENGES OF PRIVATE CITIZENS IN ORBIT

Since 2001, the ISS has been visited by a few private citizens, "space tourists" who have spent short stays largely on the Russian side of the station. In 2004, Scaled Composites (funded by Microsoft co-founder Paul Allen) developed and flew a reusable suborbital vehicle, SpaceShipOne, twice within 2 weeks (and was awarded $10 million for winning the Ansari X-Prize) [37]. Nearly a dozen privately funded launch companies are vying to develop their own suborbital vehicles to offer space tourists a 3 to 5-minute weightless experience; Bigelow Aerospace is developing inflatable habitats for an eventual orbiting space hotel [38, 39].

With private citizens of variable health status paying for the privilege to reach space, the need for defining criteria for "flight eligibility" and minimal training, particularly on launch vehicle safety considerations (especially during the critical period the tourist experiences weightlessness and possible motion sickness), will evolve in order to enable as many to fly as possible. The Aerospace Medical Association drafted initial guidance for commercial spaceflight medical eligibility [40, 41] and the Federal Aviation Administration has issued its own initial guidance on medical qualifications for space tourists [42], but clearly this aspect of space health care is in its infancy, and has the potential to become highly regulated if care is not taken in the early days of commercial flight to ensure maximal safety of the paying passenger. That is why the authors founded their practice, Space Medicine Associates (see Chapter 6). Exhibit 155 highlights some of the concerns under discussion in the emerging arena of personal spaceflight.

D.8 PART 1: CONCLUSIONS

It is difficult sometimes to realize that humans have been living and working in space for fewer than 50 years. During that time, much has been learned about the physiology of human beings and other living organisms in the microgravity environment. Nevertheless, we have a long way to go until we can transform that knowledge into viable preventative health care measures that can mitigate the effects of microgravity on the body. Solving this challenge will not only benefit future interplanetary

Exhibit 154. SPACE PSYCHOLOGY.

One of two reviews posted on Amazon.com about Valentin Lebedev's *Diary of a Cosmonaut* calls it "a profound book about what it's like to be flying in a tin can for more than half a year." ... Dr. Lebedev's mission would seem entertaining compared with a trip to Mars. The round-trip, including a stay on the surface, would take about 17 months. Which is why it is surprising that within a few days of it being advertised, more than 3,000 people have applied to take part in an experiment planned by the Russian Institute for Biomedical Problems and the European Space Agency to simulate such an outing ...

The experiment's popularity is all the more bizarre given that Dr. Lebedev is far from being the only cosmonaut whom space has driven a bit cranky. Half of all the cosmonauts developed a condition that Russias psychologists call "asthenization" (and the American ones do not even recognize). This is characterized as irritability and low energy. Crewmembers often get on badly with one another. Individuals develop "space dementia". Orbiting astronauts have also become clinically depressed and panicked at psychosomatic illnesses.

Nick Kanas of the University of California, San Francisco, and his colleagues have scrutinized seven years of interactions between crew and ground staff during missions to the International Space Station (ISS) ... The 17 crewmembers involved, and the 128 mission controllers, rated the social climate of the mission and also their emotional state through weekly questionnaires, beginning a month before launch and ending a fortnight after their return. Dr. Kanas' analysis, published in the June 2007 issue of *Aviation, Space, and Environmental Medicine*, was broadly consistent with his smaller study of Mir, the Soviet's first station. Over the course of both Mir/ISS missions, there were no overall trends in the quality of social interaction, and mood swings did not dip in the third quarter of the mission as expected, as often happens on submarines and Antarctic research stations. That it does not, implies that the psychological tactics of ground-control medics are working ... But Dr. Kanas' studies with both Mir and ISS uncovered a more worrying finding: crews coped with stress by blaming the ground team and perceiving them negatively, classic case of "transference".

This tendency to convert tensions onboard into feelings that people on Earth do not care is one reason why sending people to Mars would be as much a psychological as well as a technical challenge ... Perhaps it would be better to stick to more psychologically robust and less libidinous space explorers: robots.

Source: Adapted from "Space Psychology: Space Mood Swings," *The Economist*, June 30th, 2007, pp. 89–90 (*www.economist.com*).

explorers, but private citizens as well who are on the cusp of venturing into space. More emphasis has to be placed on the health care team in orbit, and the role of such professionals as nurses, dentists, social workers, occupational therapists, etc., as well as the training of space paramedics.

Exhibit 155. PERSONAL SPACEFLIGHT HEALTH ISSUES.

Recent professional space conferences have focused on issues involved in launching civilian passengers into space. This is encouraging because it signifies that the embryonic industry of personal spaceflight is taking seriously the all important matter of customer service and safety. Leading the effort is the Commercial Space Transportation Advisory Committee for the Federal Aviation Administration. This has resulted from a U.S. Congressional-mandated study of flight safety and standards regarding spacefarers. COMSTAC meetings increasingly are devoted to biomedical and training issues which are being addressed by member companies in the Personal Spaceflight Federation.

At such convocations, three spaceflight-training companies report on their endeavors and concerns. The NASTAR Center, for example, recommended that training people for space launch should be based on systematic collection of data about the effects of spaceflight on morphologically diverse populations. In addition to the expected physiological skills and conditioning, mindsets should be created that develop confidence and expectations among those paying for space travel ... For a safer, more comfortable experience in orbit, Paragon's specialists argue that all data collected so far on spacefarers is about being fit within the constraints of defined parameters for astronauts and cosmonauts. But now we consider civilians who may not be so fit: their height, weight, physical condition, and other such factors which require changes in the usual cabin conditioning systems and support technologies, as well as the ergonomic designs of seats, pressure suits, etc. ... Barrios Technologies would like to see the industry create a method for trainer certification in preparing paying customers for going aloft, while RLV representatives want the industry to address more training and standards issues.

Space Medicine Associates made presentations at such meetings urging the nascent industry to consider human factors for civilian customers, such as disease control, alcohol or other substance use or abuse, handling prescriptions, dealing with claustrophobia, managing medical emergencies, as well as passenger personality incompatibilities. For long-duration space travel, possibly to the Moon or beyond, SMA professionals raised settlement issues such as the effects of lunar dust on biological systems, long-term impact of exposure to hypogravity, and combinations in unpleasant mixtures of contradicting conditions requiring development of future mitigating technologies.

Source: Adaptation of NASTAR Center's Alex Howerton report on "Biomedical Considerations Addressed at Recent Space Conferences," *The Space Review*, June 28, 2007 (email: *ahowerton@NASTAR.com*).

Insufficient attention has been given in the present planning of a lunar base to subjects like emergency care, family health, sexuality, pregnancy, and even space pediatrics (yes, children may even be born on the Moon). *Astromedicine* is still in its infancy!

Exhibit 156. Space Nursing Society symbol. The origins of the nursing profession go back to Florence Nightingale in the 19th century. The concept of space nursing offworld originated with a New York Univeristy professor of nursing, Dr. Martha Rogers in the 20th century. Source: *www.spacenursingsociety.net*

PART 2 FUTURE ROLE OF SPACE NURSING IN SPACEFLIGHT ARENA

Introduction: Patricia Grace Smith, Regulator for Commercial Spaceflight from the Federal Aviation Administration, expects to see working spaceports increasing within 50 years. The first half of the 21st century will also see the opening up on a commercial scale of spaceflight allowing the average citizen a chance to experience sub-orbital (zero-*g*), orbital (Virgin Galactic's SpaceShipTwo), and possibly a hotel stay in space (Bigelow Aerospace). These commercial ventures will need health care and medical support for their customers. This creates an opportunity for nurses throughout the world to actively participate in the creation of a spacefaring society. Nursing will need to redefine and expand its professional role to include off-world healthcare delivery. Advances and spinoffs from space technology have already transformed health care delivery on Earth, so it is vital that nurses are deeply involved in innovative research and developing health care delivery offworld. Now is the time for nurses to ensure that they do indeed partake in transforming events of spaceflight (Harris, 2001; Plush, 2003).

D.9 A VISIONARY NURSING LEADER: MARTHA E. ROGERS

Professor Martha E. Rogers was not only a founding member of the Space Nursing Society, her research and writings originated the concept of nurses serving beyond Earth. A futurist, Dr. Rogers anticipated this development in the nursing profession and its specific challenges for the Space Nursing Society. She reminded nurses of the speed of accelerating change and its impact on human evolution, as well as the need to create an enhanced *image of nursing*. One that respects its nursing heritage for providing professional health care, whereever humans have lived and worked on the home planet or beyond. In her words (Rogers, 1970):

> "New concerns and new visions engage the nurses of tomorrow. Independence of thought and action, creative ideas, human compassion, and enthusiasm for the unknown abound. Diversity is a valued norm. Human health and welfare have new dimensions. As Florence Nightingale said, "No system can endure that does not march."

By discussing anticipated body changes and field patterning, Dr. Rogers struggled to integrate into her *science of unitary beings*, a sense of the emerging differences in future "spacekind". She supported the value of transcendental meditation, imagery, and therapeutic touch for both earthbound and patients offworld. Rogers even discussed how the use of color, light, and sound could help space travelers in their adjustments to orbital living. Dr. Rogers often spoke of a whole new world of transcendental unity, while defining the role of a space nurse as *to promote human/spacekind wellbeing, whatever that may be*. She challenged her colleagues who follow to ensure that nursing becomes not only a distinctive scientific discipline, but also have its distinctive body of knowledge contribute to fuller human emergence on the high frontier Subsequently, Dr. Philip Harris (2003) maintained that apart from the Shuttle program and the International Space Station, the Moon is likely to offer the next great opportunity for space-nursing skills within the next 50 years.

D.10 ROLE OF NURSES IN A FUTURE LUNAR SETTLEMENT

In an address to SNS members in 2001, space psychologist Dr. Philip Harris explored these ideas concerning the probable role of space nurses in a lunar colony. Nurses aloft will have to be multitalented, carrying out other work assignments aloft. Beside exercising their usual professional skills, he envisioned high-performing nurses enhancing the lunar dweller's sense of self and wellness by

- furtherance of interpersonal and family relations on the Moon;
- promoting unique communication and language needs within the lunar environment;
- fostering healthy values and norms for community lunar living;
- helping with personal safety and protective gear required by environmental conditions/hazards;
- responding to medical/health care emergencies and life support needs;
- developing beliefs, customs, traditions, and culture so a lunar colony has a positive quality of life;
- assisting with diet and nutritional needs, as well as feeding habits, to ensure an adequate supply of suitable food and sustenance;
- working effectively with medical robotic assistants, while being capable of maintaining such systems.

In addition, founding SNS president Linda Plush issued a scope of practice describing the space nurse as a *transfer specialist* who will assess the spacefarer's health status and environment, such as during a transfer from Earth to the Moon or Mars. The nurse will support health care needs to ensure a safe transition and passage for, during, and after space travel. In some instances, the traveler may have to go though various adjustment stages, from one environment to another. Nursing aid, with timed intervals of acculturation to various environments, may become critical in cases of gravity changes.

Dr. Barbara Czerwinski and colleagues noted usual or expected areas of concern beyond the often cited bone demineralization, space motion sickness, cardiac decomposition, and muscle wasting. Space nurses might also provide services to delimit culture shock and isolation issues; to help patients with sleep disturbances and motivational decline, diminishment of memory and/or concentration abilities; to prevent or treat radiation exposure, anxiety, and stress (mental and physical), headaches, monotony of visual landscape. Over long durations in orbit, such stress may prove to be related to dependency on an artificial atmosphere/lighting equipment for life support, home sickness, or usual problems of people trying to cope in an extreme environment (Czerwinski, Plush, and Bailes, 2000).

Furthermore, each different space environment may require a rewriting of health care procedures with innovative nursing care for its unique circumstances. However, nurses may contribute their expertise in five general categories for all space travel: (a) assessment prior to flight; (b) orientation before liftoff and on reaching orbit; (c) transitioning through environments during space travel; (d) onsite support aloft; and (e) re-entry services to homeworld.

D.11 EMERGING PERSONAL/COMMERCIAL SPACEFLIGHT NURSING

Space nurses may contribute to improved spacefaring, whether under the sponsorship of the public or private sectors. Space nursing can learn from the experience of early commercial aviation. The first 25 years of commercial passenger services in the 20th century was a time when the public preferred the support of nurses on early flights for health and safety reasons. Now in the 21st century, the emerging personal/commercial spaceflight industry will create a similar, but more lasting demand for the comfort and safety of their passengers, as well as to minimize any risk of adverse medical events. Professionals with knowledge and skills such as nurses will again be in demand (indicated in the above exhibits and Chapter 6).

D.12 ADVANCING SPACE NURSING

In a seminal article on the promising specialization of space nursing, Dr. Elizabeth Barrett, a professor at Hunter-Bellevue School of Nursing, made some insightful observations (summarized in Exhibit 157).

Exhibit 157. THE FUTURE OF SPACE NURSING.

Humankind is on the threshold of a new cosmology transcending our Earth-bound past, as Martha Rogers noted. She was acutely aware of the transition of nursing from prescience to science, as indicated in her writings on the nursing of unitary human beings. For Rogers, medicine cures, but nursing cares, whether its services are to people on this planet or in outer space. In her conceptual framework, nursing cares for the whole person constantly, acting as a link between technology and the needs of the patient and families.

In the early stages of the space program, NASA hired Air Force Nurse Delores O'Hara to set up a flight clinic for the original seven astronauts at Cape Canaveral, Florida. Eventually her responsibilities expanded to include emergency nursing care and health counseling to 55 astronauts and their families, serving as a link between them and Agency systems. Currently, nurses seek to serve in orbit and envision services aloft that range from coping with cultural differences and microgravity sex, to slower aging and death in space. The nurse will help spacefarers and their families to cope with the overview effect that causes alterations in perception, time, and space, as well as the experience of silence and weightlessness aloft. Implementing Rogers' thinking, future space nurses will accompany those on the transcendent voyage to becoming *spacekind*. Their varied services will involve new modalities for health patterning of the environment aloft, so as to provide comfort, safety, wellness, and healing. Their assistance in coping with orbital changes will include transcending the boundaries and limitations imposed by space suits, vehicles, and stations, or underground space settlements. Space nurses will assess the interpersonal atmosphere, monitor diet and fitness regimens, redesign the orbital environment and lifestyle, and use a variety of techniques to improve the quality of life beyond Earth. They will learn how to improve healthcare aloft by utilizing automation and robotics, as well as by sophisticated computer conferencing systems and *telenursing*. Nursing will be transformed as space nurses become collaborative health participants in living and working in space.

Source: Adapted from Elizabeth A. M. Barrett, Ph.D. "Space Nursing," *CUTIS*, **48**(4), October 1991, 299–303.

In its publications and conferences, the Space Nursing Society has discussed the above challenges, including strategies for the advancement of space nursing by

(a) seeking inclusion of material on space health care in various nursing schools;
(b) promoting nurse participation in various aerospace programs, corporations, and agencies worldwide;
(c) offering scholarships for nurses to attend the International Space University, as well as other space education programs;
(d) providing public information and media coverage on behalf of space nursing;
(e) encouraging inclusion of nurses in various aerospace research projects;

(f) advancing nurse membership, conference participation, and journal contributions on space nursing;
(g) forging strategic alliances for SNS with various institutions, individuals, and international organizations in nursing, medicine, health care and space;
(h) promoting grantsmanship and research about space nursing;
(i) supporting participation of space nurses in professional activities that are multidisciplinary in nature.

Since undertaking the above agenda, the Space Nursing Society has expanded from its original 11 founders, to over 450 members and is still growing. This global Society welcomes members from all nursing specializations and countries who are interested in advancing a spacefaring civilization!

D.13 PART 2: CONCLUSIONS

Nursing is among one of the oldest healing arts, and from ancient times its practitioners have been concerned about nurturing humanity. With Florence Nightingale (1820–1910) modern professional nursing emerged. With the coming of Martha Rogers in the 20th century, the concept of space nursing was born, but in the 21st century, its orbital practice will expand.

The latter will happen based on existing trends in space travel. Business and regulatory professionals predict expansion of commercial spaceports within the next 50 years. The emerging personal/commercial spaceflight movement, along with governmental support for a permanent return to the Moon, is creating a need in space exploration and development for more nurses and other health care professionals. Visionary nursing leader Dr. Martha Rogers challenged all health care professionals, especially nurses, to anticipate that need, by research and training which prepares them to effectively serve humans who live and work offworld. She understood that there would be changes in nursing along with other professions as we venture into space.

Astronurses will become a vital part of orbital health care teams, contributing to the transformation of the nursing career. The Space Nursing Society has begun to prepare for the advancing specialty of space nursing. Its members are participating in *The Vision of Space Exploration* (see Exhibit 125)! That is why the SNS newsletter is entitled, *Expanding Horizons*.

D.14 REFERENCES

Part 1 Health services aloft

[1] von Beckh, H. F. "The Space Medicine Branch of the Aerospace Medical Association," *Aviation, Space, and Environmental Medicine*, **50**(5), 1979, 513–516.

Appendix D: Health services aloft: Space nurses 595

[2] Nicogossian, A. E.; Huntoon, C. L.; and Pool, S. L. (eds) *Space Physiology and Medicine*, Third Edition. Philadelphia, PA: Lea & Feiberger, 1993.

[3] Swenson, L. S.; Grimwood, J. M.; and Alexander, C. C. *This New Ocean: A History of Project Mercury* (NASA SP-4201). Washington, D.C.: U.S. Government Printing Office, 1989.

[4] Dietlein, L. F. "Skylab: A Beginning," in Johnston, R. S.; and Dietlein, L. F. (eds.) *Biomedical Results from Skylab* (NASA SP-377). Washington, D.C.: U.S. Government Printing Office, 1977.

[5] Nicogossian, A. E.; Huntoon, C. L.; and Pool, S. L. (eds.) *Op. cit.* [2].

[6] Link, M. *Space Medicine in Project Mercury* (NASA SP-4003). Washington, D.C.: U.S. Government Printing Office, 1965.

[7] Nicogossian, A. E.; Huntoon, C. L.; and Pool, S. L. (eds.) *Op. cit.* [2, ch. 21].

[8] Grinter, K. "Project Mercury Goals," [Internet article]. Kennedy Space Center [homepage]. Florida: NASA-KSC; c2002 [updated 2000 September 27; cited 2002 November 4]. Available from *http://www-pao.ksc.nasa.gov/history/mercury/mercury-goals.htm*

[9] Kennedy, J. F. "Putting a Man on the Moon," in Podell, J.; and Anzovin, S. (eds.) *Speeches of the American Presidents*, Second Edition. Bronx, New York: H. W. Wilson Co., 2001.

[10] Mueller, G. E. "Introduction," in *Gemini Summary Conference* (NASA SP-138). Washington, D.C.: U.S. Government Printing Office, 1967.

[11] Nicogossian, A. E.; Huntoon, C. L.; and Pool, S. L. (eds.) *Op. cit.* [2].

[12] Johnston R. S. "Introduction," in Johnston, R. S.; Dietlein, L. F.; and Berry, C. A. (eds.) *Biomedical Results of Apollo* (NASA SP-368). Washington, D.C.: U.S. Government Printing Office, 1975.

[13] Nicogossian, A. E.; Huntoon, C. L.; and Pool. S. L. (eds.) *Op. cit.* [2].

[14] Nicogossian, A. E.; Huntoon, C. L.; and Pool, S. L. (eds.) *Op. cit.* [2].

[15] Graybiel, A. "Coping with Space Motion Sickness in Spacelab Missions," *Acta Astronautica*, 8(9/10), 1981, 1015–1018.

[16] Vogel, J. M.; Whittle, M. W.; Smith, M. C.; and Rambault, P. C. "Bone Mineral Measurement: Experiment M078," in Johnston, R. S.; and Dietlein, L. F. (eds.) *Biomedical Results from Skylab* (NASA SP-377). Washington, D.C.: U.S. Government Printing Office, 1977. See also interview with astronaut physician Dr. David Wolf after 119 days aboard the Russian station, Mir, in *EIR Science and Technology*, March 13, 1998, pp. 54–61.

[17] Whedon, G. D. *et al.* "Mineral and Nitrogen Metabolic Studies: Experiment M071," in Johnston, R. S.; and Dietlein L. F. (eds.) *Biomedical Results from Skylab* (NASA SP-377). Washington, D.C.: U.S. Government Printing Office, 1977.

[18] Rizzo, K. "Scientists Rave about Glenn's Health during Space Mission" [Internet article]. Detnews.com [homepage]. Detroit: *The Detroit News*; © 2000 [updated 2000 September 29; cited 2002 November 4]. Available from: *http://www.detnews.com/2000/nation/0001/29/01290070.htm*

[19] *The Vision for Space Exploration*. Washington, D.C.: National Aeronautics and Space Administration, February 2004.

[20] Brown, I. "NASA Taps Temps for Station Work" [Internet article]. UPI [homepage]. Washington, D.C.: UPI; © 2002 [updated 2002 October 28; cited 2002 November 4]. Available from: *http://www.upi.com/view.cfm?StoryID=20021028-075256-9818r*

[21] Burrough, B. *Dragonfly: NASA and the Crisis aboard Mir*. New York: HarperCollins; 1998.

[22] Stenger, R. "Shuttle brings home record-setting Astronauts" [Internet article]. CNN.com [homepage]. Atlanta: CNN, LLLP, © 2002 [updated 2000 September 29; cited 2002 November 4]. Available from: *http://www.cnn.com/2002/TECH/space/06/19/shuttle.land/*
[23] *http://www.spacetoday.org/Questions/SpaceflightEnduranceRecords.html* (accessed October 14, 2007).
[24] Burrough, B. *Op. cit.* [21].
[25] Burrough, B. *Op. cit.* [21].
[26] Burrough, B. *Op. cit.* [21].
[27] Burrough, B. *Op. cit.* [21].
[28] Halvorson, T. "Station Cost Overrun Prompts New Round of Proposed Cuts" [Internet article]. Space.com [homepage]. New York: Space Holding Corp.; © 1999–2002 [updated 2001 June 29; cited 2002 November 4]. Available from: *http://www.space.com/mission launches/missions/iss_overruns_010629.html*
[29] Dismukes, K. "The International Space Station: Where People Live, Work and Breathe" [Internet diagram]. NASA Human Spaceflight [homepage]. Washington, D.C.: NASA; © 2002 [updated 2002 October 16; cited 2002 November 4]. Available from: *http://space flight.nasa.gov/station/crew/people/issc_us_crv.htm*
[30] Berger, B. "NASA Refining Its Plan for Space Station Crew Rescue," *Space News*, **28**(6), October 2002.
[31] NASA. *Bioastronautics Roadmap: A Risk Reduction Strategy for Human Space Exploration*, NASA SP-20046113. Washington, D.C.: National Aeronautics and Space Administration, February 2005.
[32] *http://humanresearch.jsc.nasa.gov/news/neemo12.asp* (accessed October 14, 2007).
[33] *http://humanresearch.jsc.nasa.gov/elements/exmc.asp* (accessed October 14, 2007).
[34] Personal communication. George Beck, October 8, 2007.
[35] Plush, L. M. H. "Where No Surgery Has Gone Before," *OR Nurse Journal*, **1**, October 2007, 40–43 (*www.ORNurseJournal.com*).
[36] Harrison, A. A. *Spacefaring: The Human Dimension*. Berkeley, CA: University of California Press, 2001 ... Freeman, M. *Challenges in Human Space Exploration*. Chichester, U.K.: Springer/Praxis, 2000 (*www.praxis-publishing.co.uk*).
[37] *http://www.scaled.com/projects/tierone/041004_spaceshipone_x-prize_flight_2.html* (accessed October 15, 2007).
[38] *http://www.space-frontier.org/commercialspace/#comspaceanchor1* (accessed October 14, 2007).
[39] *http://www.bigelowaerospace.com/* (accessed October 14, 2007).
[40] Aerospace Medical Association Task Force on Space Travel, "Medical Guidelines for Space Passengers," *Aviation, Space, and Environmental Medicine*, **72**, 2001, 948–950.
[41] "Medical Guidelines for Space Passengers: II," *Aviation, Space, and Environmental Medicine*, **73**, 2002, 1132–1134.
[42] FAA, "Human Space Flight Requirements for Crew and Space Flight Participants," *Federal Register*, **71**, 2006, 71615–71645.

Part 2 Space nursing bibliography

Barrett, E. A. "Space Nursing," *CUTIS*, **48**(4), October 1991, 299–303.
Czerwinski, B. S.; Plush, L. M.; and Bailes, B. K. "Nurses' Contributions to the US Space Program," *American Operating Room Journal*, **71**(3), 2000, 1051–1057.

Harris, P. R. "The Nurse's Role in Moon/Mars: Personnel Deployment Systems," presented at Space Nursing Society session in conjunction with UMTB Center for Aerospace Medicine and the Center for Advanced Space Studies, Houston, TX, September 25–27, 1997, 78 pp.

Harris, P. R. "Future Role of Space Nursing in a 21st Century Lunar Colony," presented at AHNA/SPN/SRS Conference on a *Nursing Odyssey and a Tribute to Martha Rogers*, Huntsville, AL, June 22, 2001, 10 pp.

Harris, P. R. "Space Nursing on the Moon," *Launch Out: A Science-Based Novel about Space Enterprise, Lunar Industrialization, and Settlement*. West Conshohocken, PA: Infinity Publishing, 2003, ch. 13 (*www.buybooksontheweb.com*).

O'Rangers, E. A.; and Plush, L. M. "Healthcare and Safety in the U.S. Space Program," *Ad Astra*, July/August 2003, pp. 29–37.

Plush, L. M. "Origins, Founding, and Activities of the Space Nursing Society," *Journal of Pharmacy Practice*, **16**(2), 2003, 96–100 (published by Sage Publications).

Plush, L. M.; and O'Rangers, E. A. "NASA Technology Puts a Positive Spin on Nursing," *Nursing Spectrum*, June 2004, p. 38.

Rogers, M. E. *The Theoretical Basis of Nursing*. Philadelphia, PA: F.A. Davis Publishing Co., 1970.

Rogers, M. E. "Nursing Science in the Space Age," *Nursing Science Quarterly*, **5**, 1992, 27–34.

Symanski, M. E. "A Nurse on Mars? Why Not?" *American Journal of Nursing*, **100**(10), October 2000, 57–61.

For more information on the Society or to obtain copies of the above publications, go to *www.spacenursingsociety.net* (The Space Nursing Society, 3053 Rancho Vista Blvd. H377, Palmdale, CA 93551)... To contact the Space Medicine Associates, write to 4720 Water Park Drive #7, Belcamp, MD 31017 (or *www.spacemedicineassociates.com*).

VISIONING

Seeing, looking, thinking, searching
What wonders do enfold
Unbounded by time and space
Visions sail before us, unvesseled
Soaring beyond all we know
Vision—joyfully speculating
What unpredictable futures
Will become manifest
In our unitary journey
To the stars

Source: Arlene T. Farren, October 2000, *Visions of Nursing in the 21st Century: A Rogerian Nursing Journey*.

Exhibit 158. ILO 2. Early rendering #2 of what the International Lunar Observatory would look like when assembled automatically on the Moon. Source: Space Age Publishing (see Chapter 8, Exhibit 131).

Appendix E

International Lunar Observatory strategy
Steve Durst[1]

> *Support facilities on Earth will necessarily be associated, to a greater or lesser extent, for any enterprise on the Moon. The extent will greatly depend on the nature and scope of the enterprise. Initially, in early ventures and when a project is likely to be robotic, a stakeholder may have within its own organization all the technical expertise required, including research and test facilities ... However, with the passage of time, plus the increasing scope and complexity of the undertaking aloft, considerable increase and need for outside support facilities may emerge beyond the resources of a single stakeholder. In that case, advantage may be seen for collaboration by a number of stakeholders, in setting up and expending research facilities.*
>
> Michael Ross[2]

E.1 INTRODUCTION

The International Lunar Observatory (ILO) is a multi-national, multi-wavelength astrophysical observatory, power station, and communications center that is planned to be operational near the south pole of the lunar surface before 2010.

The International Lunar Observatory Association (ILOA) is the organization that supports the ILO and its follow-on missions through timely, efficient, and responsible utilization of human, material, and financial resources of spacefaring nations, enterprises, and individuals.

---

[1] For author biographies see "About the contributors" page in the prelims. See also Exhibit 131 and Chapter 10.
[2] Michael Ross' *From Footprints to Blueprints: Developing the Moon and Private Enterprise.* Bloomington, IN: Author House, 2005, p. 32 (*www.AuthorHouse.com*).

E.2 HISTORY

The ILO mission was conceived during the historic International Lunar Conference 2003 at the fifth meeting of the International Lunar Exploration Working Group (ILEWG) [1]. A distinguished panel of lunar scientists, entrepreneurs, policy makers, advocates, and others gathered to discuss the next vital step in human exploration of the Moon within the decade. What manifested was the Hawaii Moon Declaration, a one-page *Ad Astra per Luna* manifesto, written and signed by conference participants. Soon after the positive momentum, the ILO mission began to take shape.

An ILO Advisory Committee was established in 2005, consisting of about 50 supporters from the international science, commerce, and space agency communities. The Committee convened for the ILO Workshop in November 2005 on Hawaii to discuss the organizational, scientific/technical, and financial/legal direction of the ILO. A formal Founders Meeting was held November 4–9, 2007, in Waikola Beach, Hawaii. Representatives from government space agencies, astronomy organizations, and commercial space enterprises in more than seven countries gathered to advance the ILO mission. Several new strategies were reviewed, such as collaboration with the Google Lunar X-Prize teams, "Put Your Name on the Moon" endorsement campaign, and participation in the Pacific International Space Center for Exploration Systems (PISCES).

E.3 INTERNATIONAL SUPPORT

Pivotal to the success of the ILO is the cooperation and support of the world's major spacefaring powers, most notably the U.S.A., Canada, China, India, Italy, Japan, and Russia. Fundamental to ILO advancement are the private and commercial enterprises which catalyze, organize, and help attract the support of nations. The impetus behind a unified international mission is to engage the human and material resources of these nations, enterprises, and individuals into a pioneering and peaceful mission that benefits all of humanity.

E.3.1 Strategies within states and nations

In getting the International Lunar Observatory under way, efforts were made to establish connections within the global astronomical community, particularly with a view to future partnerships. Progress to date is reported below.

Canada

The ILO was presented to the Canada Astronomical Society (CASCA) in Calgary in June 2006, as well as the Canada Aeronautics Space Institute and the International Institute for Space Law conference at McGill University in Montreal in June 2006. The ILOA has neighborly relations with CFHT, allies in the Canada Space Agency, as well as Optech and MDA.

China

For several years, the ILOA has shared a strong rapport and mutual support with chief China lunar scientist Ouyang Ziyuan, who has explicitly advocated a telescope on China's first lunar lander Chang'e 2 mission. An ILO astronomy MOU was established with National Astronomical Observatories R&D Planning and Funding Director Suijian Xue in April. Additional rapport and ILO interest exist with Shanghai Astronomical Observatories, the Chinese Academy of Sciences, the China National Space Administration, and the Chinese Society of Astronautics.

Hawaii, U.S.A.

Within the State of Hawaii, the formation of the ILOA in 2006 has initiated multiple outreaches, strategies, and communications with supporters of the ILO. Within the state, these relationships include Mauna Kea Support Services, Canada France Hawaii Telescope (CFHT), West Hawaii Astronomy Club, the Onizuka Space Center, and the Hawaii Island Space Exploration Society.

India

The ILO has direct support within the Indian Space Research Organization (ISRO) and the Indian Institute of Astrophysics, most notably with ISRO PRL Council Chairman and cosmologist U. R. Rao, lunar scientist Narendra Bhandari, and current ISRO Chair Madhavan Nair.

Italy

Italy was the site of the 2005 Moonbase Symposium, which received widespread Italian industry and academia support. Italy is also the site of the 2007 ILEWG conference. The Italian Space Agency (ASI) has established agreements with China and Germany on lunar rover mission planning.

Japan

Support for ILO efforts exist within the Japan Aerospace Exploration Association, notably with lunar scientists Kohtaro Matsumoto, Susumu Sasaki, Yoshisada Takizawa, and Hitoshi Mizutani.

Russia

The ILOA has a rapport with prominent Russia lunar scientists, such as Viacheslav Shevchenko, Vladislav Ivashkin, and Erik Galimov.

E.4 FINANCING THE ILO MACROPROJECT

As a private enterprise, ILO hopes to attract support from both the public and private sectors for its long-range development program. A study performed by

ILO prime contractor, SpaceDev, Inc. of Poway, California, states that the first ILO mission can be launched, landed, and operated within a modest budget of under $50 million (USD). The strategy to fund this mission includes securing $10 million in investments (possibly at $2 million a year for five years) from various supportive government entities. These entities include science and space agencies/institutions, of Hawaii (U.S.A.), Canada, China, India, Europe, Japan, and Russia. Commercial sponsorship from high-tech companies such as Google, Cisco, Yahoo and others is also sought, as well as endowments from philanthropic groups and individuals.

E.5 ILO LUNAR LOCATION

Malapert Mountain: Located about 122 kilometers from the lunar south pole, the adjacent mare plain just north of the 4.6-kilometer high Malapert Mountain is the intended landing region for the ILO. Near-constant sunlight (thought to be 89% full, 4% partial) provides an energy-rich environment, and the lunar peak enjoys a continuous line-of-sight to Earth and direct Earth–Moon communications. The mountain dominates its surrounding area for an excellent vantage point and is near enough to expected water ice deposits (which can be utilized for oxygen, drinking water, and rocket fuel) around the lunar south pole (Shackleton Crater, Aitken Basin). Given these factors, Malapert Mountain is considered to be the most suitable location for the ILO to conduct astronomy and catalyze commercial lunar development and human lunar base build-out.[3]

In 2003, the Lunar Enterprise Corporation (LEC) hired SpaceDev to serve as the prime contractor for the ILO. A Phase A feasibility study conducted by SpaceDev concluded that "it is possible to design and carry out a private commercial lunar landing mission within the next several years." The Phase B study, conducted the following year, recommended researching a safe and accurate lunar-landing navigation system that can deliver the ILO to a "Peak of Eternal Light". Also concluded was the possibility of landing the ILO to a specific target within an accuracy of about 100 meters using "currently available commercial technology". The ILO will utilize a mix of leading-edge propulsion, inertial navigation, and celestial navigation together with established Earth-based deep-space tracking to achieve the required accuracy.

E.6 ILO'S MASTER PLAN

An International Lunar Observatory/Association (ILO/A) Master Plan was completed in February of 2006 by Optech Space Division Director and International Space University co-founder Bob Richards. The plan outlines how to build the ILO as a science, organizational, and commercial entity that operates within the scope of

---

[3] See Schrunk, D. G.; Sharpe, B. L.; Cooper, B. L.; and Thangavelu, M. "The First Lunar Base: Mons Malapert," *The Moon: Resources, Future Development, and Settlement*. Chichester, U.K.: Springer/Praxis, 2008, ch. 5, p. 84 (*www.praxis-publishing.co.uk*).

Appendix E: International Lunar Observatory strategy 603

Exhibit 159. ILO 3. Artist close-up rendering of an automated, functional observatory on the lunar surface. Source: Space Age Publishing.

investor markets, equity players, management, industry, and customers. Presently, the key participants include the following organizations: Space Age Publishing Company, Lunar Enterprise Corporation, SpaceX (of Europe), the Canadian Space Agency, the University of Hawaii and representatives from countries such as China, India, and Russia. International MOUs would be established between the ILO Association and the participants for exchanging science data and commercial communications, as well as sources of project funding.

The ILO will serve as a multi-wavelength astrophysical observatory that will utilize VLF, millimeter, submillimeter, and optical wavelengths. Scientific objectives of the ILO include (1) imaging the Galactic center; (2) analyzing interstellar molecules to determine the origin of the Solar System; (3) searching for NEOs (near-Earth objects) and Earth-like planets; (4) observations of the Earth, planets, and Solar System; and possibly (5) searching for extraterrestrial intelligence.

The ILO has the potential for interferometrical buildout, much like the Harvard-Smithsonian Submillimeter Array on Mauna Kea that operates the first such observatory (eight mobile antennas at six meters each operating interferometrically). Another example of cutting-edge interferometry is the 64-dish Atacama Large Millimeter Array being constructed at 5,000-meter altitude in Chile's ultra-dry high desert. Submillimeter and millimeter astronomy is forefront cutting-edge science and interferometry, operating in the 0.25 mm to 1.7 mm wavelengths, and offers a multitude of beneficial applications.

E.7 ILO FEATURES AND BENEFITS

The ILO program and its Association strives to innovate within several niches of the human, science, and technological endeavor. For the benefit of science, commerce, and humanity, the ILO will serve as

(1) *Astrophysics facility:* a unique and leading-edge enterprise for the next frontier of astronomy, with special emphasis on advancing Hawaii's astronomical leadership in the 21st century.
(2) *Power station:* a solar device will be required to supply power to the ILO and its instruments, and may offer additional capacities for other energy-related functions.
(3) *Communications center:* transmission of astrophysical data globally by utilizing the Space Age Publishing Company's (SPC) online publication *Lunar Enterprise Daily*, as well as global commercial media, broadcasting, advertising, and imaging. Already, a series of lunar commercial communications workshops are being conducted in Silicon Valley through the support of SPC's California office.
(4) *Lunar property rights:* ILO will seek to establish a legal precedent regarding ownership rights on the Moon in light of previous United Nations space treaties.
(5) *Site characterizer:* gather and report data of the surrounding area, including solar wind and radiation measurements, temperature, altitude, and seismic and meteoric activity.
(6) *Automated advance outpost:* ILO activities will serve as a precursor to future missions by other entities to build a functioning lunar base.
(7) *Virtual nexus dynamic website:* ILO aims to develop a website that delivers real-time astrophysical data, lunar video, Earthrise imagery, broadcast communications, and other viable information to institutions, popular media, schools, and the general public.
(8) *ILO synergy:* ILO and its Association will cooperate with the global space community in development of a fully functioning lunar base, with industrial facilities and habitats [2].

E.8 CONCLUSIONS

For those seeking to participate in private space enterprise, we invite inquiries to be among the first generation in the 21st century to successfully land and operate a scientific/commercial/robotic device on the Moon [3].

E.9 INTERNATIONAL LUNAR OBSERVATORY ASSOCIATION NEWS
(Volume 3, Issue 2, March–April 2008, Hawai'i Island)
Official organ of communication with ILO supporters

E.9.1 ILOA moves forward with precursor mission, advancing Galaxy forums, and private funding initiatives; SpaceDev completes ILO Payload Study; BoD meeting set for July in Canada

The International Lunar Observatory Association (ILOA) is developing a precursor lunar equatorial mission aboard a Google Lunar XPrize entrant's spacecraft. The mission would last 14 days and include First Light Galaxy imaging for science observation, education, and branding, as well as other observations and commercial activities. The mission would serve as a proof of concept for valid observation and communications, to lay groundwork and momentum for the ILO 1 Polar Mission and follow-on Human Service Mission. Second-round funding/investment has been made towards this initiative. A formal announcement of a partnership agreement is expected between ILOA and a Google Lunar XPrize Team on July 20, 2008.

ILO prime contractor SpaceDev recently completed its Phase 6 ILO Payload Definition study, which details various suggestions and configurations for lunar-based astronomy and communications. ILOA may contract a near-term Precursor Mission instrument definition study as a next step.

The first ILOA business plan was recently completed with the help of a Honolulu-based consulting firm. The plan was delivered along with a prospectus by ILOA Founding Director Steve Durst in ILOA's first formal funding outreach to the Durst Family/Foundation/Organization in New York City. Currently, the ILOA seeks a Hawai'i-based ILOA fundraiser/finance officer.

The 2008 Galaxy Initiative by ILOA and the Space Age Publishing Company includes forums on July 4 in California; July 26 in Vancouver, Canada; mid-October in Beijing; late 2008 possibly in New York. The purpose is to explore Galaxy consciousness and education in the 21st century.

The latest State of Hawai'i PISCES and HISES collaborations include ongoing initiatives as well as ILOA's offer and outreach of support for joint cosmic ray research between the University of Hawai'i Hilo and Tibet University, among other international institutions.

ILOA's Board of Directors has set a meeting in late July in Vancouver, Canada which offers the opportunity for all to meet freely with ILOA directors and friends from around the world, and to advance ILOA networking and support in Canada.

The ILOA is soliciting feedback on its website, please visit *http://www.iloa.org* to send a comment or email *comments@iloa.org*

E.10 REFERENCES

[1] Durst, S. M.; Bohannan, C. T.; Thomason, C. G.; Cerney, M. R.; and Yuen, L. (eds.) *Proceedings of the International Lunar Conference 2003/International Lunar Exploration Working Group 5—ILC2003/ILEWG5.* San Diego, CA: UNIVELT/AAS, Vol. 108, 2004 (*www.univelt.com*).

[2] Schrunk, D. G.; Sharpe, B. L.; Cooper, B. L.; and Thangavelu, M. *The Moon: Resources, Future Development, and Settlement.* Chichester, U.K.: Springer/Praxis, 2008 (*www.praxis-publishing.co.uk*).

[3] For further information, contact the International Lunar Observatory Association/Space Age Publishing's *Lunar Enterprise Daily* (Hawaii office: Mamalahoa Highway, D-20, Kamuela, Hawaii 96743; California office: 480 California Ave #303, Palo Alto, CA 84306; tel.: 650-324-3705; fax.: 650-324-3716; email: *info@iloa.org*; www.iloa.org or *www.spaceagepub.com*).

Exhibit 160. Concept of a lunar rover. The Japanese space agency envisions a lunar rover like this operating on the Moon's surface. Source: NASDA (now JAXA).

Appendix F

Resources

SELECT REFERENCES AND WEBSITES

Below are listed some resources which may help readers to pursue more information on the themes and references in this book, *Space Enterprise*. P.R.H.

Aerospace Medical Association For health care professionals; publishes journal *Aviation Space & Environmental Medicine* (*www.asma.org*). Related websites: *http://spacedoc/com*; females in space (*www.femsinspace.com*; *www.spacenursing society.net*); French Institute for Space Medicine and Physiology (*www.medes.fr/home.html*); National Space Biomedical Research Institute (*www.nsbri.org*).

American Institute of Aeronautics and Astronautics A professional society which assists those in the aerospace field. Sponsors conferences and workshops; publishes proceedings, books, and newsletters; provides courses and on-line services (*www.aiaa.org*). Affiliated with the International Astronautical Federation (*www.iaf.org*) which conducts annual world conferences and congresses.

Apogee Space Books, Collectors Guide Publishing Inc. A Canadian publisher of popular books and media on space activities. Among their titles of easy-to-read volumes and mission reports: *Astronautics*; *Interstellar Travel: Multigeneration Space Ships*; *Rocket Science*; *Surveyor: Lunar Exploration Program*; *High Frontier*; *Women Astronauts*; *Women of Space*; *Moonrush*; *Return to the Moon* (*www.cgpublishing.com* or *www.apogeebooks.com*).

Book stores online with astronautics/space volumes for purchase Astrobooks.com (*www.astrobooks.com*); Buy Books on the web.com (*www.bbotw.com* or *www.buybooksontheweb*, e.g., *Launch Out*, a space novel by P. R. Harris); Microcosm Press (*www.home.earthlink.net/microcosm*); Boggs Space Books, new and used on the history of space exploration (*www.boggsspace.com*).

Columbia House Producers of videos, DVDs, and CDs (e.g., HBO's *From Earth to the Moon* mini series of six video cassettes 1-800/538-7766 or *www.columbiahouse.comvl*).

Elsevier Science Ltd. Publishes journals *Acta Astronautica*, *Space Technology*, and *Space Policy* (*www.elsevier.nl/locate/*) , as well as books on business and management (e.g., Harris & Moran's *Managing Cultural Differences Series*, *www.books@elsevier. com/business*).

Extraterrestrials Contact *Cultures of the Imagination*, an interdisciplinary forum of scientists, writers, and artists exploring the problems and possibilities for human future. Holds annual conference, produces role-playing simulations, and a newsletter (*www.contact.com* or tel.: 1-650/941-4027). Also *SETI Institute*, sponsors projects that search for extraterrestrial intelligence; publishes newsletter; sponsors lectures and events (*www.seti.com*).

Google Inc. Beside their general Internet search capabilities, they have three interesting online space websites which eventually may be merged: *www.moon.google.com* (allows armchair astronauts to explore the lunar surface while zooming in and out); *www.mars.google.com* (highlights landing zones and surrounding terrain on the Red Planet filmed from automated spacecraft); *www.Earthgoggle.com* (enables the viewer to scrutinize the home planet from outer space, a standalone program that can be downloaded on your computer). Sponsor of Google Lunar X-Prize.

Greenwood Press Publishes occasional space and astronomy books (e.g., R. D. Launinus' *Frontiers of Space Exploration*, *www.greenwood.com*).

Human role in space Current books: R. L. DeHart and J. R. Davis' *Fundamentals of Aerospace Medicine*. Philadelphia, PA: Lippincott, 2002, Third Edition, pp. 604–605 ... Harrison H. Schmitt, *Return to the Moon: Exploration, Enterprise and Energy in the Human Settlement of Space*. New York: Springer/Copernicus/Praxis, 2006, ch. 13 ... A. A. Harrision's *Spacefaring: The Human Dimension*. Berkeley, CA: University of California Press, 2001 ... Marsha Freeman's *Challenges of Human Space Exploration*. Chichester, U.K.: Springer/Praxis, 2000 ... Bob Krone, Langdon Morris, and Kenneth Cox (eds.) *Beyond Earth: The Future of Humans in Space*. Burlington, Ontario: Apogee Books, 2006.

Imaginova Corp. Publishes *Space News*, a business publication on space commerce (*www.space.com*). Imaginova Studio for global education (e.g., *Starry Night*, an astronomy planetarium program in full color, *www.livescience.com* or *www.starry night store.com*, or 1/800/252-5417). Also produces the annual *Space News: Directory of Worldwide Space*.

International space agencies In addition to NASA listed below, here are some of the world space agency websites: Argentina (*www.conae.gov.ar*); Brazil (*www.cnes.br/*); Canada (*www.space.gc.ca/csa*); ESA (*www.esrin.esa.it/*); France (*www.CNES.org*); Italy (*www.asi.it/*); Germany (*www.dir.de/*); Russia (*www/federalspace.ru* or *www.rka.org* or *www.COSMOS.org*); U.K. (*www.bnsc. org* or *www.bis.org*) ... Japan (*www.JAXA.org*); China (*www.csna.gov.cn*); India (*www.ISRO.org*); Australia (*www.asri.org.au/* or *www.industry.gov.au/space*); U.N. Office of Outer Space Affairs, Vienna, Austria (*www.COPUOS.org* or *ftp://ecf.hq.eso.org/un/un-homepage.html*).

Kluwer Academic Publishers Publishes space books and journals (e.g. Wertz and Larson's *Reducing Space Mission Costs*; Ekart's *Spaceflight, Life Support, and Biospherics*; DeWitt, Duston, and Hyder's *The Behavior of Systems in the Space Environment*; Journal of Intelligent and Robotic Systems (*www.wkap.nl*).

Krieger Publishing/Orbit Series Publishes technical and professional space books (e.g. Churchill's *Fundamental of Space Life Sciences, www.krieger.com*).

National Academies of Science Scientists report on space-related research, such as *Space Biology and Medicine* or *Microgravity Research in Support of Technologies for Human Exploration* (*www.nationalacademies.org/sbb/*).

National Air and Space Museum A branch of the Smithsonian Institution in Washington, D.C. and Dulles Airport. Offers space exhibits, audio-visual presentations, and retail store, plus a membership society (*www.nasm.edu*). The Smithsonian Institution Press publishes *Air and Space* magazine and books. Leader in some 25 science and space museums around the United States, such as the San Diego Aerospace Museum (*www.sandiegoairandspace.org*).

National Aeronautics and Space Administration American space agency with more than a dozen field centers and installations around the country. Media images and complimentary magazine (*NASA Spinoff: Space Technology Innovation*) about space, available on request (*www.nasa.gov/search/index.html*). Related websites: *listserves ti.nasa.gov*; microgravity research, *http://magnwww.larc.nasa.gov.sites.html*; information services, *http://spacelink.nasa.gov*; educational programs, *www.hq.nasacodeef/ education*; human spaceflight, *http://spaceflight.gov/index-m.html*; agency library, *www.hq.nasa.gov/hqlibrary/*. NASA also publishes a number of technical reports about spaceflight (NTRS, *http://ntrs.larc.nasa.gov/search/jsp?*).

National Aerospace Training and Research Center A subsidiary of Tectronics Corporation, this private enterprise engages in research and training for both aviation and space. Dr. Richard Hamilton of Drexel University heads its Medical Team, while NASA astronauts, like Dan Barry, are presenters at the NASTAR Education Center. Sir Richard Branson, founder of Virgin Galactic, underwent training here with his staff in preparation for flights on SpaceShipTwo which is under order for space tourists. Contact NASTAR Center, 125 James Way, Southampton, PA 18966 (tel.: 866/482-0933; website: *www.nastarcenter.com*; email: *ahowerton@nastarcenter.com*).

National Space Society An international space advocacy organization with chapters throughout the U.S.A. and abroad. Sponsors the annual International Space Development Conference, conducts the NSS Online Report, and publishes *Ad Astra*, a popular space magazine (*www.nss.org* or *www.nss.org/adastra* or email: *nssmembers @aol.com*). NSS space art calendar (*www.turnerpublishing.com/* or *www.nss.org/ settlement/calendar/gallery.htm*).

Praxis Publishing U.K. publisher of professional space books in conjunction with Springer Publishing and its subsidiary Copernicus Books. Check out titles in its *Space Science and Technology Series* and its *Astronomy, Astrophysics and Space Science Series* (e.g., Harris' *Space Enterprise*; Schrunk *et al.*'s *The Moon: Resources, Development and Settlement*; Van Pelt's *Space Tourism*); websites: *www.praxis-publishing.co.uk* or *www.copernicusbooks.com* or *www.springer.com*.

Space Age Publishing Publishes *Lunar Enterprise Daily*, an online daily news; sponsors the annual International Lunar Conference, and promotes the International Lunar Observatory project (*www.spaceagepub.com*).

Space information and societies In addition to NSS above, check out these websites: *www.space.com*; for space information on Australia or other countries, *www.aero spaceguide.net/worldspace/australia_in_space* or *www.asri.org.au/ASR/index.xml*; U.S.A., *www.spaceforum.com/*; Japan, *www2.jsforum*; *www.space-generation.org/*; for the annual *World Space Industry Survey*, *www.EUROCONSULT.org* ... Space Adventures travel and tourism, including visits to the International Space Station on Russian spacecraft, *www.spaceadventures.com* ... The *Space Review* offers online commentary and essays about the final frontier (*www.thespacereview.com*) ... The Space Camp for children and adults who wish to experience the space program through simulations (*www.spacecamp.com*) ... The Challenger Center, located at varied museums around the U.S.A., is a learning center offering students an opportunity to experience space teamwork and simulations in math, science, and technology (*www.challenger.org*) ... CNN Television (*www.cnn.com/TECH/space*) ... Kids Space (*www.kids-space.org/*) ... Space Telescope Science Institute, astronomy resources (*www.stsci.edu/*) ... The Moon Society (*www.moonsociety.org* or *www.moonminders manifesto.com* or *www.lunar-reclamation.org/*) ... The Mars Society (*www.mars society.org*) ... Students for the Exploration and Development of Space (*www.seds. org*) ... The Planetary Society (*www.planetary.org*) ... Space Foundation publishes an annual Space Report, a guide to global space activity (*www.TheSpaceReport.org*) ... United Societies in Space publishes *Space Governance* journal and sponsors The Buzz Aldrin Space Library (*www.angelfire.com/space/usis* or *www.spacelaw.org*) ... Space art, including the works of Dennis M. Davidson (*www.novaspace.com*).

Space Policy Institute Elliot School of International Affairs, George Washington University (Ste. 430, 1957 E Street NW, Washington, DC 20052; tel.: 202/994-7248; fax.: 202/994-1639; email: *logsdon@gwu.edu*; website: *www.gwu.edu/-spi*). Leading graduate studies and research in emerging field of space policy headed by noted author and policy consultant Dr. John Logsdon.

UNIVELT, Inc. Premier publisher and distributor of space books. Official publisher for the American Astronautical Society conference proceedings (*www.univelt.com* or *www.aas.org*).

Universities Universities Space Research Association, a consortium of universities engaged in space research (*www.usra.edu*); International Space University, graduate degrees and research in space science (*www.isu.edu*); North Dakota State University, Masters of Science in Space Studies (*www.space.edu*); Rutgers University/Dept. of Mechanical and Aerospace Engineering (*www.lunarbase.rutgers.edu/* or *http://coewww.rutgers.edu/~benaroya/*

NOT THE END, BUT THE BEGINNING AND A TIME TO SING

*Fly me to the Moon,
Let me play among the stars,
and see what life is like
on Jupiter and Mars!*

Bart Howard, composer of the song, *Fly Me To The Moon*.

Exhibit 161. Starry nights—the Sombrero Galaxy. Source: NASA and the Hubble Heritage Team.

Index

Aldrin, Buzz, 4, 97, 298, 302, 326, 352, 472, 476
Analogs
  terrestrial, 248–254
  educational 230–231
Antarctica (polar regions), 252–253, 505–556
Anthropology, space, 121–124
Anderson, Eric, 413
Ansari, Anousheh, 413
Apollo, space missions, 11, 361–365, 380, 481, 580–582
Armstrong, Neil, 4, 302, 326, 352
Assessment, space personnel, 265–274
Asian, space (Case study), 44–54
Astronauts (cosmonauts/spacefarers throughout the text), 13–24, 28–35, 183–242, 276
Astronomy (observatory, *see* Hubble), 266, 346–347, 599–605
Australia Space Program, 53
Automation & Robotics (A&R), 222–224, 227, 437–440

Behavioral science, space, 103–147
Bennett, Steve, 413
Bezos, Jeff, 413
Bigelow, Robert, 80, 130, 236, 412–413, 417–418
Biosphere, 2, 258–261, 290

Bluth, B. J., 133–134, 221, 286–287, 299–300
Brady, Joseph, 144–145
Branson, Richard, 244, 355, 414
Brin, Sergey, 244
Bush, George (U.S. presidents)
  Senior 18, 20
  George W., 22
Business, NewSpace (*see* Space commerce), 416–422

California Space Institute (note references to *Space Resources NASA SP–509*), li, 283
Canadian Space Agency (Case study), 25–26, 83
Carmack, John, 413
Cheston, Stephen, 143–145
China Space Agency (CNSA Case study), 22, 45–48, 83, 435–436, 481–482
Christenson, J. M (and Talbot, J. M.) 127–128
Colon, Angel (and Patricia), 213
Cordell, Bruce, 247, 283
Commerce, space (space business/industrialization), 407–433, 460–463, 569–576
Criswell, David, xxxiv, 119, 224, 330, 479, 545–565

Culture
  space, 153–180
  organizational 367–368

DLR, German Space Agency
  (*see* Germany), 40
Deployment, space personnel, 264–303
Diamandis, Peter, 523, 567
Douglas, William, 208–209, 229, 294
Durst, Steve (Space Age Publishing), xxxix,
  460, 490–491, 599–605

Earth–Moon system, xlvii, 3, 52, 469–516,
  536
Education, space (learning/training), 168,
  232, 277–279, 281–283, 345–346,
  382–384, 386–388, 400–401,
  463–464
Eisenhower, Dwight, 14
Energy, space-based, 545–565
Enterprise, space (entrepreneurs), 1, 56,
  355, 407–463, 469–515, 523,
  569–575, 567, 577
Environment, space, 106–107, 114, 119,
  132, 138–142
Ethos, space, 9–11, 27, 35, 43, 47, 50, 52,
  54, 58
European Space Agency (European
  Union), 254–258, 269, 275, 361
  337–338, 434–435, 454–455,
  482–484
European space (Case study), 37–44
Exploration, space, 1, 65–99
Extra Vehicular Activity (EVA), 152, 154,
  196, 219, 312

Feeny, Brian, 413
Female spacefarers, 204–206, 270, 303–304
Finney, Ben, 120–121, 124, 131, 136, 379,
  503
France, space program (CNES), 37–39,
  328–329
Freeman, Marsha, 579

German Space Agency (DLR) 38, 40
Glaser, Peter, 553–554, 561
Global Exploration Strategy (GES),
  349–351

Global Positioning System (GPS), 405
Goldman, Nathan, 407–458, 597
Governance, space (space law), 446–451,
  535–542
Great Britain space ethos (U.K.) 40–41

Habitation, space (habitability, *see*
  settlement), 79, 81–86, 103–115, 119,
  121–124, 125–143
Hall, Stephen, 113, 199, 207, 210
Harrison, Albert A., 9–10, 24, 104–107,
  120–121, 127, 130, 136, 138–143,
  175–177, 193–194, 199, 225, 234,
  253, 272, 286, 295
Health, space, 210–211, 579–597
Heinlein, Robert, 427–28
Human behavior, space, 103–147,
  153–180, 183–312
Human Resource Development, spacefarers
  (HRD, *see* education), 277, 297–231
Hubble Observatory (*see* Astronomy), 226

IDEEA (International Design for Extreme
  Environments Assembly), 261–263
India Space Program, 51–53
Italy Space Program, 41
International Luna Observatory, 346–347,
  469, 600–606
International Space Station (Case study),
  385–400
International Space University, 91, 167,
  215–216, 225, 346, 446, 483, 539

Japan Space Program (JAXA), 48–50, 83,
  244, 435, 460, 479–480
Johnson, David, 413
Johnson, Stewart and Martha, 87–88

Kennedy, John F., 14
Kistler, Walter (Kistler Aerospace), 413,
  575
Kluckholm, C., 122
Knight, Eric, 413
Knowledge, space workers/management/
  culture, 86–90, 365–367
Krafft, Ehricke, 140, 350, 497, 517

Launches, space, 440–49; 535–541
Laurie, Dennis (Transorbital), 490, 523
Law, space (*see* Governance), 440–449, 535–541
LEDA (Lunar Economic Development Authority), 497–503
Libration points (Exhibit Pi), xlvi
Life sciences, space 32, 213, 239
Living systems theory in space (*see* J. G. Miller), 136–141
Livingston, David (The Space Show), 57, 459,
Logsdon, John 452–453, 531
Lunar activities (Moon), 10, 18, 126, 139, 141, 187, 214, 233, 238, 289–291, 337–338, 347, 422, 455–460, 468–516, 545–565, 585, 591–592, 599–605

MacDaniel, William E., 134, 238–239, 311, 360
Macrothinking/planning/management (large-scale enterprises), 317–403
Management culture, space, 371–373
Managers, Earth-based for space, 373–377
Mars, 263, 301, 327, 378, 518, 534, 585–586
Matrix management, space, 363–364
Matula, Thomas, xxxv, 81, 523, 569–576
Medical services for space (*see* Healthcare), 579–597
Mexico Space Program, 26
Miller, James Grier, space living systems, 105–111, 137–141, 176, 202, 228, 265, 285, 294–295, 503
Moon Treaty, U.N. (*see* Lunar), 443–508
Morrison, David, 146–147
Musk, Elton, 413, 415–416

National Aeronautical Space Administration (NASA, cited throughout book), xlvii, 13–25, 107, 173–176, 181, 193, 316, 336–338, 489, 609
National Aerospace Training & Research Center (NASTAR), 411–412, 609
National Biomedical Research Institute, 101

National Space Commission, xlviii, 16–17
National Space Society, 17–18, 78, 478, 610
North American Space Case Study (U.S.A./Canada/Mexico), 11–27

Offworld (beyond Earth) 11, 407–463
O'Neill, Gerard, 114, 508, 516, 525
O'Rangers, Eleanor, xxxviii, 579–597
Orbital deployment (of people), 245–303
Orbital research, 42, 60, 89, 200, 213
Orbital Science Corporation, 572
Organizational culture (Aerospace), 170–175, 367–372

Paine, Thomas (*see* National Commission on Space), 140
Palinkas, Lawrence, 253
Pearle, Kim (solar civilization), 55
Performance (space productivity), 183–239
Pierce, John, 359
Plush, Linda, xxxvi–xxxvii, 579–597
Policy, space, 443–445, 451–458
Praxis, (space publications), 610
Psychology, space, 125–130, 588

Reagan, Ronald, 16
Robinson, George (*see* Space law), xxxiii, 23–24, 57, 288, 378, 407, 447, 535–542, 571
Rocketplane XP (Chuck Lauer/Kistler Aerospace), 575–576
Russian Space Case/Program, 28–36, 173, 295–296, 299–300, 484–485, 530
Rutan, Bert (SpaceShipOne), 355, 411, 414, 418, 523, 575

Salk, Jonas, 5
Satellites, space, xvii, 184
Schmitt, Harrison (Astronaut Scientist), 448, 476, 496, 508, 553
Schrunk, David, xv–xvi, xxxi, 440–441, 417, 510–513
Settlement, space (*see* Habitation), 72–81, 115–121, 446–449
Sheenin, Geoff 413
Shuttle, space, 28, 184, 296, 343, 584–585
Shuttleworth, Mark, 407

Sociology, space, 131–137
Space authorities (*see* LEDA), 499–503
Space Nursing Society (SNS), 590–594
Spaceports, 330–336, 584–585
Space Services of America (David Hannah), 570–571
Space Stations (Skylab, Mir, ISS), 113, 140, 388–399
  Case study, 405
Suedfield, Peter, 107, 131
Synergy, space, 90–97

Team, space culture, 219–221
Terraforming (planets), 349, 384, 503–506
Thagard, Norm (M.D., Astronaut), 288
Tourism, space, 304–311, 334–335, 344–345
Tsiolkovsky, Konstantin, 30, 36, 59, 65
Turner, Ron, 114

United Nations (UNESCO), 441, 443, 446, 451, 456, 477–478, 501, 508

United Societies in Space, 339–341, 450–451
U.S.S.R. (Soviet Space Program, *see* Russia) 30–32
Universities Space Research Association 463, 611

Vakjch, Douglas, 74
Visions, space, 3–10, 523–527
Vision for Space Exploration (VSE, *see* Space policy), 2, 6–9, 22–25, 60, 69, 73, 468
Von Braun, Wernher, 2, 6, 351, 516
Von Puttkamer, Jesco, 2, 3, 22–25, 104, 106

Webber, Derrek, xxxii, 330–337

X-Prizes, 523

Printing: Mercedes-Druck, Berlin
Binding: Stein+Lehmann, Berlin